TID-26241

NUCLEAR POWER PLANT DESIGN ANALYSIS

Alexander Sesonske

Professor of Nuclear Engineering
Purdue University

1973

Published by
Technical Information Center, Office of Information Services
UNITED STATES ATOMIC ENERGY COMMISSION

Available as TID-26241 for $10.60 from
National Technical Information Service
U. S. Department of Commerce
Springfield, Virginia 22151

International Standard Book Number 0-87079-009-9
Library of Congress Catalog Card Number 73-600245
AEC Distribution Category UC-80

Composed by the Technical Publications Department
Oak Ridge National Laboratory
Oak Ridge, Tennessee

Printed in the United States of America
USAEC Technical Information Center, Oak Ridge, Tennessee

November 1973

Preface

This book evolved from material used as background in several design-oriented courses for students who have had an introduction to nuclear engineering. A primary objective is to develop an additional perspective of engineering areas important to reactor design which interplay with one another. Emphasis is given to such areas as engineering economics, thermal-hydraulics, safety analysis, and fuel systems, which in an introductory course are often not covered with sufficient breadth to provide a student with enough "tools" to participate in a meaningful group design effort. Since many parameter interplays are brought out, the book is also intended for the inexperienced engineer interested in improving his background for reactor design.

Significant technological progress has been made in many of the areas described during the years that were required to prepare the book. Although continual updating occurred, compromises were necessary to achieve publication. The reader should therefore use the book as a source of ideas, not as a reference for the latest technique or design value.

In general, notation used in equations is that of the original source as a convenience to readers who are already familiar with such material. As a result, some notation inconsistencies do occur between various sections of the book.

This book is a direct result of the stimulation received during my association with Dr. Samuel Glasstone in the preparation of *Nuclear Reactor Engineering*. In addition, I am very grateful to Dr. Glasstone for his continual encouragement and inspiration.

The contributions of Roger Stover and Owen Gailar to Chapter 5 are gratefully acknowledged. I am also indebted to the several generations of students who used the material in draft form and commented on it. Although it is not feasible to list the many reviewers of the draft manuscript who contributed valuable suggestions, I want to express my gratitude to them. Since it was not possible to incorporate all the suggestions made, I accept full responsibility for the material appearing in the book.

Thanks are due to R. F. Pigeon of the AEC Office of Information Services, who was the administrator for the writing and the production of the book. Invaluable editorial assistance was provided by W. F. Simpson, Jr., Margaret L. Givens, and Dee Jared of the AEC Technical Information Center.

Although this book was prepared under its auspices, the Atomic Energy Commission allowed me complete editorial freedom. Any opinions expressed in the book are therefore mine and do not necessarily represent those of the Commission.

Alexander Sesonske

August 1973

Contents

Introduction 1

1.1 This book is primarily intended to provide nuclear engineering graduate students and inexperienced engineers with background material on several topics that play a vital role in the design of commercial nuclear power plants. As part of this introduction, a major objective is to bring out the interplays among engineering disciplines that affect the design. Since many of the topics considered are described at length only in the report literature, an attempt is made to at least discuss the appropriate highlights. A broad coverage consistent with the need to provide perspective is essential to meet these objectives. Some corresponding sacrifice in detail is therefore justified since the reader can consult reference material once he knows how such material fits into the overall picture.

1.2 As a way of limiting the presentation, it is assumed that the reader is familar with general references such as *Nuclear Reactor Engineering.*[1] Some duplication with such sources is necessary, of course, in order to retain topic unity.

ENGINEERING DESIGN

INTRODUCTION[2,3]

1.3 The engineer's primary function is to create a structure, device, process, etc., that will meet a practical requirement. The creative process required, known as *design,* can therefore be considered the very heart of engineering practice. Although this book is concerned primarily with those efforts, whatever they may be called, which result in the construction and reliable operation of a nuclear power plant, it is useful first to consider in a general way the processes required to reach this goal and the terms that are applied.

1.4 The desired goal can be considered a "problem," the solution of which proceeds through a number of logical steps.-The initial step is to *define* the problem in a total way. This step includes the sorting out of irrelevant information and of presently available solution approaches from the true nature of the problem to be solved. The next step is to *analyze* the problem, wherein the effects of various parameters and restrictions are evaluated. The establishment of a model that may involve computer simulation is likely to be required for this stage.

1.5 Subsequent design stages include the creative effort required to actually search for various alternate "solutions" to the problem and to evaluate them in order to choose the best solution. Characteristically, each path to a solution is likely to be unsatisfactory in some way, and a compromise between advantages and disadvantages may be necessary. A systematic procedure for accomplishing this balancing operation is *optimization.* Finally, after a solution has been chosen, it is necessary to work out its complete details, or to prepare *specifications.*

1.6 Nuclear power design problems can be extremely varied. The conceptual design of the entire power plant, the selection of a fuel-management scheme, or the development of a small component can all be pursued along the lines of the general format described. This book, however, devotes primary attention to the *analysis* phase as applied to the reactor portion of the power plant.

DESIGN RESPONSIBILITIES IN NUCLEAR REACTOR ENGINEERING PROJECTS

1.7 In the evolvement of a power-reactor system, from conception, through the construction stage, and finally to acceptance, the design engineering responsibilities are normally shared by several different groups.[4] The *user,* the *architect–engineer,* the *nuclear equipment manufacturer,* and the *engineer–constructor* each plays a different role. The user, or customer, owns the proposed plant and therefore desires primarily to obtain a system that will provide energy in a reliable manner over the lifetime of the reactor at minimum cost. User design is therefore oriented toward the establishment of specifications and the evaluation of proposals. A great deal of conceptual or preliminary design effort may be necessary during the early stages of the project to reach a decision regarding the size and general features of a nuclear plant as well as the relative merits of a fossil-fuel plant. The specification procedure may include studies of alternate nuclear-plant concepts. *Consultants* hired by the user may play an important role in this design effort.

1.8 The concept of a product known as the *nuclear steam supply system* (NSSS) is important in the establishment of design responsibilities. This concept evolved from an initial "turnkey" project approach in which all equipment and subsystems were supplied by the reactor vendor. Next, the scope of the reactor vendor's responsibility was reduced by deleting the turbine generator and much of the field work, leaving a "nuclear island" consisting of the reactor, containment, and all systems within the containment structure. Finally, the NSSS concept received acceptance by both utilities and vendors as the preferred package for buying and selling nuclear steam-generating facilities.

1.9 The manufacturer of the nuclear steam supply system designs the subsystems, manufactures the fuel and most of the hardware directly connected with the reactor vessel, inside and outside, and specifies and builds or procures the equipment directly associated with the reactor coolant and the principal safety subsystems, warranting the output and performance of the system. In fact, a major reason for the evolvement of the NSSS package is that it provides a basis for warranties.

1.10 The primary responsibility of the architect–engineer is to design and prepare specifications for the nonreactor portion of the power plant, which includes buildings, auxiliary services, etc. Such firms also have the usual architectural responsibility for the coordination of engineering details, supervision of construction, and inspection of the work of contractors. Considerable variation is possible in the areas of design responsibility, which may be assigned by the user to the architect–engineer, depending on the engineering resources available to each and the relative role of consulting organizations. Such a responsibility could include the nuclear steam supply system itself. Some engineering firms serve as general contractors for the entire project; they subcontract the reactor portion to a reactor manufacturer. A separate engineer–constructor is often given responsibility for supervising the construction of the plant. The division of responsibilities tends to vary. In fact, some utilities act as their own architect–engineers and engineer–constructors.

1.11 It is important to remember that the architect–engineer and engineer–constructor act only as agents for the user; they carry out no manufacturing or actual construction. They merely provide a professional service. In the design of an advanced system, the architect–engineer is therefore limited by the ability of various contractors to meet the specifications that he establishes. As a result, the performance specifications are likely to be in the range of existing practice. On the other hand, the equipment manufacturer, in preparing a design for a similar system consisting of a number of components, can take advantage of his own experience and development capability by introducing innovations and improved-performance requirements. Engineering responsibility for advanced experimental reactors may therefore be assumed by the user.

1.12 Design effort by the equipment manufacturer is oriented toward supplying a *product,* perhaps a complicated system, which meets the customer specifications. In designing the product, he can take advantage of the advanced technology that he may have developed internally. The manufacturer, as do contractors for other services, normally submits a bid in competition with others. Consequently there is incentive to prepare a design that will permit a bid low enough to win the contract but high enough to allow the manufacturer to make a profit.

1.13 The detailed design of reactor components, including the pressure vessel, core, and all its internals, is therefore the responsibility of the manufacturer. This is in contrast to the design effort of the user, who is concerned primarily with the overall requirements of the plant and their effect on the utility system. Since there are many interplays between what might be called the "internal" plant design and the "external" requirement design, both are considered in this book.

DESIGN ANALYSIS

1.14 At a number of locations around the world, electrical generating plants are being built which use nuclear fuels instead of coal, oil, or gas. What is the basis for this choice? In practically every case the decision was made after a study which concluded that the nuclear plant would be more economical or would be preferred from the viewpoint of atmospheric pollution (§ 1.47) or other national-goal considerations (§ 1.37).

1.15 More is involved, however, than a simple comparison between the proposed nuclear plant and a fossil-fuel plant. Nuclear plants of different types might meet the economic and engineering criteria for the specific proposed plant and hence must be studied and compared. Questions of safety and fuel-resource utilization may be important in the selection. In fact, a substantial effort is involved in making planning decisions, studying the effects of many parameters on the behavior of the system under design, and thoroughly understanding the frequently complicated interplay between these parameters. It is also common for variables to operate at cross-purposes so that a compromise is necessary to achieve a workable design.

1.16 The systematic consideration of the important individual parameters that contribute to a reactor design is known as *design analysis*. The design itself, of course, consists of the synthesis of many individual considerations into a working system. This synthesis, however, is only one step in the so-called *engineering design process* (§ 1.3), which also includes analysis, evaluation, and iteration steps.

1.17 Before applying a formal engineering design process, however, the experienced design engineer can do much to "sort out" parameters which have an important influence on the concept from those which do not. Basing his work on past experience and reasonable assumptions, he may be able to establish a rough or preliminary design which is quite satisfactory for many purposes and which involves a combination of parameters "not far" from the optimum. Furthermore, this design model may be useful for the systematic study of the effects of changing parameters on reactor characteristics. Various approaches are also available (§1.29 and Chap. 9) for the optimization of the design with respect to some criterion, usually economic in nature.

SYSTEMS ENGINEERING

INTRODUCTION

1.18 Systems engineering[5] is closely associated with engineering design. In fact, the terms are sometimes used interchangeably. Operations research, concerned with the analysis of mathematical models of systems, is also related. Since many of the processes of both operations research and systems engineering provide powerful tools useful in the design of power-reactor systems, a brief introduction to the terminology and approaches is given here. A further survey of optimization techniques is given in Chap. 9.

1.19 The word "system" is used in many different ways, even by engineers. According to one definition, a system is a group of things that make up the whole in a regular, interrelated, organized manner. From the viewpoint of engineering design, a system can be considered a device or process of some complexity, the behavior of which can be accurately described. The description can be presented by means of a mathematical model, which includes provision for various "inputs," system "states," and "outputs." Inputs can be considered various restrictions or design requirements imposed on the designer, states as the actual conditions of the mechanism, and output as a product. For a nuclear reactor design, these terms can be used rather loosely.

1.20 A large system can normally be divided into smaller units, each also having inputs, outputs, etc. Such *subsystems* may be closely interrelated, with the output of one serving as the input to another. Of course, additional subdivision is possible. The coarseness of subdivision, however, depends on the needs of the problem. Subsystems that are not relevant to the problem under consideration are called *components.*

1.21 The systems approach is therefore applicable to the design of any engineering mechanism consisting of a number of interrelated components for which a mathematical model describing the behavior of the individual components can be established. An optimized design can then be achieved through formalized mathematical approaches (Chap. 9).

1.22 Although in theory such procedures can also be applied to the design of a complete nuclear power system, an exact mathematical representation, including economic parameters, would be too complex for the practical application of existing optimization procedures. In addition, the uncertainty that may be associated with some physics and economics parameters limits the accuracy of representation and the corresponding degree of mathematical sophistication that is appropriate. It is possible, however, to apply the systems approach to portions of the plant complex. With future improved confidence in parameter representation as well as improvements in computer facilities, so that very complex problems can be treated economically, it is likely that the systems approach can be applied to the design of a complete nuclear plant.

SYSTEM ENVIRONMENTAL FACTORS

1.23 A useful systems engineering concept is associated with the *environment* or decision space external to what has been defined as the *system.* Various considerations outside of the system which affect system design in some way can thus be identified. The concept of a boundary between the system and the environment across which internal–external influence factors can be applied is a helpful way to sort out the variables relevant to the system itself from those which may only affect the interactions between the system and its subsystems.

1.24 For a nuclear power plant, both physical and economic "external" factors influence the overall system requirements. Such factors are established at the initial stage of the design. For example, the so-called requirements analysis must consider existing systems, or, in the case of a power reactor, the *functional requirements* for the reactor in relation to the utility system in which it will become a part. Interactions with the public on such questions as safety and ecological effects of the plant site require analysis. In addition, consideration must be given to the *economic environment,* which includes management policy, relations with government, and the price and cost structure for the planned system design. For the projected plant to be profitable, the cost of electricity produced must be consistent with costs from other utility system facilities. Other environmental factors might include the various items that contribute to the cost of fuel and all the complicated accounting requirements necessary to determine an actual manufacturing cost (§3.5).

1.25 In one sense the state of existing technology could be considered an environmental factor. Design practice, based on such existing technology, provides a background upon which a great deal of the system description can be based, including the influence of various limitations imposed by materials, reliability of theoretical predictions, etc. An important phase of the existing technology environment concerns standards and codes (Chap. 6). For example, safety standards, establishing design specifications that result in desired high levels of reliability in plant performance, have much to do with the nature of the design.

1.26 In a broad sense, all phases of systems design are affected by these environmental-decision-space factors. The specifications, including necessary boundary conditions for the new system, are derived from the environment. Once designed, the system must be evaluated with reference to its projected environment. This requirement should be kept clearly in mind by the designer in considering various portions of the systems design, e.g., the thermal transport system, control and neutronic design, cost estimation. A power-reactor plant is so complex that it is essential for the systems engineer to concentrate on functional variables that affect, or are affected by, the preceding environmental factors and to avoid being submerged in the large amount of design detail not related to the environment "interface." Although such design-detail problems must be solved before the reactor system is operable, they need not be given high priority in the early phases of the design effort when it is important to fix the general characteristics of the system. The proper identification of pertinent environmental factors is therefore important at this early design stage.

DESIGN FEEDBACK AND ITERATION

1.27 Although the term "feedback" is normally applied to communication and control systems, it is also applicable to the design process. A closed-loop-design approach for either a complete reactor system or a subsystem is shown in Fig. 1.1. This feedback loop is a result of introducing evaluation into the design

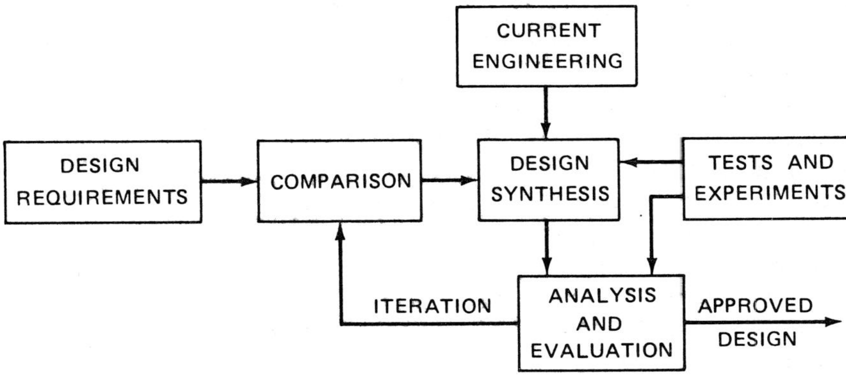

Fig. 1.1 Closed-loop-design approach.

process. Such evaluation may be based on information derived from either planned experiments or a consideration of parameters external to the system that were not included in the original design. It is normal for the evaluation process to bring to light the need for design adjustment, the resolution of inconsistencies, or, in some cases, a change in the original specifications. Such problems can be solved by "trips" around the feedback loop with appropriate adjustments each time as needed.

THE ROLE OF MODELS, SIMULATION, AND COMPUTERS

1.28 If an accurate mathematical model of the system to be designed could be prepared, computer techniques could be used to search for a combination of specifications that would meet the desired design constraints, i.e., to prepare the design. Similarly, an analog or analog–digital hybrid approach could be used to simulate the dynamic behavior of the reactor system. These methods have received considerable attention but, of course, are limited in their application to the complexity of representation that is justified considering the accuracy of the input data. Another consideration is the trade-off between the expense of computation and the accuracy of results that may be desired for a given purpose (Chap. 8).

OPTIMIZATION

1.29 Optimization[6] is the final stage of system design. Optimization is the process of seeking a combination of conditions that results in some "most favorable" situation with respect to some criterion such as maximum cost or minimum weight. An introduction to design-optimization approaches is given in Chap. 9. A few introductory comments are given here, however, as part of the discussion of the design process.

1.30 In a classical approach from elementary calculus, a maximum or a minimum can be found by setting equal to zero the first derivative of the criterion function with respect to an independent variable. Such an approach is applied to the following typical nuclear design problem:

Example

Determine the height-to-radius ratio (H_c/R_c) for a buckling corresponding to a minimum critical volume for a right-circular cyclinder. For a cylinder,

$$B_c^2 = \left(\frac{2.405}{R_c}\right)^2 + \left(\frac{\pi}{H_c}\right)^2$$

but taking $V_c = \pi R_c^2 H_c$ and multiplying by $(\pi R_c^2 H_c / V_c)^{2/3}$ gives

$$B_c^2 = \left(\frac{\pi}{V_c}\right)^{2/3} \left[2.405^2 \left(\frac{H_c}{R_c}\right)^{2/3} + \pi^2 \left(\frac{H_c}{R_c}\right)^{-4/3}\right]$$

Now, differentiating B^2 with respect to H_c/R_c and setting the derivative equal to zero gives

$$\left(\frac{H_c}{R_c}\right)^2 = \frac{2\pi^2}{2.405^2} \quad \text{and} \quad \frac{H_c}{R_c} = 1.85$$

1.31 When additional independent variables are involved, it is necessary to differentiate with respect to each of them and set the derivatives equal to zero.

When the relations between the independent variables are not continuous or cannot be easily expressed analytically, graphical procedures and trial-and-error methods are desirable. Such procedures are useful, however, only for fairly simple systems containing two independent variables. Because optimization has been necessary for the design of complicated systems in the space and process industries, special mathematical techniques have been developed, as discussed in Chap. 9. For each method, however, the relations between the system variables must be completely understood.

1.32 Some of the simple functional relations can be clarified by considering a geometric analogy. Assume that there exists some mathematical function of the variables $x_1, x_2, x_3, \ldots, x_k$ which is equal to the cost of power on a unit-energy basis, C.

$$C = \phi(x_1, x_2, \ldots, x_k) \tag{1.1}$$

The function is sometimes called a *response* function. For only two variables, x_1 and x_2, C can be represented by a surface. Some constraints may, of course, be effective. The designer, having previously arrived at some combination of variables, using his knowledge of similar systems or engineering judgment, is interested in finding the direction in which changes in the value of the variables will lead to an optimum. Furthermore, it is necessary to determine the position of the maximum (or minimum).

1.33 In the method of *steepest descent,* one of the *gradient* methods, a small area of the curved surface around a point remote from the minimum can be represented by the best-fitting sloping plane. The direction toward a minimum is determined from the slope of the tilted plane. Then, by repeated trials, the optimum can be approximately located in the region in which the slope becomes small. In practice, the procedure is usually carried out analytically since more than two variables, as required for a geometric surface, are usually of interest. The partial derivatives of the cost function indicate its rate of change with respect to the variables involved. The gradient, ∇C, of the cost function can then be determined:

$$\nabla C = \mathbf{v}_1 \frac{\partial C}{\partial x_1} + \mathbf{v}_2 \frac{\partial C}{\partial x_2} + \mathbf{v}_3 \frac{\partial C}{\partial x_3} + \ldots \tag{1.2}$$

where $\mathbf{v}_1, \mathbf{v}_2, \mathbf{v}_3, \ldots$ are unit vectors in the direction of the axes corresponding to the variables in multidimensional space. The gradient, ∇C, in turn, is a vector that can be directed along a path in which the response of the function, C, might increase or decrease rapidly, much as one might explore the tilt of a plane about a point in the case of a two-variable surface. A minimum condition is where $\nabla C = 0$.

1.34 The procedure in Eq. 1.2 is valid only if one of the partial derivatives of the function vanishes within the range of interest of the variables and not at one of the boundary conditions. Such a situation, where the optimum would be at the boundary, occurs when the function is linear. Should this be the case, a trial-and-error approach is necessary to search for an optimum of the function of a number of variables satisfying linear inequalities or equalities as boundary conditions. Such an approach is known as *linear programming.* Systematic procedures utilizing digital computers have been developed for the trial-and-error operations, as discussed in Chap. 9.

1.35 Mathematical and computational difficulties arise in the case of multistage processes in which each stage operates in sequence. In this case an optimum must be obtained for each stage involving, let us say, M independent variables. If there are N interconnected stages in the system, the problem is of magnitude MN. *Dynamic programming* is a technique utilizing a high-speed computer for reducing such an MN-dimensional optimization problem to a case of M variables considered N times. This procedure is also described more fully in Chap. 9.

1.36 Since the usefulness of a "formalized" optimization approach is limited by the ability to describe the system, the amount of sophisticated mathematical treatment that is appropriate for a given problem is subject to evaluation. In general, these formalized approaches are most useful for preliminary surveys in which a small number of parameters are studied over a wide range and high accuracy is not needed. For final designs for which parameter variation may cover only a narrow range, however, secondary effects are likely to become important. Additional parameters are thereby introduced which may be difficult to accommodate in the computational procedure of a formal approach.

NUCLEAR POWER PROJECTS

INCENTIVES

1.37 An engineering design project is undertaken to meet a practical need. Before discussing nuclear projects, therefore, and some of the design-parameter interplays in such projects, we should consider some of the reasons why such projects are undertaken. The *incentive* for the specific project may strongly affect its nature and determine the design compromises that may be necessary. With utility company central-station plants, for example, favorable economics is probably the most important incentive, but other factors, such as freedom from air pollution and uncertain fossil-fuel supplies, may contribute. An incentive for a nuclear system for space propulsion is improved performance.

1.38 A number of countries, including the United States, have "national programs" wherein nuclear power activities are encouraged or subsidized for reasons other than the potential for producing electrical energy economically.

Such encouragement is justified by contributions that nuclear power makes toward the attainment of national goals.[7]

1.39 One such goal is to utilize energy resources efficiently. Although the known reserves of coal, oil, and gas may vary considerably from one country to another, clearly such fuels will not be adequate to supply energy requirements for many years at reasonable cost.

1.40 Another aspect of such energy-resource conservation is concerned with the need to make use of fertile nuclear materials, such as ^{238}U and thorium, rather than depend entirely upon naturally occurring fissile material. World energy needs are accelerating rapidly with electric power generation doubling about every 10 years in the United States. Although various projections have been made of the proportion of this growth expected from nuclear power, uranium-ore requirements can be estimated if an anticipated light-water-reactor capacity of about 150,000 Mw(e) by 1980 is realized. As discussed in Chap. 7, a requirement of about 260,000 tons of U_3O_8 is anticipated, which is greater than the assured U.S. reserves at \$5 to \$10 per pound. With consideration of uncertainties in the projections, it seems clear that breeding reactors must share part of the load if nuclear fuels are to make a major contribution toward meeting the energy needs estimated for the United States.

1.41 At present there is little economic incentive for reactors designed for the more efficient utilization. Furthermore, in many countries such as the United States, where fossil fuels are comparatively adequate, there is little pressure from the conservation viewpoint alone to promote use of nuclear fuels instead of fossil fuels. Since the need does exist for a logical pattern of growth that will result in the efficient use of nuclear-fuel resources to meet future energy requirements, the necessary encouragement and planning for such a pattern becomes part of a national program.

1.42 On a long-term basis, however, an economic incentive for developing breeder reactors has been demonstrated by a cost-benefit analysis.[8] A number of studies were made with varying parameters to compare the benefits accrued from an economy with a breeder with those from an economy without a breeder. A breeder economy saves in uranium-fuel costs, particularly considering the rising cost of ore as lower cost reserves are depleted. Another saving is in the cost of enrichment. With reasonable interest rates, the dollar savings anticipated some years hence if a breeder economy is available can be converted to present-day dollar values (§2.24) and compared with the cost of developing such reactors. Such a comparison shows that the savings to be realized tend to be greater than the estimated cost of development.

1.43 For some countries without adequate conventional fuel resources, nuclear fuels offer a way to self-sufficiency. Even if the country in question has no nuclear-fuel mineral resources, the long lifetime and comparatively low weight of a nuclear reactor core, compared with an equivalent amount of coal or oil, can provide some strategic advantages. A fuel supply not dependent on long supply routes that could be vulnerable under wartime conditions may be important. The United Kingdom's nuclear power program, for example, received encouragement after the Suez crisis in 1956.

1.44 National policies on nuclear weapons can affect nuclear power programs in two rather opposite ways. Should a nuclear disarmament program become effective, civilian nuclear power is a means of recovering some of the financial investment in weapon fissile materials and associated production facilities. If a country is seeking to build up its weapon capabilities, nuclear facilities of various kinds may allow production of nuclear power as a by-product effort with economic benefits and may thus strengthen the overall program.

1.45 An active nuclear program may receive encouragement in order to promote national prestige. Both prestige and economics may be involved in the desire of a given country to export reactors, components, and materials. Such export is possible, of course, only if the country has a strong program.

1.46 Certain secondary economic benefits, from the national-goal viewpoint, may result from the encouragement of a nuclear power industry. For example, in an economy where a choice is to be made between the use of nuclear fuels and either oil, gas, or coal, competitive pressures tend to encourage technological advances, which, in turn, can lead to lower costs for all fuels. We can appreciate the "stakes" involved in deciding the type and location of a new power plant of the size now frequently considered when we realize that a market for 1 to 4 million tons of coal per year or its equivalent in some other fuel may be involved. In the lifetime of a plant, this could represent a total value of the order of a half *billion* dollars. The substantial competitive pressures involved, therefore, can lead to savings for the consumer which, in turn, represent an economic benefit to a nation.

AIR POLLUTION

1.47 Growing public concern over air pollution has encouraged the choice of nuclear-fuel plants instead of fossil-fuel plants in urban areas planning new electrical generating capacity. In most areas power plants must comply with new regulations prohibiting smoke emission.[9] Although modern, efficient fossil-fuel plants no longer emit a significant amount of unburned carbon, "fly ash" must be removed from the flue gases. Various mechanical devices, such as cyclones and electrostatic precipitators, can reduce particulate emission by over 99%. Costs for such devices vary from about $14 per kilowatt for a 360-Mw(e) plant to less than $1.50 per kilowatt for a modern 950-Mw(e) unit. This corresponds to a power-cost increment between 0.2 and 0.02 mill per kilowatt-hour. Since nuclear power plants cause no air pollution, the economic increment may be significant, particularly for smaller size plants.

1.48 Much more significant, however, is *gaseous* air pollution from fuel combustion, which cannot be removed at a reasonable cost. In the Los Angeles area, for example, where a major cause of smog is derived from sulfur oxides from liquid fuels, steam–electric plants have had to switch to natural gas during the 7-month period from April to November, when weather conditions lead to the greatest smog formation.

1.49 For a suggested ground-level-concentration limit of 0.1 ppm of sulfur oxides, coal must contain less than 1% sulfur. Since only 10% of the coal used in power plants meets this standard, there is a substantial interest in developing economical processes for removing SO_2 from flue gas. One method, the alkalized alumina process, adds a cost penalty equivalent to about $1.00 per ton of coal burned, provided credit can be obtained from the sale of sulfur or sulfuric acid by-product.[10] Since coal has a value of the order of $8 per ton, the additional $1 charge represents a severe cost burden when nuclear fuels are being considered on a competitive basis.

1.50 Tall stacks can reduce the level of air pollutants at ground level. Since this approach does not limit emission to the atmosphere or prevent the dispersion to the ground from a plume many miles from the tall stacks, it is not really a satisfactory solution to the air-pollution question.

1.51 The preceding comments indicate how the air-pollution question can contribute to a decision to use nuclear fuel. Actually, the subject of air pollution and the choice of a suitable fuel to meet air-control standards involve many factors in addition to those mentioned, particularly those concerned with the selection of a site. Acceptable air-quality levels can vary at different locations. A trade-off between energy-transmission costs and pollutant-removal-process costs may therefore also affect a fuel choice.

1.52 The reliability of the fuel supply is very important in a decision for a new power plant. In recent years, for example, many utilities have been unable to obtain desired long-term contracts for coal. Coal shortages have resulted from a combination of circumstances, including a lack of labor, some reluctance by operators to open new mines, and uncertainties about the sulfur problem. Such uncertainties tend to provide an incentive for utilities to choose nuclear fuel.

HEAT-DISSIPATION CONSIDERATIONS

1.53 A major consideration in selecting a site for either a nuclear-fuel or a fossil-fuel power plant is the effect on the environment of the heat that must be dissipated in the thermodynamic cycle. For example, the discharge of condenser cooling water can cause temperature changes in receiving waters which might affect aquatic life. Allowable temperature changes are generally specified as part of water-quality standards enforced by individual states.

1.54 The effluent from power plants in the 1000-Mw(e) size range may well cause temperature changes in some major streams during summer months which exceed the quality standards. It may therefore be necessary to include cooling towers or cooling ponds in the design, which, of course, adds to the capital cost. Such design considerations associated with the site are discussed in Chap. 6.

1.55 Light-water nuclear plants have a thermodynamic efficiency of about 32% compared with 40% for a fossil-fuel plant. It is therefore necessary to dissipate about 42% more energy as heated condenser cooling water from the nuclear plant than from the more thermally efficient fossil-fuel plant for the same electrical output. When the choice between a nuclear-fuel and a fossil-fuel plant is on a near-equal basis from other considerations, the heat-dissipation factor may therefore result in a decision for the fossil-fuel plant. Nuclear plants using high-temperature reactor coolants, such as liquid metals, helium gas, or molten salts, however, are as thermally efficient as modern fossil-fuel plants and hence dissipate no more energy to the environment.

DESIGN PARAMETERS

DESIGN–PARAMETER INTERPLAY

1.56 The design of a power plant is affected by many parameters that operate in many interrelated ways. It is common to have trade-offs where the beneficial effect of a change in one parameter is offset by a detrimental effect of an accompanying change in another parameter, with a need to reach a design compromise. Although several areas of concern, such as economics, heat removal, safety, and the fuel system, are treated in subsequent chapters, it is useful to first become acquainted with the nature of some of the common interplays.

1.57 Although each reactor concept presents a different framework for the interplay of design variables, it is helpful to consider in general classifications some of the variables that can be considered semi-independent in the sense that the designer has some freedom to adjust them. The degree of dependency, however, will vary from one concept to another. A few general types of design choices are given in the following sections as an introduction to some specific-concept examples.

Fuel—Moderator—Coolant Combination

1.58 Although in some cases the choice of the fuel–moderator–coolant combination is made when the concept is selected, in other cases some options may be available. At any rate, it is well to recognize how a choice will affect other parameters. The nuclear and physical properties of a moderator–coolant introduce a "package" of design features that the experienced engineer will bring to mind when considering various possibilities. For example, the moderator-to-fuel ratio for a D_2O-moderated system will be much larger than that for a light-water system, primarily because of low D_2O neutron absorption. Other parts of the package include the possibility of a pressure-tube core as a result of the large fuel-pin pitch and the good neutron economy with need for only very slightly enriched fuel, which, however, is capable of only low burnup. Similarly, the choice of coolant has a marked effect on the choice of other materials, the heat-removal-system parameters, and many details of the core design.

Structural Materials

1.59 Corrosion and other performance limitations of structural materials selected for the system are likely to affect pressure and temperature specifications and, indirectly, the power density. An important design limitation for pressurized-water-reactor pressure vessels, for example, is the fast-neutron dosage that can be accepted without undesirable changes in the mechanical properties of the steel. Although structural materials are normally specified to meet design requirements that have been established from other parameters, some compromise may be possible by adjusting these requirements to accommodate materials that may be preferred because of cost, fabricability, etc.

Core-Lattice and Thermal-Transport Parameters

1.60 The fuel-to-moderator ratio is an important physics design parameter, particularly in light-water-cooled and -moderated reactors, which tend to be somewhat epithermal. In such reactors, increased moderation will increase reactivity for a given fuel enrichment. The degree of fuel subdivision, or fuel-rod diameter, normally has a strong effect on the core power density obtainable since it determines the ratio of fuel volume to fuel-element surface area available for heat removal. Similarly, the coolant cross-sectional area through the core, fixed by the fuel-to-moderator ratio if the moderator is also the coolant, has a marked effect on the heat-removal-system circulation specifications, although for a given nonboiling-reactor power level the product of the core temperature rise and coolant circulation rate remains constant. Consideration must also be given to the core pressure drop, heat-transfer coefficient, etc., which, of course, are sensitive to the coolant velocity and, in turn, to the cross-sectional area.

Neutronics Parameters

1.61 Any one of a number of parameters that affect the neutron balance can contribute to the transient and steady-state reactivity and flux distribution.

Most of these, however, are not independent and therefore need not be considered as separate variables. A fuel enrichment is required to achieve criticality under the conditions that may have been established with the fixing of heat-removal requirements etc. The principal design challenge therefore tends to be the neutron-behavior description rather than the parameter choice. Local enrichment in the core, however, may indeed vary with the degree of power flattening desired and the fuel burnup obtained. The possibility of several fuel-management schemes also introduces some design flexibility in the selection of neutronic specifications.

Safety Parameters

1.62 Since a reactor design is acceptable only if it meets established safety standards, parameters associated with these standards may indeed prove to be the independent variables that fix the design. Transient characteristics are particularly important. Permissible excess reactivity for a water-moderated reactor, for example, may limit available burnup and establish limits for fuel-management specifications. For a large fast reactor (see Chap. 8), the need for a combination of reactivity coefficients and thermal–hydraulic characteristics may establish important limits to the nature of the fuel specified and the acceptable configuration of the core.

1.63 The safety standards consist of a combination of criteria, codes, and regulations as described in Chap. 6. In the United States, reactor projects must be approved by licensing authorities who must be convinced that the facility will not result in hazard to the health and safety of the public.[11] Various criteria have been established in the course of numerous reactor licensing reviews which may serve as guides for the designer. In addition, codes, such as the American Society of Mechanical Engineers code, which applies to pressure vessels, and a series of industrial standards compiled by the American National Standards Institute provide design criteria.[12]

EXAMPLES OF PARAMETER INTERPLAY

1.64 As a further introduction to the interdependency of parameters, let us consider a few examples of design characteristics for specific reactor concepts. A discussion of some of the effects in light-water-moderated systems, heavy-water-moderated advanced converters, and sodium-cooled fast reactors covers an adèquate range for the present purpose. For each concept the preceding interplay classifications are considered. All the effects pertinent for each concept are not discussed here. Only a few of the topics important to the designer are mentioned. Furthermore, it is assumed that the reader is familiar with the general characteristics of the concepts.

Pressurized Light-Water and Boiling-Water Reactors

1.65 The properties of available structural materials place an important limitation on the maximum thermodynamic cycle temperature, which, in turn, limits the thermodynamic efficiency for the usual water-cooled system. As an example, the maximum system pressure and temperature are determined by pressure-vessel design limitations. For fuel material, the available burnup could be limited by either the ability of the fuel to withstand irradiation or reactivity changes that may be acceptable. Actually, an optimum fuel lifetime can be determined by an analysis of fuel-cycle economics (Chap. 7).

1.66 The moderator-to-fuel ratio can be considered both a core and a neutronic parameter. If the moderator is also the coolant, the power density obtainable in the core depends on this parameter as well as on the degree of fuel subdivision. The degree of subdivision can be considered a second important lattice parameter. As the fuel-element diameter is decreased, the surface-to-volume ratio is increased, tending to give a high power density and specific power but an increasing fabrication cost.

1.67 The cladding design affects the thermal design of the fuel element, the neutronics of the core, and, to some extent, fuel utilization. For high-burnup fuels (>3 at. %) generating significant amounts of fission-product gases, containment of the gases by the cladding must be considered. Though perhaps better categorized under "materials," cladding interplays strongly with the preceding factors. Optimization is again possible. In a choice between stainless-steel and zirconium-alloy cladding, for example, the zirconium absorbs fewer neutrons parasitically but is more difficult to fabricate than stainless steel. In the past stainless steel has been the preferred material because of lower cost and better-understood characteristics.

1.68 The thermal transport parameters of the light-water-reactor core are similar to those for other reactor concepts. Previous mention of some of these emphasizes a strong interplay with other parameters. The power density, for example, a parameter of much economic significance, is normally limited by either "burnout" (Chap. 4) or the volumetric heat-generation rate permissible within the fuel element. Although the characteristics of the materials used determine the maximum temperature that can be tolerated, the temperature pattern itself is very sensitive to the diameter of the fuel, the temperature of the coolant, and the fluid heat-transfer coefficient. The temperature and flow characteristics of the coolant through the core therefore affect the power density.

1.69 The desire for a maximum coolant velocity past the fuel-element surface, which would tend to give a high surface heat-transfer coefficient, must be compromised since the coolant pressure loss (or pumping power) is generally proportional to the square of the coolant velocity. Another important factor, however, is the coolant temperature rise through the core, which depends on the mass flow rate. For a large pressurized water reactor, water velocities are normally in the range of 15 to 20 ft/sec, corresponding to a total pressure loss across the reactor vessel of about 50 psi.

1.70 For a given amount of power to be extracted from the core, the coolant circulation rate (and pumping power) is likewise inversely proportional to the permissible temperature rise of the coolant while passing through the core. In this case the maximum temperature depends on the limits imposed by the materials used. A low inlet temperature, desirable to reduce coolant flow, is governed by the compromise in cycle thermodynamic efficiency that is considered acceptable. Although details of cycles vary, the higher the temperature of extraction from the fuel to the coolant, the higher is likely to be the thermodynamic efficiency.

1.71 Bear in mind that the importance of attaining a high thermodynamic efficiency is derived not only from the need to extract as much useful energy as possible from a given amount of fuel consumed, as for fossil-fuel reactors, but also from the desire to minimize the capital costs required for thermal-power capacity to obtain the desired electrical capacity. A lower capital cost would not result, however, if increased temperatures required more-expensive materials and engineering features. Another important advantage of a higher thermodynamic efficiency is a decrease in the waste heat dissipated to the environment.

1.72 The neutronic, or physics, parameters affect the degree of fuel enrichment needed, which, in this case, is considered a dependent variable. Both the moderator-to-fuel ratio and the actual fuel diameter are important in establishing the so-called ^{238}U infinite resonance integral, which affects the plutonium production rate. This, in turn, is significant in evaluating the fuel-cycle costs.

1.73 The reactivity, or "worth," associated with the coolant is important in the transient and safety analysis. The variation of neutron flux, both as a function of core geometry and fuel depletion, is a neutronics parameter that has a marked effect on the maximum-to-average power ratio in the core. Since the thermal power is normally limited by the maximum power conditions, this parameter has a major effect on power costs.

1.74 Any reactor is acceptable only if it can meet established safety standards. All previously mentioned parameters should be evaluated from the viewpoint of safety requirements. A water reactor, fueled with zirconium alloy or stainless-steel-clad UO_2, has characteristics requiring special attention. Between the cold, shutdown condition and various operating modes, a large amount of reactivity is normally present, compensated by control rods or other methods. The design of a control system and the possibility of loss of coolant flow must therefore receive considerable attention.

1.75 Depressurization of the primary circuit must be analyzed and a containment structure designed accordingly. Such design criteria will be treated in greater detail in Chap. 6. For present purposes, however, the interplay of safety considerations with all the other design parameters should be recognized.[1]

D_2O-Moderated Reactors

1.76 In examples of design-variable interplays and parametric effects mentioned here, the reactor concepts moderated with D_2O but cooled with D_2O, organic coolants, or gas are considered as a single category. Some major engineering differences are involved, of course, between one concept and another, depending on the coolant chosen. The main reason for using D_2O as a moderator is to provide good neutron economy with a resulting high conversion and efficient fuel utilization. When design parameters are considered, therefore, neutron-economy effects are emphasized. Core materials are therefore limited to low neutron absorbers, such as aluminum and zirconium, for fuel cladding. The "boundary conditions," as determined by the structural, temperature, chemical, and radiation limitations of such materials, may therefore differ markedly from those for other types of reactors.

1.77 The excellent moderating ratio of heavy water permits either natural uranium or very slightly enriched uranium to be used as a fuel in a reactor having a much smaller core than that for a graphite-moderated system. Since keeping the D_2O moderator cooled enhances the neutron economy, most designs provide for the separation of coolant and moderator. However, this separation of a hot and cold fluid using only low-neutron-absorber materials is an engineering challenge. In one approach pressure tubes contain the coolant.

1.78 Although many of the general features of core design are similar to those of light-water reactors, a very large moderator-to-fuel ratio for heavy-water systems provides space for comparatively large-diameter fuel-pressure-tube construction if desired. Thermal transport variables, though similar to those described for light-water reactors, of course depend on the characteristics of the coolant specified in the concept. For the gas-cooled reactor, heat must be removed at the necessary high rate without undue expenditures of energy to circulate the coolant. Here the pressure of the coolant gas tends to be an important heat-transfer variable not present in systems with liquid coolants. The degree of pressurization is also important in water-cooled systems, but it affects the design in a different manner.

1.79 Neutronics, or physics, parameters are concerned with variables that influence the reactivity or neutron balance. An interplay exists between the burnup, the arrangements for flattening the flux, and the fuel enrichment. The enrichment lends itself to optimization, taking into consideration the cost of an incremental amount of enrichment, balanced against the advantage of longer fuel exposure or a smaller reactor volume or both. A small volume reduces structural expenses, particularly those associated with the containment vessel. On the other hand, an increase in reactor volume, with the necessary increase in capital investment, is likely to decrease the required fuel enrichment, with consequent savings, since the larger volume has an improved neutron economy. The optimization depends on lattice parameters, i.e., the moderator-to-fuel ratio, the cross-sectional area of the fuel element, and the average temperature of the moderator. On-load refueling, a feature of D_2O-moderated reactors, must be considered in planning the fuel-element spacing. As for other concepts, a fuel-management scheme must consider depletion, control, enrichment, neutron balance, and economic parameters.

1.80 Safety considerations introduce practical limits to the permissible range of many of the preceding variables. Questions of stability depend on the choice of coolant. For example, the coolant-void coefficient may be positive for an organic-cooled system. In any case, stability factors depend on the complex interaction between a number of variables, including the reactivity coefficients, coolant-channel hydraulics, and pressure control in the steam system.

Fast Sodium-Cooled Reactors

1.81 Let us consider here a large, sodium-cooled fast breeder reactor with ceramic fuel. Since fast reactors are justified primarily for their breeding potential, design parameters affecting fuel economy are extremely important. The need for a chain reaction using fast neutrons, however, introduces some unique problems.

1.82 Preventing the degradation of the neutron spectrum becomes important in the choice of core materials. Since the fertile material normally used, ^{238}U, has a large inelastic scattering cross section, its presence, desirable for fuel utilization as a source of bred fuel, leads to a reduction in neutron energy, with a corresponding reduction in the neutrons available for the breeding process. A fairly high atom concentration of fissile material must be used since the fission cross section is low at high neutron energies. Therefore high power densities are usually required to conserve valuable fissile atoms. An ideal coolant, therefore, must have excellent heat-transfer properties, must be nonmoderating, and must not affect the neutron economy.

1.83 Fast reactors having the most favorable economics tend to be large in size and to have large, dilute cores. Safety and stability considerations for such systems are very different from those for thermal reactors. Accidental fuel consolidation must be avoided since a dilute system may contain much more fissile material than that required for a critical mass in compact geometry. The analysis of kinetic behavior is complicated by short neutron lifetime, differences in neutron yield, and nonlinear contributions to reactivity feedback. Since a reactor must be safe to be acceptable, the most significant design interplays are associated with the need to satisfy safety requirements and yet obtain a satisfactory low-cost fuel cycle.

1.84 The transient characteristics of large-core plutonium-fuel fast reactors depend on two important parameters, the Doppler coefficient and the sodium-void coefficient, as described in Chap. 6. In the case of the Doppler coefficient, broadening of fission resonances in fissile materials increases the reactivity, whereas broadening of capture resonance in both fissile and fertile materials decreases the reactivity. For large fast reactors, in which the ratio of

^{238}U to ^{239}Pu is high, the broadening, with temperature increase of the fuel, results in a significant negative value for the Doppler reactivity coefficient and hence an enhancement of a feedback term leading to reactor stability.

1.85 The sodium coolant both scatters neutrons and moderates them. The reduction in sodium density with an increase in temperature decreases the scattering effect and thus increases neutron leakage from the core and reduces the reactivity. The accompanying loss of neutron moderation, however, leads to a more reactive situation since fewer neutrons are scattered below the ^{238}U fission threshold.

1.86 Although fuel costs depend on a number of parameters, the fissile (plutonium) fuel inventory and the breeding gain are the most significant. A large fuel inventory is costly, although replacement of the fuel by breeding, as it is consumed, lowers the costs. For very large cores with low leakage, the core may be diluted with such a large amount of fertile ^{238}U that the internal breeding ratio may approach unity.

1.87 A fuel material, therefore, must be selected in view of its effects on the Doppler and sodium-void coefficients, the breeding ratio, lifetime in the core, and fabrication costs. As in other reactor types, the core lifetime has an important effect on the fuel-cycle cost. Mechanical, physical, and chemical characteristics also have a marked effect on the core design and on the thermal transport system, similar to other reactor situations. There is greater freedom in the choice of structural materials, however, for fast reactors than for thermal reactors, because of the generally higher concentration of fissile material and the trend of all materials to have small capture cross sections at increased neutron energies. This permits long irradiation times for the fuel, with the possibility of generally severe environments, from the materials viewpoint. In fact, swelling of fuel cladding and other core components due to fast-neutron irradiation can cause serious dimensional changes of great concern to the designer.

1.88 The disposition of fission-product gases generated in the fuel element becomes a problem as long irradiation times become possible. It has been suggested that the fuel be vented to the sodium coolant to solve the otherwise intolerable stress conditions in the cladding if the accumulated gases are contained.

1.89 Fast reactor core-design parameters are concerned principally with the interplay between fuel-element thermal-design details and the thermal-transport-system variables. High heat fluxes are usually an objective but may be limited by cladding thermal stress and unacceptably high temperature gradients in the fuel.

1.90 The use of a liquid metal, such as sodium, for a coolant leads to possible mass transfer from stainless-steel constituents to the sodium at temperatures above 1200°F, particularly if the oxygen content of the sodium is not maintained at a low level. Coolant-system parameters also include the usual balance between core temperature rise and circulation rate. Since pressurization is not a problem, however, a core temperature rise as great as 400°F may be suitable, compared with a 50°F rise for a water-cooled system. The severe activation leading to ^{24}Na does lead to a requirement for primary-system shielding, however.

1.91 Fuel-cycle parameters are worthy of special mention. Refueling methods must be uncomplicated to avoid lengthy shutdown periods and technical complications. The need for handling fuel elements safely under sodium, with the ever-present danger of melting from decay heat from the previous high-power-density exposure, is a substantial design challenge. Also, the fuel-element design should not necessarily complicate subsequent fuel-processing operations.

1.92 In addition to neutronics parameters, which contribute to the Doppler and sodium-void effects, other important parameters include those which contribute to power flattening and to various concentration and spectral effects that could change the breeding ratio, inventory, and fuel-management scheme. In fact, the essential design challenge for fast power reactors is to achieve a compromise between parameters which improve fuel economics at the expense of favorable Doppler and sodium-void effects and those which have the opposite effect.

1.93 Safety questions are an important part of fast reactor design. The various transient effects must be provided for in both the engineering and the nuclear designs. Since substantial heat is emitted by the high-power-density fuel elements after shutdown, emergency-cooling capability must be provided, a requirement that complicates the coolant-system design. Parameters that could affect the consequences of an unlikely accident should be considered and such information included in the overall design picture.

CONCLUSION

1.94 A reactor system is quite complex, with design parameters interacting in ways that challenge analysis. The designer must accept this challenge and consider all effects resulting from the combination of parameters that he selects for his system specifications. It is emphasized that the preceding examples are not complete, but they do indicate the nature of the design challenge.

REFERENCES

1. S. Glasstone and A. Sesonske, *Nuclear Reactor Engineering,* D. Van Nostrand Company, Inc., Princeton, N. J., 1963.
2. Morris Asimow, *Introduction to Design,* Prentice-Hall, Inc., Englewood Cliffs, N. J., 1962.
3. E. V. Krick, *An Introduction to Engineering and Engineering Design,* John Wiley & Sons, Inc., New York, 1965.
4. Arthur D. Little, Inc., Competition in the Nuclear Power Supply Industry, USAEC Report NYO-3853-1, December 1968.

5. A. D. Hall, *A Methodology for Systems Engineering,* D. Van Nostrand Company, Inc., Princeton, N. J., 1962.
6. D. J. Wilde and C. S. Beightler, *Foundations of Optimization,* Prentice-Hall, Inc., Englewood Cliffs, N. J., 1967.
7. E. M. Brandes and L. H. Towle, Evaluation of Economic Benefits from a Continuing Civilian Nuclear Power Program, USAEC Report SRIA-65, Stanford Research Institute, 1962.
8. U.S. Atomic Energy Commission, Cost–Benefit Analysis of the U. S. Breeder Reactor Program, USAEC Report WASH-1126, April 1969.
9. Office of Science and Technology, Considerations Affecting Steam Power Plant Site Selection, 1968.
10. J. R. Garvey, *Chem. Eng. News,* **44**(21): 56 (1966).
11. W. B. Cottrell and A. W. Savolainen (Eds.), U. S. Reactor Containment Technology, USAEC Report ORNL-NSIC-5, pp. 2.1 et seq., Oak Ridge National Laboratory and Bechtel Corporation, August 1965.
12. J. P. Blakely, Compilation of United States Nuclear Standards, 6th ed., USAEC Report ORNL-NSIC-76, Oak Ridge National Laboratory, July 1969.

2 Electric Utility Economics

INTRODUCTION

2.1 Cost parameters play a key role in the design of nuclear power plants. Although special types of nuclear power systems may be designed for propulsion, space-vehicle energy sources, process heat, etc., the requirements for central-station electrical generating plants currently predominate and are emphasized here. Since the type of ownership also affects the influence of economics on design, the investor-owned utilities, which predominate in the United States, as shown in Table 2.1, also are emphasized here.

2.2 The operating economics of an electrical utility company is very complex, and the analysis of costs associated with meeting different types of power requirements for a variety of customers is important in the determination of rates.[1,2] This phase of electric utility economics, however, is of only secondary importance to the designer of the electric generating plant. On the other hand, a number of interrelations between the electrical-load requirements and the financial structure of the company affect capital costs and are important to the reactor-plant designer. Therefore only the features of electric utility economics that play a major role in the design of the generating station are considered here.

2.3 Many economic considerations important to the nuclear-power-plant designer are those which also frequently apply to engineering projects in general. Therefore some description of engineering economics principles is included in

TABLE 2.1

Private and Public Electrical Utility Capital Spending

Classification	Percent
Investor-owned companies	76
Cooperatives	2
Federal agencies	12
Public power districts, municipal and state projects	10
Total	100

this chapter as background material for a subsequent discussion of electric utility operational economics pertinent to the design of the system.

NATURE OF ELECTRIC UTILITY SYSTEMS

INTRODUCTION

2.4 Public utilities differ from the usual business enterprise in that they do not have direct competition but they must provide service to all who apply in their area of operation. For example, it would clearly be uneconomical, and indeed wasteful, for two or more electric utility companies to compete with one another in a given area, providing a duplication of lines etc. Regulation, usually by state government, is substituted for the pressure of competition to provide customer service at reasonable cost.

2.5 Because of regulation, both by state commissions and by federal agencies, many of the economic practices of the company are subject to public scrutiny, and the necessary rulings are subject to review by the courts. An approved rate structure is therefore the result of a rather complicated economic analysis that considers the costs, the need for a fair return, and the details of various customer requirements. Many of the unique economics features of electric utility companies are associated with this combination of public regulation and service to a variety of customer types. Table 2.2 lists the classes of service and the energy-generation percentage for each.

2.6 The actual generation of electricity in a power plant constitutes only a portion of the effort expended by an electric utility company. In addition to the generating activity, the operation of an electric utility includes a bulk transmission system in which substantial energy is transmitted at high voltage to large-size users or other load centers, a secondary distribution system, and

TABLE 2.2

Electric Service Classes

Class of service		Percent
Residential		30
Commercial		21
Industrial		45
Other		4
	Total	100

TABLE 2.3

Electric Utility Cost Functions

Functions		Percent
Production system		55
Bulk transmission system		7
Secondary distribution system		25
Customers' activities and sales		6
General administration		7
	Total	100

substantial efforts associated with customer billing, customer relations, and general administration. Table 2.3 gives an indication of the cost requirements for these various functions.

2.7 The nuclear engineer is concerned primarily with the reactor portions of the nuclear power plant. This necessary nuclear engineering activity constitutes less than half the overall engineering effort required for the entire power plant. Energy production, in turn, is responsible for only about half the operating expense of the utility company. However, design decisions by the nuclear engineer involve expenditures of many millions of dollars since the electric utility industry is of considerable economic importance and has a substantial growth rate.

2.8 Plant investment in the United States represents about 2% of the national wealth, and gross revenue is about 5% of the total U. S. value added by manufacture. The growth rate each year amounts to about 8%. Projected growth values and an estimate of energy sources[3] are shown in Table 2.4.

2.9 Coal accounts for about 65% of the electric energy production from thermal sources, whereas the cost of the coal itself accounts for about 35% of the total cost of producing electrical energy from coal-fired steam–electric plants. Since coal mining is still labor intensive despite much mechanization of mining in recent years, the cost of producing power from coal-fired generating plants can be appreciably affected by rising labor costs.

TABLE 2.4

U.S. Electric Utility Requirements and Supply,* 1965–1990

	1965	1970	1980	1990
Energy requirement, 10^9 kw-hr	1.06	1.52	3.07	5.83
Peak demand, 10^6 kw	188.0	277.0	554.0	1051.0
Total installed capacity, 10^6 kw	236.1	344.0	668.0	1261.0
Hydroelectric capacity, 10^6 kw	41.7	51.4	68.0	83.0
Pumped-storage capacity, 10^6 kw	1.3	3.6	24.0	65.1
Internal-combustion and gas-turbine capacity, 10^6 kw	4.9	16.2	27.0	42.0
Fossil-steam capacity, 10^6 kw	187.5	261.2	399.0	562.0
Nuclear capacity, 10^6 kw	0.7	11.6	150.0	509.0

*Excludes Alaska and Hawaii.

TABLE 2.5

U.S. Electric Utility Power Projections Relating to Population and Consumption

	1950	1968	Estimated for 1980	Projected for 2000
Population, millions	152	202	235	320
Total power capacity, 10^6 kw	85	290	600	1352
Kilowatt capacity/person	0.6	1.4	2½	~4¼
Power consumed/person/year, kw-hr	2000	6500	11,500	~25,000
Total consumption, kw-hr	325 billion	1.3 trillion	2.7 trillion	~8 trillion
Nuclear power capacity, 10^6 kw	0	2.7	150	941
Nuclear power capacity, % of total	0	<1	29	69

2.10 The projected large increase in U. S. electrical needs is due to population growth and much greater use of electricity on a per capita basis. During recent years, for example, electrical air conditioning and home heating have been increasing major contributors to electrical-load requirements, trends that are expected to continue. Table 2.5 summarizes the population and consumption projections. Tables 2.4 and 2.5, as well as Fig. 2.1, which illustrates energy-source projections, indicate trends rather than accurate values since such

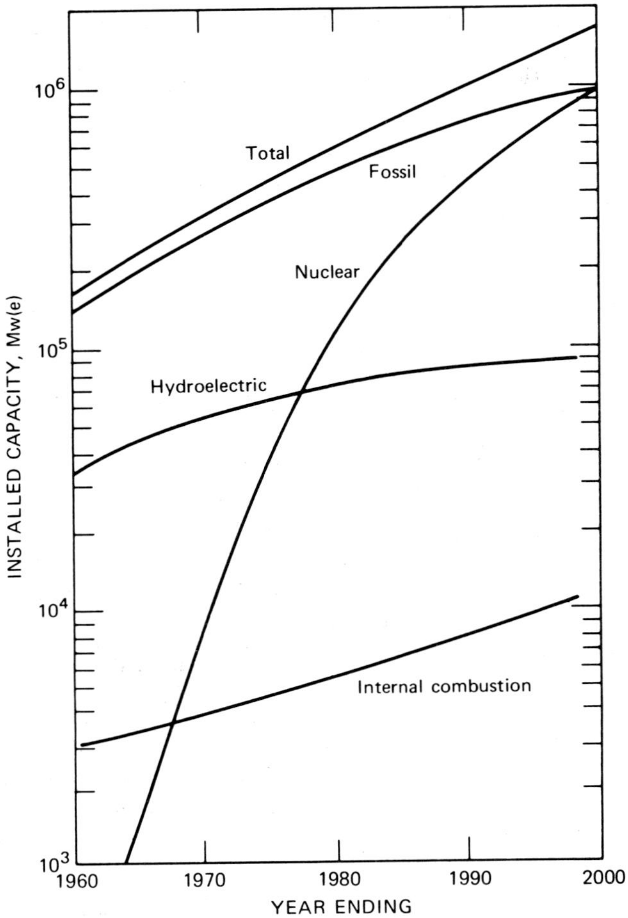

Fig. 2.1 Projected electric utility generating capacity.

estimates tend to change from year to year in response to changing economic conditions.

SYSTEM SIZE

2.11 The size of a system is an important design parameter since it determines the size of the individual generating unit that can be accommodated. Large units are normally preferred since the capital charges for such units expressed on a unit-power basis tend to be less than capital charges for smaller units (§3.33). The size of a given unit, however, is limited by the need to meet the varying demands of the system and yet provide adequate reserve capacity for emergency outages, scheduled maintenance work, fuel changes, and some contingency. Therefore the larger the unit, the greater the reserve requirement is. A unique feature of the electric utility business is the requirement for providing a product to the customer at any time and at the instant he desires it. If the supply is not dependable, users may complain to regulating agencies. Users of a large supply of electricity may also seek other energy sources or may generate their own. Therefore the capacity of the system must be adequate to meet the maximum demand.

2.12 In general, a given unit should constitute no more than 10 to 15% of the total system capacity so that, when it is shut down, no serious problems in meeting energy demand result. Although only a few systems in the United States have individual capacities over 7000 Mw,* interconnections provide a means of improving the reliability of the power-supply systems through mutual assistance, and hence medium-size utilities can operate large units [1000 Mw(e)]. The United States has a number of so-called pool arrangements with a combined capacity in some cases as great as 40,000 Mw. However, although interconnection arrangements do exist, the fine structure of transmission-line capacity of the systems involved could limit their flexibility. Furthermore, the cost of purchased energy required during peak-load periods may be relatively high (§2.15).

2.13 A trade-off does occur between the cost of transmission and the savings realized through use of large generating plants. In addition to providing flexibility for economic reasons, however, strong transmission networks now being built will provide a service flexibility to prevent the recurrence of major power failures and to avoid localized deficiencies. Thus there is considerable incentive to develop lower cost methods of transmitting electrical energy. As discussed in Chap. 6, such costs are important in the selection of generating-plant sites. In fact, since environmental considerations tend to limit the number of plant sites available close to load centers, low-cost transmission could make the necessary use of somewhat remote sites more economically acceptable.

LOAD STRUCTURE

2.14 The characteristics of the utility company's load structure[4] represent an important design parameter for the generating plant. Load changes are of two types: first, changes that occur over a fairly long time period, such as hours of the day, days of the week, and seasons of the year, and, second, momentary surges that require very rapid changes in the generating capacity of the system. Although it is clearly the obligation of the utility company to provide service instantaneously as these changes in demand are made, the pooling of a large number of diverse requirements provides a degree of averaging.

---

*Although there are about 200 investor-owned companies in the United States, the ten largest account for about 35% of the total capacity.

2.15 In a given community a pattern is normally established for electrical service demands which depends on personal habits and commercial and industrial schedules. Variations in this pattern are generally predictable. Weather and other climatic conditions, however, have a marked effect on the load requirements. Such changes are somewhat predictable on an average basis from weather records.

2.16 Although the load-structure pattern may vary from one utility company to another, the daily load-variation pattern shown in Fig. 2.2 is typical. The system requirements are minimal during the early morning hours. A peak is reached between midmorning and noon, and a drop occurs toward the latter part

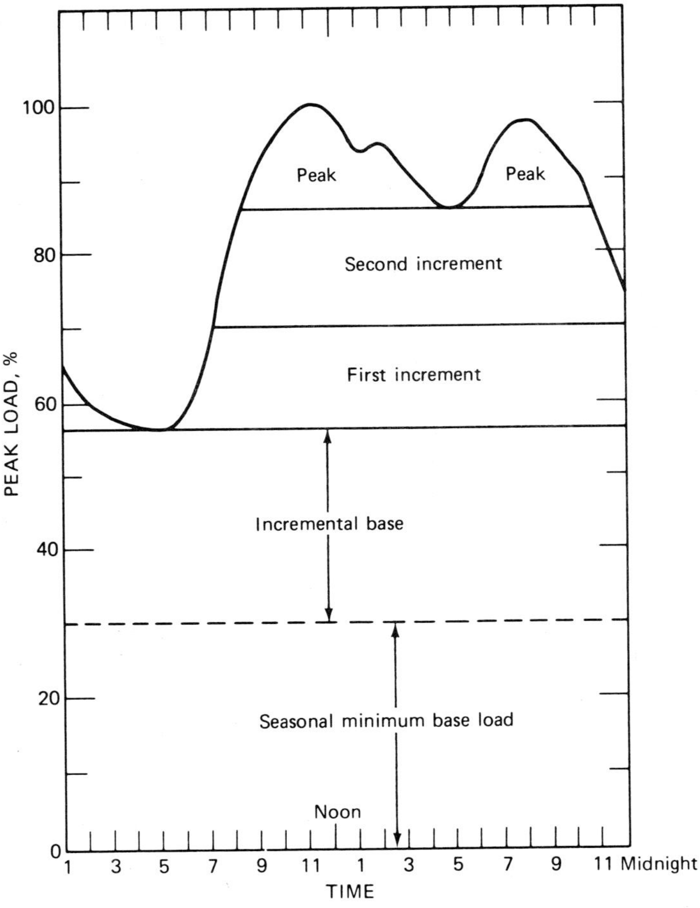

Fig. 2.2 Typical daily-load-variation curve.

of the afternoon. When darkness falls while offices and factories are still operating, an increase occurs, which can be substantial during the winter months. As industrial and business loads decrease in the evening, a decline follows and continues until the early-morning "low" is reached.

2.17 Various other types of graphs useful for load analysis include integrated loads on a daily and monthly basis and, more important, the seasonal variation integrated on a 1-hr basis, shown in Fig. 2.3. The seasonal off-peak periods in spring and autumn may be useful for the planned plant shutdowns necessary for routine maintenance and for fuel change in a nuclear plant. The nature of seasonal variations, however, varies considerably from one part of the country to another. In such states as Illinois, Missouri, and Iowa, for example, peak loads tend to occur during the summer months as a result of air-conditioning requirements. In neighboring Minnesota, however, the peak occurs in winter. Therefore energy exchanges between utilities serving these areas prove of mutual benefit.

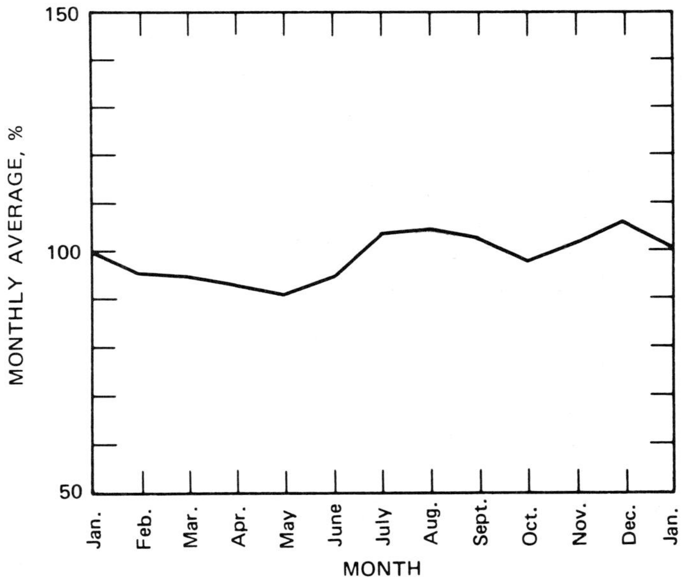

Fig. 2.3 Typical monthly-load-variation pattern.

2.18 A load-duration curve as shown in Fig. 2.4 is useful for showing the probable reduction in capacity factor for an individual unit as it becomes older, and presumably less efficient when compared with newer units. In any system, units with lower *incremental* operating costs will be assigned a larger share of the available load. For a given plant some cost components are load dependent whereas others are not. The fixed charges associated with the plant investment

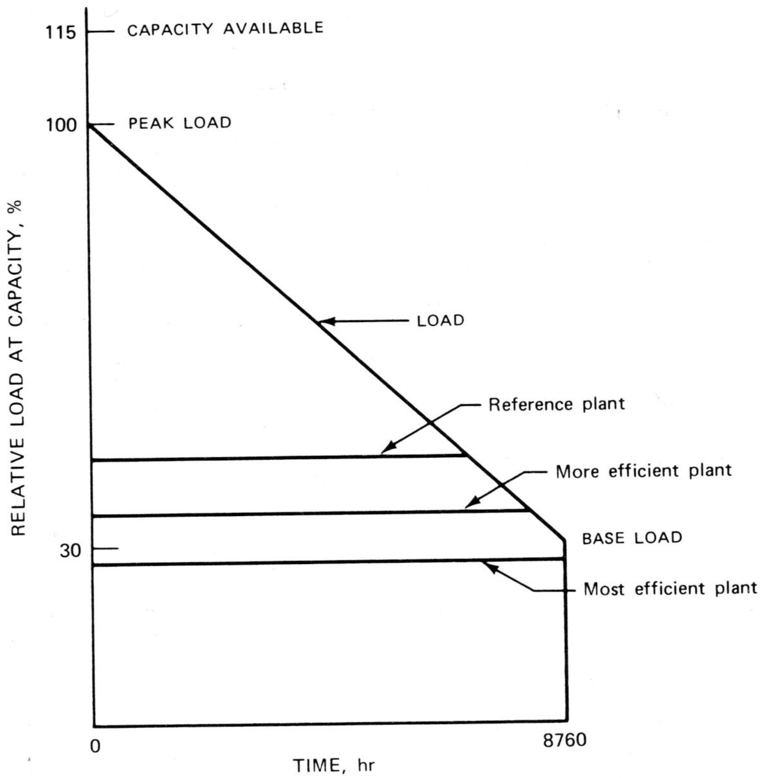

Fig. 2.4 Typical load-variation curve for a 1-year period.

must be paid, for example, whether or not the machine is loaded, but fuel costs depend on the energy produced. The change in operating expense with energy produced, approximately 80% of which is due to the increased fuel requirement, is known as the incremental cost. The plant that produces the most energy with the least expenditure of fuel is the most efficient plant and therefore tends to have the lowest incremental costs.

2.19 Consider a system in which the load is divided among the generating units in such a manner that the total cost of operating the system is minimal. If the load demand is increased, a new load assignment between plants must be made to operate at minimum cost at the new level. In a simple analysis the increase must be provided by the plant that exhibits the smallest increase in unit cost, i.e., the smallest incremental cost. The various incremental loading regions shown in Fig. 2.2 represent the planned assumption of load requirements above the so-called "base" load by units having progressively higher fuel costs.

2.20 In any load distribution, therefore, the plant with the higher incremental costs will be assigned load only after the plant with lower incremental cost is at capacity. Consequently, new plants with lower incremental costs will displace older plants (see bottom of Fig. 2.4). This picture is somewhat idealized, however, since there will be system-capacity additions specifically designed for peaking service; base-load plants will therefore tend to retain their high-capacity factors. In a mixed system containing both fossil-fuel and nuclear-fuel plants, nuclear-fuel plants are likely to have low incremental operating costs in comparison with newer fossil-fuel plants and hence remain at base-load service.

2.21 Several types of *operating factors* are related to the load requirements and are of major importance in cost analysis (§3.24). A *capacity factor* (or *plant factor* or *use factor*) is defined as the actual energy production during a given period divided by the energy production during the same period if the machine operated continuously at the manufacturer's rated capacity. A *load factor* is the ratio of the average load during a specified time period to the peak load. A *plant availability factor* is defined as the ratio of the integrated megawatt-hour output capability for a given period to the total rated megawatt-hours during the period. Planned outages for maintenance and refueling and forced outages caused by equipment breakdowns will decrease the availability. The availability factor will always be equal to or greater than the capacity factor since a plant that is ready to run may be held in reserve, depending on the load demand of the system. Therefore the availability factor is important in an analysis of a system's reserve-capacity requirements.

ENGINEERING ECONOMICS PRINCIPLES

2.22 The design engineer must have a thorough understanding of the economic parameters[5,6] affecting the selection of the "best" path to his design objective. Since many nuclear engineering graduate students and practicing designers have not received formal training in engineering economics, some background material is presented here for the subsequent discussion of nuclear-plant economics.

CAPITAL

2.23 The nature of capital funds is of considerable importance to the engineering designer. He may, for example, need to decide on a reduction in the required plant investment which would be compensated by an increase in other costs. To analyze the effects of changes in capital requirements, he must consider the factors that depend on the sources of the investments. An analysis of such sources can be quite complex since a combination of debt (bonds) and common-stock equity is normally involved along with important secondary effects contributed by income taxes. The "time value of money" is also pertinent. An introduction to these concepts is presented here.

Fixed-Charge Costs

2.24 Fixed-charge costs for engineering projects consist of the return on investments, income taxes, and depreciation. These are sometimes called *indirect charges.* In addition, total costs include such nonfixed charges as operating expenses (labor supplies and maintenance), certain types of property taxes, and insurance. For power plants, particularly nuclear plants, a fuel-cost item is also included. For a nuclear system, part of the fuel cost is in the fixed-charge category. An analysis of fixed charges is generally important in the study of engineering-project alternatives but tends to be complicated because of the time factors involved. The accounting terms described in §2.37, such as the present-worth factor, are useful for the purpose.

Capital Sources

2.25 "Capital" funds needed for new construction, purchase of fuel, and other purposes are obtained from the sale of stocks and bonds as well as from certain internal company sources, such as depreciation-charge recoveries and retained earnings. The most important sources of capital are equity funds (stocks) and borrowed funds (bonds). Equity funds are provided by the owners, the stockholders of the corporation. Funds may also be obtained by long-term borrowing through the sale of bonds. The relation of the bond holders to the corporation is quite different from that of the stockholders since bonds represent an agreement to pay interest and principal on a definite basis.

2.26 Capital funds are generally more readily available for privately owned public utility companies than for private industry because the necessity for rapid expansion is recognized. The "return on investment" is normally based on the results of rate negotiations with regulatory agencies, which recognize that new capital for necessary utility company expansion can be obtained only if the prospect of earnings is sufficient to attract such capital. Rates are established such that charges to the public are reasonable and that the earnings from the sale of energy cover operating expenses plus the fixed charges, including depreciation, income taxes, and a return on the investment. The allowed return takes into consideration a reasonable dividend rate needed to attract equity capital and to pay interest on borrowed money. It depends on the ratio of borrowed to equity capital, on the past financial management of the company, and on the general financial market existing at the time.

2.27 In 1969 the typical return on investment was estimated to be 8.0%. The total investment was assumed to be 52% funded debt and 48% common stock. These figures, which are typical of public utilities, show a much higher fraction of debt than is customary for so-called competitive industry. An interest rate of 6% was assigned to the debt capital, and a 10% return was assumed for the common stock after taxes. The use of bonds for financing capital needs is generally favored by utilities since the cost of such capital tends to be lower than equity capital and such bonds tend to be attractive to investors because of the inherent stability of a regulated company with earnings assured through the rate structure. An important reason for the lower cost of capital derived from bonds is that the interest on bonds is tax deductible whereas the funds that pay stock dividends are taxed.

Effect of Income Tax

2.28 Taxes on net income are another expense to be added to the expenses for operation etc. An analysis of factors affecting the amount of tax is quite complex, however, since depreciation accounting and other matters are involved. Only some effects of the tax itself will therefore be considered here.

2.29 Since funds to be distributed to stockholders as their "returns" (§2.26) are considered profit and thus are subject to tax, a substantial difference exists in the rate of return before taxes and after taxes. If the state regulatory agency determines that the allowable rate of return is 10%, for example, cost estimates must provide for a sufficiently higher fixed-charge rate so that, when the approximately 50% federal corporate income tax is applied, the desired 10% rate of return is still left for the stockholders. This applies, of course, only to the equity portion of the capital requirement and not to debt capital. Interest paid on bonds is, in fact, a deductible expense for income-tax purposes.

2.30 Economic studies for regulated utility companies are somewhat different from those for industry. Normally a utility's "revenue requirements" are made up of current operating expenses, an allowance for depreciation, income taxes, and an allowable return on the depreciated book value of the investment. As a result, changes in requirements for capital investment which could reduce taxable income do not really have a saving effect, since any investment saving would be accompanied by some decrease in allowable revenues. This is quite different from the normal industrial situation where the price of a competitive product is based on market conditions. A considerable incentive therefore exists in industry to reduce all manufacturing expenses, including taxes.

TIME FACTORS

2.31 The time value of money in any economic analysis can be most easily expressed in terms of interest. This is true even for equity funds where a "rate of return" is really meant. In the broad sense, therefore, interest can be considered money paid for the use of other money, either borrowed or part of the equity of a company. The interest concept must be applied whenever time is a variable whether or not interest is actually paid. This concept can be applied in various ways. For example, capital investments in equipment may not be made at the same time when the equipment is put into service or when "revenue" is produced as a result of the equipment purchased. Therefore interest charges must be added to the investment cost.

2.32 A somewhat different time effect occurs for plutonium produced in a reactor. Since this plutonium will yield revenue (salvage value) at a future time when the fuel is reprocessed, the bred plutonium may be carried on the books as an asset with interest applied to account for the time factor involved. Other approaches are possible, however. In fact, the accounting procedures for plutonium have not really become standardized.

Interest Relations and Terms

2.33 The concept of interest is very simple; it is the application of a fixed rate applied to a "principal" sum of money over a period of time, but a number of equations and special terms are useful, particularly in dealing with compound interest. If a sum of money, P, is invested at an interest rate, i, for a year, the interest earned at the end of the period is Pi. In compounding, this amount is added to the principal so that the sum invested for the second year is now $P + iP = S_1$ and the new interest earned at the end of the second year is $(P + iP)i$, or more systematically,

$$\begin{aligned} S_0 &= P \\ S_1 &= P + iP = P(1 + i) \\ S_2 &= P(1 + i) + Pi(1 + i) \\ S_2 &= P(1 + i)^2 \end{aligned} \tag{2.1}$$

At the end of n years, therefore, the total sum accumulated is $S_n = P(1 + i)^n$.

2.34 The quantity $(1 + i)^n$ is known as the *single-payment compound amount factor.* The reciprocal, $1/(1 + i)^n$, the *single present-worth factor*, is useful in finding a principal, P, that will give a required total amount, S, in n years.

$$P = S \left[\frac{1}{(1 + i)^n} \right] \tag{2.2}$$

P may then be considered as the *present worth* of the total, S, anticipated after n years.

2.35 Another type of accumulation results from the investment of a fixed sum, R, at the end of each year for n years. The total is the summation of the individual subtotals compounded over the years applicable for each payment.

$$S = R + R(1 + i) + R(1 + i)^2 \ldots R(1 + i)^{n-1} \tag{2.3}$$

where the amount $R(1 + i)^{n-1}$ is the accumulation from the payment made at the end of the first year and the first term, R, is the final payment, which earns no interest.

Multiplication by $(1 + i)$ and then subtraction of Eq. 2.3 yields

$$iS = R[(1 + i)^n - 1]$$

or

$$S = R\left[\frac{(1 + i)^n - 1}{i}\right] \tag{2.4}$$

This expression is known as the *uniform-annual-series compound factor.* Its reciprocal, $i/[(1 + i)^n - 1]$, is called the *sinking-fund-deposit factor.* This factor is used to determine the regular payment required to produce a desired amount at the end of a given period of time.

$$R = S\left[\frac{i}{(1 + i)^n - 1}\right] \tag{2.5}$$

The sinking-fund concept is used in one type of depreciation accounting to provide for the replacement of an asset at the end of its useful life.

2.36 A variation of this principle is the determination of the uniform end-of-the-year payment needed to repay a debt. From the viewpoint of the lender, this is the same as making a single present investment, P, which is returned, with interest, as a series of end-of-the-year payments. The payment scheme is the same as that for the sinking-fund deposit except that the lump-sum payment at the beginning is related to the alternate accumulation, S, by

$$S = P(1 + i)^n$$

Then substituting in Eq. 2.5 yields

$$R = S\left[\frac{i}{(1 + i)^n - 1}\right] = P(1 + i)^n\left[\frac{i}{(1 + i)^n - 1}\right] = P\left[\frac{i(1 + i)^n}{(1 + i)^n - 1}\right] \tag{2.6}$$

The final expression in brackets in Eq. 2.6 is known as the *capital-recovery factor.* It is equal to the sinking-fund factor plus the interest rate. The reciprocal, which would be useful if P were expressed in terms of R, is known as the *uniform-series present-worth factor.*

Present-Worth Concept

2.37 In the evaluation of alternate engineering projects involving the expenditure of funds, incurring of costs, and receipt of revenue, all at different times, a systematic treatment of the effect of the time variable of money is useful. The value of money can be considered to change as it is moved through time. In the present, for example, money has a greater value than it would have at some time in the future because it can be put to a useful purpose in the interim. Some of the terms discussed in the previous section are useful in design comparisons involving expenditures.

2.38 The *present-worth concept*, for example, provides for the shifting of money from one time level to another with a corresponding shift in value. If r is the effective earning rate or interest rate, then the present worth of 1 dollar *due* 1 year in the future is $1/(1 + r)$. This corresponds to shifting backward in time. Similarly, the present worth of 1 dollar invested a year previously would be $(1 + r)$, corresponding to a shift forward. In electric utility economics, it is useful to know the present worth of revenue requirements for future years of plant operation. This corresponds to the case of the backward shift. In the case of simple interest (single payment), the present value, P, can be expressed as

$$P = \frac{1}{(1 + r)^n}$$

where n is the number of years involved. The *compounding* of interest results in the equations mentioned in §2.25. Present-worth tables for various compounding schemes are available.[6]

Example 2.1

An electric utility is planning an orderly expansion of generation and distribution facilities to meet projected increases in load over the next 20 years. The first of two alternate plans being considered requires an initial investment of \$60 million followed by an additional investment of \$20 million at the end of 10 years. During the first 10 years, an expenditure of \$10 million per year will be required. This will rise to \$15 million per year for the last 10 years. Finally, a salvage value of \$5 million is estimated at the end of the 20-year period.

In the second alternate plan, an initial investment of \$20 million is required, followed by additional investments of \$30 million at the sixth and twelfth years. Annual expenditures of \$12 million per year are required over the 20-year period with no salvage value at the end. Compare the present worths of the two plans on a 6% interest-rate basis.

SOLUTION

	Millions of dollars	
	Plan 1	**Plan 2**
Initial cost	60.0	20.0
Additional investment		
1. 20(single-payment present-worth factor,* 6%, 10 years) = 20(0.5584)	11.2	

SOLUTION (Continued)

2. 30(single-payment present-worth factor, 6%, 6 years) = 30(0.7050)		21.2
30(single-payment present-worth factor, 6%, 12 years) = 30(0.4970)		14.9
Annual payments		
1. 15(uniform-series present-worth factor, 6%, 20 years) − 5(uniform-series present-worth factor, 6%, 10 years) = 15(11.470) − 5(7.360)	135.2	
2. 12(uniform-series present-worth factor, 6%, 20 years) = 12(11.470)		137.6
Salvage, 5(single-payment present-worth factor, 6%, 20 years) = 5(0.3118)	−1.5	
Totals	204.9	193.7

*See Eq. 2.2.

DEPRECIATION

2.39 Depreciation is an important component of the cost of capital. Any differences in calculating depreciation that affect the cost of capital are therefore significant to the designer. *Depreciation accounting* is a systematic approach to the distribution of tangible capital assets, less salvage value, over the useful life of the asset. Different methods of determining this *allocation* can result in different year-by-year "charges" and, as an important secondary effect, significant differences in income taxes, which are a separate component of cost of capital. Depreciation-accounting methods may provide a uniform write-off during the service life, a greater write-off during the early years than during the final years, or a smaller write-off earlier than later.

2.40 Although the uniform, or straight-line, method is easiest to apply, it fails to consider the practice of using assets for standby or "inferior" uses during the final years of life. This is particularly true for electric utilities where older, less-efficient generating units may provide only peak-load service (§2.20). Therefore methods giving a greater write-off during early years of asset lifetime are important here.

2.41 Two common methods providing for early write-off are *double-rate declining-balance accounting and sum-of-the-years'-digits accounting.* In the first method an annual depreciation rate computed as 200% of the estimated life in years results in a write-off of about two-thirds of the cost of an asset in the first half of the service life. In the second method the digits corresponding to the number of years of estimated life are added together. Considering a 30-year asset, for example, the sum of the digits is 466. During the first year 30/466 of the first cost − salvage value would be depreciated; during the second year, 29/466; during the third year, 28/466; and so on. The charge therefore decreases by 1/466 each year. By this method about 75% of the depreciable cost is written off during the first half of the asset lifetime.

2.42 Another method of depreciation accounting is based on the establishment of an "imaginary" sinking fund by uniform end-of-the-year deposits throughout the life of an asset. The accumulation, with interest, is just sufficient to equal the cost of the asset less its salvage value at the end of its estimated life. The so-called book value at any time can be determined from the difference between the accumulation and the net cost. Since the book values tend to be above the values obtained from the straight-line method, the sinking-fund approach tends to give a less favorable depreciation for tax purposes than do methods that provide a greater write-off during the early life of the asset. A company can legally use one method of depreciation for tax purposes and another method for its own accounting. Similarly, for tax purposes the depreciation lifetime may be different from the "book" lifetime. The effects of such complications are beyond the scope of this introductory treatment, however.

ECONOMIC ANALYSIS

Cash Transactions

2.43 Although many variations are used in the economic analysis of engineering projects, the *cash-flow concept* is very helpful for orientation. In simple terms all expenses, such as fuel purchases, interest payments, and supplies expenditures, are considered as cash transactions at a specific time, whereas revenues are similarly considered as cash transactions in the opposite direction. By shifting the value of money through time by the present-worth concept, the entire cash-flow model can yield equivalent costs at any specific time in the history of the project or provide an average cost picture for a desired time period.

Direct and Indirect Costs

2.44 In the economic analysis of engineering systems, it is often very useful to separate the so-called indirect costs from the direct costs. Direct costs are generally for services or materials of a cash-flow nature and include no provision for the time value of money or present-worth considerations. On the other hand, indirect costs, or fixed charges (§2.24), usually include such items as (1) interest on borrowed money, (2) earned surplus (profit), (3) federal and state income taxes, (4) other taxes, and (5) other costs associated with the time value of money.

2.45 The exact calculation of such costs can be quite complicated and would require a detailed cash-flow analysis. The relation between direct and such indirect costs tends to be nonlinear. For many design purposes, however, it is satisfactory to treat indirect costs in a manner similar to an interest charge on borrowed money whereby a fixed percentage per unit time is applied to the principal sum. The rate in this case includes not only the composite cost of money but also an allowance for taxes and other indirect costs that are related to the investment.

2.46 With a simple-interest approach, therefore, the indirect cost can be expressed as the product of the outstanding debt, an annual "interest rate," and the time duration in years. The debt may vary with time, however, particularly with nuclear fuels.

2.47 Figure 2.5 illustrates in simplified form the time behavior of an outstanding debt for one of the direct-cost components.[7] In this example, N, the number of time periods, equals 3. The sloping line from t_a to t_b approximates

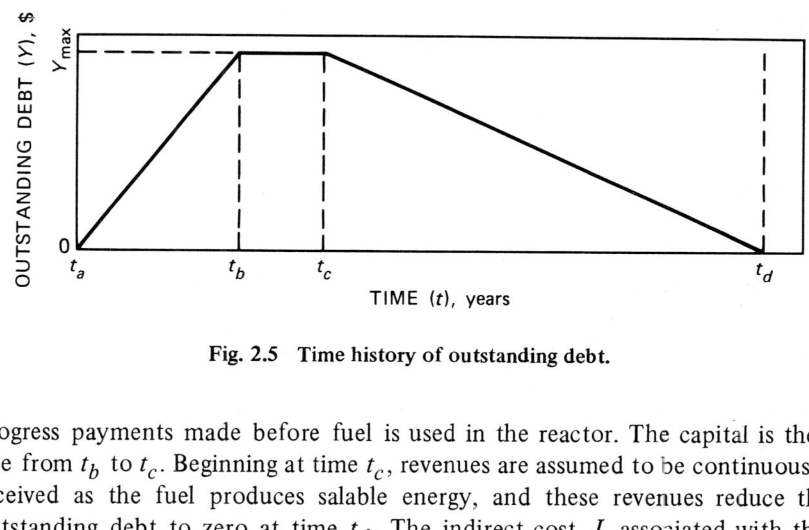

Fig. 2.5 Time history of outstanding debt.

progress payments made before fuel is used in the reactor. The capital is then idle from t_b to t_c. Beginning at time t_c, revenues are assumed to be continuously received as the fuel produces salable energy, and these revenues reduce the outstanding debt to zero at time t_d. The indirect cost, J, associated with this direct-cost component is therefore

$$J = Y_1 i_1 t_1 + Y_2 i_2 t_2 + Y_3 i_3 t_3$$

where Y is the average value of the outstanding debt during the time period, t, and i is the interest rate.

Working Capital

2.48 In addition to the investment, or "outstanding debt" associated with the physical plant or other major direct-cost component, an engineering project requires *working capital,* or a *working fund.* This category includes the funds that must be borrowed to provide *cash flow* to pay due accounts of the project before corresponding revenues are received. This working-fund investment is not subject to depreciation. In the operation of a nuclear power plant, for example, cash is needed to pay for various fixed charges associated with the fuel, particularly the cost of fabricating the core, before revenue is received. Other working capital is required for materials and supplies, etc., as listed in §3.66. An economic analysis, therefore, must include the indirect charges, or "interest," associated with the working capital.

Considerations of Capital Expenditure in Project Evaluation

2.49 In project analysis emphasis is often given to capital expenditures. Even allowing for the time value of money, we must remember that capital funds are by no means available in unlimited amounts. The determination of availability can be a rather complex matter since funds for future capital expenditures may come from both internal and external sources. Internal sources may include the charges allowed for depreciation and a certain fraction of the earnings that are retained rather than distributed to the stockholders of a corporation. External sources include the proceeds from the sale of bonds or stock issues. The availability of external sources depends on many factors, including the financial health of the corporation and general economic conditions. At any rate, capital is generally considered precious, with proposed expenditures worthy of careful analysis.

2.50 Two somewhat different criteria are used to evaluate proposed capital expenditures. Both consider the time value of money. The *present-worth* approach is similar to that shown in Example 2.1, where the expenditures required for the alternatives being considered are converted to a consistent basis by shifting them through time to some reference period and using as an interest rate a so-called *minimum rate of return.* The minimum rate of return, or discount rate, may be taken as the overall cost of money to the enterprise; but it is sometimes taken at a somewhat higher value, with the argument presented that the rate of return must be above the cost of the funds or there would be no point in the investment. Furthermore, an allowance must be made for certain intangible factors. In many corporations a rate of return for new investments is established by management.* With the use of an appropriate discount rate, the

*In detailed analyses the effective interest rate and rate of return prove to be quite complicated parameters that depend on such factors as income tax, the effect of issuing new equity shares on the earnings of existing shares etc. Also, the engineering economist and the accountant may have some differences in viewpoint. Such matters are beyond the scope of this discussion, although some discount-factor considerations are given in §2.59.

present-worth method can be used to place proposed capital-expenditure plans in economic order by comparing the present value of the stream of expenditures with the corresponding stream of benefits (revenues).

2.51 In the *rate-of-return* method, an iterative procedure determines the unknown discount rate that is needed to balance the stream of expenditures and benefits. It is similar to the present-worth method except that now the rate of return, or interest rate, is the unknown quantity. The calculated rate of return can then be compared with the minimum rate of return to determine how attractive the proposed expenditure is. The project that provides the highest rate of return tends to be the most attractive when alternates are considered.

2.52 Table 2.6 illustrates a cash-flow comparison for a project evaluation of proposed improvements, known as the Cascade Improvement Program (CIP), to the U.S. gaseous diffusion plants. According to the study,[8] made in 1968, an expenditure of $477 million on the CIP during the period 1969 through 1978 would increase the separative capacity of the three gaseous diffusion plants from 17,213 to 22,112 metric tons per year. The power level, the waste assay, and the separative work produced, sold, and preproduced *without* (case 1072) and *with* (case 1073) the CIP are all described in the study but are not given here. The study also shows that, starting in 1976 and continuing through 1990, greater quantities of separative work would be sold if the CIP were installed (case 1073) than if it were not installed (case 1072).

2.53 Table 2.6 gives the cash flows without the Cascade Improvement Program and with the CIP starting in FY 1973. Column 10 gives the differences in the net cash flows between these two cases. Note that there are *net disbursements* each year starting in FY 1969 and continuing through FY 1978. From FY 1979 through the end of the study period, there are net revenues due to the greater quantities of separative work being sold.

2.54 The interest rate that makes the present worth of the cash-flow stream in column 10 of Table 2.6 zero is the rate of return earned on the $477 million. This interest rate is determined by a trial-and-error process (§2.51). In this case, 17.3% is the rate of return earned on the $477 million investment. This capital expenditure is made over a period of years and has the effect of increasing the net revenues in future years. In other words, the increases in the net revenues caused by the capital expenditure are sufficient to (1) pay 17.3% interest on the unrecovered capital investment outstanding at the end of each year and (2) pay off the capital investment by the end of the specified period, 15 years ending in 1990.

2.55 During the years before additional net revenue is received, compound interest at 17.3% accrues on the capital investment. The stream of annual net revenues then pays the interest on the unrecovered capital balance (original investment plus compound interest minus repayments), and the unrecovered balance is reduced by the amount by which the annual net revenue exceeds the annual interest payment. By the end of the period, the net revenues pay off in total the unrecovered capital balance.

TABLE 2.6

Cash Flows Without Cascade Improvement Program Compared with Cash Flows with Cascade Improvement Program Starting in Fiscal Year 1973

(Most Likely Requirements for Separative Work)

	Without CIP (case 1072), 10^6 $				With CIP starting in FY 1973 (case 1073), 10^6 $				Case 1072 vs. case 1073 (increase in revenue due to FY 1973 CIP installation),* 10^6 $
FY	Revenue	Cascade operating cost	Capital cost	Net cash flow*	Revenue	Cascade operating cost	Capital cost	Net cash flow*	
(1)	(2)	(3)	(4)	(5)	(6)	(7)	(8)	(9)	(10)
1969							1	−1	−1
1970	120	125	4	−9	120	125	12	−17	−8
1971	182	152	3	27	182	152	38	−8	−35
1972	245	178	1	66	245	178	47	20	−46
1973	279	214	0	65	279	214	81	−16	−81
1974	344	247	0	97	344	247	90	7	−90
1975	430	268	0	162	430	267	78	85	−77

1976	480	270	0	210	496	272	68	156	−54
1977	561	273	0	288	579	277	50	252	−36
1978	653	276	0	377	673	281	20	372	−5
1979	411	278	0	133	815	284		531	398
1980	412	276	0	136	581	283		298	162
1981	448	276	0	172	575	283		292	120
1982	448		0						
1983	448		0						
1984	448		0						
1985	448		0						
1986	448		0						
1987	448		0						
1988	448		0						
1989	448		0						
1990	448	↓	0	↓	↓	↓		↓	↓
				3272			485†	4599	1327

Present value (7-1-1969) of increase in revenue due to FY 1973 CIP installation (column 10) at 5% = \$540 million, at $7\frac{1}{2}$% = \$334 million, at 17.3% = \$0 million. Rate of return is 17.3%.

*Disbursements are preceded by minus.

†Includes \$8 million for power plant restoration and \$447 million for CIP installation.

2.56 Another approach, the *present-worth method,* is shown in Table 2.7. In this method the present worth of the annual cash disbursements and revenues resulting from the \$477 million capital expenditure is determined at a discount rate, which is the minimum attractive rate of return, taken here at 5%, the assumed cost of money to the U.S. Government. The present worth so determined represents the present value of the net revenues, which are the annual amounts remaining after the operating costs and capital charges (also at 5% interest) are paid. Column 2 of Table 2.7 is therefore identical to column 10 of Table 2.6.* Thus the present worth might be considered "profit" since it is the present value of the revenues which are in excess of those required to pay full cost of the CIP, including the cost of money.

2.57 The annual net cash flows in column 2 of Table 2.7 represent the annual net disbursements (preceded by minus) and the annual net revenues that are realizable as a result of the capital expenditure of \$477 million on the Cascade Improvement Program. Column 3 of Table 2.7 gives the present-worth factors for 5% interest, and column 4 gives the present worth of the annual net cash flows. A summation of the values in column 4 is therefore the present worth of the total stream, amounting to \$540 million. In other words, the present worth of the streams of net disbursements and net revenues realizable as a result of the capital expenditure of \$477 million is \$540 million when a discount rate of 5% is used.

2.58 The benefits from a capital improvement have therefore been determined in two different ways. Additional revenues are equivalent to a rate of return of 17.3% earned on the investment or, on a present-worth basis, a \$540 million gain.

Considerations of the Discount Factor

2.59 A picture of the interrelation between parameters that affect the discount factor and a unit selling price of energy for an electric utility is given in a simplified model developed by Vondy.[9] He assumed that revenue from an investment over a certain period will retire all associated indebtedness as well as cover all costs. In actual practice, since a utility company normally increases its services with time and makes new investments that bring about a total increase in debt, the debt is not really retired. Although "retirement of debt" actually means "freeing the money for new investment," the calculation is not altered. The indebtedness was assumed to be in a fixed ratio of stock to bonds and the interest on bonds to be tax deductible whereas return on stock was not.

2.60 Although careful consideration is given here to the payment of income tax, many complications, such as local taxes, are avoided to preserve clarity. Since the less favorable sinking-fund method of depreciation and the more favorable sum-of-the-digits method add complexity, the more elementary

*Since the CIP is being evaluated, the amounts considered are in excess of those which would apply without the CIP.

TABLE 2.7

Present-Worth Determinations of the Cash-Flow Streams Caused by $477 Million Capital Expenditure at 5% Discount Rate

(5% is the Assumed Cost of Money to the U.S. Government)

End of fiscal year	End of year net cash flows,* 10^6 $	Present-worth factor at 5%	Annual present worths, 10^6 $
(1)	(2)	(3)	(4)
1969 (7-1-69)	−1	1.00	−1.000
1970	−8	0.95238	−7.619
1971	−35	0.90703	−31.746
1972	−46	0.86384	−39.736
1973	−81	0.82270	−66.639
1974	−90	0.78353	−70.517
1975	−77	0.74622	−57.458
1976	−54	0.71068	−38.376
1977	−36	0.67684	−24.366
1978	−5	0.64461	−3.223
1979	398	0.61391	244.334
1980	162	0.58468	94.716
1981	120	0.55684	66.819
1982	120	0.53032	63.637
1983	120	0.50507	60.607
1984	120	0.48102	57.721
1985	120	0.45811	54.972
1986	120	0.43629	52.354
1987	120	0.41552	49.861
1988	120	0.39573	47.487
1989	120	0.37689	45.225
1990	120	0.35894	43.071
		Total present worth	540.124

*The net disbursements (−) in column 2 do not equal the capital expenditure, $477 million, because of differences in the amount of power used (disbursement) and in separative work sold (revenue) between the case where CIP *is not* installed and the case where the CIP *is* installed. The difference in revenue is due to a small difference in waste assays.

straight-line method (fixed periodic depreciation) is used. In the analysis "operating costs" are costs that are immediately tax deductible, whereas "investment" or "capitalized expenditures" are costs that can be deducted only as they are depreciated.

2.61 Income and outlay are assumed to occur at the end of each accounting period. There will be an outstanding debt at the end of each period which is to be eliminated at the end of the history. The unit price of electricity required to retire this debt is taken to be constant over the plant life.

The following list defines symbols used for an accounting period, n:

$Q(n)$ = amount of energy sold during period.
$Y(n)$ = outstanding indebtedness before income and outlays during period are considered.
$Z(n)$ = investment (capitalized expenditure).
$V(n)$ = income from other-than-energy sale.
$D(n)$ = depreciation.
$O(n)$ = deductible operating costs.
$T(n)$ = income taxes.
$R(n)$ = net retirement income after costs and taxes.
C = direct cost before interest.
I = interest charge, which includes real cost of indebtedness and taxes.
P = unit selling price of energy to return all investment costs.
X = discount factor defined by the development.
N = history life.
r = tax rate on taxable income.
i = required return on stock.
j = required return on bonds.
b = fractional indebtedness in bonds.
m = fixed charge or interest on an investment.

Income tax is given by the applicable fraction of taxable income:

$$T(n) = r\,[P\,Q(n) - D(n) - O(n) - jb\,Y(n)] \tag{2.7}$$

Net income is income remaining after costs:

$$\begin{aligned} R(n) &= P\,Q(n) + V(n) - O(n) - [jb + i(1-b)]\;Y(n) - T(n) \\ &= (1-r)\,P\,Q(n) + V(n) - (1-r)\,O(n) + r\,D(n) \\ &\quad - [j(1-r)b + i(1-b)]\;Y(n) \end{aligned} \tag{2.8}$$

The outstanding debt is reduced by applying the net income:

$$\begin{aligned} Y(n+1) &= Y(n) + Z(n) - R(n) \\ &= [1 + j(1-r)b + i(1-b)]\;Y(n) + Z(n) \\ &\quad - (1-r)\,P\,Q(n) - V(n) + (1-r)\,O(n) - r\,D(n) \end{aligned} \tag{2.9}$$

Equation 2.9 is one of recurrence in the outstanding debt. If the terms other than Y are recognized as being independent of Y, Eq. 2.9 can be simplified to

$$Y(n+1) = (1+X)\,Y(n) + A(n) \tag{2.10}$$

where

$$X = j(1-r)b + i(1-b) \tag{2.11}$$

For an initial investment and indebtedness of $Y(1) = Z(0)$, the solution to Eq. 2.10 is given by the expression

$$Y(a) = \sum_{n=0}^{n=a-1} (1+X)^{a-n-1}\,A(n)$$

and retiring all indebtedness, $Y(N+1) = 0$, is given by the expression

$$\sum_{n=0}^{N} (1+X)^{N-n}\,A(n) = 0 \tag{2.12}$$

In terms of the primary variables, the solution is given by

$$\sum_{n=0}^{N} (1+X)^{N-n}\,[Z(n) - (1-r)\,P\,Q(n) - V(n) + (1-r)\,O(n) - r\,D(n)] = 0 \tag{2.13}$$

The solution of Eq. 2.13 for an unknown unit selling price of energy is

$$P = \frac{\displaystyle\sum_{n=0}^{N} (1+X)^{N-n} \left[\frac{Z(n) - V(n)}{(1-r)} + O(n) - \frac{r}{1-r}\,D(n)\right]}{\displaystyle\sum_{n=1}^{N} (1+X)^{N-n}\,Q(n)} \tag{2.14}$$

Equation 2.13 discounts all items to the end of the history, i.e., future-value discounting; it is generally more flexible to work with present-value discounting, obtained by multiplying the numerator and denominator of Eq. 2.14 by $(1+X)^{-N}$:

$$P = \frac{\displaystyle\sum_{n=0}^{N} (1+X)^{-n} \left[\frac{Z(n) - V(n)}{(1-r)} + O(n) - \frac{r}{1-r}\,D(n)\right]}{\displaystyle\sum_{n=1}^{N} (1+X)^{-n}\,Q(n)} \tag{2.15}$$

2.62 A single expression is therefore developed for the unit selling price of energy needed to return all investment cost. Although the expression appears somewhat complex, it can easily be evaluated with a digital-computer approach.

Levelized Equivalent Uniform Annual Values

2.63 Year-by-year economic studies of the cost of capital are frequently necessary for the high degree of precision required for management decisions on specific projects. Slightly less exact methods, however, are frequently useful; these make use of percentage rates or levelized equivalent values representing the average value of year-by-year changes taken over the lifetime of the project. The various techniques for converting economic information to a uniform annual basis are described in engineering economics texts.[6]

ELECTRIC UTILITY COST ANALYSIS

INTRODUCTION

2.64 Rates for a regulated utility company are determined by analyzing the revenue requirement (§2.5), considering various operating expenses and a fair return on the investment. In other words, the rate depends on the cost of energy produced. Since various types of customers are served, however, the apportionment of total costs in a fair manner is complicated.

2.65 The cost of energy produced can be analyzed by using a model, in which variables for the different customer load requirements are considered.[1] The annual operating expense, for example, can be expressed as a summation of four elements: (1) elements that are constant and independent of installed capacity, peak effects, or amount of energy generated; (2) a capacity element; (3) an element associated with the need to be ready to carry load without necessarily producing energy; and (4) an element proportional to the energy produced. Elements 3 and 4 are significant in arrangements for operating capacity and energy interchanges associated with power pooling. On the average, about 25% of the cost of energy can be attributed to the constant and capacity elements, 50% to the peaking capability, and 25% to the energy itself. Element 4, the so-called incremental cost of energy production, is also the key variable in determining base, semibase, and peaking types of production units.

2.66 This model can be described by a four-term equation,

$$C_T = A + B\,(\text{kw capacity}) + C\,(\text{kw peak}) + D\,(\text{kw-hr}) \qquad (2.16)$$

where C_T is the total cost for the specified kilowatt-hours produced and A is constant for all expenses unrelated to the energy produced or the energy-demand characteristics. The second and third terms are related to the demand but each in a different way. Constant B accounts for all expenses directly *proportional* to installed capacity, and C accounts for the additional expense associated with the necessity of being *ready* to carry a peak load without actually producing energy. In each case nonlinear effects could be described with additional terms.

2.67 The installed-capacity term, B, is a function of the individual customer's peak-load requirements, but constant C depends on the peak-load requirements as determined at the generating station. The difference is based on the "smoothing" of irregular requirements by combining needs of different types of customers in a large system. The generating station, of course, provides the needs of the entire system.

2.68 A detailed study of the components of cost analysis, particularly as they affect the "cost to serve," is not appropriate here. We shall, however, discuss the determination of unit costs and the allocation of unit costs to the customer.

UNIT COSTS

2.69 Equation 2.16 serves as the basis for the determination of unit costs. The first and last terms, those associated with fixed effects and with the energy produced, are straightforward. The two middle terms, involving capacity and peaking, are more difficult to evaluate, however. One approach is to study the effect on cost of incremental changes in system loading. Loading may have different effects on different portions of the utility system. For example, it is convenient to consider separately the production system, the bulk transmission system, and the various primary and secondary voltage distribution systems. Although the most important measure of capacity is the kilowatt rating, the power factor, and hence the kilovolt-ampere rating, is also important, particularly in the distribution system. Since this phase of the analysis is to determine the cost of changes in the load structure, certain parameters describing the load structure (e.g., various types of peaking ratios) also become useful. An indication of the relative cost contribution of model elements to each of the various functions of a utility system is given in Table 2.8.

PARAMETERS FOR ALLOCATION OF COST TO CUSTOMERS

2.70 The allocation of costs to the customer is important in setting up a rate structure. Actually, such an allocation can go hand in hand with the establishment of a unit-cost structure, which also depends on the different load requirements of various types of customers.

2.71 Coincidence factors are important in allocating the demand-related portion of costs to different classes of customers. Load changes of the two types

TABLE 2.8

Typical Utility Cost Distribution

Function	Constant	Customer peak	Generating peak	Energy
	Cost contribution, %			
Generating station			60	40
Bulk transmission			100	
Distribution system	34	16	50	
Fixed (billing, sales, metering)	95	5		
Administrative overhead	40	5	50	5
System total	20	5	50	25

mentioned in §2.14 affect the cost analysis and determine the transient specifications for the generating system. Since the details of these variations may differ from one customer to another, some "smoothing out" of system fluctuations is obtained as more and more customers of different types are connected to the system. In other words, each customer has a varying load requirement and demands a certain reliability of service to meet it. He could obtain this from an individual facility. However, if the requirements among the customers are substantially different, the load changes can be averaged out to provide service at less cost than would be possible under an individual basis. Also, the use of larger generating units to meet the needs of many customers provides an economic advantage. The expense of distribution, however, reduces this advantage to some extent.

2.72 Since irregular demand requirements increase the cost of providing service when demand is expressed on a unit-energy (mills per kilowatt-hour) basis, a pattern of use by different types of customers which reduces such irregularities through averaging, therefore reduces the unit cost. Load-research studies have provided empirical relations between the energy use and demand requirements of different classes of customers, such as residential, manufacturing, and commercial users. Such information can be used to develop cost-to-serve relations, including the effects of a number of parametric cost components summarized in Eq. 2.16.

INFLUENCE OF UTILITY SYSTEMS ON DESIGN REQUIREMENTS

2.73 Since the generating plant, whether nuclear or fossil fueled, is a subsystem or a portion of the utility system (§1.18), a number of design interactions apply between the requirements for the generating plant and the system characteristics. This follows in a straightforward manner from the concepts of system engineering design (§ 1.16) and is particularly important in optimization procedures.

2.74 The load pattern, for example, and the resulting interaction with various fuel-system economic factors can lead to decisions regarding peaking or base-load service and, in turn, to the establishment of control-system specifications. As a second example, the system size and interconnections with other systems can have an important effect on the maximum size of a single generating plant that can be accommodated in the system. Seasonal load changes can determine the periods appropriate for generating-plant downtime for routine maintenance. Since fuel changes in a nuclear system can best be accomplished during such downtimes, fuel-management specifications are also affected.

REFERENCES

1. C. W. Bary, *Operational Economics of Electric Utilities,* Columbia University Press, New York, 1963.
2. P. J. Garfield and W. F. Lovejoy, *Public Utility Economics,* Prentice-Hall, Inc., Englewood Cliffs, N. J., 1964.
3. F. S. Brown, *Trends and Growth Projections of the Electric Power Industry,* Federal Power Commission Statement, Hearings Before the Joint Committee on Atomic Energy, Oct. 28, 1969, Superintendent of Documents, U.S. Government Printing Office, Washington, D. C.
4. J. M. Vallance, Nuclear Power Briefing for the Coal Industry, September 29–30, 1966, Oak Ridge, Tennessee, USAEC Report CONF-660930, p. 152, Oak Ridge Operations Office (AEC), February 1967.
5. E. L. Grant and W. G. Ireson, *Principles of Engineering Economy,* 4th ed., The Ronald Press Company, New York, 1964.
6. N. N. Barish, *Economic Analysis for Engineering and Managerial Decision Making,* McGraw-Hill Book Company, Inc., New York, 1962.
7. NUS Corporation, Guide for Economic Evaluation of Nuclear Reactor Plant Designs, USAEC Report NUS-531, pp. 5–19, January 1969.
8. S. G. Tucker Arnold, Three Methods of Evaluating Capital Expenditures in the Public Sector in which the Time Value of Money Concept is Used, USAEC Report ORO-674, Oak Ridge Operations Office, December 1969.
9. D. R. Vondy, Basis and Certain Features of the Discount Technique, in A Comparative Evaluation of Advanced Converters, USAEC Report ORNL-3686, M. W. Rosenthal et al. (Eds.), Oak Ridge National Laboratory, January 1965.

3 Nuclear-Plant Economics

INTRODUCTION

3.1 The cost analysis of a nuclear power plant is a special case of an engineering economy study. Analyses can provide a basis for business decisions involving alternate projects, such as a nuclear- or fossil-fuel plant or the selection of one nuclear-reactor concept from among several. Cost analysis is indeed essential to the engineering design of a nuclear power plant. Since the plant must produce energy at costs competitive with those for other energy sources, economic parameters are extremely important in achieving an optimum design. The nuclear-plant designer must therefore be completely familiar with reactor costs and the nature of an economic analysis for a proposed plant.

3.2 Economic studies are of two types. One concerns a specific project to determine the projected costs under known conditions but perhaps with several alternatives. "Classic" examples are the 1963 study by the Jersey Central Power and Light Company of a proposed 620-Mw(e) generating station[1] at Oyster Creek, N. J., and a 1966 Tennessee Valley Authority (TVA) study.[2] Both organizations decided to build a nuclear plant after studying both nuclear- and fossil-fuel alternates. The economic analyses were based on each organization's accounting practice and their experience in such matters as investment return and taxes.* Since such an analysis is complicated and oriented toward the user's

*There are marked differences in fixed-charge rates for an investor-owned utility and for TVA, a U. S. Government agency (§3.36).

(§1.8) requirements, it is of limited use at the conceptual design stage in determining the effects of parametric changes on costs.

3.3 The second type of economic study consists of *concept* comparisons or parametric studies and makes much greater use of average conditions. To provide for a uniform treatment of cost parameters, the U. S. Atomic Energy Commission (AEC) has issued a standard procedure, *Guide for Economic Evaluation of Nuclear Reactor Plant Designs,*[3] which is useful for preliminary cost analyses but may not be appropriate for actual cost determinations. This procedure is emphasized here since it is more straightforward than the first approach for the handling of economic-parameter contributions to a systems design analysis.

3.4 Consistent with the growth in use of digital computers for all types of engineering calculations, numerous economic codes have been developed to meet a variety of needs. Since quite complicated economic calculations can be handled easily by machine methods, codes for survey studies normally permit quite detailed descriptions if desired. Therefore they tend to fall within the first of the preceding categories with options to accommodate the user. Examples of codes for computing nuclear power costs are POWERCO[4] and PACTOLUS.[5]

TYPES OF CHARGES

3.5 An economic analysis of a nuclear power plant includes a study of *all* contributions to the cost of generating electricity. These range from items based on fixed charges, which apply whether or not there is any production, to costs that on an annual basis depend directly on the energy generated. Since questions that affect the uniformity of energy requirements, the nature of capital funding, fuel options, and a wide variety of engineering parameters are thereby introduced, a systematic accounting basis for these contributions is essential.

3.6 In the *Guide for Economic Evaluation of Nuclear Reactor Plant Designs,*[3] to be referred to subsequently as the *Guide,* power generating costs are subdivided into *plant-investment fixed charges, fuel costs,* and *operation and maintenance costs.* A suggested format for summarizing such costs is shown in Table 3.1. In addition to the *investment* required for the reactor plant itself, the fixed charges are also associated with fuel and "working capital" (§2.48). Items such as depreciation, insurance, and taxes are normally included in the fixed-charge rate applied to plant investment. The determination of actual costs, normally expressed on an annual basis, depends on the nature of the financing practices appropriate for the owner, either private or public. For investor-owned utilities, for example, corresponding fixed charges each year may amount to about 14% of the investment required for plant construction. A somewhat lower rate would apply to capital that does not depreciate, as explained in §3.36.

3.7 Several features of the fixed charges should be emphasized: Since the fixed charges are a function of investment, or capital cost, any design parameters

TABLE 3.1

Format for Summarizing Electric Energy Generation Costs

Cost component	Capital cost, 10^3 $	Fixed-charge rate, %	Annual cost (or revenue), 10^3 $	Unit energy cost (or revenue), mills/kw-hr
Plant investment				
Depreciating assets	———	———	———	
Nondepreciating assets	———	———	———	
Subtotal			———	———
Fuel				
Unit direct cost			———	———
Unit indirect cost	———	———	———	———
Subtotal				———
Operation and maintenance				
Direct cost			———	
Working capital	———	———	———	
Subtotal				———
Total electric energy generation cost			———	

that affect the investment will, in turn, affect the fixed charge. As reactors are made larger, for example, the investment required, when expressed on a unit-energy basis (dollars per kilowatt-hour), tends to decrease markedly. Similarly the contribution of fixed charges to the generation cost will decrease as the magnitude of the fixed charge decreases with increased size, again when expressed on a unit-energy basis.

3.8 A second important feature is the relation between capital investment and the fixed charge. For example, differences in the financial characteristics of utility companies which could affect the cost of money could result in differences in the percentage rate applied to the capital investment to determine the fixed charge. Thus, to obtain an optimum combination of engineering specifications for which structural features requiring capital investment must be evaluated, the designer must consider the derived fixed charge, not the investment. As another example, reactor concepts requiring comparatively large capital investments may be more attractive for a public-owned agency, for which a low fixed-charge rate would apply, than for a privately owned utility (§3.36).

3.9 A third consideration is that fixed charges must be paid whether or not the reactor is producing energy and hence earning revenue. In the apportionment of the fixed charges to the energy produced, to determine the cost on a unit-energy basis, the fraction of time for which the reactor is able to run at rated power becomes very important. As a result the fixed charges are inversely proportional to the load factor. The reactor load, in turn, depends on the requirements of the utility company grid and the relative efficiency of the unit.

3.10 Fuel costs (see §§3.41 to 3.63) normally include all charges associated with the fuel, fixed as well as direct. These costs are very significant to the designer not only because they may make up about 35% of the total production cost but also because they may vary over a wide range and are sensitive to various engineering parameters. The third category, operation and maintenance (see §§3.64 and 3.65), concerns the cost of such items as salaries and labor, maintenance materials, and operating supplies. Although these charges represent only a second-order contribution to the power cost and comprise perhaps 10% of the annual operating expense, activities necessary to keep the reactor plant in service are important since low plant "availability" (§2.21) will reduce revenue and, in turn, increase the fixed charges on a unit-energy basis. On the other hand, these operation and maintenance charges are relatively insensitive to engineering parameters.

PLANT-INVESTMENT FIXED CHARGES

3.11 The principal nonfuel fixed charges are derived from the capital investment required for the construction of the nuclear power plant. Construction costs, in turn, consist of the direct construction costs (labor, materials, and equipment) and indirect construction costs (§3.14). For identifying the various cost categories and the interrelation that may be involved, the *Guide*[3] is very useful.

DIRECT CONSTRUCTION COSTS

3.12 Direct construction costs consist of the actual cost of the plant equipment, materials, and the labor required for installation. Fine points of categorization, such as the purchase of subsystems as a package or labor overtime, are not considered here. A *uniform system of accounts* is convenient for estimating and reporting direct construction costs. Such a system, in which various items are identified by numbers in a standard manner, is similar to the one used by electric utilities established by the Federal Power Commission. For illustration, the direct construction costs estimated for a 1000-Mw(e) pressurized-water reactor are given in Table 3.2. The values listed are significant only in relation to one another since actual costs change from year to year. In fact, direct construction costs for plants to be started in the mid-1970s have been estimated at about $250 million.[6]

3.13 Although subdivisions exist under each account number in Table 3.2, they are given only in Account No. 22, which concerns the nuclear-reactor engineer. The table includes only the direct construction costs; indirect costs are considered in §§3.20 to 3.32.

TABLE 3.2

Direct Construction Costs* for 1000-Mw(e) Pressurized-Water Reactor

Account number and description			Capital cost
20 Land and land rights			$ 1,000,000
21 Structures and improvements			34,000,000
211 Ground improvements		$ 2,000,000	
212 Buildings			
Turbine and auxiliary buildings		5,000,000	
Control, service, and office		2,000,000	
219 Reactor containment and building		18,000,000	
Other account 21		7,000,000	
22 Reactor-plant equipment			81,000,000
221 Reactor equipment		22,000,000	
Vessel	$11,000,000		
Control rods etc.	6,000,000		
Miscellaneous	5,000,000		
222 Heat-transfer system		32,000,000	
Reactor coolant system	3,000,000		
Steam generators	17,000,000		
Miscellaneous	7,000,000		
223 Fuel-handling and -storage facilities		4,000,000	
224 Fuel reprocessing and refabrication			
225 Waste disposal		2,000,000	
226 Instrumentation and control		6,000,000	
227 Feedwater supply and treatment		6,000,000	
228 Steam condenser and feedwater piping		9,000,000	
23 Turbogenerator plant			65,000,000
24 Accessory electrical equipment			10,000,000
25 Miscellaneous power-plant equipment			4,000,000
		Total direct construction costs	$195,000,000

*Contingency allowances are included.

3.14 The major accounts listed in Table 3.2 are as follows:

1. *Land and land rights.* The cost of land for the plant site and for an adjacent exclusion area belongs to this account. A "standard" 370-acre typical site along a river and 25 miles from a city is described in the *Guide*.[3] Since the estimate of $1 million for the site is an almost insignificant proportion of the

total reactor-plant construction cost, it need not be considered in detail. In actual cases, however, the cost of land and site development may amount to about \$2 million.[7] Since the land cost is not subject to depreciation, it is handled separately in the calculation of the annual fixed charges.

2. *Structures and improvements.* The cost of preparing the site and constructing all the buildings for the reactor plant, including the reactor building, shielding, and the containment vessel comprise this account. Depending on the type of reactor system, the account may range from \$20 to \$40 million for a 1000-Mw(e) facility.

3. *Reactor-plant equipment.* This account includes the reactor proper, reactor vessel (and thermal shields), coolant, moderator, foundation, and fuel-handling system. The costs of the heat-transfer system, comprising heat exchangers, auxiliary piping, tanks, etc., as well as of various support systems, such as instrumentation, water treatment, hot cells, remote maintenance, waste disposal, and special ventilation equipment, are in this category. The costs of the plant equipment are not very sensitive to the reactor type, at least for water-cooled, heavy-water-cooled, high-pressure gas-cooled, and sodium-cooled reactors.[8] A total cost of \$70 to \$90 million for a 1000-Mw(e) plant is appropriate for this category.

4. *Turbogenerator plant.* The turbogenerator, its foundations, and all its accessories are in this account. Differences in cost depend on the quality of steam available from the given reactor system, but the variations are not large. The cost for a 1000-Mw(e) nuclear plant ranges from \$50 to \$70 million, a substantial proportion of the total. Since the condenser heat-rejection systems would be in this account, a requirement for cooling towers could add about \$15 million to the total.

5. *Accessory electrical equipment and miscellaneous power-plant equipment.* These accounts do not involve a large fraction of the capital expenditure. Their total costs amount to about \$15 million for the 1000-Mw(e) installation. Included are (1) switchgear, wiring, and conduit work; (2) "other equipment," such as compressed air and refrigeration systems; and (3) the main power transformer.

6. *Special materials,* such as moderator and coolant, comprise Account 26, which is not applicable to the pressurized-water-reactor listing in Table 3.2. Although heavy water is a valuable material that can be recovered at the end of nuclear-plant life, its salvage value is uncertain. Accounting practice is therefore to apply depreciating fixed-charge rates to the heavy-water investment, as is done with all other equipment items.

3.15 Although the actual dollar values in an estimate of this type are likely to change yearly, the size of the relative cost contributions is a useful guide for the designer. Major items, for example, justify considerably more design attention than items representing only a small fraction of the total construction cost.

3.16 For a pressurized-water-reactor the containment is a major structural item. The reactor vessel and controls represent the major items of "reactor equipment." Remaining major items are the heat-transfer system and the turbine generator. Since comparatively low pressure steam is fed to the turbine in this concept, the turbine-generator cost tends to be high. Similarly, a substantial and an expensive pressure vessel is required.

CONTINGENCY ALLOWANCE

3.17 For each of the plant-investment items listed, an allowance should be made for "contingency" to provide for unforeseen or unpredictable costs. Often the designer can determine an appropriate allowance item by item from experience and knowledge of the uncertainties that may be involved. Little or no contingency allowance may be necessary, for example, for items priced on a firm basis, whereas a generous allowance may be needed for installation in anticipation of adverse weather, strikes, and delays in equipment delivery. Some guidelines are given in Table 3.3 for contingency allowances.

TABLE 3.3

Recommended Values for Contingency Allowances

Item	Allowance, %
Labor (all accounts)	10
Equipment and material	
Accounts 21, 24, 25, and 91	5
Account 22:	
Proven design	3
Extrapolated design (scale-up of a plant for which a construction permit has been received and equipment costs are known)	10
Novel design	20
Account 23	
Proven design	3
Extrapolated design	5
Novel design	10
Account 26	3

3.18 In Table 3.3 differences in uncertainties that may be appropriate for different types of reactors as well as for different account categories are considered. In the case of the reactor-plant equipment (Account 22), for example, the contingency allowance has three categories. The first (proven design) applies when a firm price for a complete package of equipment, which includes all the more critical or expensive items, or both, is available from a nuclear steam supply system (NSSS) vendor. The second (extrapolated design) applies when the design of the reactor plant is based on a somewhat similar one that is smaller by a factor of 3 or less, has received a construction permit, and also has known component costs. The third category (novel design) applies for all other evaluation studies, particularly for the expected economic performance of a concept yet being developed.

3.19 The equipment in Account 23 also has three categories. Here the important item is the turbine–generator unit since it accounts for approximately three-quarters of the total equipment cost. In most instances, however, the power-plant concept probably will not require other than a "proven design" unit. Since the cost of the turbine–generator can be closely fixed by a single contract and since designs are proven and changes are infrequent, a smaller contingency factor for Account 23 is justified than for Account 22.

INDIRECT CONSTRUCTION COSTS

3.20 A number of indirect construction costs must be added to the direct construction costs. These include the costs of contracting, design, engineering, inspection, start-up, and interest during construction. Such indirect construction costs often vary widely according to differences in accounting systems and to the experience of the contractor in building a specific plant.

3.21 The reactor station may be constructed and all initial equipment installed under a fixed-price or "turnkey" contract. In this case indirect costs are included in the fixed price. Many utilities, however, prefer to contract for the reactor portion and other components separately. In either case it is common practice for the utility to employ a firm of engineering consultants (§1.7) to review proposals not only for the station and accessory equipment but also for indirect services, such as licensing and employee training (§3.27).

3.22 For study and evaluation, indirect costs can be estimated according to standard procedures. One approach is to follow the *Guide,*[3] which prescribes percentages of the direct construction costs for each of several indirect cost accounts. Utilities compute indirect costs in a variety of ways. One approach, useful for study and evaluation, is described in the *Guide,* which breaks down such costs into four major accounts:

Construction facilities, equipment, and services, Account 91.
Engineering services, Account 92.
Other costs, Account 93.
Interest during construction, Account 94.

A discussion of these accounts follows. The results of applying the *Guide* to a pressurized-water reactor direct construction cost are shown in Table 3.5.

Construction Facilities, Equipment, and Services (Account 91)

3.23 This account includes buildings and other facilities that are removed after construction has been completed, the cost of construction equipment, and the cost of services, such as utilities, labor training, and maintenance. In the absence of detailed information, Fig. 3.1 may be used to obtain a value useful for evaluation studies.

Engineering Services (Account 92)

3.24 The various engineering services for the project may be grouped into two accounts: reactor engineering and plant engineering, with the architect–engineer and construction management services combined in the latter account.

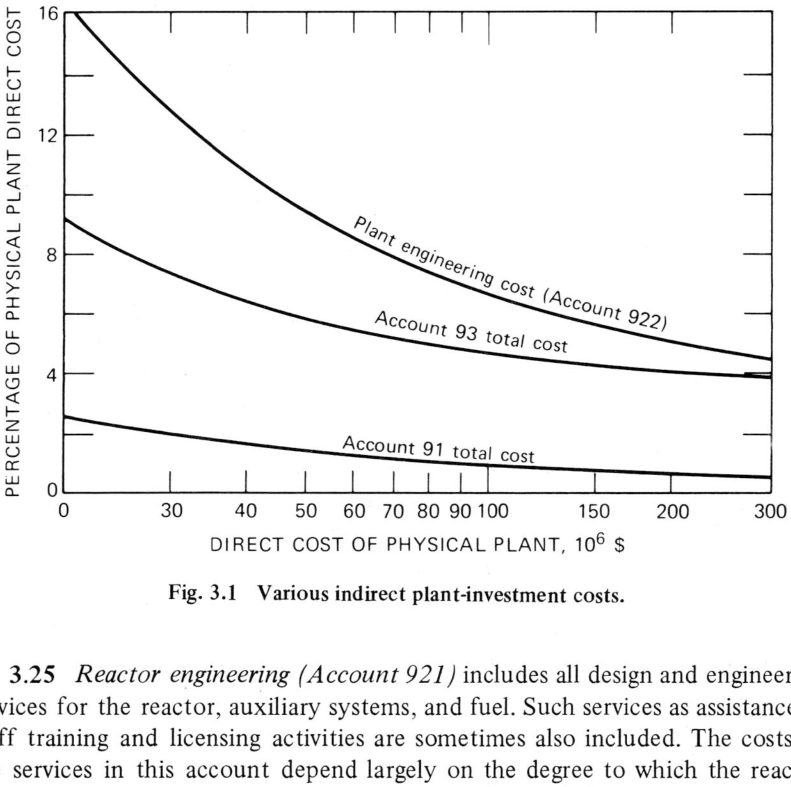

Fig. 3.1 Various indirect plant-investment costs.

3.25 *Reactor engineering (Account 921)* includes all design and engineering services for the reactor, auxiliary systems, and fuel. Such services as assistance in staff training and licensing activities are sometimes also included. The costs of the services in this account depend largely on the degree to which the reactor design has been proven. An additional complication is the engineering development that may be required for new reactor comcepts (novel design), which really should not be included directly in a cost estimate. A cost basis for the nondevelopment effort is given in Fig. 3.2 for estimates as a function of the Account 22 cost. An estimating basis is also given for an intermediate category, an "extrapolated design" in which the design of the plant is based on one that is similar but probably different in size. Plants of proven design may be considered as a third category, which, indeed, may have their engineering costs included by

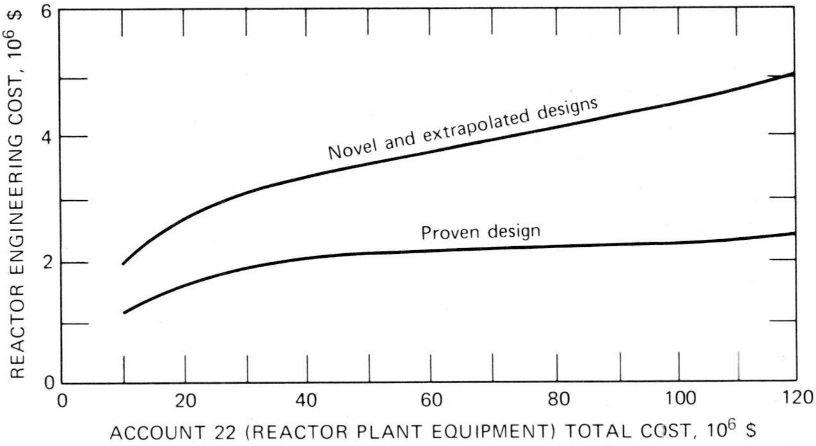

Fig. 3.2 Reactor engineering cost (921).

the vendor as part of Account 22. An estimation basis is included in Fig. 3.2, however. This approach is similar to that suggested for contingency allowances (§3.17).

3.26 *Plant engineering (Account 922)* includes all other engineering services associated with the project. Examples of such services are the preparation of specifications and evaluation of proposals for major equipment packages, the management and direction of construction activities, and preoperational activities. A cost basis for estimation is given in Fig. 3.1.

Other Costs (Account 93)

3.27 This account covers a variety of expense items that are part of the overall task of building a nuclear power station and preparing for its operation. According to the *Guide*, the account may be broken down into three component subordinate accounts, as follows:

Taxes and Insurance (Account 931): This account covers both property and all-risk insurance with nuclear rider, state and local property taxes on the site and improvements during the construction period, and sales taxes on purchased materials and equipment. Insurance and tax rates vary widely with geographical location, even within a given state. For evaluation studies, it is often assumed that the total of the insurance and tax charges prior to full-power commercial operation is equivalent to 3.0% of the total physical plant direct cost.

Staff Training and Plant Start-up (Account 932): These two indirect-cost categories may be combined because most of the costs are associated with staff salaries, and the training and start-up phases overlap. Final testing of the entire plant is included. Costs in this subcategory are of the order of $1 million.

Owner's General and Administrative (Account 933): This account is intended to cover all the miscellaneous costs incurred by the owner in carrying through the project. Included are licensing costs not included elsewhere, costs associated with public relations programs, and the general and administrative overheads assignable to the project. General and administrative costs are also of the order of $1.0 million.

3.28 The preceding three subcomponent cost categories tend to be insensitive to the type of reactor and can be correlated with the total direct cost of the plant. Therefore a total Account 93 cost for evaluation purposes may also be obtained from Fig. 3.1.

Interest During Construction (Account 94)

3.29 This account covers the net cost of the funds used to finance the design and construction of the plant. The interest during construction is therefore the sum of the interest charges for each expenditure made in connection with the accounts listed previously. These interest charges are a function of three quantities: the amount of the expenditure, the time period for which funds are borrowed, and the interest rate. The time when the plant achieves full-power commercial operation is taken as the end of the construction period since this marks the beginning of the period over which energy produces revenue and the fixed charges are included as a component of the energy-generation cost (§3.5). Table 3.4 lists time periods from the *Guide* which may

TABLE 3.4

Guidelines for Design and Construction Period

	Design and construction period, months		
Plant rating, Mw(e)	Proven design	Extrapolated design	Novel design
150	50	56	62
300	56	61	67
500	61	65	72
750	64	68	75
1000	66	70	77
1500	68	72	78

be assumed if no specific information is available. In 1972, however, such values appeared to be about 20% too low. The total interest during construction cost includes three components:

Physical Plant and Associated Indirect Costs: When specific information is not available, the percentage may be taken from Fig. 3.3 as a function of the design and construction period and interest rate. The curves are based on the spending rate shown in Fig. 3.4.

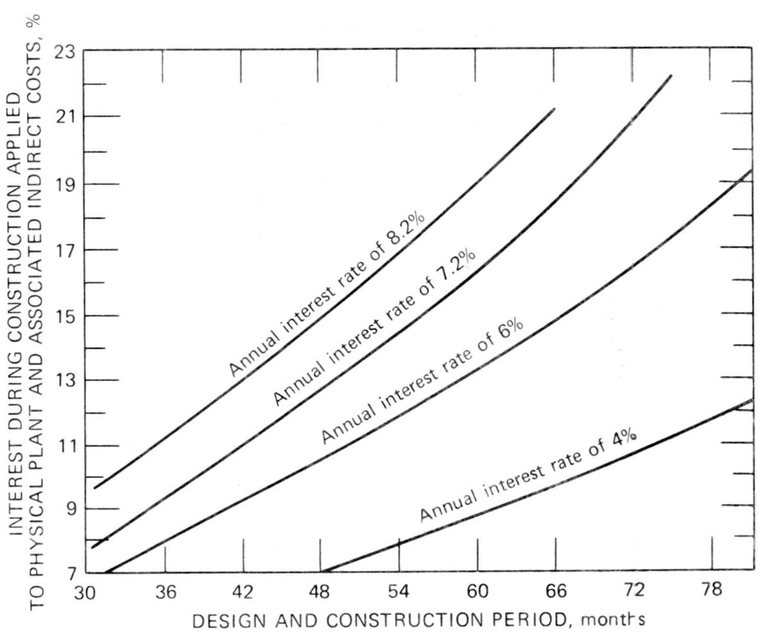

Fig. 3.3 Interest during construction.

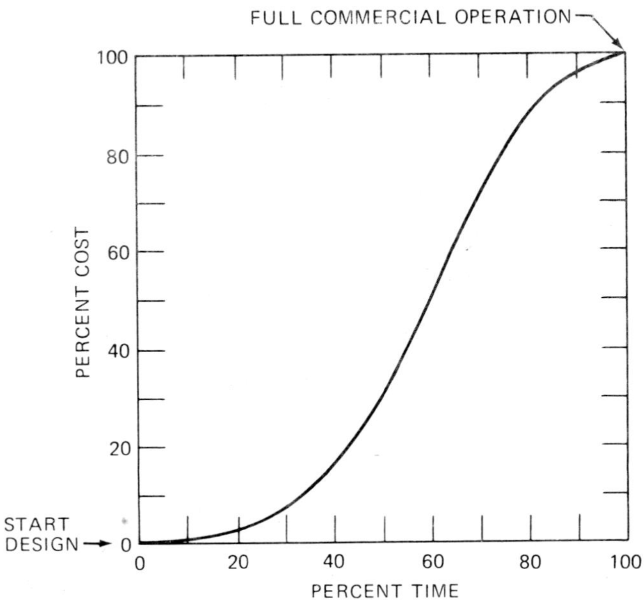

Fig. 3.4 Plant capital investment expenditures vs. time.

Land and Land Rights: For evaluation studies, when an actual land purchase date is unknown, it may be assumed that the land is owned for the full design and construction period plus 6 months. This interest during construction cost component is then found by computing the compound interest charges on the Account 20 amount for this time period.

Special Materials (as listed in Account 26): Materials to be installed during the construction of the reactor are assumed to be purchased 3 years prior to full-power commercial operation, whereas materials to be installed after all the reactor systems have been completed (such as fluids) are purchased 9 months ahead of project completion.

OTHER INDIRECT COSTS

3.30 Several other indirect costs not included in the preceding AEC categories should be considered if estimates are made on an individual project basis.[9] Local property taxes are frequently applied as construction progresses. If land is initially paid for, construction taxes may be expressed as:

$$\text{Taxes during construction} = \frac{rT}{100}(L + 0.45Q) \tag{3.1}$$

where $r(\%)$ is the annual tax rate, L is the land cost, and Q is the portion of plant costs to which the property tax is applicable. The constant $0.45 = (1 - 0.55)$ is a time-averaging factor.

3.31 An allowance for escalation in costs during the period of construction is sometimes provided. This is separate from the contingency allowance. The amount that should be applied will, of course, depend on inflationary trends prevalent at the time. One approach suggests that the turbine-generator price be considered firm for 3 years after the order is placed but that it increase at 6% per year thereafter. Other direct and indirect costs may increase at 7.5% per year.* Such escalation, amounting to about a $50 million[6] addition to the total construction cost listed in Table 3.5, may therefore be a substantial cost item.

3.32 Although provision is made for licensing expenses in Account 92, possible public acceptance problems requiring additional expenditures may need special attention. Additional allowances of $500,000 for licensing are common.

EFFECT OF SIZE AND CONCEPT

3.33 As the electric capacity of the power plant is increased, the total construction cost in dollars per kilowatt shows a significant decrease. For example, an increase in capacity from 500 Mw(e) to 1200 Mw(e) decreases unit construction costs by a factor of about 2.0. Therefore most nuclear-fueled plants

*The expenditure rate shown in Fig. 3.4 is applicable.

TABLE 3.5

Total Construction Costs for 1000-Mw(e) Pressurized-Water Reactor

Account number and description	Cost
Total direct construction costs (from Table 3.2)	$195,000,000
91 Construction facilities, equipment, and services (0.8% of direct cost)	1,500,000
92 Engineering Services	
921 Reactor engineering	2,300,000
922 Plant engineering (5.0% of direct cost)	10,000,000
93 Other costs (4.0% of direct cost)	8,000,000
94 Interest during construction at 8.2% per year (66-month period, 21% of direct cost)	41,000,000
Total construction cost	257,800,000
Allowance for escalation during construction*	55,000,000
Grand total construction cost (rounded)	$313,000,000

*See §3.31. The applicability of this indirect cost strongly depends on inflationary trends during the construction period.

planned for the 1970s have a rating of at least 800 Mw(e). Whether savings are likely to be realized at the same rate for capacities larger than 1200 Mw(e) is not clear, since increased construction periods, the need for greater reserve generating capacity, and operating differences tend to complicate the picture.

3.34 Although this trend is based on data for highly developed pressurized-water and boiling-water reactors, similar size–effects probably apply to other concepts. However, capital costs for other concepts cannot be estimated with the same degree of reliability as those for light-water reactors, although it is possible to compare design features and thereby predict probable cost variations from the "known" system.* Since a large fraction of the construction costs for the entire reactor-plant system arises from items that are external to the reactor proper, however, total costs do not vary markedly from one concept to another at the same respective stage of development. Moreover, the sensitivity to size is likely to be similar.

3.35 Several factors contribute to the decrease in construction costs on a unit-power basis. Some features of the design, such as instrumentation and fuel handling, need not be much more expensive or more complicated for a large plant than for a small one. Consequently there is a substantial decrease in the cost of "auxiliary systems" on a unit-power basis for the large system. Another

*The high-temperature gas-cooled reactor (HTGR) is quite well developed and has many design features quite different from those used in light-water reactors. It should therefore be considered separately.

factor is the possibility of increasing the design capacity of process-type equipment (e.g., pumps and heat-exchange systems) by using additional materials but with only modest additional cost for fabrication and manufacturing overhead. In fact, a useful rule of thumb in the chemical and petroleum industries is that the total investment will vary as the 0.6 power of throughput or capacity. Finally, many of the indirect costs are relatively insensitive to size. About the same engineering effort is required for a small plant, for example, as for a large one. An additional savings is realized when several plants of the same design are ordered. The manufacturer may then distribute the indirect costs over several plants. This is particularly true when several plant units are built on the same site with some facilities shared.

FIXED-CHARGE RATES

3.36 The annual charges for the capital represented by the total construction costs consist of such items as interest on borrowed money, return on equity, depreciation, taxes, and insurance. Although the exact calculation of these items can be fairly complicated (§2.24), usually for preliminary estimates the annual charge for each item can be determined by applying a constant fixed percentage to the initial total construction cost. The rate should be such that the present-worth grand total of these constant charges will equal the total of charges calculated in an exact manner on a yearly basis. The constant percentages to be applied to a new estimate will change, of course, from year to year as pertinent economic factors change. As an illustration the basis for an annual fixed-charge rate of 14.7% on depreciating capital for investor-owned power plants in the United States is shown in Table 3.6.

3.37 Fixed charges for working capital are associated with funds necessary to operate the plant. For example, in addition to a working fund for daily operations, working capital (§2.48) is required to pay for work done before revenue is received from the sale of electricity, e.g., fabrication of the core. Since no depreciation is involved in such a fund, an annual fixed-charge rate lower than that for depreciating capital applies. Actual rates vary, of course, depending on market conditions, the ratio of debt to equity capital (bonds to stocks), taxes, etc. (§2.27). For preliminary estimates an annual rate of 13.8% can be assumed. Land and land rights should also be included in the nondepreciating-capital category. According to the *Guide* the fixed charges related to the fuel, such as fabrication working capital, are included with the fuel charges (§3.53) whereas the plant-operation working capital is included in operation and maintenance expense (§3.64).

3.38 In the United States the annual fixed-charge rate for depreciating capital for government-owned utilities, municipal power plants, and plants constructed with the aid of loans from the Rural Electrification Administration may be 6 to 9% because of favorable interest rates and tax exemption. Similarly,

TABLE 3.6

Utility System Economic Parameters

(Plant Capacity Factor: 0.80; Plant Life: 30 years; Salvage Value: Zero)

	Fixed-charge rates, %			
	Publicly owned		Investor owned	
Component	Depre-ciating	Nondepre-ciating*	Depre-ciating	Nondepre-ciating*
Interest or return investment	5.00	5.00	8.20†	8.20†
Depreciation (30-year sinking fund)	1.78		1.02	
Interim replacements	0.35		0.35	
Property insurance	0.25		0.25	
Federal income taxes‡			2.04	4.80
State and local taxes	1.00	1.00	2.84	0.80
Total	8.4	6.0	14.7	13.8

*Use indicated total for land and operation and maintenance working capital.

†52% bonds at 6.5%, 48% equity at 10%.

‡Based on sum-of-the-year's-digits depreciation for tax purposes and federal taxes at 50% of taxable income.

nondepreciating-capital annual fixed-charge rates range from 4 to 6.5%. Fixed-charge rates also vary from one country to another, which may account for the different power-cost estimates for plants of the same type in different countries. This difference is significant because the fixed charges for nuclear power represent 60 to 80% of the final cost of the electricity produced. Since the cost of electricity is determined by the expenses during the period of generation, including the fixed charges, the fixed-charge rate is very important to the designer in selecting the most economical concept for a given service and in determining optimum operating conditions. When operating costs can be reduced, e.g., by spending additional money on construction, additional capital requirements can be more easily tolerated under public ownership, where a low fixed-charge rate applies, than under private ownership.

3.39 Since fixed costs are constant per unit of time and are independent of electrical output, the contribution of the fixed costs to the cost of energy, expressed in mills per kilowatt-hour, depends on the plant factor (or use factor). This is defined as the actual energy produced during a given period divided by the energy that could have been produced during the same period if the machine had operated continuously at the manufacturer's rated capacity (§2.16). These considerations are important in a utility's economic analysis and operational planning (§2.15).

TRENDS

3.40 Much of the previous discussion of capital costs is based on *estimates* for plants to be completed in the mid-1970s rather than on actual costs. Over a period of only a few years around 1970, such estimates have also shown a marked increase due to escalation of labor and material costs, increase in the cost of money, lengthening of construction time, and additional costs to meet safety and environmental requirements. Safety-related costs have included stringent requirements for quality assurance and design changes after an advanced stage of construction has been reached.[10] Although consideration has been given to these trends in the values cited here, it was not possible at the time this was written (1972) to forecast behavior during the 1970s.

FUEL COSTS

INTRODUCTION

3.41 Fuel expenses are the second major component of nuclear power costs. In fact, since costs can be reduced by improving both fuel-cycle technology and fuel design during the lifetime of the plant, the analysis of fuel costs is of continuing interest to the electric utility, whereas fixed charges do not change after the plant has been built. The analysis of fuel-cycle expenses is complicated, however, since a number of operations are involved and economic factors are sensitive to irradiation time in the reactor as well as to numerous financial and process variables. Furthermore, with private ownership of fuel, opinion varies on accounting methods for some of the charges. Our primary objectives here are to identify the principal cost components and to examine how they are affected by changes in the reactor design. In this chapter the calculation of fuel cost as a component of the energy cost is emphasized. The fuel system and the analysis of economic parameters are considered in greater detail in Chap. 7. It is therefore helpful to consider separately four major items affecting fuel-cycle economics: (1) the nature of the fuel itself; (2) the time required for fuel operations; (3) "fuel management"; and (4) manufacturing and processing costs.

3.42 The importance of these items can be seen from Figs. 3.5 and 3.6, which show the time dependence of the expenditures required for the various fuel-cycle operations. Since operations are performed on a valuable material, fixed charges depend on both the value involved and the time required. Any strategy or management options therefore require time and value considerations. Finally, the cost of carrying out the individual operations themselves must be included.

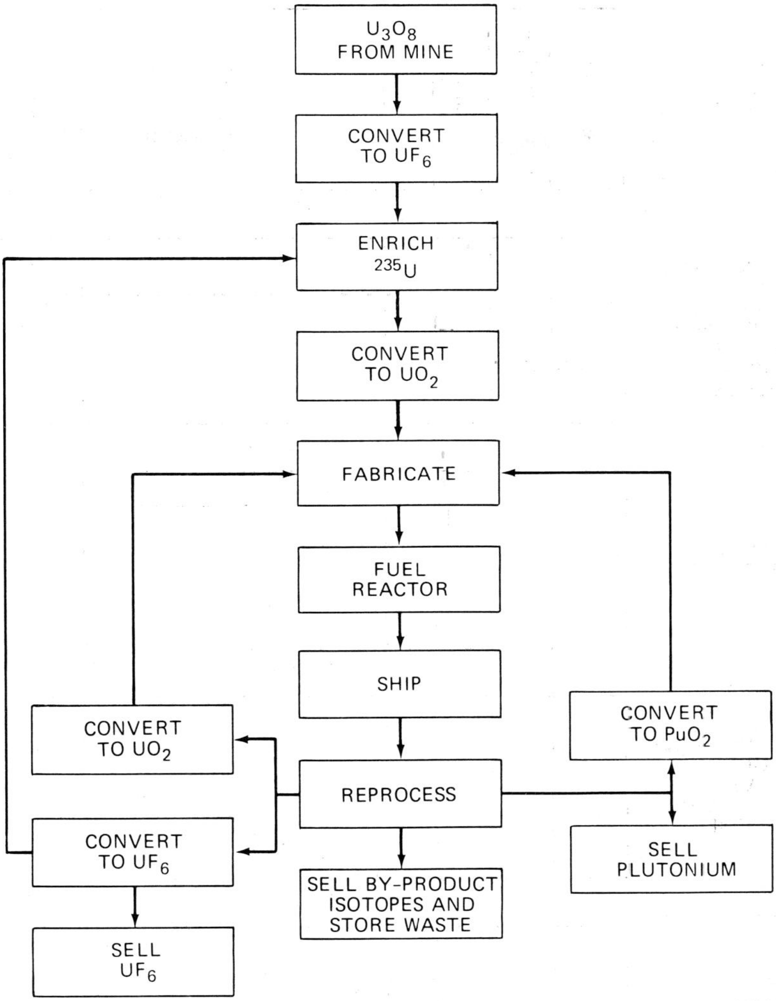

Fig. 3.5 The nuclear-fuel cycle.

Fuel Value

3.43 The value of the fissile atoms in the fuel is one important contribution to the nuclear-fuel-cycle cost. For partially enriched uranium the value depends on both the cost of the original ore and the cost of enrichment. The value of plutonium, on the other hand, can be related to that of uranium, but it also depends on the type of reactor using the plutonium (§3.46).

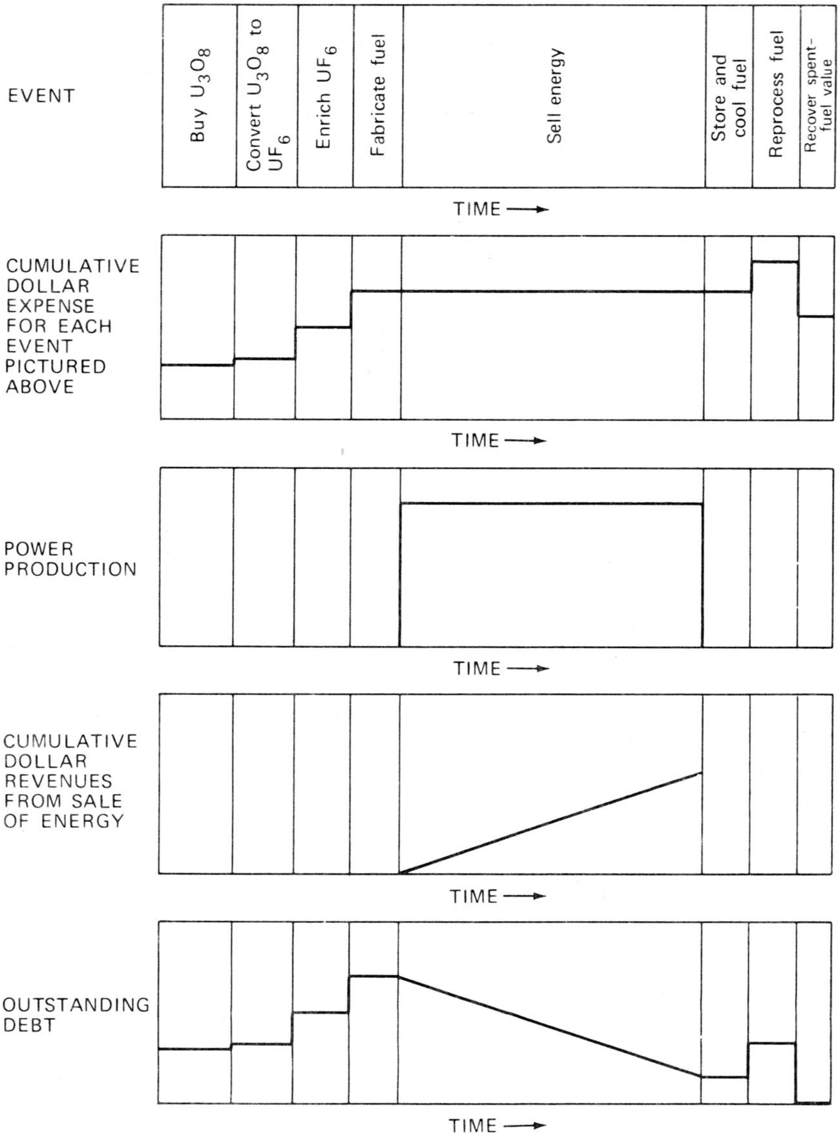

Fig. 3.6 Economic picture of nuclear fuel.

3.44 The cost of uranium ore, which is determined by market conditions, serves as a reference for the value of all uranium fuels, enriched as well as nonenriched. Although contractual prices have covered a wide range, as more power reactors are built and as they absorb existing supplies of low-cost uranium, a rise in the market price[11] of U_3O_8 to about $7 to $8 per pound is expected in the 1970s. Beyond this the availability of new supplies and the introduction of fast breeder reactors could influence the market price. Since a 30-year operational period is anticipated for reactors starting up in the early 1970s, such projections are very important to the reactor designer concerned with planning an optimum fueling strategy for the reactor lifetime.

3.45 The expenses of enrichment in a gaseous-diffusion plant must also be considered part of the value of partially enriched uranium fuel. An analysis of the gaseous-diffusion cascade[12] provides a value of enriched-uranium product in terms of the cost of raw material, the cost of separative work, and the optimum waste-stream composition as follows:

$$C_p = C_s \left[(2x - 1) \ln \frac{x(1 - x_w)}{x_w(1 - x)} + \frac{(x - x_w)(1 - 2x_w)}{x_w(1 - x_w)} \right] \tag{3.2}$$

where C_p is the value in dollars per kilogram of enriched uranium with a weight fraction, x, of ^{235}U, C_s is the unit cost of separative work in dollars per kilogram of uranium,* and x_w is the weight fraction of ^{235}U in the waste stream. The "optimum" waste-stream composition is the composition obtained from this equation when C_F, the unit cost of natural-uranium fuel, is substituted for C_p and when $x_F = 0.00711$, the composition of natural uranium, is substituted for x. This waste-stream composition corresponds to material that has zero value. The derivation of this relation and the effect of variations in the parameters on the calculated cost of enriched ^{235}U are considered in Chap. 7.

3.46 The value of plutonium is important, both in converter reactors producing some plutonium and in plutonium-fuel fast breeders. Although a useful value for estimation has been the "buyback" price of $9 to $10 used by the AEC, a future market value will be responsive to supply and demand. This should, in turn, depend on the type of reactor using the plutonium. Although plutonium has a higher value than ^{235}U in a fast reactor because of its more favorable physics in a fast spectrum, in a thermal reactor its value is about equal to that of fully enriched ^{235}U on a dollar-per-gram basis.[13] The market price of plutonium will therefore be related to that of enriched uranium, which is a function of ore and separation costs (§3.45). Although the economics of plutonium fuels is complicated and is treated at greater length in Chap. 7, a value of $8 to $10 per gram is useful for preliminary cost estimation.

Time Requirements

3.47 Since the time required for all the operations shown in Fig. 3.5 may total over 5 years and the material in process may have a value of about $200

---

*Also in dollars per separative work unit (see Chap. 7).

per kilogram, significant fixed charges are involved. Important savings can therefore be realized by reducing some of the times. There is, for example, considerable interest in fuel-recovery processes that can handle fuel after only short cooling times.

Fuel Management

3.48 Strategies for fueling the reactor at intervals which might include various options for shifting the fuel within the core at refueling can strongly affect the fuel cost, principally as a result of changes in the percentage of total fuel atoms fissioned. The corresponding study of the enrichment requirements for various core zones is also pertinent to the cost. Efficient use of the fuel, which might include breeding and high conversion as well as the recycling of fuel from one type of reactor to another, may be considered in the cost analysis. Fuel utilization is treated in Chap. 7.

Fuel Manufacturing and Processing

3.49 The costs of manufacturing the fuel and processing the irradiated material after its removal from the reactor are the third major influence on the economics of fuel cycles. Processing costs are affected, of course, by the nature of the operations required. Of great interest, however, is the throughput. Since there is a minimum plant size required for efficient operations, unit costs for volumes less than those appropriate for this size tend to increase rapidly. In fact, this type of analysis is helpful as a way of concentrating attention on areas that do or do not depend on the maturity and size of a nuclear power industry. Although the technology for fuel manufacturing and processing is now reasonably well developed, lower unit costs would be possible, for example, if larger amounts were required with resulting economies in efficient high-throughput operation (§7.20).

FUEL-COST CALCULATIONS

3.50 The energy cost component attributable to the fuel can be determined either by relatively exact computer programs or by somewhat approximate hand calculations. Before we turn to the details of possible approaches, however, it is useful to consider six parameters that affect fuel-cycle costs: (1) fabrication, (2) inventory or use, (3) depletion, (4) processing, (5) plutonium credit, and (6) transportation. For discussion purposes all fuel charges may be included in one or another of these categories, which have been used for many years under both private and government ownership of fuel. In some cases the choice is fairly arbitrary, but the following discussion indicates the considerations involved. Steps in the fuel cycle are also discussed in Chap. 7.

Fuel-Fabrication Costs

3.51 The direct and indirect fuel-fabrication charges together are important in fuel-cycle costs. The direct cost includes charges for converting uranium hexafluoride into the oxide, metal, alloy, or other form to be used as fuel, as well as costs for shipping and actual fabrication. The fabrication costs vary, depending on the choice of materials (including cladding), the difficulty of fabrication, and the complexity of the design. Rigid manufacturing tolerances, which are usually required, add to the cost because of the need for inspection and the high percentage of rejections. Recycle fuel containing plutonium may require remote handling, which also increases the cost. As is true in most manufacturing activities, efficiencies are realized as the throughput is increased. Unit fabrication costs, normally expressed as dollars per kilogram of heavy metal, therefore tend to decrease as the requirement (kilograms per day) is increased. Estimated fabrication costs for fuel elements in several types of power reactors[14] are given in Table 3.7. Since the core of a central-station reactor may contain some 150,000 kg (metal basis), the direct cost of fabricating the fuel material for a complete core may range from \$10 million to \$15 million. Estimates of unit fabrication costs can be obtained from bid information from fuel manufacturers or from standard methods involving cost equations set up for computer solution.[15] One such code is FABCOST.[16]

TABLE 3.7

Estimated Fuel-Fabrication Cost and Average Burnup

Reactor type	Fuel	Cladding	Fabrication cost,* \$/kg of uranium	Average burnup, Mwd/tonne
Pressurized water	UO_2	Zircaloy	75	25,000
Boiling water	UO_2	Zircaloy	70	22,000
Heavy water–natural uranium	UO_2	Zircaloy	50	10,000
Fuel	PuO_2–UO_2	Stainless steel	170†	75,000
Blanket	UO_2	Stainless steel	50†	

*Based on manufacturing throughput of 1000 kg/day.
†Updated.[17]

3.52 Although published cost equations and computer codes are based on operation-by-operation studies of hypothetical plants, the resulting estimates do not necessarily agree with prices quoted by commercial fabricators.[11] Studies of hypothetical plants tend to vary on the approach used to treat amortization, return on investment, and indirect costs, which may be rather specific to an individual manufacturer. Pricing information obtained from fabricators must also be viewed with caution by the estimator since commercial bids tend to be quite complex; variations are common in such conditions as warranty terms, escalation provisions, and possible discounts for second cores that might be sold to a utility.

3.53 In addition to the direct costs, allowance must be made for indirect costs on the capital invested in fabricated fuel. Working-capital requirements for nuclear power plants are much greater than those for fossil-fuel power plants, largely because of the considerable investiment in fabricated fuel both before and during reactor operation. Hence an annual charge on working capital is assigned as an indirect cost of the fuel cycle. The average investment by the utility in fabricated fuel may be about 60% of the fabrication cost for one full reactor loading. The annual fixed charge for working capital is the same as other rates used for nondepreciating capital. Although this rule of thumb was useful when the utility could not own the fissile material in the fuel, with the advent of private ownership, all the fuel-cycle capital charges are probably more appropriately considered as separate items. Furthermore, depreciation would amount at least to the value of one core over the plant lifetime. For the present purpose, however, these secondary considerations will be disregarded.

Fuel-Inventory Fixed Charge

3.54 A charge associated with the value of the fuel (§3.43) must be applied. The *Guide* procedure was originally applied to fuel owned by the AEC for which a lease charge of 4.75% per annum was applied. When the fuel material is privately owned, fixed charges based on the funds invested in the material are applied. To simplify the desired identification of cost components, however, capital charges for privately owned fuel will be treated in the same manner as the lease charge but with an annual rate appropriate for nondepreciating capital.* The annual capital, or use, charge is applied to the cost of enriched uranium as uranium hexafluoride (UF_6) according to an official schedule of prices.[3] These prices are based on the cost of isotope separation in the gaseous-diffusion plants, the value of normal uranium hexafluoride being taken as $23.50 per kilogram of uranium.

3.55 The use charge is payable not only on the fissile material in the core but also on the fuel being fabricated, stored, cooled, and reprocessed. Since one approach to fuel-cost analysis is to calculate separately all costs associated with each step, the time elapsed and value of material are necessary to determine fixed charges. Although enriched material is privately owned, the corresponding fixed charges are determined throughout the fuel cycle by applying a carrying

*The percentage to be applied may vary from that shown in Table 3.6 since it depends not only on the cost of money but also on the tax structure.

charge to the average investment rate. Plutonium is treated in the same manner as enriched uranium; on the other hand, natural uranium, thorium, and depleted uranium are treated in the same manner as ordinary purchased materials, the charge being included in the fabrication cost.

Fuel-Depletion Cost

3.56 The depletion cost is the value of fissile uranium consumed during reactor operation. The charge is the difference between the value of the uranium loaded into the reactor and that discharged. Consumption of ^{238}U during reactor operation by fast-fission and radiative capture also causes the total weight of uranium discharged to be slightly less than that charged. If depletion is determined from the energy produced, the contribution of fissioning plutonium that may have been formed in place from fertile ^{238}U must be considered.

Other Costs

3.57 Of the other three costs, chemical processing and plutonium credit tend to be about equal in importance, at least for nonbreeder reactors, whereas transportation is generally an incidental cost. Spent-fuel processing costs can be determined from a rate scale published by commercial processors. Although economies of scale might tend to lead to future reductions as fuel-processing needs increase, cost forecasts are complicated by inflation and other factors, which could have an opposite effect. Current processing-cost scales are therefore appropriate for estimating.

Calculation Methods

3.58 It is helpful to keep in mind a pattern that is common to most calculation methods. For each fuel-cycle step, as shown in Fig. 3.5, the cost may be divided into three categories: the cost of carrying out the operation in question, capital charges, and losses. The capital or carrying charges depend both on the value of the material being handled in the step and on the time required for the operation.

3.59 In practice, fuel costs are normally calculated with various computer codes, which include the "fine structure" of present worth and other second-order effects. On the other hand, hand-calculation methods are more useful for illustrating the principles involved and the influence of changes in design parameters. One such procedure for pressurized-water reactors follows a so-called "seven-page" format[18] in which costs of the significant steps of the nuclear-fuel cycle are presented in chronological order.

3.60 Fixed charges are determined from net capital depreciation and net carrying charges. Net capital depreciation is the difference between initial and final fuel investments, whereas net carrying charges are the cost of supporting the time-varying capital investment in the fuel region. Carrying charges include return on investment, federal income taxes, and state and local taxes. The approach can best be described by the following example:

Example

The unit fuel cost for a boiling-water reactor is desired, and the following data are available:

Thermal output, Mw	3293
Electrical conversion efficiency, %	32.8
Load factor	0.80
Number of fuel assemblies	764
UO_2 per assembly, lb	487.4

Equilibrium core averages (per kilogram of uranium initially charged)

Charged, ^{235}U, g	25.6
Discharged	
Burnup, Mwd	27.5
^{235}U, g	6.19
^{236}U, g	3.31
^{238}U, kg	0.953
^{239}Pu, g	4.61
^{240}Pu, g	2.07
^{241}Pu, g	0.93
^{242}Pu, g	0.36

Fuel management, 25% of the core removed per cycle at equilibrium

Assume \$70 per kilogram of uranium as the unit fabrication price and annual carrying charge at the following rates:

Uranium and plutonium, %	13.2
Working capital (fabrication), %	12.8
Spent-fuel shipping and reprocessing (nontaxable operating costs), %	7.2

Nuclear-Fuel-Cost Estimate (Basis: One Reload Cycle)

1. Reference design parameters
   a. Region power, Mw(t) (3293/4) — 823.25

b. Initial enrichment, wt.% $\left(\frac{25.6}{1000} \times 100\right)$	2.56
c. Final enrichment, wt.% $\left(\frac{6.19}{953 + 6.19 + 3.31} \times 100\right)$	0.643
d. Initial uranium weight, kg	37,189

$$\left[\begin{aligned} M_{UO_2} &= 764 \times 487.4 \times 0.454 \text{ kg/lb} \\ &= 168{,}871 \text{ kg of } UO_2 \\ M_U &= 168{,}871 \times \frac{(0.0256)(235) + (0.9744)(238)}{(0.0256)(235) + (0.9744)(238) + (1.0)(32)} \\ &= 148{,}755 \\ M_U &= 148{,}755/4 \text{ for region} \end{aligned}\right]$$

e. Final uranium weight, kg $\left(\frac{953 + 3.31 + 6.19}{1000} \times 37{,}189\right)$	35,794
f. Fissile plutonium produced, kg $\left(\frac{4.61 + 0.93}{1000} \times 37{,}189\right)$	206
g. Discharge burnup, Mwd/tonne	27,500
h. Full-power hours $\left(\frac{27{,}500}{1000} \times \frac{37{,}189}{823.25} \times 24\right)$	29,814
i. Lifetime at 80% capacity factor, months $\left(\frac{29{,}814}{0.8 \times 24 \times 30}\right)$	51.76
j. Thermal-energy output, 10^{11} Btu $(823.25 \times 3413 \times 10^3 \times 29{,}814)$	837.77
k. Plant thermal efficiency, %	32.8

2. Monthly carrying charges

a. Uranium and plutonium (13.2/12), %	1.1000
b. Working capital (fabrication) (12.8/12), %	1.0667
c. Spent-fuel shipping and reprocessing (7.2/12), %	0.6000

3. Average time periods (assumed)

Months prior to commercial use	
a. Procure U_3O_8	12
b. Complete conversion	10
c. Complete enriching	7
d. Complete fabrication	2
Months subsequent to refueling	
e. Complete spent-fuel cooling	4
f. Complete spent-fuel shipment	7
g. Complete uranium and plutonium reprocessing	10

4. Initial investments

Uranium investments required	
a. Uranium as enriched UF_6, kg	37,189
b. Uranium as natural UF_6 feed to enriching facility, kg $\left(37{,}189 \times \frac{2.56 - 0.20^*}{0.711 - 0.20}\right)$	171,754
c. Units of separative work per kilogram of enriched uranium as UF_6 (from Table 7.4)	3.356
d. Total units of separative work (37,189 × 3.356)	124,806
e. Natural uranium feed for U_3O_8 conversion facility, kg (171,754/0.995) (loss during conversion assumed to be 0.5%)	172,617
Uranium investment	
f. U_3O_8 (172,617 × 2.6 lb U_3O_8/kg uranium × \$7.50 per pound of U_3O_8, assumed ore price)	3366
g. Conversion (171,754 × \$2.50 per pound of uranium, assumed conversion charge)	429
h. Enriching (124,806 × \$32.00 per unit of separative work, assumed)	3994
i. Initial uranium investment (3366 + 429 + 3994), 10^3 \$	7789
Fabrication investment	
j. Unit fabrication price, \$ per kilogram of uranium	70
k. Fabrication investment (37,189 × 70), 10^3 \$	2603
l. Total initial investment (7789 + 2603), 10^3 \$	10,392

5. Final investments and credits

Spent-fuel shipping investment	
a. Spent-fuel shipping investment (37,189 × \$6.00 per kilogram of uranium initial, assumed shipping charge), 10^3 \$	223

*From a material balance, Feed/Product = $(x_p - x_w)/(x_f - x_w)$.

Reprocessing investment

b.	Recovery of uranium and plutonium in nitrate form (35,794 × \$30 per kilogram of uranium, assumed processing charge), 10^3 \$	1074
c.	Conversion of uranium nitrate to UF_6 (35,794 × 0.99 × \$5.60 per kilogram of uranium, assumed conversion charge), 10^3 \$ (loss assumed to be 1.0%)	198
d.	Reprocessing investment (1074 + 198), 10^3 \$	1272

Equivalent uranium requirements for uranium credit

e.	Uranium as reprocessed UF_6 (35,794 × 0.987), kg (loss during reprocessing assumed to be 1.3%)	35,329
f.	Uranium as natural UF_6 feed to enriching facility, kg $\left(35{,}329 \times \frac{0.643 - 0.2}{0.711 - 0.2}\right)$	28,489
g.	Units of separative work per kilogram of enriched uranium as UF_6 (from Table 7.4)	−0.066
h.	Total units of separative work (35,329 × −0.066)	−2332
i.	Natural uranium feed for U_3O_8 to UF_6 conversion facility (28,489/0.995), kg (loss assumed to be 0.5%)	28,632

Uranium credit

j.	Equivalent U_3O_8 investment (28,632 × 2.6 × \$7.50 per pound of U_3O_8), 10^3 \$	558
k.	Equivalent conversion investment (28,489 × \$2.50 per kilogram of uranium conversion charge), 10^3 \$	71
l.	Equivalent enriching investment (−2332 × 32.00), 10^3 \$	−75
m.	Uranium credit (558 + 71 − 75), 10^3 \$	554

Plutonium credit

n.	Fissile plutonium credit (206 × 0.99 × 8000), 10^3 \$ (recovery loss assumed as 1.0% and plutonium value as \$8.00 per gram)	1631
o.	Total final investment (554 + 1631 − 223 − 1272), 10^3 \$	690

6. Uranium depletion and carrying charges

a.	Uranium depletion (7789 − 690), 10^3 \$	7099
	Carrying charges prior to commercial use	
b.	On U_3O_8 investment, 10^3 \$ $\left(3366 \times 12 \times \frac{1.1}{100}\right)$	444
c.	On conversion investment, 10^3 \$ $\left(429 \times 10 \times \frac{1.1}{100}\right)$	47

d. On enriching investment, 10^3 \$ — 308

$$\left(3994 \times 7 \times \frac{1.1}{100}\right)$$

e. Subtotal (444 + 47 + 308), 10^3 \$ — 799

Carrying charges during commercial use

f. On average uranium investment, 10^3 \$ — 2375

$$\left(\frac{7789 + 554}{2} \times 51.76 \times \frac{1.1}{100}\right)$$

Carrying charges after refueling

g. On uranium credit, 10^3 \$ — 61

$$\left(554 \times 10 \times \frac{1.1}{100}\right)$$

h. Total uranium carrying charges, 10^3 \$ (799 + 2375 + 61) — 3235

7. Fabrication depreciation and carrying charges

a. Fabrication depreciation (2603), 10^3 \$ (no salvageable value from fabrication process) — 2603

Carrying charges

b. Prior to commercial use, 10^3 \$ — 56

$$\left(2603 \times 2 \times \frac{1.0667}{100}\right)$$

c. During commercial use, 10^3 \$ — 719

$$\left(\frac{2603}{2} \times 51.76 \times \frac{1.0667}{100}\right)$$

d. Total fabrication carrying charges (719 + 56), 10^3 \$ — 775

8. Spent-fuel shipping and reprocessing

Capital accumulation and carrying charges

a. Spent-fuel shipping accumulation (223), 10^3 \$ — 223

b. Reprocessing accumulation (1272), 10^3 \$ — 1272

c. Total accumulation (1272 + 223), 10^3 \$ — 1495

Carrying charges displaced by accumulations

d. During commercial use, 10^3 \$ — 232

$$\left(\frac{1495}{2} \times 51.76 \times \frac{0.6}{100}\right)$$

e. After refueling, 10^3 \$ — 86

$$[(223)(7) + (1272)(10)]\left(\frac{0.6}{100}\right)$$

f. Total displaced carrying charges (232 + 86), 10^3 \$ 318

Since the capital accumulations and carrying charges for fuel shipping and reprocessing are collected in advance from revenues obtained from the sale of energy during the "commercial use" period, carrying charges appear as a credit to the fuel-cycle cost.

9. Plutonium appreciation and carrying charges

a.	Plutonium appreciation (1631), 10^3 \$	1631
	Carrying charges	
b.	During commercial use, 10^3 \$	464

$$\left(\frac{1631}{2} \times 51.76 \times \frac{1.1}{100}\right)$$

c.	After refueling, 10^3 \$	179

$$\left(1631 \times 10 \times \frac{1.1}{100}\right)$$

d.	Total plutonium carrying charges (179 + 464), 10^3 \$	643

Note: In this approach bred plutonium is an investment on which carrying charges are applied and a fuel-cycle expense.

10. Summary

a.	Uranium depletion (6a), 10^3 \$	7099
b.	Fabrication depreciation (7a), 10^3 \$	2603
c.	Spent-fuel shipping and reprocessing capital accumulation (8c), 10^3 \$	1495
d.	Plutonium appreciation (9a), 10^3 \$	1631
e.	Net capital depreciation (10a + 10b + 10c – 10d), 10^3 \$	9566
f.	Uranium carrying charges (6h), 10^3 \$	3235
g.	Fabrication carrying charges (7d), 10^3 \$	775
h.	Spent-fuel shipping and reprocessing carrying charges displaced (8f), 10^3 \$	(318)
i.	Plutonium carrying charges (9d), 10^3 \$	643
j.	Net carrying charges (10f + 10g – 10h + 10i), 10^3 \$	4335
k.	Total fuel cost (10e + 10j), 10^3 \$	13,901
l.	Thermal-energy output (1j), 10^{11} Btu	837.77
m.	Total unit energy fuel cost (10k/10l), ¢/10^6 Btu	16.59
n.	Plant thermal efficiency (1k), %	32.8
o.	Total unit energy produced fuel cost, mills/kw-hr	1.73

$$\left(10\text{m} \times 3.412 \times \frac{1}{10\text{n}}\right)$$

FUEL-CYCLE ANALYSIS CODES

3.61 For operational analysis and planning, the operator of the nuclear power station needs accounting procedures more exact than those provided by hand-calculation methods such as the one described. Digital-computer codes have therefore been developed for this purpose. Although the mathematics of such codes are relatively simple compared with those used for nuclear design, these codes can perform a great deal of otherwise laborious data manipulation. Two widely used codes are CINCAS[19] and FUELCOST.[20]

3.62 Such codes offer a wide variety of options, which provide for both accounting and engineering economy needs. Costs may be allocated on either an energy or an elapsed-time basis. For example, some of the features provided by CINCAS are:

1. Monthly calculation of dollar costs and mass inventory on a batch, core, and case basis for each month that the fuel is considered to be in service.
2. Variable monthly batch heat-production rates and plant efficiencies.
3. Allocation of costs by heat production.
4. Fuel weights of uranium and plutonium on a burnup-dependent basis.
5. General formula for calculation of enrichment costs.
6. Flexible output editing.
7. Results in dollars, mills per kilowatt-hour, and cents per 10^6 Btu.
8. Burnup-averaged costs.
9. Present-worth and levelized costs.
10. Package fuel purchase or sale options.
11. Cash-payment-schedule editing.

The calculations are generally based on a model that follows the economic picture given in Fig. 3.6.

3.63 Several other methods also widely used for comparing alternate concepts lend themselves to computer programming. For example, Dragoumis, Cademartori, and Milioti[21] consider just two components in fuel-cycle costs: the fuel-cycle operating expense and the capital charges on the investment of materials in the cycle. Flow of cash, revenues, expenditures, and the amortization of the investment receives primary attention. The development of a single, levelized, total fuel-cycle cost over the lifetime of the nuclear plant is an important feature. The approach equates the 30-year present-worth sum of fuel-cycle costs with the 30-year present-worth sum of revenues, assuming a lifetime levelized revenue requirement, or cost, L, for electrical energy. Thus

$$L = \frac{\sum_{i=1}^{n} P_i F_i K_i}{\sum_{i=1}^{n} P_i K_i} \tag{3.3}$$

where P_i is the present-worth factor for the ith irradiation period based on the composite rate of return on investment, F_i is the calculated power cost (mills/kw-hr) for the ith irradiation period, and K_i is the net electrical power (kw-hr) produced during the ith irradiation period. The product F_iK_i is given by

$$F_iK_i = 1000V_i$$

where

$$V_i = V_{fe} + V_f + V_s + V_{FE}(i + a)\,\Delta t + V_F i + (V_F - V_S)a\,\Delta t \qquad (3.4)$$

The components in Eq. 3.4 denote values, times, and interest rates for a given irradiation period and are defined as follows:

V_{fe} = operating expense related to the new fuel elements added to the core for each irradiation period

V_f = operating expense for the replacement fuel supply required at the beginning of each irradiation period

V_s = operating expense associated with fuel recovery after each irradiation period and is equal to the difference between the recovery cost for the batch reprocessed and credit received by the utility for net recovered uranium and plutonium

V_{FE} = initial value of the investment in the total number of fuel elements comprising the first complete core

i = annual fixed-charge rate

a = amortization rate over the lifetime of the investment given in this study as 0.0116 (30 years, 6.50%) sinking-fund factor

V_F = initial value of the fuel in the first complete core

Δt = length of the irradiation period in years

V_S = net scrap value of the recovered fuel from the entire core after the final irradiation period and is equal to the credit received for the recovered fuel less the recovery cost

OPERATION AND MAINTENANCE COSTS

3.64 Operation and maintenance expenses, the third category of nuclear power costs, include the direct and indirect payroll, the cost of materials, coolant makeup (including D_2O), and general supplies. Miscellaneous operation and maintenance costs include such items as public relations, new-staff training, rents, and travel. The category also includes liability insurance and the fixed charges for the working capital to pay for the items in the category. Nuclear liability insurance combines commercial coverage limited to \$82 million with

TABLE 3.8

Summary of Annual Direct Costs for Plant Operation and Maintenance*

	Annual cost		
	750 Mw(e)	1000 Mw(e)	1500 Mw(e)
Staff payroll and fringe benefits	$709,000	$726,000	$752,000
Consumable supplies and equipment	340,000	400,000	500,000
Outside support services	120,000	140,000	175,000
Miscellaneous	70,000	80,000	95,000
General and administrative (15% of total of four items above)	186,000	202,000	228,000
Nuclear liability insurance	300,000	340,000	430,000
Property damage insurance	350,000	410,000	470,000
Total annual direct costs (excluding fuel)	$2,075,000	$2,300,000	$2,650,000

*For boiling-water and pressurized-water reactors.

TABLE 3.9

Working-Capital Requirements for Operation and Maintenance*

Item	Capital basis
Average net cash required	
2.7% of annual direct operation and maintenance cost excluding nuclear liability insurance	$ 42,000
50% of insurance annual cost	375,000
Materials and supplies in inventory	
Consumable supplies and equipment	$340,000
25% of annual cost of materials and supplies	85,000
Total operation and maintenance working capital	$502,000
Annual cost (13.8% rate)	$ 69,000

*For a 1000-Mw(e) pressurized-water reactor.

government indemnity having a yearly premium charge of $30 per megawatt of heat to provide a total coverage of $560 million.

3.65 Typical values for the direct-cost items for several sizes of light-water-reactor plants are shown in Table 3.8. For developing the working-capital contribution (§2.48), the *Guide*[3] suggests the procedure in Table 3.9 for the 1000-Mw(e) example. Although some variations in manpower requirements, maintenance supplies, and other items are likely to exist for different types of plants, the costs involved represent only a minor component of the annual charges for power production. Furthermore these charges may be assumed to be only moderately dependent on reactor size. Interestingly, estimated operation and maintenance costs for 1000-Mw(e) nuclear plants are not very different from those for coal-fired plants of the same capacity. Although a higher level of training is required for nuclear-plant personnel than for coal-plant personnel, the substantial size of the staff necessary for coal handling and maintenance is eliminated. Operating and maintenance costs of about $2.4 million per year for a 1000-Mw(e) plant correspond to about 0.34 mill per kilowatt-hour, or about 5% of the cost of energy generation.

TOTAL ENERGY PRODUCTION COSTS

3.66 The total energy cost on an annual and a kilowatt-hour basis is obtained by adding the amounts in the various categories as shown in Table 3.10 for a 1000-Mw(e) pressurized-water reactor and following the procedure given in the *Guide*. Actual values for such a system may vary, of course, depending on design, economic, and other parameters. Attention has been given here primarily to one applicable approach. Other approaches are equally satisfactory as long as they are consistent.

TABLE 3.10

Format for Summarizing Costs of Electric Energy Generation

Cost component	Capital cost, 10^3 \$	Fixed-charge rate, %	Annual cost or revenue, 10^3 \$	Unit-energy cost or revenue, mills/kw-hr
Plant investment				
Depreciating assets	312,000	14.7	45,864	
Nondepreciating assets	1,000	13.8	138	
		Subtotal	46,002	6.08
Fuel				
Total fuel cost			12.89	1.70
Operation and maintenance				
Direct cost			2,300	
Working capital	502	13.8	69	
		Subtotal	2,369	0.34
		Total electric energy generation cost		8.12

TECHNICAL–ECONOMIC INTERRELATIONS

3.67 Cost components and the many parameters affecting them influence the engineering design of the reactor system markedly. Although a number of these interrelations will be brought out later, some of the parameters associated with fuel-cycle cost components are mentioned here. It is important to realize also that a change in one cost category may well affect another category. Since reduction in fuel-cycle costs could also conceivably be offset by an increase in capital costs, the design variables related to the fuel cycle must be chosen to provide the lowest nuclear energy cost, not the lowest fuel-cycle cost.

3.68 As a basis for discussion, the general behavior* of light-water-reactor cost components (on a unit-energy basis) as a function of burnup is shown in Fig. 3.7. Several competing factors contribute to the shape of the curves. As burnup is increased, the cost of fabricating the core and reprocessing it can be allocated to a larger number of kilowatt-hours produced during the longer exposure time. Hence such costs, which apply to a batch of fuel elements when expressed on the basis of unit energy produced, tend to decrease. On the other hand, if reactivity of the reactor is to be maintained during the longer exposure period, fuel having a higher enrichment than that required for a shorter exposure

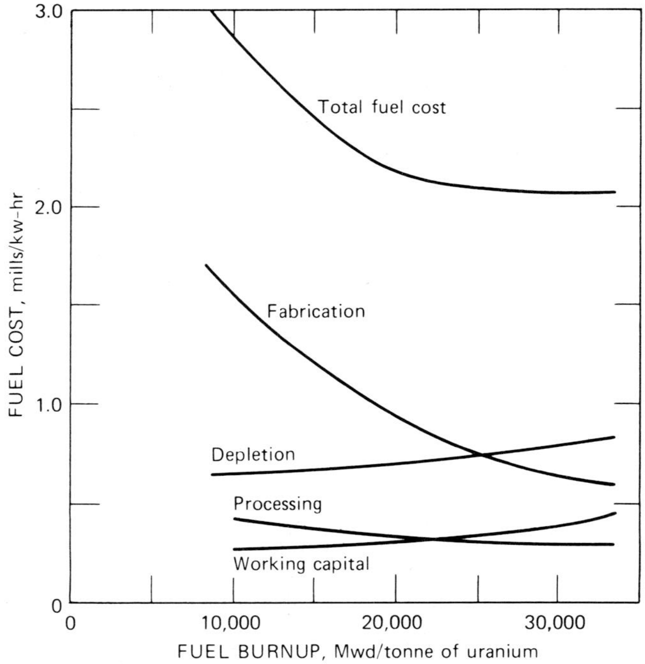

Fig. 3.7 Fuel-cycle cost components as a function of burnup.

*Costs shown are only relative and are not necessarily representative of current practice.

is necessary. Because of the increased cost of this enrichment, the atoms of ^{235}U that fission in the longer burnup fuel are more expensive. The necessary higher investment per fissile atom of fuel also results in higher interest charges. This somewhat simplified picture is complicated, however, by the influence of the bred plutonium. Although the amount of plutonium discharged and available for sale does not increase much with increased fuel exposure, more plutonium atoms are fissioned while they remain in the core. This sharing of the energy-production load by plutonium therefore results in somewhat less depletion of ^{235}U than would otherwise be the case. Despite this effect the depletion cost rises slightly with increased exposure.

3.69 This preliminary picture does not consider the probable need for certain design changes in the fuel element to enable it to withstand the desired increased exposure. For example, it may be necessary to increase the thickness of the cladding to contain the increased quantity of fission gases released. In addition to increasing the cost of fabricating a fuel element, such a change, particularly if the cladding is stainless steel, would result in an increase in neutron absorption, which, in turn, would have to be compensated for by fuel with higher enrichment. Similarly, the need to provide for compensation for the reactivity change associated with long-burnup fuels could result in inefficient use of neutrons by capture in control rods, burnable poisons, and accumulated fission products with consequent economic penalties. The conflicting trends in Fig. 3.7 show that it is possible to design for an optimum fuel burnup. An analysis of all the economic, neutronic, and engineering factors contributing to the determination of this optimum point is by no means simple. Changing economic conditions (e.g., interest rates, separative-work charges, and uranium-ore costs) as well as shifts in reactor load requirements, all occurring during the projected lifetime of the reactor, further complicate the determination of such an optimum.

POWER-SUPPLY ECONOMICS

3.70 An estimate of total energy costs for a particular nuclear plant may represent only one part of a larger economic analysis concerned with power-supply planning. Management decisions concerning proposed new generating facilities require a complete picture of available options and parametric effects.

3.71 Fossil-fuel costs required for a break-even point with nuclear power can be studied, for example, as a function of fixed-charge rates. Compared with coal plants, nuclear stations tend to have higher construction costs but lower fuel costs. A rise in fixed-charge rates therefore tends to favor the coal system. On the other hand, a general rise in labor costs would affect the cost of coal much more than that of nuclear fuel.[22]

3.72 Since it is generally necessary to build steam power plants in large sizes to achieve favorable unit costs, factors such as the availability of cooling water, siting problems, load patterns, and transmission costs may make it desirable to consider as an alternate a number of smaller-size gas-turbine facilities that could meet local needs. It is important to remember that transmission facilities have associated with them substantial fixed charges that are also sensitive to changes in economic conditions.

3.73 From the viewpoint of utility planning, the economics of power supplies is therefore quite complex with many parameters that lend themselves to analysis by systems-engineering techniques. Some of the optimization approaches that may prove useful are considered in Chap. 9. However, in developing a system model for optimization, which includes both technical and economic parameters, one should keep in mind the probable relevance of both advancing technology and economic trends.

REFERENCES

1. Jersey Central Power and Light Company, Report on Economic Analysis for Oyster Creek Nuclear Electric Generating Station, 1964.
2. Tennessee Valley Authority, Comparison of Coal-Fired and Nuclear Power Plants for the TVA System, 1966.
3. NUS Corporation, Guide for Economic Evaluation of Nuclear Reactor Plant Designs, USAEC Report NUS-531, January 1969.
4. Royes Salmon, Revision of Computer Code POWERCO (Cost of Electricity Produced by Nuclear Power Stations), USAEC Report ORNL-4116, Oak Ridge National Laboratory, August 1969.
5. C. H. Bloomster, J. H. Nail, and D. R. Haffner, PACTOLUS: A Code for Computing Nuclear Power Costs, USAEC Report BNWL-1169, Battelle–Northwest, January 1970.
6. *Nuclear Industry*, **18**(10 and 11, Part 1): 23 (1971).
7. R. L. Ashley (Ed.), Nuclear Power Reactor Siting, A National Topical Meeting held February 16–18, 1965, Los Angeles, Calif., Report CONF-650201, American Nuclear Society, 1965.
8. M. W. Rosenthal et al., A Comparative Evaluation of Advanced Converters, USAEC Report ORNL-3686, Oak Ridge National Laboratory, January 1965.
9. L. Geller, J. F. Hogerton, Jr., and S. M. Stoller, Analyzing Power Costs for Nuclear Reactors, *Nucleonics*, **22**(7): 64 (1964).
10. W. Steigelmann, The Outlook for Nuclear Power-Station Capital Costs, *Reactor Technol.*, **13**(1): 14 (1970).
11. A. D. Little, Inc., Competition in the Nuclear Power Supply Industry, USAEC Report NYO-3853-1, December 1968.
12. M. Benedict and T. H. Pigford, *Nuclear Chemical Engineering*, McGraw-Hill Book Company, Inc., New York, 1957.
13. E. A. Eschbach and M. F. Kanninen, Uranium Price Schedules and Bred Fuel Values, USAEC Report HW-72219, General Electric Company, December 1964.
14. Kaiser Engineers, Guide to Nuclear Power Cost Evaluation, USAEC Report TID-7025, Mar. 15, 1962.

15. NUS Corporation, Guide for Economic Evaluation of Nuclear Reactor Plant Designs, USAEC Report NUS-531, p. D.4-1, January 1969.
16. A. L. Lotts, T. N. Washburn, and F. J. Homan, FABCOST-9: A Computer Code for Estimating Fabrication Costs for Rod-Bundle Fuel Elements, USAEC Report ORNL-4287, Oak Ridge National Laboratory, August 1968.
17. W. R. Gee, Jr. (Comp.), Conceptual Plant Design, System Descriptions, and Costs for a 1000 Mw(e) Sodium-Cooled Fast Reactor, Task II, Report of 1000 Mw(e) LMFBR Follow-On Work, USAEC Report GEAP-5678, General Electric Company, December 1968.
18. J. F. Bader, K. J. Kitzke, and D. A. Nordman, Estimating Average Nuclear Fuel Costs, *Power Eng.*, **73**(12): 50 (December 1969).
19. T. W. Craig, P. J. Fulford, and Richard Swanson, Engineering Economy in Nuclear Cost Evaluations, *Trans. Amer. Nucl. Soc.*, **10**: 438 (1967) (CINCAS is available through the Argonne National Laboratory Code Center).
20. N. B. McLeod, FUELCOST-II, The Calculation of Fuel Costs and Working Capital Requirements in Multi-Batch-Fueled Nuclear Power Reactors, Report NUS-446, NUS Corporation, April 1968.
21. P. Dragoumis, J. J. Cademartori, and S. J. Milioti, Estimating Nuclear Fuel-Cycle Costs, *Nucleonics*, **24**(1): 40 (1966).
22. F. M. Swengel, A New Era of Power Supply Economics, *Power Eng.*, **74**(3): 30 (March 1970).

4 Thermal-Transport Systems and Core Design

INTRODUCTION

4.1 Since power reactors exist primarily to produce useful energy from heat generated in the fuel, heat-transfer and thermal-transport parameters are important in the design of the reactor plant. In a typical system the thermal-transport path proceeds from a point of fission energy deposition within a solid fuel, through layers of fuel, through gas at the interface of fuel and cladding, and then through the cladding to the interface with a fluid coolant. The heat transported into the body of the flowing coolant then causes the temperature of the coolant to rise. The coolant transports absorbed energy to a heat exchanger in which steam is generated, and the steam is then expanded in a turbine generator to produce electricity. Since there are also spatial variations, transient effects, and a dependency upon neutronic parameters, the actual thermal–hydraulic picture is complex.[1]

4.2 Thermal-transport considerations strongly influencing the design of the nuclear steam-supply system are emphasized here rather than all the relevant heat-transfer principles, which are discussed elsewhere.[2,3] To pinpoint the thermal-transport topics, we shall consider the balance between design *objectives* and *restrictions* that limit the specifications. For example, a high core power density (kilowatts/liter) reduces the reactor vessel size, whereas a high fuel specific power (kilowatts per kilogram of uranium) reduces the fuel inventory needed. In ceramic-fueled water reactors, this desire for a minimum amount of fuel in a minimum core volume is restricted both by the temperature that can be tolerated in the fuel and by the heat-transfer behavior at the cladding–coolant interface. Therefore it is important to examine the thermal performance of a fuel element and the nature of the thermal-flux limitations at the cladding–coolant interface.

4.3 Two other types of effects are of major importance to the designer: (1) variations in power distribution spatially and with time and (2) transients from abnormal operating situations. The power distribution may vary with time as the fuel is depleted and also as the result of different fuel reloading strategies. The thermal consequences of start-up, shutdown, and inadvertent operating situations must be considered in the safety analysis as well as the detailed effects of various postulated accidents. Many of these possibilities are considered in Chap. 6.

4.4 Significant in fuel-element design but only partially related to the thermal behavior is the maintenance of element integrity during burnup and in the event of a power transient. Fission-product accumulation, fuel swelling, and cladding deformation must all be considered in the design analysis of the core.

REACTOR THERMODYNAMIC SYSTEM

INTRODUCTION

4.5 The temperature levels of the coolant system provide a basis for the core-temperature pattern and hence represent a good starting point in the discussion of thermal transport. This coolant-temperature behavior, however, depends on the system thermodynamic cycle.

4.6 The use of the thermal energy released by the fuel and the corresponding rejection of waste heat to the environment depend, of course, upon the thermodynamic efficiency of the heat power cycle. Thus thermodynamic-cycle parameters also concern the designer from this point of view. Optimization may be in order if capital costs must be increased to improve the thermal efficiency. In fact, steam cycles have been studied for many years as components of fossil-fuel power plants, and quite complex systems have been developed to achieve efficiencies as high as possible for a given set of steam conditions. However, to examine some of the basic features of the steam cycle, we shall consider first a very idealized system such as that shown in Fig. 4.1.

4.7 The coolant fluid, which is heated in the reactor, is pumped to a separate steam generator, in which the heat is absorbed by the thermodynamic-cycle fluid. Useful energy, usually as electricity, is then produced by expanding the steam in a turbine, condensing it, and recycling the condensate to the steam generator, as might be done for a conventional fossil-fuel power plant. However,

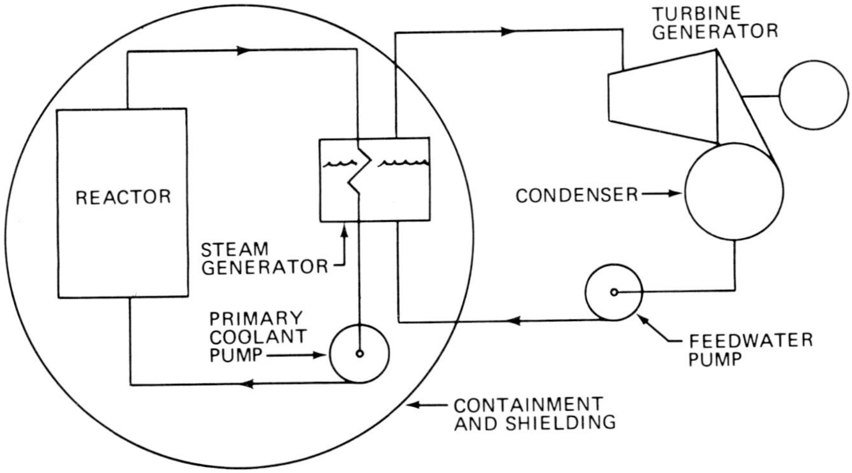

Fig. 4.1 Simple indirect steam cycle.

a practical cycle is much more complicated than that shown since some efficiency can be gained by the reheating and partial expansion approaches developed by the power industry over many years.

4.8 The nature of the temperature-driving forces associated with thermal transport is shown in the temperature–entropy plot of Fig. 4.2. Heat from the fuel element is absorbed by the primary coolant. The coolant and the fuel element are each assumed at constant temperature, with a thermodynamic irreversibility necessary for the transfer of heat from the fuel-element temperature, T_1, to the coolant temperature, T_2. The heat absorbed in the primary loop is then rejected in an intermediate heat exchanger (steam generator) to the working fluid in the secondary loop. For the idealized case no loss in temperature in the primary loop is assumed, but, again, an irreversibility is necessary to accomplish the heat transfer from the primary loop temperature, T_2, to the secondary loop temperature, T_3, assumed as constant. Mechanical work is then extracted from the secondary coolant in the usual way, as indicated by the cycle, A, B, C, D, E, with the irreversibility associated with the turbine expansion indicated. A heat-transfer irreversibility is, of course, associated with the heat rejection from the cycle to T_5. Since temperature T_5 depends upon the heat-sink temperature available, efficiency is increased if temperature T_3 is made as high as possible.

4.9 By contrast, the temperature–entropy pattern for a variable-temperature heat source is shown in Fig. 4.3. A nonvaporizing primary coolant increases in temperature as it absorbs energy from the fuel, as shown. In sodium systems an intermediate loop is normally used between the primary coolant loop and the vaporizing working fluid. Such a system is also illustrated in Fig. 4.3, where a counterflow heat exchanger minimizes the irreversibilities associated with the necessary heat transfer. An important design parameter is the point of minimum temperature difference between the two lines, called the pinch point. If the temperature difference at the pinch point is small, the process tends to be less irreversible, but a greater investment in steam-generator heat-transfer area is needed. The trade-off between efficiency and investment therefore requires optimization. In practice, the pinch point tends to be between 20 and 40°F.

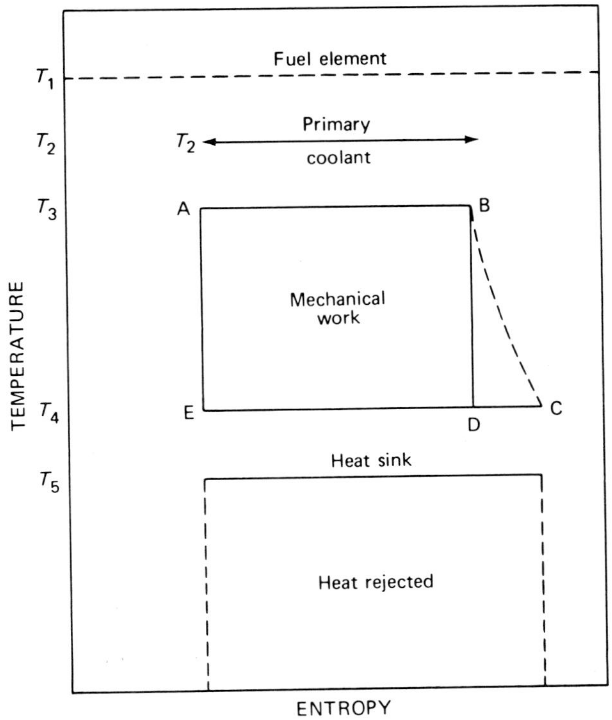

Fig. 4.2 Idealized thermal-transport and power-cycle system.

4.10 Practical steam cycles include more features than those shown in Fig. 4.3. For example, a temperature–entropy diagram for a pressurized-water reactor is shown in Fig. 4.4, still on a somewhat idealized basis. Cycle efficiency is improved without increasing the maximum pressure or temperature by partially expanding high-pressure steam from state 1 to 2 and reheating it from state 2 to 4 before expanding lower pressure steam from state 4 to 5. The reheat steam cycle is similar to that for a conventional fossil-fuel power station except that moisture separation must be provided for since the steam entering the high-pressure turbine is saturated rather than superheated. Figure 4.5, a complete flow diagram for the steam and power conversion system of a pressurized-water reactor, is quite complex in contrast to the simplified cycles discussed.

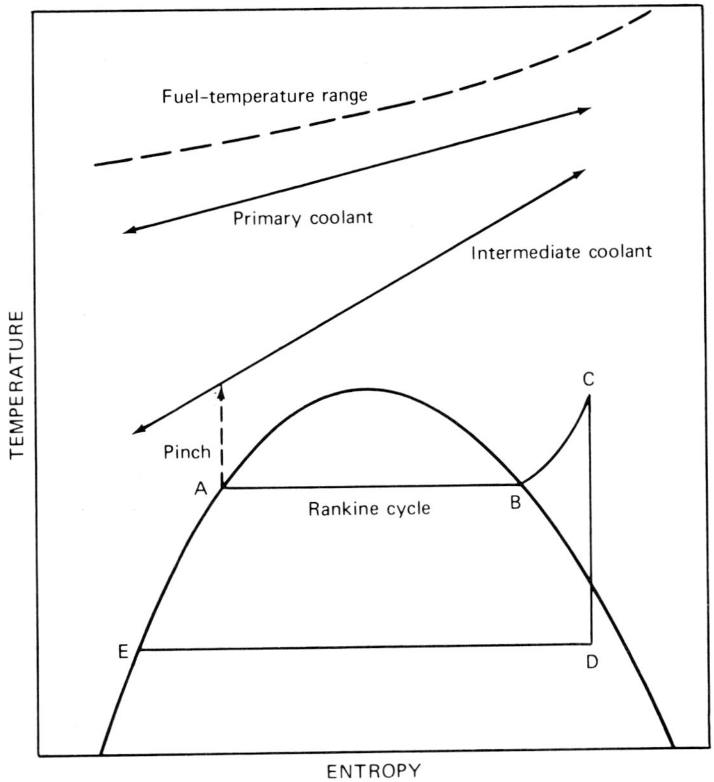

Fig. 4.3 Coolant-loop and working-fluid temperature–entropy pattern.

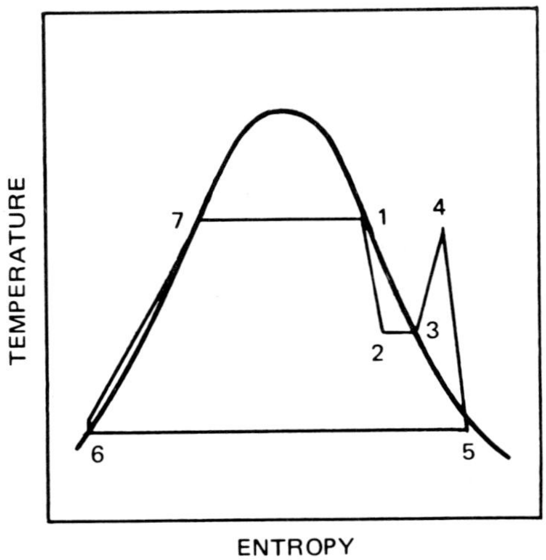

Fig. 4.4 Steam cycle for a pressurized-water reactor.

TEMPERATURE PATTERN

4.11 The thermal-transport and energy-conversion system temperatures serve as bench marks for much of the reactor system design. These are not independent, however, but are subject to iterative adjustment. The nature of the coolant, of course, has an important effect on the pattern. Liquid-metal and, to some extent, gaseous coolants characterize high-temperature high-power-density systems, but water coolants operate at temperatures of the order of 500°F.

4.12 A number of factors must be considered in choosing a coolant-temperature rise. In addition to being influenced by the interplay between circulation rates and the fuel-element temperature pattern, the higher coolant-temperature limit is governed by the properties of the containment materials, and the lower limit depends on practical circulation temperatures. The introduction of cold coolant past the reactor core under "scram" conditions must also be considered. Thermal shocks, possible if the temperature of the entire coolant became as low as the inlet temperature, might well introduce structural design problems. Generally, the higher the differential temperature through the reactor, the more severe this requirement is and the greater the cost. An optimization study is therefore appropriate. It should be mentioned that sodium-cooled reactors can involve temperature rises as high as 400°F compared with rises of 40°F which are common in pressurized-water systems.

4.13 The temperature conditions may be determined as in Fig. 4.6 for a typical sodium system. First, assume that a steam pressure of 1200 psia has been found appropriate. Then select the reactor-outlet temperature, considering the materials available and the primary system design with 1000°F assumed here. Next, by choosing a 50°F driving force across the intermediate heat exchanger and steam generator as reasonable, the intermediate-heat-exchanger outlet is fixed at 950°F and the steam temperature at 900°F. On a temperature–enthalpy diagram, a vertical line can then be drawn from the steam conditions of 900°F and 1200 psi (point 1) to the 950°F temperature line. A line may then be drawn on the diagram describing the path of the secondary sodium temperature from a point 40°F higher (pinch point) than the water-saturation line at the steam pressure (point 3). If the temperature rise through the reactor has been determined as 400°F, the intermediate loop return temperature will be at 550°F. The diagram line is then extended to point 4, where it intersects with the 550°F line. A vertical line is then drawn to the water-saturation line which fixes the feedwater temperature at 430°F.

SOME COOLANT-SYSTEM DESIGN CONSIDERATIONS

The Primary Coolant System

4.14 In addition to transporting heat generated in the core to a heat exchanger, the primary coolant system contains any fission products released

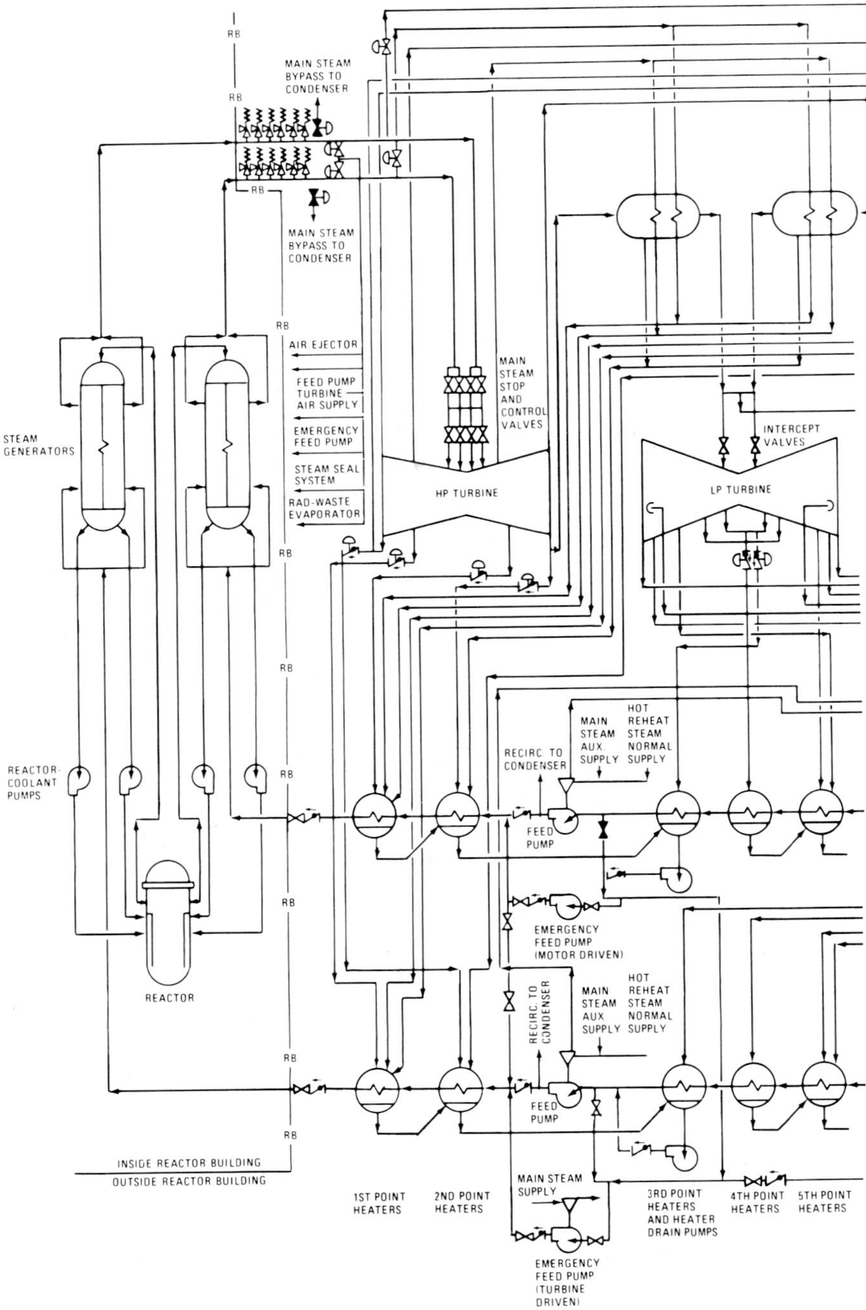

Fig. 4.5 Steam and power conversion system.

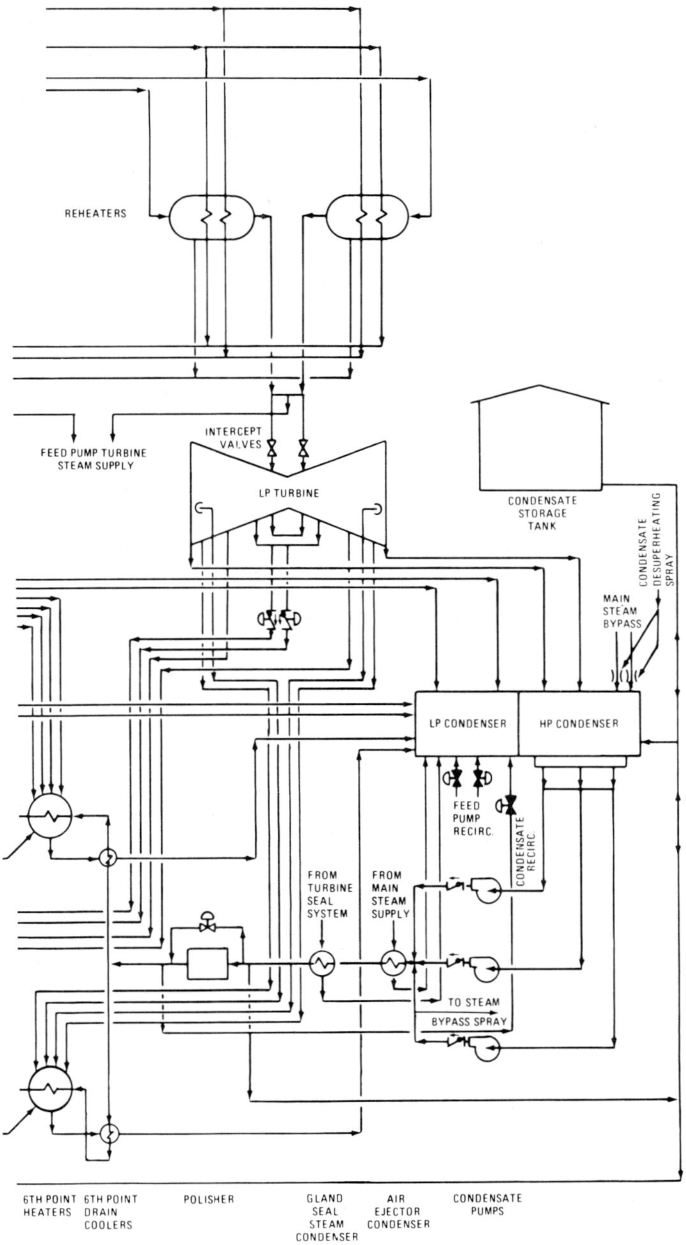

REHEATERS
FEED PUMP TURBINE STEAM SUPPLY
INTERCEPT VALVES
LP TURBINE
CONDENSATE STORAGE TANK
CONDENSATE DESUPERHEATING SPRAY
MAIN STEAM BYPASS
LP CONDENSER
HP CONDENSER
FEED PUMP RECIRC.
CONDENSATE RECIRC.
FROM TURBINE SEAL SYSTEM
FROM MAIN STEAM SUPPLY
TO STEAM BYPASS SPRAY
6TH POINT HEATERS
6TH POINT DRAIN COOLERS
POLISHER
GLAND SEAL STEAM CONDENSER
AIR EJECTOR CONDENSER
CONDENSATE PUMPS

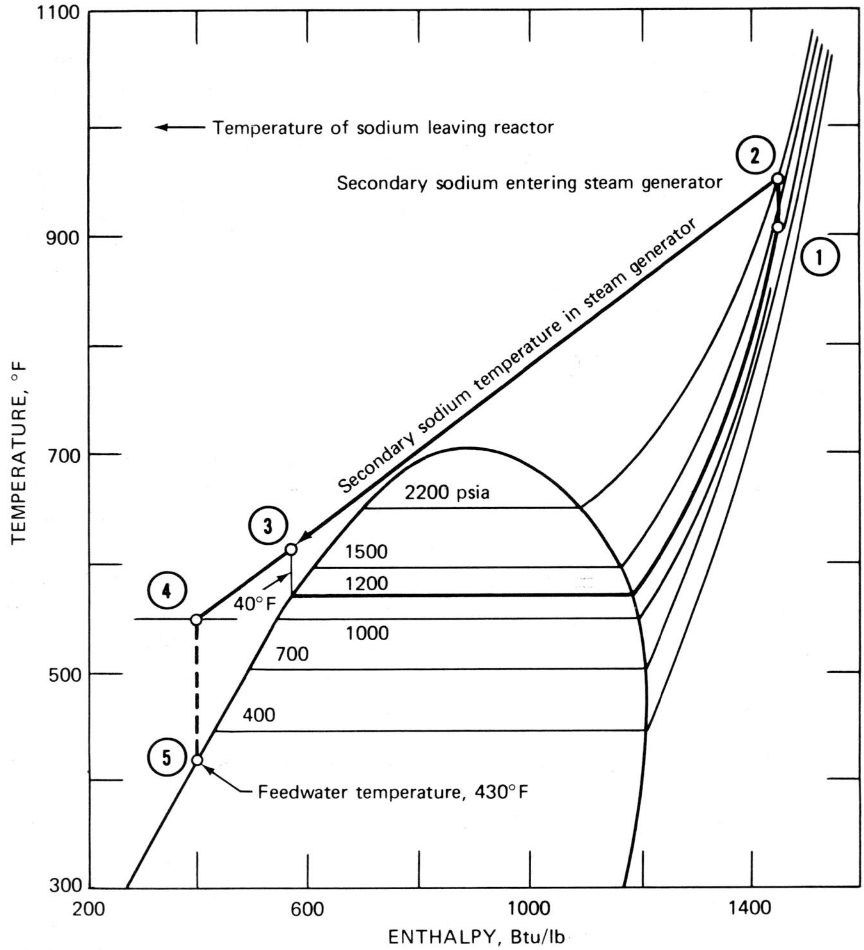

Fig. 4.6 Temperature condition selection.

from the fuel elements. It also cools the fuel in the core in the event of a shutdown or reactor transient. The coolant can flow either upward through the core or downward. The upward flow pattern, by far the most common, provides the advantage of consistency with natural circulation. Furthermore, an important disadvantage of a downward flow pattern is the need for sealing the reactor plug, or cover, against the maximum pump-discharge pressure. Hydrodynamic problems also occur with the upward flow pattern, however, since the core must be kept from rising as a result of the pressure differential between the inlet at the bottom and the outlet at the top. Such a differential, a result of the friction loss through the core, can be of the order of 35 psi and produces substantial upward forces.

4.15 Generally, the smaller the number of coolant loops, the lower the component cost is. This may not be true, however, if such major components as circulating pumps are of undeveloped size or of special design. The same consideration holds true for heat exchangers. Efficient use of building space is also important. For example, three loops can generally be spaced more efficiently than one or two loops. Emergency cooling also requires consideration, perhaps using only one loop with a special type of circulation arrangement. Generally, four loops can be accommodated in a building smaller than that necessary for an equivalent two- or three-loop design, since the components are smaller. Either three or four loops are ordinarily appropriate. However, for large reactors, five, six, or more loops may be desirable.

Primary and Secondary Coolant Loops

4.16 The need for a secondary loop system depends on the coolant fluid and thermodynamic working fluid. The simplest system using steam is the direct-cycle boiling reactor in which the steam is formed directly in the reactor core and then expanded in the turbine generator. Although some boiling may take place in the core of a pressurized-water reactor, an external steam generator produces the steam expanded in the turbine. The type of system normally used for sodium is more complicated; both a primary and a secondary sodium circulation system are used to transport energy to the steam generator, and three independent coolants are therefore involved.

4.17 Although a primary coolant system used alone in a sodium system would cost less to install, an intermediate-loop system is usually considered necessary for safety and reactor containment. If an intermediate loop is used, the reactor-containment structure can be designed to accommodate only the primary coolant system, which includes no water-containing components. Steam-generator failures, with water possibly entering the sodium system, could then be isolated from the primary loop containing radioactive sodium activated in the reactor. The containment-vessel design is also simplified by the addition of an intermediate loop since attention need be given only to the liquid metal–air reaction, and not to the liquid metal–water reaction, possible if the steam generator were contained therein. Also, a steam generator external to the containment vessel is accessible for maintenance.

FUEL-ELEMENT THERMAL PERFORMANCE

4.18 Since the fuel element is where energy is released by fission, its design depends on the way generated heat can be transported through the element to a circulating heat-transfer medium without producing temperatures that could cause failure of the element. The word "failure" in this sense can mean a change in physical properties, or dimensions, as well as melting or corrosion. In a general sense, therefore, "failure" is a loss in the functional abilities of the fuel element. The thermal design of a fuel element is therefore concerned with the effect of temperature and thermal transport on the fuel element. Although such parameters as thermal input, fuel-rod-material properties, rod diameter, and coolant characteristics can lead to a workable thermal design, economic and neutronic parameters are also pertinent to an optimized design of a fuel element.

4.19 The thermal design of a fuel element must meet certain criteria:

1. The maximum temperatures must not lead to deterioration of the materials.

2. Stresses from the combined effects of thermal gradients and accumulated fission gases should lie within limitations of design practice. Thermal stresses depend on the temperature gradient, and the rate of gas evolution depends on temperature and fission rate. The possibility of cladding creep or embrittlement can also limit the burnup.

3. The thermal flux must be below that which would lead to the so-called *critical heat flux,* in which coolant boiling instabilities occur.

4. The thermal and hydraulic design can be related to nuclear-stability criteria through coolant-void limitations.

The heat flux and temperature profile depend on each other, of course. The determination of the temperature profile is therefore the "key" to fuel-element thermal analysis.

FUEL-ELEMENT TEMPERATURE PROFILE

4.20 A review of the sequence of thermal flow is useful as a guide to the parameters that affect a temperature profile such as that in Fig. 4.7 for an annular fuel pellet. Energy is generated in the fuel volume at a rate depending on the fission rate and not on inherent combustion limitations, which apply, for example, to fossil-fuel plants. From the point of generation, the energy flows through fuel material, through a possibly gas-filled region between fuel and cladding material, through cladding material to an interface with a coolant, and finally through a portion of the coolant which will transport the energy from the core.

4.21 At a given power level, the temperature in the fuel depends both on the temperature *gradient* through the various materials in the flow sequence and on the bulk temperature of the coolant at the point along the length of the fuel element being studied. Starting with the coolant temperature, which we may consider a reference temperature, and using known temperature gradients, we can determine the actual temperature profile in the element.

4.22 As shown by the Fourier heat-flow equation (Eq. 4.1), the respective temperature gradients depend on the heat-flow rate, which, in turn, depends on the power being generated and the thermal conductivity,

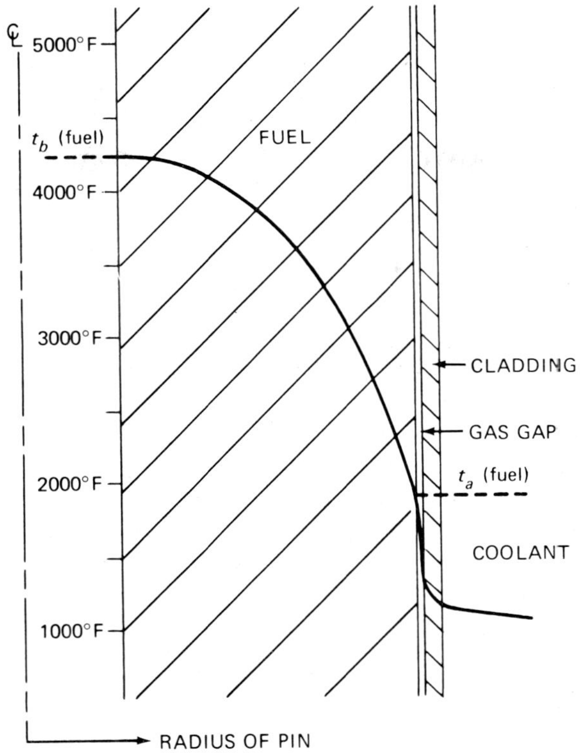

Fig. 4.7 Temperature distribution in fuel pin.

$$\frac{dt}{dx} = -\frac{1}{k}\frac{q}{A} \tag{4.1}$$

where dt/dx is the gradient in the x direction, k is the thermal conductivity, and q/A is the heat flux. At the fluid–cladding interface,

$$t_{\text{clad}} - t_{\text{fluid}} = \frac{1}{h}\frac{q}{A} \tag{4.2}$$

where h is the heat-transfer coefficient. The coolant reference temperature depends, of course, on the temperature rise through the core, which, in turn, depends on the coolant circulation rate, coolant properties, etc. In the actual calculation, a number of other factors must be considered, such as accounting for the volumetric nature of the heat generation, geometric effects, and parameters that affect the heat-transfer coefficient.

4.23 For most purposes the fuel-element radial temperature distribution can be obtained by assuming a uniform volumetric heat source in a conduction problem. Although solutions of the pertinent equations are readily available for a number of different geometries, the cylinder case is given here since it is so widely applicable.

Consider a thin cylindrical section of thickness dr at the radius r. In the steady state the heat-rate balance requires that

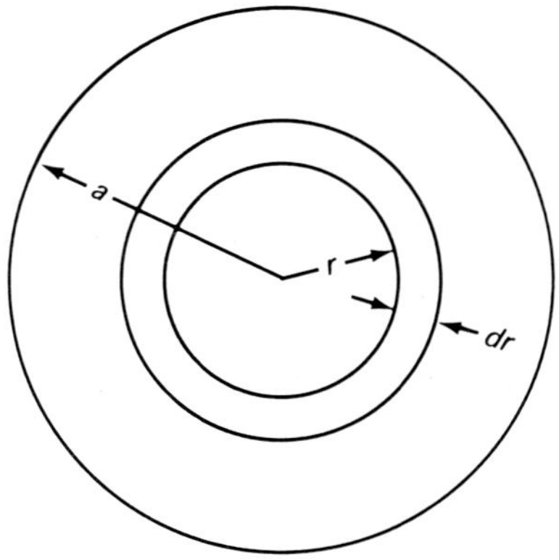

Heat conducted out of cylindrical section − heat conducted into cylindrical section = heat generated in cylindrical section of thickness dr

or for unit length in the axial direction $q_{out} = q$ at $(r + dr)$ and $q_{in} = q$ at r, or according to the heat-conduction, or Fourier, equation

$$q = -k(2\pi r)\frac{dt}{dr} \tag{4.3a}$$

$$q_{in} = -k(2\pi r)\frac{dt}{dr} \tag{4.3b}$$

and

$$\begin{aligned} q_{out} &= -\left[k(2\pi r)\frac{dt}{dr} + \frac{d}{dr}\left(k2\pi r\frac{dt}{dr}\right)dr\right] \\ &= -2\pi k\left[r\frac{dt}{dr} + \left(\frac{dt}{dr} + r\frac{d^2t}{dr^2}\right)dr\right] \end{aligned} \tag{4.3c}$$

then

$$q_{out} - q_{in} = -2\pi k\left(\frac{dt}{dr} + r\frac{d^2t}{dr^2}\right)dr = Q(r)\ 2\pi r\ dr \tag{4.3d}$$

Since $q_{generated} = Q(r)\ 2\pi r\ dr$ where $Q(r)$ is the volumetric heat source, the equation for the cylinder in the steady state reduces to

$$-k\left(\frac{dt}{dr}+r\frac{d^2t}{dr^2}\right)dr = Q(r)\,r\,dr \qquad (4.4a)$$

or

$$-k\left(\frac{d^2t}{dr^2}+\frac{1}{r}\frac{dt}{dr}\right) = Q \qquad (4.4b)$$

This is a particular form of the general equation for the steady state in a system with an internal source; namely, $-k\ \nabla^2 t = Q(r)$, where ∇^2 is the Laplacian operator.

The general solution for Eq. 4.4b is

$$t = -\frac{Qr^2}{4k} + C_1 \ln r + C_2 \qquad \text{for } 0 \leqslant r \leqslant a \qquad (4.4c)$$

The boundary conditions in the present case are $dt/dr = 0$, r = C, and $t = t_1$, $r = a$. Therefore the appropriate solution is

$$t - t_1 = \frac{Q}{4k}(a^2 - r^2) \qquad \text{for } 0 \leqslant r \leqslant a \qquad (4.4d)$$

where t is the temperature at the radial distance r. If t_0 is the temperature along the central axis, where $r = 0$, then

$$t_0 - t_1 = \frac{Qa^2}{4k} \qquad (4.5a)$$

for the temperature drop across the fuel itself.

4.24 Where the thermal conductivity, k, changes with temperature, it is convenient to express Eq. 4.5a as

$$\int_{t_1}^{t_0} k(t)\,dt = \frac{Qa^2}{4} \qquad (4.5b)$$

This parameter is characteristic of the material for a given temperature range. Another useful parameter is the heat-generation rate per unit length of cylindrical fuel element, q_L. Since $q_L = Q(\pi a^2)$,

$$\int_{t_1}^{t_0} k(t)\,dt = \frac{Qa^2}{4} = \frac{q_L}{4\pi} \qquad (4.5c)$$

The temperature difference between axis and surface of round rod with uniform heat generation therefore depends only on the linear heat load and is independent of the rod diameter, even when thermal conductivity varies with temperature. Experimental measurements may be used to determine the value of

$$\int_{t_1}^{t_0} k\,dt$$

Experimental determination of the thermal conductivity itself for ceramic fuel materials is difficult since substantial temperature gradients are present in a sample under study and the property desired is dependent on the temperature.[4] Furthermore, some fuel materials such as oxides tend to fracture and change characteristics during heating and in-pile irradiation, which makes generalized correlation difficult. The designer should therefore consult current literature for needed values.

4.25 Design requirements for some reactors fueled with oxide, for example, permit no center melting of the fuel in the hottest spot in the core. Experimental evidence in the range of fuel-rod diameters under consideration may show that center melting does not occur until the linear heat load, q_L, exceeds 20 kw/ft. If the design specification is 16 kw/ft, the ratio 16/20 provides a measure of the design conservatism. Tabulations are frequently made of $\int k\, dt$ which also may have the units of kilowatts per foot. The conversion, $\int k\, dt \times 4\pi = q_L$ should not be overlooked, however.

4.26 Solutions can be developed for various other geometrics as well as other types of cylindrical systems. For a hollow cylinder of radius a, inner radius b, externally cooled, with internal heat generation, the equation is

$$\int_{t_a}^{t_b} k(t)\, dt = \frac{q_L}{4\pi}\left[1 - \frac{2b^2}{a^2 - b^2} \ln\frac{a}{b}\right] \tag{4.6}$$

The second term in brackets on the right-hand side of Eq. 4.6 can be interpreted as a measure of the reduction in center to outer surface-temperature difference obtained by using a hollow configuration instead of a solid cylinder at the same linear heat rate.

4.27 Another useful parameter is the heat flux, the heat rate per unit area, which is obtained from the power per unit length by the relation

$$q_A \text{ (at radius } a) = \frac{q_L}{2\pi a} \tag{4.7}$$

The heat flux is very useful in determining other components of the temperature distribution, such as the temperature drop across the gas gap between the cladding and the fuel and the temperature drops associated with the cladding and coolant film. In each case the radius appropriate to the resistance must be used. For example, the relations for the temperature changes across the gas gap,* cladding, and coolant film as shown in Fig. 4.7 can be written as follows:

$$\Delta t_{\text{gap}} = \frac{q_L}{2\pi a h_{\text{gap}}} \tag{4.8a}$$

$$\Delta t_{\text{clad}} = \frac{q_{L}(\text{clad thickness})}{2\pi r_{\text{LM}}\, k_{\text{clad}}} \tag{4.8b}$$

In the cladding a logarithmic mean radius (LM) is appropriate. However, the arithmetic mean radius is normally an adequate approximation.

*A typical value for UO_2 pellet-clad gap conductance with fission gas in the gap is 1000 Btu/(hr)(sq ft)(°F).

4.28 In thermal reactors the power generation is actually not uniform across the diameter but is depressed at the center because of self-shielding, in which absorption by outer layers of fuel tends to deplete the supply of neutrons of a given energy reaching the center. Therefore, in light of the central temperature limitation, a uniform power distribution is a conservative assumption for design.

4.29 The designer must also consider the effect of burnup on fuel-material physical properties. For example, for UO_2 both the melting point and the thermal conductivity apparently decrease as burnup increases. Therefore the permissible linear heat rating must be reduced as the burnup specification is increased if central-pin melting is to be avoided.

OXIDE PIN BEHAVIOR

4.30 Since most current designs for light-water thermal reactors and sodium-cooled fast reactors use mixed uranium and plutonium* oxide fuel pins, some of the special thermal characteristics of such pins are considered here. Light-water reactors generally use rods of UO_2 in the form of pellets, powder, or particles, clad within Zircaloy-2 or -4. Fast reactors use stainless-steel cladding.

Thermal Conductivity

4.31 Although many measurements have been made of the thermal conductivity of UO_2, the results tend to be only in fair agreement since many parameters affect the nature of the material used in a given experiment. These include the density, method of fabrication, and oxygen content. Thermal-conductivity measurements at high temperature are also very difficult. The designer is therefore faced with a range of values rather than a definitive thermal-conductivity standard in terms of the independent variables of interest, which are temperature and the extent of radiation. A range of representative results is shown in Fig. 4.8 for unirradiated material. One expression[5] for the thermal conductivity of polycrystalline UO_2 to about 1300°C is

$$k = \frac{1}{11.75 + 0.0235t} \tag{4.9}$$

where k is the thermal conductivity [watts/(cm) (°C)] and t is the temperature (°C). The density is 95% of theoretical.

4.32 Above 1300°C, uncertainty increases since measurements become more difficult. However, there appears to be a consensus that the thermal conductivity tends to increase with increasing temperature with a minimum near 1600°C.

*Emphasis is given here to characteristics contributed by the UO_2 present.

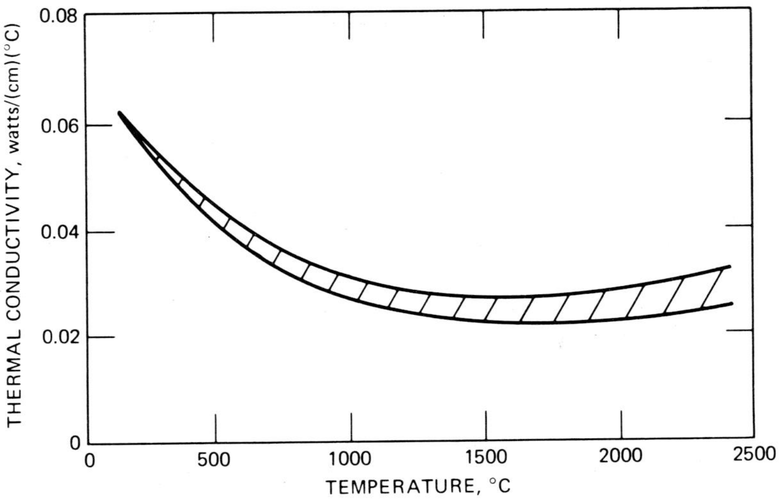

Fig. 4.8 Thermal conductivity of unirradiated UO_2 (95% of theoretical density).

4.33 The actual mechanism of heat transfer in uranium dioxide is not yet completely understood. Thermal transport through crystalline UO_2 is considered to be the result of the sum of at least three mechanisms, phonon or lattice conductivity, radiative or photon conductivity, and electronic conductivity. Although phonon conductivity is significant at lower temperatures and radiative transport becomes important as the temperature is raised, as expected, the picture is apparently complex. Electronic conduction is also a factor but is not clearly understood.[5]

4.34 Measurements of thermal conductivity made under irradiation conditions involve additional parameters and uncertainties, and therefore a graph here is not appropriate. However, it does appear clear that the conductivity of UO_2 decreases upon irradiation at temperatures under 500°C. On the other hand, irradiation appears to have little effect on the thermal conductivity at higher temperatures.[5]

4.35 Integral measurements, in which the value of $\int k\, dT$ for a specific temperature range is developed, tend to provide a more satisfactory basis for design work than differential measurements. The experiment is inherently easier since only terminal temperatures and a heat flux need be measured. The effect of such structural changes as columnar grain growth (§4.43) which are temperature dependent can also be studied. A typical integral conductivity graph[6] for mixed-oxide fuel is shown in Fig. 4.9. Note the change in slope at approximately 1800°C, the threshold temperature for grain restructuring (§4.42).

4.36 An approach even closer to the expected service conditions is to merely carry out a family of measurements with linear heat rate as the parameter

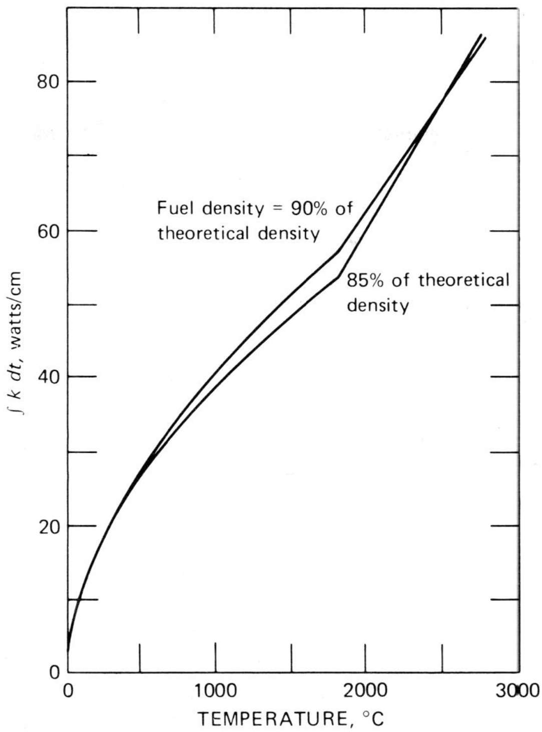

Fig. 4.9 Integral conductivity of mixed oxide fuel as a function of fuel temperature. It is assumed that sintering has occurred.

and experimentally establish the heat-rate value at which unsatisfactory pin performance can be anticipated as indicated by central melting or other criteria. An operating condition can then be specified at a lower linear heat-rate value with a safety factor provided. With this approach it is common to reduce the specified heat rate somewhat as burnup progresses to correspond with an anticipated decrease in thermal conductivity of the oxide, even though laboratory measurements of conductivity indicate little change. Typical design limitations as a function of both initial density and burnup for an oxide-fueled liquid-metal-cooled fast breeder are shown in Fig. 4.10.

Dimensional Changes and Pin Integrity

4.37 Light-water fuel rods designed for burnup levels exceeding 30,000 Mwd/tonne at linear heat rates of over 20 kw/ft must also meet stringent quality-assurance requirements so that the probability of pin failure during

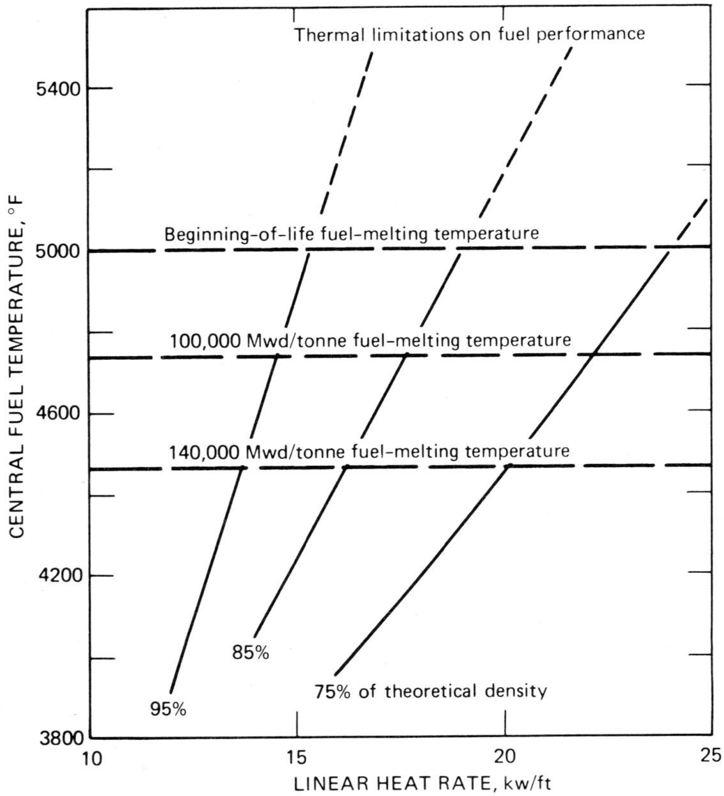

Fig. 4.10 Effect of initial fuel density and burnup on allowable linear heat rate.

Assumptions

	Columnar grain formation	Equiaxed grain formation
Temperature, °F	3200	2700
Density, %	98	94
	Gap coefficient, 1500 Btu/(hr) (sq ft) (°F)	

exposure is minimal.[7] Many of the design considerations concern the interplay of thermal- and material-property parameters.

4.38 Important in fuel-rod design is fuel swelling, dependent on both temperature and fission-product inventory. Accommodation of growth induced by irradiation has been attempted by including void volumes within the rod. Various suggested schemes include the use of high-density pellets with large fuel–cladding gaps, low-density pellets, annular pellets, deep dished-end pellets, and a combination thereof.

4.39 Additional irradiation-performance considerations are rod bowing, fuel–cladding chemical reactions, fatigue, cladding strains arising from steady-state and transient power operation, hydriding of the cladding material, power distortion, and other dimensional distortions. Generally these effects can be controlled by establishing design limits. For example, rod bowing is caused by radial thermal gradients and can be partially controlled by modifying the fuel-rod support system or by limiting the radial gradient. Fuel–cladding reactions can result from an attack by molten fuel or fission products on the cladding. The possibility of fuel melting can be reduced by limiting the peak power level of the rod. Fission-product reactions tend to be temperature dependent and can therefore be controlled by limiting the peak power level of the rod.

4.40 Interactions between fuel and cladding and, to a lesser extent, fission-gas release affect the performance and subsequent integrity of the fuel rod significantly. Considerations include fatigue from plant swingload, pressure, or power cycles, dimensional strain within the fuel, and circumferential clad strain. These effects are all dependent on temperature, time, fuel thermal expansion, irradiation, and fuel–cladding diametral-gap dimensions.

4.41 In fast reactors, where the fuel is normally subjected to several times the burnup that is practical for thermal reactors, dimensional changes of the fuel cladding and structural components become a major design consideration. Fuel swelling is caused primarily by the accumulation of fission-product gas and the buildup of solid fission products. However, stainless-steel cladding also tends to swell under irradiation, presumably owing to the formation of voids caused by the aggregation of irradiation-induced vacancies. The amount of swelling is a function of fluence (flux–time product), irradiation temperature, and cold work. Mechanical deformation of the cladding results from the stress pattern developed by the interaction between the cladding and the swelling fuel. However, the interaction may be due to thermal expansion of the fuel and cladding during power or temperature changes of the reactor rather than the steady-state pressure of the fuel.[8] At any rate, the amount of mechanical deformation is a function of cladding temperature, mechanical properties, and load pattern.

Restructuring

4.42 In the central regions of oxide fuel pins at temperatures above about 1700°C, the pores migrate toward the hottest point at the center.[9] This tends to enhance the formation of a central void as well as densify the hotter fuel region. Pore migration is thought to be due to vaporization on the hot side of the pore followed by condensation on the cold side in response to a thermal gradient. As shown in §4.26, an annular fuel pin can be subjected to a higher linear heat rate than a solid pin. Densification in the central region from pore migration also increases the apparent thermal conductivity that enhances this effect.

4.43 In this region subjected to temperatures above 1700°C, a columnar grain structure about 97% of theoretical density forms but tends to have extensive radial cracks. In an outer, intermediate-temperature region (1450 to 1750°C), equiaxed grains may also be formed to become somewhat denser than the as-fabricated state. Restructuring may therefore result in three distinct radial regions: an outer, as-fabricated region; an intermediate equiaxed-grain region; and a central columnar-grain-structure region.

Fission-Gas Release

4.44 In fuel designed for long burnup, particularly fast-reactor fuel insensitive to the poisoning effects of fission products, pressures from fission-product-gas accumulation impose important mechanical-based design limitations. Although at first glance the calculation of such pressures appears to be a straightforward application of the gas laws, the selection of an average representative temperature and a proper volume is a problem. For an ideal gas the pressure may be expressed as

$$P = \frac{NR}{\sum_i (V_i/T_i)} \tag{4.10}$$

where N is the number of moles of gas and V_i/T_i is the contribution of an interconnected volume element V_i at a temperature T_i. Although the amount of fission gas produced can be calculated from the fission rate,* the amount actually released from a ceramic lattice structure depends on a variety of factors. For UO_2 the release appears to be by a combination of "solid-state" mechanisms, with the temperature determining which one predominates.[10] Below 600°C, fission-gas release is independent of temperature and is apparently caused by the "knock-out" process of secondary ejection when a recoil fission fragment leaves the UO_2 surface. Above 600°C the rate-controlling process appears to be the release from defect traps of gas that previously diffused rapidly through the UO_2 matrix.

4.45 For engineering purposes in the case of UO_2, Lewis[11] suggests a model that assigns a maximum percentage gas release to certain temperature bands:

---

*For most fissile atoms, approximately 27 atoms of gas are produced per 100 atoms fissioned.

Percentage release $\leqslant$

$$\frac{0.5 \int_{\text{surface}}^{1000^\circ\text{C}} k\,dt + 10 \int_{1000^\circ\text{C}}^{1300^\circ\text{C}} k\,dt + 60 \int_{1300^\circ\text{C}}^{1600^\circ\text{C}} k\,dt + 95 \int_{1600^\circ\text{C}}^{\text{center}} k\,dt}{\int_{\text{surface}}^{\text{center}} k\,dt} \tag{4.11}$$

More gas tends to be released as the oxide density is decreased for a given $\int k\,dt$ since k is lower and more surface is available.

4.46 Another more approximate approach assumes that all the fission gases are released at temperatures in the fuel above 1700°C. The vclume fraction of fuel below 1700°C may be assumed to release only 50% of the fission gases produced. The temperature profile (longitudinal and lateral) in the fuel must be known, however, before this rule of thumb can be applied.

4.47 An additional complication in calculating the pressure is the dependency of $\sum_i (V_i/T_i)$ on reactor power since the volume available for the gas tends to change as a result of fuel plastic flow, cracking, etc. Notley[12] provides some calculation guides and reports experimental pressures ranging from 300 psi at a thermal reactor heat load q_L of 125 watts/cm to 600 psi at a q_L value of 625 watts/cm.

Analysis Models

4.48 For design and analysis a pin thermal-transport model that takes into consideration the preceding characteristics and lends itself to computer representation is useful. Since experimental information is not completely consistent, reasonable assumptions must be made as needed. One example of such a model[9,13] developed for sodium-cooled fast reactors follows and is based on three fuel regions and a center void region as shown in Fig. 4.11. Material in the outer region retains its fabricated density and is assumed to move outwardly from thermal expansion. Fuel in this region is assumed to be nonplastic and brittle. Material in the middle region is above the temperature for sintering and forms equiaxed grains. It is in a plastic condition and is restrained from expanding by the outer region. Thermal expansion in this region therefore moves the material inwardly.

4.49 The inner region, Region I, has a steep thermal gradient with porosity migration that causes columnar grains to form. A void forms in the center of the fuel from the increase in density (decrease in porosity) of Regions I and II. This center void has an important effect on the thermal behavior of the fuel pin.

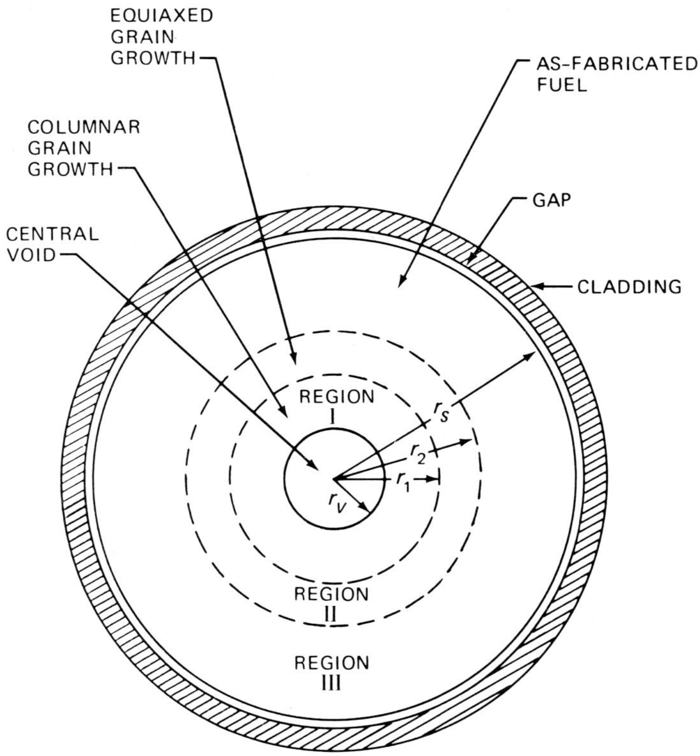

Fig. 4.11 Three-region analytical fuel model.

4.50 In fast-reactor fuel rods, there is very little flux depression, and the heat generated is proportional to the mass of the fuel. The volumetric heat rate in each of these three regions is therefore

$$Q_{\mathrm{III}} = \frac{q_L}{\pi r_s^2} \tag{4.12a}$$

$$Q_{\mathrm{II}} = Q_{\mathrm{III}} \frac{\rho_{\mathrm{II}}}{\rho_{\mathrm{III}}} \tag{4.12b}$$

$$Q_{\mathrm{I}} = Q_{\mathrm{III}} \frac{\rho_{\mathrm{I}}}{\rho_{\mathrm{III}}} \tag{4.12c}$$

where q_L is the lineal power, r_s is the surface radius of the pellet, and ρ is the density. Subscripts used for temperatures correspond to those used for radii.

4.51 Solutions for the heat-transport equation with appropriate boundary conditions in each region are given in the following equations:

For Region III:

$$\left(\frac{r_2}{r_s}\right)^2 = 1 - \frac{4\int_{T_s}^{T_2} k_{III}\, dT}{Q_{III} r^2 s} \tag{4.13}$$

For Region II:

$$\left(\frac{r_1}{r_2}\right)^2 + \left(1 - \frac{Q_{III}}{Q_{II}}\right) \ln\left(\frac{r_1}{r_2}\right) = \frac{4\int_{T_2}^{T_1} k_{II}\, dT}{r_2^2\, Q_{II}} \tag{4.14}$$

For Region I:

$$\int_{T_1}^{T_v} k_I\, dT = \frac{Q_I}{4}\left\{r_1^2 - r_v^2\left[1 - 2\ln\left(\frac{r_v}{r_1}\right)\right]\right\} \tag{4.15}$$

From a mass balance and values of r_s, r_1, and r_2, the radius of the void is

$$r_v^2 = r_2^2 \frac{\rho_{II} - \rho_{III}}{\rho_I} + r_1^2 \frac{\rho_I - \rho_{II}}{\rho_I} \tag{4.16}$$

4.52 The temperature of the fuel surface can be found from the cladding temperature and the conductance across the fuel–cladding gap. The conductance across this gap is related to the differential expansion of the fuel and the cladding and to the thermal conductivity of the cover gas. From these relations the power to cause melting with a center void can also be expressed as

$$q_L = \frac{4\pi \int_{T_s}^{T_m} k\, dT}{1 - \left(\frac{r_v}{r_s}\right)^2 \left[1 - 2\ln\left(\frac{r_v}{r_s}\right)\right]} \tag{4.17}$$

4.53 Other models are available which describe the effects of cladding swelling on fuel-rod performance, a behavior important in fast-reactor analysis. OLYMPUS is a computer code of this type,[14] and SWELL and LIFE are codes[15] designed to provide a detailed picture of the pin performance for its complete lifetime. These and similar fuel-performance codes may also be used in conjunction with more general thermal-hydraulic design codes for the thermal analysis of the entire core (§4.139).

CRITICAL HEAT FLUX

4.54 In liquid-cooled* reactors the heat flux must be kept below a condition that produces instabilities in the forced-convection boiling process. This flux,[16,17] referred to as the *critical heat flux* (CHF), causes a marked decrease in the heat-transfer coefficient, with a resulting increase in the fuel-element surface temperature at constant heat-generation rates. Critical heat flux is normally considered the same as *dryout* [or DNB (Departure from Nucleate Boiling)], which refers to the breakdown of the liquid film covering the hot surface in two-phase flow. This breakdown causes small but rapid rises in surface temperature corresponding to the appearance and disappearance of the dry areas on the heated surface. *Burnout* refers to the failure of the heating surface due to high surface temperatures caused by the poor heat transfer through the vapor film that covers the heater beyond dryout. The corresponding flux is referred to as the *burnout heat flux. Boiling crisis* is another term applied to the temperature excursion following critical heat flux.

4.55 Although critical heat flux is one of the most important considerations in the design of a reactor core, it is the subject of controversy among investigators.[17,18] Since the mechanism is quite complex, attempts to develop models for prediction have met with only limited success. Empirical data correlations for predicting conditions that will lead to critical heat flux must also be carefully examined for applicability. Therefore we shall provide the student a guide to some of the highlights of the subject without recommending design procedures or evaluating different approaches. Thus the designer should consult current literature when he needs to determine critical-heat-flux conditions.

POOL BOILING

4.56 Boiling heat transfer under forced convection is a very complicated process with many variables. For orientation we shall review the characteristics of a simple system in which a heated surface is submerged in a pool of boiling liquid (pool boiling). The familiar boiling regimes that occur in such a nonflow system are shown in Fig. 4.12. At low values of the parameter ΔT_x, the difference between the heated-surface temperature and the saturation temperature, the heat-transfer mechanism is primarily free convection with only liquid (no vapor) contacting the heating surface. At higher heating-surface temperatures, nucleate boiling occurs, in which bubbles form at the surface. In this process a high rate of heat transfer is obtained from intense convection currents

*Although "boiling crisis" is possible in systems cooled with any liquid, the critical heat flux is normally a major design limitation in water-cooled reactors and not in sodium-cooled cores.

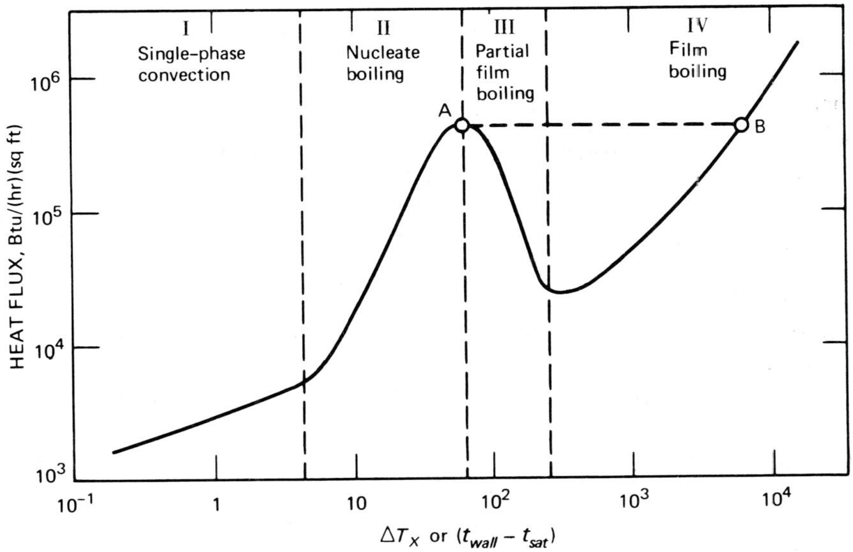

Fig. 4.12 Pool-boiling regimes.

at the heating surface and from bubble agitation and liquid movement to the surface as vapor bubbles leave the heated area. Experiments have shown that most of the heat transferred across the surface interphase is absorbed by local superheating of the fluid rather than by the formation of a vapor bubble immediately.

4.57 In pool boiling the surface temperature rises further, the thermal driving force to the fluid increases, and bubbles form at a greater rate and tend to coalesce. Finally the vapor column formed from the coalesced bubbles is so great that columns of liquid can no longer move toward the hot surface to replace the liquid lost by evaporation in the bubbles. On subsequent increases in the difference in surface and saturation temperatures, the behavior of this system depends on which variables are independent and which are dependent. In the case of nuclear fuel elements, for example, the heat flux is controlled and the surface temperature is dependent. Therefore, if the heat flux is increased above the vapor-blanketing point (A), a sudden transition in surface temperature to the corresponding ordinate (point B) on the film-boiling curve will occur, as shown. Since this temperature can be above the capabilities of the fuel material, the term "burnout" has been applied to such a case.

4.58 Numerous investigators[16,18,19] have studied boiling and have applied various theories to the nucleation process and the subsequent growth and motion of the bubbles formed. Discussions of such approaches are available elsewhere[16] and are not considered appropriate here. Such work, however, does provide a basis for various semiempirical correlations of experimental data using dimensionless groups. Since bubble agitation plays a major role in the mechanism, various "bubble Reynolds numbers" have been postulated. Using a dimensionless-group approach, Rohsenow[19] was able to correlate much pool-boiling data by the following equation:

$$\frac{c_l\,\Delta T_x}{h_{fg}\mathrm{Pr}_l^{1.7}} = C_{sf}\left[\frac{q/A}{\mu_l h_{fg}}\sqrt{\frac{g_c\sigma}{g(\rho_l-\rho_v)}}\right]^{0.33} \tag{4.18}$$

where c_l = specific heat of saturated liquid [Btu/(lb$_\mathrm{m}$) (°F)]
q/A = heat flux [Btu/(hr) (sq ft)]
h_{fg} = latent heat of vaporization (Btu/lb$_\mathrm{m}$)
g_c = conversion factor [4.17×10^8 lb$_\mathrm{m}$-ft/(lb$_\mathrm{f}$) (hr^2)]
g = gravitational acceleration (ft/hr^2)
ρ_l = density of the saturated liquid (lb$_\mathrm{m}$/cu ft)
ρ_v = density of the saturated vapor (lb$_\mathrm{m}$/cu ft)
Pr_l = Prandtl number of the saturated liquid
μ_l = viscosity of the liquid [lb$_\mathrm{m}$/(hr) (ft)]

The term C_{sf} depends on the surface roughness and the heating-surface material–fluid combination. For water–stainless steel, $C_{sf} = 0.014$.

4.59 Pool-boiling correlations are not of major interest to the nuclear-core designer. They are mentioned here merely to indicate the methods that can be applied and the parameters that are pertinent. Photography has been helpful in attempts to describe the conditions existing at critical heat flux and in the development of a model. Various analytical descriptions have been applied, including that of a Helmholtz instability, at the vapor–liquid interface as a result of the relative phase velocity. There has been some disagreement, however, among investigators describing the mechanism and attempting a correlation. A typical critical-heat-flux correlation based on a model describing the coalescence of bubbles from neighboring sites leading to vapor blanketing is that of Rohsenow and Griffiths[19]:

$$\frac{(q/A)_{\mathrm{max}}}{\rho_v h_{fg}} = 143(g^{1/4})\left(\frac{\rho_l-\rho_v}{\rho_v}\right)^{0.6} \tag{4.19}$$

The quantities $[(q/A)_{\mathrm{max}}/\rho_v h_{fg}]$ and $g^{1/4}$ also appear in other equations, although other terms differ. A semiempirical correlation of Kutateladze,[20] for example, describes many experimental data quite well:

$$\frac{(q/A)_{\mathrm{max}}}{\rho_v^{1/2} h_{fg}} = 0.14[\sigma g_c(\rho_l-\rho_v)g]^{1/4} \tag{4.20}$$

4.60 For confined systems (tubes, channels, rod bundles, etc.), where the liquid is under forced convection, critical-heat-flux behavior is somewhat different from that for pool boiling.

4.61 Observations of the behavior differ in the literature, with considerable speculation regarding mechanisms involved. Experimental work does indicate, however, that different mechanisms may exist for the subcooled flow region, the low-quality region, and the high-quality region and that each mechanism is influenced differently by such system parameters as the density, mass flow rate, and the channel length to diameter ratio. Although some progress has been made using hydrodynamic models, a satisfactory theoretical correlation is not yet available. However, we shall examine some of the ideas proposed for describing the mechanism.

4.62 Consider the types of liquid and vapor distributions that can occur under two-phase flow conditions. Visualization studies show that a number of different flow regions can exist. The type of flow that predominates in a tube in a given situation depends on the value of certain hydraulic parameters as first measured by Baker[21] for water–air mixtures.

4.63 For the flow-boiling mechanism, the most important regimes are *bubble flow* and *annular flow*. These can best be visualized with the aid of Figs. 4.13 and 4.14 by considering progressive boiling along a heated vertical channel. For illustration consider a channel or tube such as that in Fig. 4.14 in which the various regimes are encountered in turn. Initially, the incoming water is heated by convective heat transfer with no boiling occurring. As the temperature of the flowing water increases, a point in the channel is reached where the temperature

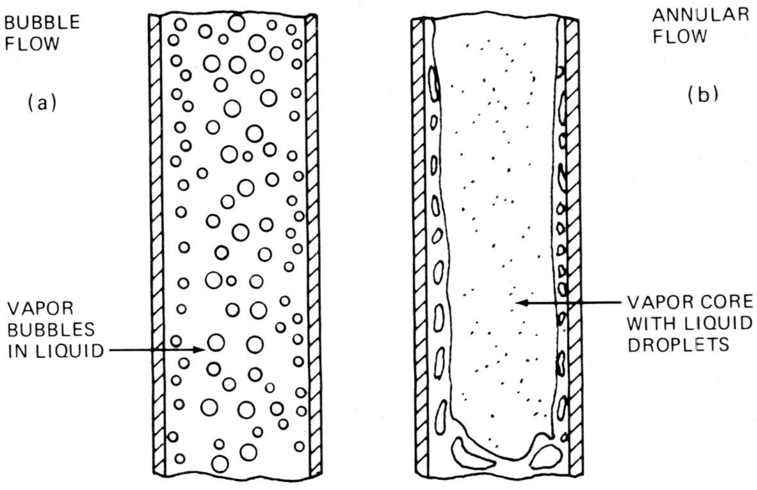

Fig. 4.13 Boiling flow patterns.

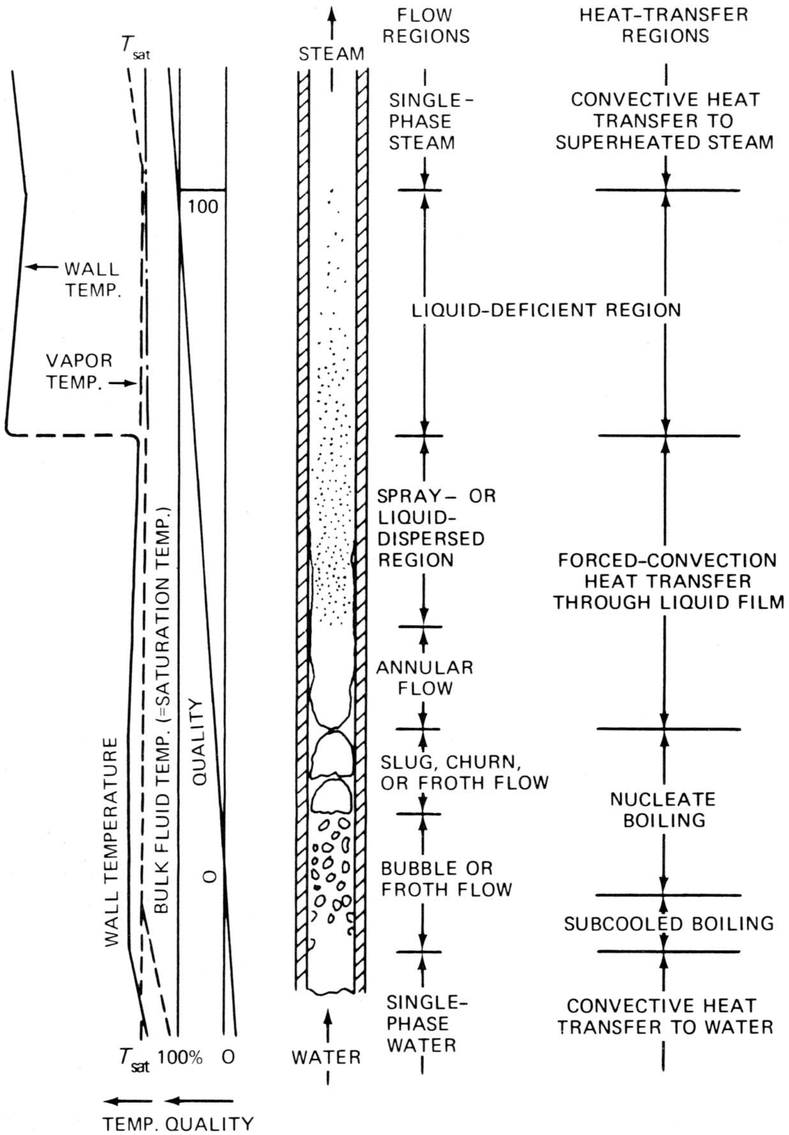

Fig. 4.14 Regimes of two-phase flow.

of the slowly moving fluid adjacent to the wall is above the saturation temperature although the temperature at the same level in the more rapidly moving core is still below saturation.

4.64 Initial vaporization therefore occurs along the wall under *subcooled* (local boiling) conditions. Bubbles grow and are carried along in the superheated layer close to the wall, but they condense on being mixed with the subcooled liquid core. After the central core reaches saturation conditions, however, the vapor bubbles being fed from the superheated layer near the surface no longer collapse but are carried along in the stream. *Bubble flow* is therefore characterized by a "regular" distribution of vapor bubbles in the continuous liquid phase. With increased vaporization and a resulting higher void fraction (higher quality), a transition to *annular flow* occurs as shown in Fig. 4.13b. This is characterized by a continuous vapor phase in the central core in which some liquid drops may be dispersed and by a continuous liquid along the heated wall in which vapor bubbles are dispersed.

4.65 Attempts to explain flow boiling behavior in terms of a model are very qualitative. In the low-quality or subcooled region, a high heat flux can cause a very large number of bubbles in the layer next to the heated surface; this, in turn, can prevent access to the surface by liquid from the bulk stream. Under high-flow-velocity high-pressure conditions, however, a viscous layer of small bubbles flows parallel to the heated surface between a superheated liquid layer immediately next to the surface and the central core.[16] Such a pattern usually occurs in pressurized-water reactors. A boiling crisis probably starts with the formation of a vapor area next to the surface which grows and contracts in an oscillatory manner. The onset of this condition depends primarily on the heat flux but also on flow and thermodynamic parameters.

4.66 With normal heat flux the transition point from bubble flow to the annular flow regime is not well established but depends on pressure, mass velocity, geometry, and perhaps other factors as well as the quality. Now the liquid layer moving along the heated surface tends to suppress bubble nucleation. Most of the heat is therefore transferred by pure conduction through the liquid film, and no boiling takes place. As a result most of the evaporation occurs at the interphase between the outer liquid column and the inner vapor core. An annular flow pattern usually occurs in boiling-water reactors. With increasing rates of evaporation, however, and hence increasing vapor velocities, the vapor tends to "drag" portions of the surrounding liquid. Such intermittent shearing of the liquid away from the wall and replacement by vapor can result in wall-temperature oscillations that can become divergent and lead to a boiling crisis. When the liquid layers disappear from the heated surface, the term "dry-out" applies.

4.67 In the higher quality region, the approach to dry-out depends on hydrodynamic considerations. Predictions of a boiling crisis are therefore based on a critical enthalpy rise rather than in terms of a critical heat flux. Note that an important difference between boiling two-phase flow and an adiabatic system such as air–water is the increase in void fraction along the channel due to vaporization. Therefore, at any point in the channel, the *previous history* or *upstream condition* of the flow is an important variable.

4.68 After the critical heat flux has been exceeded, as shown in Fig. 4.14, the flow regime consists of entrained droplets in a central vapor core. Droplets

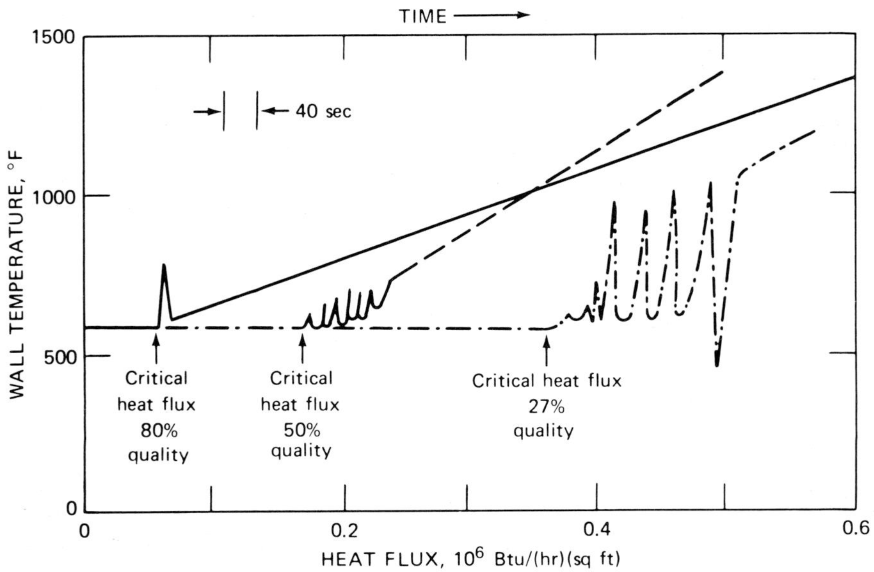

Fig. 4.15 Temperature history as heat flux is increased in steam–water system. $G = 1.4 \times 10^6$ lb/hr/sq ft at 1000 psia.

striking the wall are evaporated. However, if the wall temperature exceeds the *Leidenfrost point,* no wetting of the wall occurs, and heat is transferred directly from the wall to suspended droplets.[22] In this case the momentum of the rapidly evaporating vapor between the liquid droplet and the hot surface forms a steam cushion to support the droplet and prevent the liquid from wetting the surface. This is *stable film boiling,* a mode of heat transfer in which the heat-transfer coefficients are low but stable. Flow is in the *liquid-deficient regime.* Since the heat generation in nuclear fuel elements continues at about a constant rate even though the heat-transfer coefficient is markedly reduced, a corresponding increase in surface temperature results with material failure likely.

4.69 Some experimental observations[23] are shown in Fig. 4.15; wall-temperature behavior is indicated as the heat flux from a vertical heated rod is increased with time for three different initial steam–water mixtures. In the low-flux region for each case, stable and very high heat-transfer rates occur in the nucleate-boiling region, characterized by evaporation at the heated surface or at the liquid film–vapor core interface. The surface temperature is just slightly above saturation. The onset of critical heat flux is indicated at the point of the first substantial temperature rise, or oscillation of the temperature trace, to points well above saturation. In two cases the oscillations first increase, then pass through a maximum in the transition boiling region where segments of the heated rod may be covered intermittently by water and steam, and finally decrease when stable film boiling prevails. Three boiling regions are therefore observed: an initial nucleate boiling region, a transition oscillatory region, and a stable film-boiling region.

4.70 Since the reactor fuel-element–heat-generation rate is independent of changes in the heat-transfer coefficient, the surface temperature will oscillate erratically. An additional complication, however, is the disturbance in a parallel-flow system which will tend to reduce the coolant flow. Such a "choked" flow would lead to high coolant temperatures and, in turn, a rise in the fuel temperature, which could result in failure.

PREDICTIONS AND CORRELATIONS

4.71 Limiting boiling conditions are normally of greater interest to the reactor-core designer than the prediction of steady-state heat-transfer coefficients. Should heat-transfer coefficients be desired for determining surface temperatures or other reasons, a useful first approximation is to merely "superposition" (or directly add) the heat-transfer rate obtainable from pool boiling to the single-phase forced-convection heat flux, each for the same ΔT_x driving force.

4.72 For subcooled or local boiling, the relation

$$\frac{q}{A} = \left(\frac{e^{p/900}}{1.9} \Delta T_x \right)^4 \tag{4.21}$$

was found to be applicable[24] where q/A is in Btu/(hr) (sq ft), p is the pressure (psia), and ΔT_x is the temperature difference between the heated surface and the saturation value (°F).

4.73 Although attempts have been made to develop models that can describe critical-heat-flux conditions, design correlations are usually a representation of laboratory data and use dimensionless quantities or parameters derived from the model work. The approach is primarily empirical and depends on the validity of the measurements used. It is therefore important for the designer to evaluate the correlation and supporting measurements in terms of his own geometry, flow conditions, and heat-flux distribution.

4.74 Data interpretation can be complicated by test-system instabilities not directly related to the boiling process itself. The presence of compressible fluid upstream of the test section can cause flow oscillations and an apparent critical heat flux much lower than that obtained from a stable system. Apparently, much reported data are of this nature.

4.75 The axial heat flux along a reactor channel is not uniform; yet most laboratory critical-heat-flux data have been taken in electrically heated, uniform-heat-flux test sections. Investigators disagree on the conclusion that the critical heat flux is a function of local conditions and does not depend on the integrated heat transferred or the exit enthalpy. Some measurements with a cosine-heated test section show a 20% reduction in the critical heat flux compared with uniform flux results. Tong[16] includes an F-factor correction in his correlation (§4.78) to account for nonuniform heat flux.

4.76 The more satisfactory empirical critical-heat-flux correlations include a large number of terms and are complicated to use. However, they can readily be incorporated into computer procedures needed for industrial design. Only a few of the more important approaches will therefore be mentioned here. Macbeth,[16,25] confining his attention to water, developed a correlation for critical heat flux in tubes, for mass velocities above certain low values, of the form

$$\left(\frac{q}{A}\right)_{\text{crit}} = A - \frac{C}{4} d G \chi_e h_{fg} \tag{4.22}$$

where A and C are functions of mass velocity, diameter, and pressure; χ_e is the quality; and h_{fg} is the enthalpy of vaporization. This correlation has proved quite successful for a vast amount of data in the bubble-flow as well as the annular-flow regime. For uniformly heated round tubes, the critical heat flux in the *low*-mass-velocity region is given by

$$\left(\frac{q}{A}\right)_{\text{crit}} \times 10^{-6} = \frac{(G \times 10^{-6})(h_{fg} + \Delta h_i)}{158 D^{0.1} (G \times 10^{-6})^{0.49} + 4(L/D)} \tag{4.23}$$

4.77 For pressurized-water reactors a critical-heat-flux correlation developed by Tong[26] and coworkers is normally used. Designated as "W-3," it replaces earlier correlations (W-2) developed by the same group. The correlation may be applied to a nonuniform-flux distribution. In contrast to earlier approaches, which yielded a critical flux in the subcooled region and a critical enthalpy rise in the quality region, with a discontinuity at the saturation temperature, the W-3 correlation is continuous over an exit quality range of ±15%.

4.78 In the W-3 correlation an equivalent uniform DNB (Departure from Nucleate Boiling) flux is first calculated. An F-factor is then used to convert the uniform-flux value to one applicable to a nonuniform shape.

The equivalent uniform (EU) DNB flux $q''_{\text{DNB,EU}}$ is calculated from the W-3 equivalent uniform-flux DNB correlation as follows:

$$\begin{aligned} \frac{q''_{\text{DNB,EU}}}{10^6} &= [(2.022 - 0.0004302p) + (0.1722 \\ &- 0.0000984p)e^{(18.177 - 0.004129p)\chi}] \\ &\times [(0.1484 - 1.596\chi + 0.1729|\chi|)\, G/10^6 + 1.037] \times [1.157 - 0.869\chi] \\ &\times [0.2664 + 0.8357e^{-3.151 D_e}] \times [0.8258 + 0.000794(H_{\text{sat}} - H_{\text{in}})] \end{aligned} \tag{4.24}$$

The heat flux is in Btu/(hr) (sq ft). The other units for the ranges of parameters used in developing the correlation follow:

System pressure (p) = 800 to 2300 psia
Mass velocity (G) = 0.5×10^6 to 5.0×10^6 lb/hr/sq ft
Equivalent diameter (D_e) = 0.2 to 0.7 in.
Quality (χ_{loc}) = −0.15 to +0.15
Inlet enthalpy (H_{in}) = 400 Btu/lb
Length (L) = 10 to 144 in.
$\dfrac{\text{Heated perimeter}}{\text{Wetted perimeter}}$ = 0.88 to 1.00
Geometries = circular tube and rectangular channel
Flux = uniform and equivalent uniform flux converted from nonuniform data by F-factor

The local uniform $q''_{\mathrm{DNB,N}}$ is calculated from

$$q''_{\mathrm{DNB,N}} = \frac{q''_{\mathrm{DNB,EU}}}{\mathrm{F}} \tag{4.25a}$$

where

$$F = \frac{C}{q''_{\mathrm{local}} \text{ at } l_{\mathrm{DNB}} \times (1 - e^{-Cl_{\mathrm{DNB}}})} \times \int_0^{l_{\mathrm{DNB}}} q''(z)\, e^{-C(l_{\mathrm{DNB}}-z)}\, dz \tag{4.25b}$$

where l_{DNB} = distance from inception of local boiling (in.) and

$$C = \frac{0.15(1 - \chi_{\mathrm{DNB}})^{4.31} \text{ in.}^{-1}}{(G/10^6)^{0.478}} \tag{4.25c}$$

4.79 The General Electric Company determined critical heat flux for boiling-water reactors somewhat differently.[27] Careful experiments with rod bundles which simulated reactor conditions as closely as possible were emphasized. The data were treated statistically, and "design-limit" lines were established. Finally, design correlation and prediction were obtained through the use of a hydraulic and thermodynamic analytical model that determined flow distribution in a parallel-channel assembly and critical heat flux in terms of design parameters (§4.80). The computation model here was closely related to experimental results for both input information and confirmation of predictions.

4.80 Some features of the analytical model are of interest. The critical heat flux is assumed to depend only upon local conditions. To apply test data obtained from four- and nine-rod laboratory assemblies to reactor assemblies containing many fuel rods, one must therefore know the flow and enthalpy distribution for each case. Each flow channel is divided into a number of nodes. The pressure drop is calculated for each of the channels and flow is redistributed to maintain the same pressure loss in each channel for the first node. The calculation then proceeds up the channels in an interative manner to determine a consistent flow-pressure loss pattern. Important to the calculation is an energy balance with provision for thermal mixing across adjoining surfaces as determined experimentally.

4.81 With the steam quality and local flow rate known, existing correlations, such as Macbeth's for pipe flow (§4.76), are used to introduce the critical-heat-flux calculation into the analysis. An important feature of the entire procedure is the opportunity to cross-check results with planned experimental measurements as parameters are varied and then to make suitable adjustments in the model.

ALKALI METALS AND ORGANIC FLUIDS

4.82 Although the foregoing discussion pertains primarily to water systems, boiling conditions and the possibility of critical heat flux are also important to other reactor coolants. This is particularly true in accident analysis and safety evaluation (§6.151). A good deal of work has been done on net boiling in liquid-metal systems. Only a moderate amount of research has been devoted to critical heat flux, however. In general, the mechanism appears to be similar to that proposed for water systems with correlations of the same *form* applicable. Similar comments apply to organic-cooled systems. Detailed review of either system is not appropriate here. Since the thermal conductivity of liquid metals is high, other design considerations, such as fuel melting, are limiting rather than the possibility of critical heat flux for sodium-cooled oxide-fueled fast reactors.

4.83 For boiling binary (two component) liquid systems, either organic or liquid metal, there is some evidence that the heat-transfer coefficient may be lower than that of either component. Similarly, the critical heat flux may be higher. Such behavior is consistent with Scriven's theory[28] of bubble-growth coefficients.

BOILING BEHAVIOR IN REACTOR CORES

4.84 An identification of critical flux conditions is only one step in the design of a reactor core. To the designer the ratio between the critical heat flux and the actual heat flux (DNB ratio) along a channel is more important than the absolute value of the limiting critical heat flux. Typical profiles for a boiling-water-reactor "hottest channel" are shown in Fig. 4.16. Since the critical

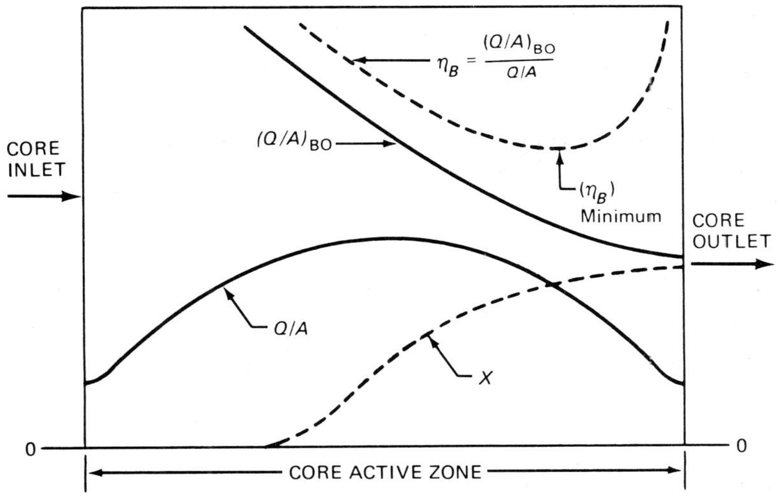

Fig. 4.16 Heat flux and steam distribution relating to burnout conditions. X, steam quality; Q/A, maximum operating heat flux; $(Q/A)_{BO}$, burn-out heat flux.

heat flux tends to decrease along a channel while the heat flux itself can peak at various positions depending on the control-rod position, fuel burnup, and degree of subcooling, an understanding of the channel boiling behavior is an operating requirement as well as an important part of core design.

4.85 Core geometry may affect both flow distribution and thermal patterns. Fuel-rod spacers, for example, can introduce complex flow-passage geometry that can cause poor mixing and low critical heat fluxes. Rod-surface obstructions may also affect the applicability of dryout correlations by disrupting the liquid-flow pattern. Cross-flow mixing between parallel channels is another important design challenge and has been the subject of computer analysis (§4.80).

4.86 Transient effects are also important, particularly as part of the safety analysis (§6.122). Therefore transient heat transfer and its effect on boiling has received much research.[29] A boiling system itself must also be examined for hydrodynamic instabilities that can cause premature burnout of the fuel elements.[30]

ALLOWANCES FOR POWER PEAKING AND HOT CHANNELS

INTRODUCTION

4.87 Specifications for core performance averaged over both volume and lifetime of the core are useful as a reference in the design of the core.

Performance limitations, however, are based on maximum, not average, values of such parameters as heat flux and fuel temperature. In other words, when a reactor is in the early design stages, the core is assumed to be a perfect structure; i.e., lattice parameters are nominal in dimension, fuel parameters are nominal in all respects, reactor coolant flow satisfies idealistic flow patterns, and neutron flux behaves ideally. Knowing that these conditions cannot actually exist, the designer must find a tool to aid in developing a core design that acknowledges these basic uncertainties, with specifications that will indeed ensure performance within limitations.

4.88 Upper thermal limits can be predicted from average conditions, or, conversely, the fixing of average design conditions can be predicted from maximum permissible values by using factors relating the maximum to the average values. A reactor core is complex, however, and deviations from average values result from a variety of causes. Therefore a systematic analysis is necessary first to evaluate a separate factor associated with each cause and then to combine the factors in a way that will give the desired ratio between average and maximum conditions. Since design methods have not been standardized, the subsequent discussion provides guidance to possible approaches rather than outlining a specific procedure.

4.89 Variations in heat flux, material temperatures, and other parameters are derived from two types of sources, nuclear and engineering. The local rate of *heat generation* depends on the fission rate, which, in turn, is a function of the product of the macroscopic fission cross section and the neutron flux, $\Sigma_f \phi$, which is neutron-energy dependent. Neutron diffusion results, of course, in both axial and radial variations in neutron flux. Local power variations also result from changes in such parameters as fuel, moderator, and poison concentration, as well as in control-rod effects. Long-term transient effects and transient effects caused by operational problems must also be considered.

4.90 *Engineering factors* are concerned primarily with the thermal and hydrodynamic behavior of the coolant. Variations and uncertainties in coolant flow, instrumentation, heat-transfer coefficient, etc., as well as deviations from nominal dimensions of fuel-rod spacing, cladding thickness, etc., can lead to local hot spots.

4.91 The power margin of reactor systems has frequently been determined by the "hot-channel" concept, where a so-called hot channel is identified from radial peaking information and then local or "hot spot" effects are determined along the channel.

4.92 Factors are therefore defined which show how uncertainties in each variable affect enthalpy rise, heat flux, film drop, etc. It may then be assumed that all the effects occur in the pessimistic direction at the same place and same time in the core. The individual hot-channel factors are multiplied together to give a conservative estimate of combined factors. The estimates are used in the thermal and hydraulic design calculations to yield the information of interest.

4.93 This approach, though simple, is unnecessarily conservative. Since random effects apply to many of the factors, efforts have been made to apply probability theory to the analysis. With the development of computer programs for this type of analysis, statistical effects can be incorporated into the program easily.

STATISTICAL FACTORS

4.94 It is important to distinguish between hot-channel factors of design variables subject indeed to random-type uncertainties and those introduced to account for inadequate knowledge. The random uncertainties derived primarily from manufacturing tolerance experience lend themselves to statistical analysis. It is not valid, however, to attempt statistical treatment of factors that describe uncertainties in performance.*

4.95 Some examples of random-type uncertainties related to manufacturing operation which can be treated statistically are:

- Fuel-pin diameter
- Fuel-rod pitch
- Rod-surface roughness
- Variation in fuel density and enrichment
- Cladding thickness
- Cladding eccentricity
- Variations in coolant flow area

Some uncertainties that should not be treated statistically are:

- Heat-transfer coefficients
- Material mechanical properties
- Radiolytic decomposition of coolants
- Variations in coolant velocity
- Void distributions
- Various flux-peaking effects
- Instrumentation efficiency
- Thermal conductivities
- Erosion and corrosion of fuel elements
- Flutter and vibration due to high-velocity coolant flow
- Irradiation effects on fuel rods and structural materials
- Interchannel mixing of coolant
- Effect of turbine trip-out
- Operator efficiency

Note that some of these uncertainties, such as operator error, instrumentation problems, and turbine trip-out, are related to operations outside of the core. In

*Experienced designers can sometimes establish *confidence limits* by subjective judgment and then use statistical procedures (see §§4.98 and 6.226).

addition, the variation of these parameters over the lifetime of both the core and the reactor should be included in the analysis.

4.96 Identification of a hot-channel factor with a manufacturing uncertainty does not mean that the subsequent statistical analysis is necessarily simple or even straightforward. Although manufacturing quality-control theory is a separate subject and will not be treated here,[31] some simple considerations are mentioned.

4.97 It is reasonable to assume that deviations from a mean value are random and hence follow a normal, or Gaussian, distribution.* Such a distribution is symmetrical about the mean. Certain tests, such as the chi-square test, are available for determining whether a given set of data is indeed normally distributed when a visual comparison with the normal curve is not conclusive. The standard deviation is a measure of the dispersion of the data:

$$\sigma = \sqrt{\frac{\Sigma(x - \bar{x})^2}{N}} \tag{4.26}$$

where σ is the standard deviation and $\bar{x}$ is the arithmetic mean of N values of the quantity x.

4.98 A normal distribution curve will have 68.27% of the values between $\bar{x} \pm 1\sigma$, 95.49% between $\bar{x} \pm 2\sigma$, and 99.73% between $\bar{x} \pm 3\sigma$. When a given variation may be caused by several factors, the total variance (σ^2) can be obtained from the sum of the squares of the individual standard deviations. Uncertainties that are not related (correlated) to one another can also be combined through the use of the standard deviation. If x is an arbitrary function of independent random variables $x_1, x_2, x_3, \ldots, x_n$,

$$dx = \frac{\partial x}{\partial x_1} dx_1 + \frac{\partial x}{\partial x_2} dx_2 + \ldots \tag{4.27}$$

For a normal variance, σ^2 of such a function can be expressed[32] as

$$\sigma^2 = \sum_{i=1}^{n} \left(\frac{\partial x}{\partial x_i}\right)^2 \sigma_i^2 \tag{4.28}$$

Therefore core-design criteria, such as a maximum fuel-center-line temperature, can be related to design variables that can be expressed as probability functions.

*The "nominal" specification may not always be identical to a measured mean. For example, manufacturing tolerances are sometimes given in the form $A^{+0\%}_{-1\%}$. Skewed distributions may also sometimes result from the nature of the process.

If the temperature, t_c, depends on a surface temperature, t_s, a peaking flux factor, P, and a thermal resistance, R, the standard deviation of the temperature can be expressed as

$$\sigma_{t_c} = \left[\left(\frac{\partial \overline{t}_c}{\partial t_s} \right)^2 \sigma_{t_s}^2 + \left(\frac{\partial \overline{t}_c}{\partial P} \right)^2 \sigma_P^2 + \left(\frac{\partial \overline{t}_c}{\partial R} \right)^2 \sigma_R^2 \right]^{1/2} \qquad (4.29)$$

The combined standard deviation can then be used for an appropriate design-uncertainty evaluation.

4.99 A number of different procedures exist for combining individual factors. With such a procedure the probability that a variable exceeds a given limit can be estimated from tolerance specifications. In practice, however, the methods used and the hot-channel factors obtained are part of the digital-computer programs used for the thermal design of the core. In considering transient effects and some of the details of boiling systems, we must use computer programs since the mathematical relations required for the statistical treatment are quite complicated. Methods specifically designed for computer processing of the data have therefore been developed. In one such method[33] a second-degree polynominal is developed relating the variables subject to uncertainty to thermal criteria. The theory of experimental design is used to determine polynominal coefficients. Finally, probability calculations are performed to determine how deviations may affect performance.

4.100 Sophisticated methods for statistically combining factors have little justification without a body of manufacturing data to provide meaningful variance information. The designer, therefore, must frequently assume that tolerance-specification limits are equal to the 3σ values of a normal distribution. Table 4.1 illustrates such an approach for a pressurized-water reactor.

4.101 Statistically significant uncertainties also contribute to such design limitations as the fuel-pin temperature. For example, Fig. 4.17 schematically depicts the interrelation of pin power, peak temperature, and frequency of occurrence.[34] In this model the direct contributors are assigned a relative frequency of occurrence of 1.0, and the statistical contributors are individually chosen to represent 3σ confidence limits and are normally distributed. Their combined effect therefore also represents a 3σ limit.

4.102 Another type of representation of the influence of hot-channel factors on core-temperature trends is shown in Fig. 4.18. The relative temperatures of fuel, cladding, and coolant will vary, of course, depending on the specific design and coolant–fuel selection. Hot-channel factors are therefore the link between the nominal calculated temperatures in the core and the probable and maximum expected temperatures. Since the upper operational design limits are fixed, the hot-channel factors establish the *nominal* operating conditions, including the nominal or allowable average pin linear power. Although large hot-channel factors are desirable for safety and reliability, it is

TABLE 4.1

Engineering Subfactor Statistical Components for a 700-Mw(e) Pressurized-Water Reactor*

Subfactor contributing to engineering heat-flux factor	1σ value	2σ value	3σ value†
Pellet density	1.008	1.016	1.024
Pellet diameter	1.0009	1.0018	1.0027
Pellet enrichment	1.0065	1.013	1.020
Cladding diameter	1.004	1.008	1.012
Subfactor contributing to rod-pitch and bowing and cladding-diameter factor	**1σ value**	**2σ value**	**3σ value§**
Pitch and bowing	1.015	1.030	1.045
Cladding diameter	1.014	1.028	1.042

*From Consumers Power Company, Palisades Plant Preliminary Description and Safety Analysis Report, Jackson, Mich., 1966.

†Combination of the 3σ values by taking square root of the sum of the squares yields 1.034.

§Combined 3σ value is 1.062.

preferable to use factors which will give only a reasonable degree of confidence that the design limits will not be exceeded. Otherwise, core performance will be penalized, resulting in an unnecessarily large core volume and a proportionately low flux for a given core power.

DESIGN PRACTICE

4.103 As examples of the hot-channel-factor approach, typical methods used for pressurized-water reactors and to some extent for sodium-cooled fast reactors are discussed in the following sections.

Water Reactors

4.104 Although details differ, practice for water-cooled reactors is to consider a heat-flux factor and a coolant-enthalpy-rise factor. Each is defined as the ratio of the maximum quantity to the "core average." These total hot-channel factors are each the product of a nuclear hot-channel factor describing the neutron-flux distribution and the engineering hot-channel factor to allow for variations from design conditions. The neutron-flux distribution can be obtained from computer calculations. Engineering hot-channel factors accounting for the effects of flow conditions and fabrication tolerances are made

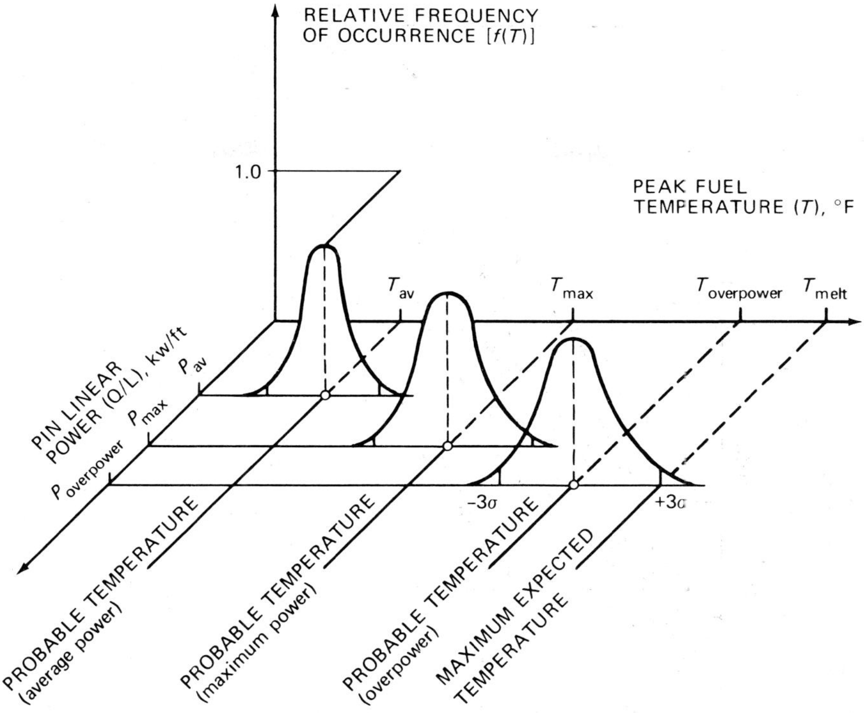

Fig. 4.17 Typical graph of pin power vs. peak fuel temperature vs. frequency of occurrence.

up of subfactors that introduce the influence of variations in fuel-pellet diameter, density and enrichment, fuel-rod diameter, pitch and bowing, inlet-flow distribution, flow redistribution, and flow mixing.

4.105 Fuel-assembly manufacturing-tolerance data are used to establish statistical limits for subfactors depending upon fuel diameter, density, enrichment, pitch, and bowing.[35] The subfactors related to flow distribution and mixing, however, do not lend themselves to statistical distribution and are determined from flow tests and computer studies.

4.106 The nuclear-power-density factor in a core containing fuel of several enrichments depends on many variables, such as fuel-management schemes and control-rod positioning, and therefore can be considered a design limit. A value of about 3.0 is normal for pressurized-water reactors. An engineering hot-channel factor of about 1.05 accounts for local variations in oxide-pellet density, diameter, and enrichment.

4.107 Some typical values for enthalpy-rise factors are shown in Table 4.2. Note that a nuclear factor to account for variations in heat input in the axial direction is also included. The calculated enthalpy rise is then used in a DNB calculation (§4.78). Tabulations such as this should be used with caution,

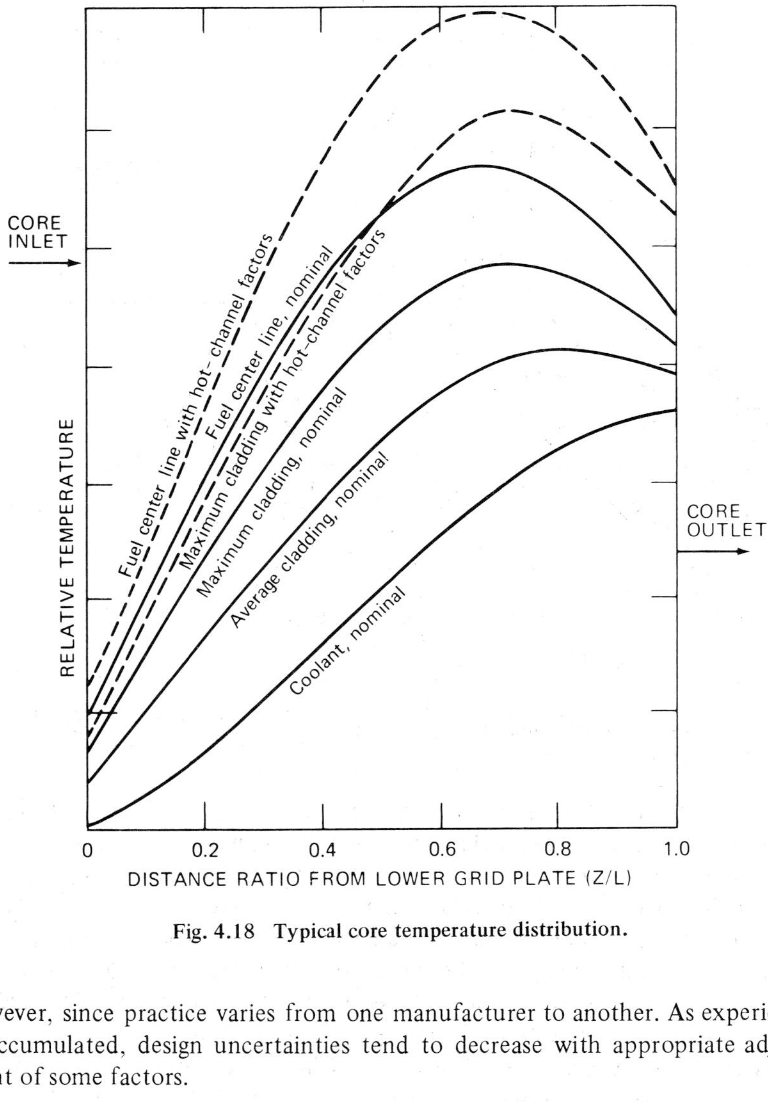

Fig. 4.18 Typical core temperature distribution.

however, since practice varies from one manufacturer to another. As experience is accumulated, design uncertainties tend to decrease with appropriate adjustment of some factors.

Computer Codes

4.108 Computer programs are useful in calculating and combining both engineering and nuclear hot-channel factors for core design and analysis. Such programs can avoid some of the limitations of both the statistical and the "multiplied together" combination methods. One example of this approach, a code used to evaluate the enthalpy rise for pressurized-water reactors,[36] is discussed here (§§4.109 to 4.112). Note that several of the subfactors are somewhat different from those listed in Table 4.1, since practice varies among

TABLE 4.2

Typical Pressurized-Water-Reactor Enthalpy-Rise Factors*

Heat-input factors	
Nuclear enthalpy-rise factor (axial power)	1.83
Engineering factor on hot-channel heat input	1.05
Total-heat-input factor	1.92
Flow factors	
Inlet plenum maldistribution (input)	1.05
Rod pitch, bowing, and cladding diameter (input)	1.065
Flow mixing (input)	0.85
Internal leakage and boiling-flow redistribution (code output)	1.25
Total flow factor	1.20
Total enthalpy-rise factor $\left(\dfrac{\Delta H_{\text{hot channel}}}{\Delta H_{\text{core}}}\right)$	2.304

*From Consumers Power Company, Palisades Plant Preliminary Description and Safety Analysis Report, Jackson, Mich., 1966.

manufacturers. Although various values are mentioned for orientation purposes, some of these values are likely to decrease as experience in operating the reactor is gained. Such codes may be included in comprehensive computer approaches for the thermal–hydraulic analysis of cores (§4.139).

4.109 The code applies to an open-channel reactor core with attention given to flow distribution and mixing. Applicable subfactors follow:

1. A lower plenum subfactor that accounts for local flow nonuniformity at the core inlet. A value of 1.07 is often used.

2. A flow redistribution subfactor that accounts for reduced flow in a hot channel owing to an increase in pressure drop caused by nucleate boiling. A value of 1.05 was used for reactor core I at the Yankee Nuclear Power Station at steady state. This value increases at overpower conditions when nucleate boiling intensifies in the hot channel.

3. Fabrication-tolerance subfactor. This subfactor accounts for the local flow slowdown between fuel rods due to a reduced pitch and the possibility of bowing of the fuel rods. A value of 1.10 was used for Yankee core I.

4. Fuel variation subfactor. A subfactor that accounts for the change in enthalpy rise due to variations in pellet diameter, density, enrichment, and eccentricity. A value of 1.04 was used for the Yankee reactor.

5. A flow-mixing factor that accounts for the enthalpy change caused by the flow mixing between channels. A value of 0.95 is often assigned to this subfactor.

4.110 Each of these subfactors is evaluated individually. Such evaluation may include statistical methods, if applicable. A challenge is how to apply them in a core design without introducing repetition. For example, a flow decrease at the inlet of a hot channel reduces the pressure drop in this channel. It would be a repetition to apply simultaneously the flow-redistribution subfactor to avoid excessive pressure drop in the same hot channel. A decreased flow at the hot-channel inlet could also result from a reduced pitch and bowing of fuel rods. The effect on enthalpy rise can then be charged to either of the these reasons but not to both. A constant mixing subfactor assigned to all reactors with different power distributions is also not valid, because the effect of mixing on enthalpy rise depends on the radial thermal gradient between the channels (i.e., on power distribution) as well as on flow conditions.

4.111 Another way to account for various hydraulic and nuclear effects on the enthalpy rise is to calculate the "hot-assembly factor" by superimposing the power distribution among the assemblies on the inlet-flow distribution with flow mixing between assemblies. Then the "hot-cell factor" is calculated by superimposing the normalized local power distribution within the hot assembly on the flow areas of the unit cells; these are formed by the fuel-rod lattice and the boundary of the fuel assembly with flow mixing between the channels. The total hot-channel factor is then the product of a hot-assembly factor and a local hot-cell factor. This calculation is quite complicated but can be accomplished with a digital-computer code. In this code variations in coolant properties along a group of heated fuel assemblies are calculated with Δp, ΔV, and ΔP in each fuel assembly chosen as the independent variables. Fluid enthalpy is related to density with the use of coolant properties. Using experimentally derived relations, one obtains the cross flow between adjacent assemblies from the pressure difference between the assemblies.

4.112 An inlet density and pressure distribution are required for the code, and the inlet velocity distribution is estimated. The code then determines the density, velocity, and pressure stepwise up the core and in general indicates a nonuniform outlet pressure distribution. Inlet velocities are then adjusted in an attempt to arrive at a uniform outlet pressure distribution by using the outlet pressure distribution as a guide and maintaining the same total inlet mass flow. The process is repeated until a uniform outlet pressure distribution is approached. The code prints out the fluid velocity, density, enthalpy, mass velocity, and static pressure at each-length step for each fuel assembly or group of assemblies. The mass and heat balance are also checked. Within an assembly the lateral resistance between the fuel-rod lattice is very small, and the effect of local cross flow on the exit pressure distribution becomes negligible.

Other Effects

4.113 The effect of both fuel burnup and schedules for fuel loading on the detailed radial and axial power distribution will, in turn, affect the parameters described by the engineering hot-channel factors. Whether or not such changes will be described by so-called nuclear factors, included in the appropriate engineering factor, or merely described in detail by a computer calculation will depend on specific design procedure.

4.114 It is also important to keep in mind the influence of secondary effects on the power-distribution factor. For example, if a nuclear enthalpy-rise factor for a water-cooled reactor is based on a power distribution, the value can be multiplied by (1) a local peaking factor to account for the presence of control-rod followers, water slots, etc., and (2) a transient hot-channel factor corresponding to the motion required for small step changes (~5%) in power.

4.115 In the study of core-design parameters, it is important to consider the dependence of power distribution on core height, enthalpy rise, and burnup. As an example, assume that a reactor with a core height of 10 ft has been designed to produce a given power output. For a new reactor with a 10% higher output, the designer might propose an increase in the core height to 11 ft and maintain the same average power density. However, the axial hot-channel factor might be greater in the 11-ft core. If the increase in the channel factor is 10%, the power capability would not increase at all with the 11-ft core height, assuming a burnout heat-flux limitation. In fact, the power distribution will probably become worse as the enthalpy rise increases since the water density becomes more nonuniform.

Fast Reactors

4.116 The hot-channel-analysis approach for fast reactors is essentially the same as that used for light-water thermal systems. In fact, differences in practice among manufacturers tend to be greater than differences between hot-channel methods. For light-water reactors, however, operating experience tends to reduce the value of some of the factors used, a trend not yet possible with fast reactor systems. One approach for designing fast reactor cores is described here to provide additional perspective.

4.117 Major hot-channel factors considered in a sodium-cooled fast breeder are those associated with fuel-pin temperatures and enthalpy rise. Eight enthalpy-rise subfactors in two groups demonstrate the techniques used. The first group includes nuclear-uncertainty subfactors that affect the axially integrated neutron flux and, in turn, the power generation and enthalpy rise for a particular coolant channel in the core.

4.118 One source of uncertainty is the approximate physics calculational methods used. If flux values are obtained via the diffusion approximation in two dimensions, an error of about ±5% is expected for the axially integrated flux for a partially spoiled core. A second flux-peaking uncertainty that affects the axially integrated flux occurs at the local core–zone interface from the enrichment ratio between the two adjacent core zones. This ratio can vary at different zone boundaries since fuel pins are loaded by batch methods. An uncertainty of about ±5% on the nominal peaking value and consequently in the axially integrated flux is introduced by this effect.

4.119 Since these uncertainties are statistical and independent, the total nuclear uncertainty for the integrated axial flux is

$$(3\sigma)^2 = 3^2\sigma_1^2 + 3^2\sigma_2^2 = (0.05)^2 + (0.05)^2 \tag{4.30}$$

$$3\sigma = \sqrt{0.0025 + 0.0025} = 0.071 \tag{4.31}$$

The nuclear hot-channel factor is therefore 1.0 + 3σ, or 1.071, for the enthalpy rise in a given coolant channel.

4.120 Subfactors in the second group are of an engineering nature, associated with the following physical effects:

1. Maldistribution of coolant, f_{mal}.
2. Mixing of the coolant, f_{mix}.
3. Measurement instrumentation and control, f_{cont}.
4. Manufacturing tolerances in fuel-pin pitch and outside diameter, f_{dim}.
5. Fuel mass per pin tolerances, f_{mass}.
6. Fissile isotope variation per fuel pin, f_{fis}.

The maldistribution of coolant, mixing of coolant, and instrumentation subfactors are taken as being nonstatistical and are combined by the factor-product method. The maldistribution of coolant, referring to the variation in the coolant flow through individual channels, includes the effect of the plenum geometry. In this case f_{mal} is taken to be 1.05 on the basis of engineering judgment.

4.121 Interchannel mixing promotes energy transfer from the hot channel and, consequently, reduces the enthalpy rise of the coolant in the hot channel. Current literature reflecting experimentation suggests a value of 0.95 for f_{mix}. The subfactor f_{cont} (equal to about 1.03) is used because instrumentation possibly may not read the true thermal and neutron characteristics of the core.

4.122 The engineering subfactors reflecting uncertainties of the pin pitch, pin diameter (and consequently the flow area), fuel-mass loading per pin, and fissile-isotope variation per pin are statistical in nature. Experience with light-water systems indicates that both the pin pitch and the fuel-pin outside diameter vary normally about the nominal design value, with a 3σ confidence that the hot-channel flow area will be within 0.97 of the nominal area. The second uncertainty of interest is that of fuel-mass quantity per pin which reflects variations in both fuel density and diameter. Experimental data show a normal distribution of fuel mass per pin which results in a 3σ subfactor of 1.004. Another effect on the heat generated per pin is the variation of the fissile-isotope concentration per pin. Since in this example the PuO_2 and UO_2 materials are blended mechanically in small batches to make up the fuel-pin load, a variation of plutonium per batch is expected not to exceed 1.7% of the nominal or specified value with a resulting subfactor of 1.017.

4.123 The preceding subfactors were combined through the use of a computer code, which calculated the enthalpy rise in a nominal hot channel and then ran parallel cases corresponding to the deviations expressed by the factors. A less sophisticated approach is to combine factors either statistically or by the product method as appropriate for each type.

EXAMPLE OF A FUEL-PIN TEMPERATURE PROFILE

4.124 A number of the design considerations contributing to the maximum temperature of a fast reactor fuel pin are illustrated by the following example, in which an annular oxide pellet is used, a design choice that is gaining favor since it has the effect of anticipating the formation of a central void (§4.48). A typical temperature profile for a pin of this type is shown in Fig. 4.7.

Example 4.1

The objective is to check on published thermal-analysis data* for an oxide-fueled fast reactor having a core thermal power of 2125 Mw; the core consists of 498 hexagonal fuel assemblies with 123 pins per assembly. Other specifications follow:

Total coolant flow rate, 1.14×10^8 lb/hr
Overall coolant-temperature rise, 250°F
Coolant-temperature rise through core, 263°F
Coolant-temperature rise through radial blanket, 263°F
Coolant inlet temperature, 950°F
Fuel Pin:
- Outer diameter of cladding, 0.300 in.
- Inner diameter of cladding, 0.244 in.
- Length, 5 ft
- Cladding thickness, 0.028 in.

Fuel Pellet:
- Outer diameter, 0.2405 in.
- Inner diameter, 0.0998 in.
- Active fuel length, 4 ft

Fuel Performance:
- Average linear heat generation, 8.6 kw/ft
- Average heat flux, 0.373×10^6 Btu/(hr) (sq ft)

---

*Allis-Chalmers, Large Fast Reactor Design Study, USAEC Report ACNP-64503, January 1964.

Power Distribution:

$(P_{max}/P_{av})_{axial}$, 1.3

$(P_{max}/P_{av})_{radial}$, 1.18

P_{max}/P_{av}, 1.52

Maximum Fuel Temperature at the Beginning of Fuel Cycle (as published):

Nominal, 3424°F

Hot Spot (multiplicative): 4615°F

Hot Spot (statistical, 99.7% confidence) (see §4.94): 3861°F

SOLUTION:

1. *Hydraulic Parameters*

Fuel-assembly side: 2.455 in. (given)

Fuel-assembly hexagonal-section area: 15.6 sq in.

Coolant area/assembly: total area − pin area
$= 4.81 \times 10^{-2}$ sq ft

$$\text{Equivalent diameter: } \frac{4(\text{coolant area})}{\text{wetted perimeter}}$$

$$= \frac{4 \times (4.81 \times 10^{-2} \text{ sq ft})}{10.89 \text{ ft}} = 1.77 \times 10^{-2} \text{ ft}$$

2. *Flow Velocity in Hot Channel*

Total coolant-mass-flow rate: 2.158×10^5 lb/hr

$$\text{Velocity: } \frac{\text{mass-flow rate}}{(\text{density})(\text{coolant area})}$$

$$= \frac{2.158 \times 10^5}{(51.2)(4.81 \times 10^{-3})(3600)} = 24.4 \text{ ft/sec}$$

3. *Heat-Transfer Coefficient*

$$\text{Nu} = 5.0 + 0.025\ \text{Pe}^{0.8} = 9.375 \qquad (\text{at Pe} = 631)$$

$$h = \frac{\text{Nu} \times k}{D_{eq}} = \frac{9.375 \times 37.8}{1.77 \times 10^{-2}} = 20{,}050 \text{ Btu/(hr)(sq ft)(°F)}$$

4. *Hot-Channel Values as Given:*

Coolant rise = 1.46

Film-temperature drop = 1.49

Cladding temperature drop = 1.44

Temperature drop through gas = 1.44

Fuel-temperature drop = 1.52

$P_{max}/P_{av} = 1.52$

5. *Hot-Channel Temperature Profile*

Coolant bulk temperature at half-channel height: 950 + ½ (263) (1.46) = 1142°F

$$t_{\text{film}} = \frac{(\text{film } \Delta t \text{ factor})(P_{\text{max}}/P_{\text{av}})}{h} \times \frac{q}{A}$$

$$= \frac{1.49 \times 1.52}{20{,}050} \times 3.75 \times 10^5 = 42.3°\text{F}$$

Cladding surface temperature: 1142 + 42 = 1184°F

$$\Delta t_{\text{clad}} = (1.52)(1.44)\left(\frac{q}{A}\right)\left(\frac{\text{thickness}}{k}\right) = 191°\text{F}$$

$$\Delta t_{\text{clad gap}} = \frac{(1.52)(1.44)(3.75 \times 10^5)}{1000} = 821°\text{F}$$

where a heat-transfer coefficient of 1000 Btu/(hr) (sq ft) (°F) is assumed for the cladding gap

Outside-fuel-surface temperature: 1184 + 191 + 821 = 2096°F

For Δt_{fuel} apply

$$\int_{t_a}^{t_b} k(t)\, dt = \frac{q_L}{4\pi}\left(1 - \frac{2b^2}{a^2 - b^2} \ln\frac{a}{b}\right)$$

(see §4.26). Average linear-heat-generation rate = 8.6 kw/ft; k = 1.6 Btu/(hr) (sq ft) (°F/ft).

$$\Delta t_{\text{fuel}} = \frac{(1.52)(1.52)(8.6)(3412)}{4\pi \times 1.6}\left[1 - \frac{2(0.0998)^2}{(0.2405^2 - 0.0998^2)} \ln\frac{0.2405}{0.0998}\right]$$

$$= 2110°\text{F}$$

Inner-fuel-surface temperature: 2096 + 2110 = 4210°F

This result of 4210°F compares with the published value of 4615°F for multiplying hot-spot factors. A measure of the uncertainty of this type of calculation is seen by comparison with the published value of 3861°F obtained by statistical treatment of the hot-spot factors.

SAFETY-RELATED THERMAL-DESIGN CONSIDERATIONS

INTRODUCTION

4.125 Analysis associated with safety considerations is important to reactor design. Thermal transport, in turn, is basic to accident-propagation models for safety analysis (§§6.124 and 6.151). For example, in water reactors, a feature of the anatomy of a loss-of-coolant accident is the very transient thermal picture of the decreasing heat-removal capability of the primary coolant as it vaporizes during blowdown, the continuing heat input from the fuel and the heat-withdrawal capability of emergency core cooling and other engineered safety features. As another example, the thermal analysis of the fuel assembly of a liquid-metal-cooled fast breeder reactor after a coolant-flow blockage includes a description of the coolant flow as it progresses through various two-phase flow regimes as well as a consideration of the corresponding transient temperature pattern of the fuel.

4.126 Analysis of abnormal operating conditions differs somewhat from accident analysis. The partial loss of coolant flow due to an inadvertent pump shutdown, for example, must be studied by the designer but is not classified as an accident with associated fast transients.

4.127 A transient thermal analysis can be applied to any part of the thermal-transport system that is relevant to safety considerations. The general approach is likely to be similar to that for steady state with the important addition of capacitive terms into the descriptive equations as well as time as a variable. If an accident or abnormal operating situation is considered, such conditions as temperature and pressure may also be very different from those considered in the steady-state case. Safety-related thermal considerations are very important to design and cover a wide range of possible cases. Two such cases, sodium-boiling initiation in the fuel channel of a liquid-metal-cooled fast breeder and the loss-of-coolant accident in a thermal reactor, follow as illustrations.

EXAMPLES OF THERMAL ANALYSIS

Sodium-Boiling Initiation

4.128 In a sodium-cooled channel, the initiation of boiling after the sodium has been superheated for several hundred degrees can have rather serious consequences. The very rapid formation of a large volume of sodium vapor in the channel will severely restrict the coolant flow. In fact, analytical models[37] of the process show that the coolant flow may even be reversed. Reduced heat-removal capability is likely to cause overheating of the fuel and then failure.

4.129 Whether or not superheating is included in the model is a basic difference in approaches.[38] For example, if superheat is not considered, vapor growth is described according to thermodynamic equilibrium. In equilibrium models the system is normally defined in terms of the conservation equations of mass (continuity), momentum, and energy for a two-phase fluid flowing upward in a vertical channel. An additional relation provides the volume fraction of vapor in the two-phase system. Computer representation is essential for this approach. Even when superheat is not considered, simplifying assumptions are necessary to keep the model from becoming too complex. For example, in some early work a constant saturation temperature was used during the transient throughout the channel. Subsequent models, however, provided for variation of the reference pressure (or saturation temperature) along the channel according to the steady-state profile during the two-phase transient.

4.130 Nonequilibrium models that take into account superheating normally empirically relate nucleation to superheat with the use of experimental results. Variables that appear to affect superheat behavior include the heat flux, nature of the heating surface, fluid velocity, and physical properties. Since considerable uncertainty is involved, it is often useful to consider the superheat as a parameter of the study.

4.131 In either analysis approach the next need is an accurate description of the point-to-point loss of coolant flow along a channel and the resulting temperature pattern. An autocatalytic situation exists since decreased coolant flow tends to enhance bubble formation that further restricts the flow. In fact, the rapid increase in vaporization could cause a downward pressure pulse that is likely to reverse the normal upward flow of coolant. Without heat removal the cladding temperature rises rapidly and failure results unless countermeasures are taken.

4.132 Since, in a sodium-cooled fast reactor, either the formation of a void or an increase in fuel temperature (Doppler effect) affects the reactivity, a realistic model for the thermal behavior should be coupled with a model for reactor kinetics. Calculation models of this type have been developed in which a description of the spatial and temporal distribution of sodium-vapor void in a reactor channel during power and flow-coastdown transients includes the effect of Doppler reactivity and void feedback on the reactor power vs. time-driving function.[39]

4.133 An important objective of this type of analysis is to develop a knowledge of how rapidly the cladding and fuel temperatures will rise as a function of various design parameters as an aid in the design of the control system and engineered safety features (§6.166). Some typical results developed by this type of computer model are shown in Fig. 4.19. Temperature profiles for both cladding and coolant are given with elapsed time from the voiding as a parameter.

Light-Water-Reactor Transient Fuel-Pin Temperatures

4.134 In the design of light-water reactors, the transient-temperature response of fuel pins to the design basis (§6.122) loss-of-coolant accident is an important feature of the safety evaluation. A performance picture can be developed with the aid of a computer model. In one analysis[40] for a pressurized-water reactor, a one-dimensional radial representation is used for

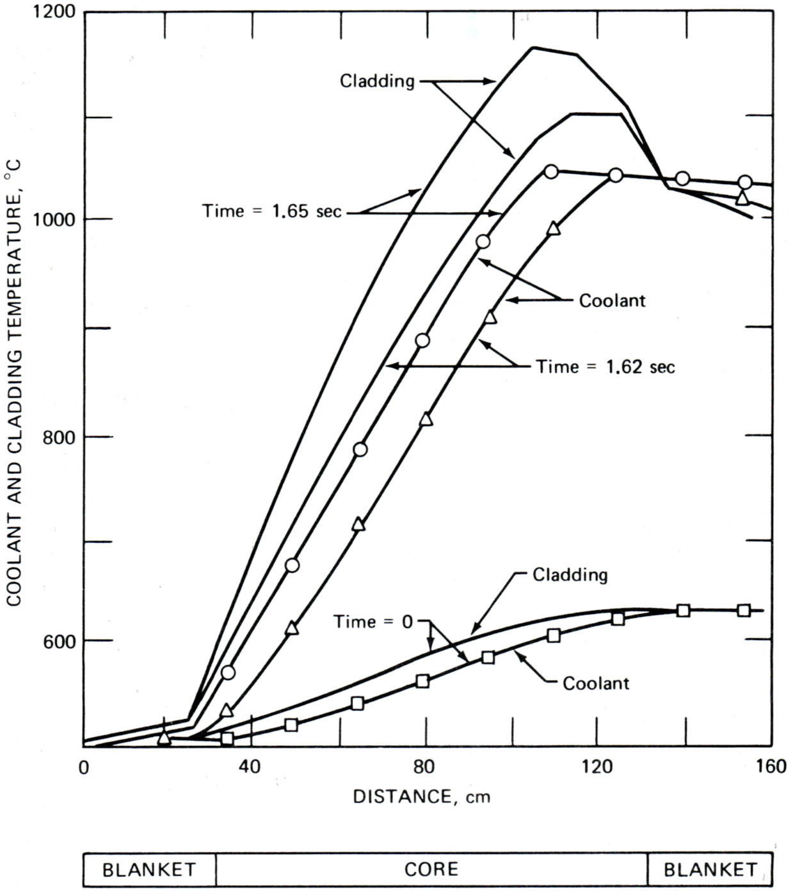

Fig. 4.19 Cladding and coolant temperature profiles at various times after initiation of boiling.

fuel, gap, cladding, and coolant. Decay power and bulk coolant temperature as functions of time were provided by other codes. The surface-heat-transfer coefficient during blowdown was assumed to be 20,000 Btu/(hr) (sq ft) (°F) until critical heat flux was reached, when it would drop immediately to 175 Btu/(hr) (sq ft) (°F) and then decrease exponentially to zero in about 8 sec.

4.135 A typical cladding-surface-temperature history after a loss-of-coolant accident is shown in Fig. 4.20. In this figure the realistic case is the one showing the effects of decay-heat input to the fuel pin and heat removal to the coolant during blowdown. Other curves reflect the importance of the three factors involved: (1) internal energy initially stored in the fuel pin during steady-state operation, (2) decay heating of the fuel pin during the transient, and (3) energy removed from the fuel pin by blowdown heat transfer.

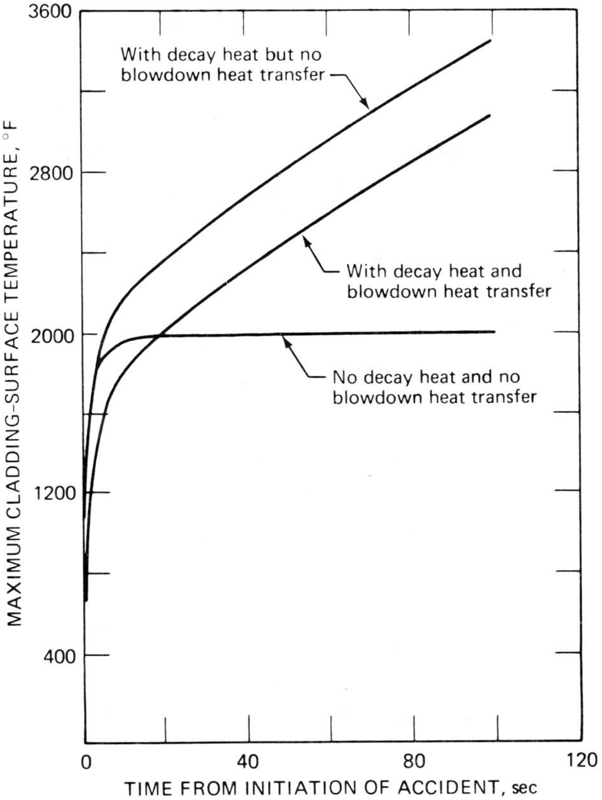

Fig. 4.20 Effects of decay heating and blowdown heat transfer on fuel-pin-cladding surface temperature for a large pressurized-water reactor.

4.136 In a different analysis for a similar situation, the influence of core quenching is considered.[41] As part of the backup protection in designs of large pressurized-water reactors for the unlikely event of a blowdown, provision is made to inject borated water to flood the reactor core and to provide additional cooling to prevent fuel failure. Figure 4.21 shows the hot-spot cladding-temperature history after the double-ended rupture of a main coolant pipe. As an aid to the design of the quenching system, the heat-transfer coefficient between the cladding and the injected fluid is considered a parameter. The quenching liquid initially vaporizes as it comes in contact with the hot cladding.

4.137 A delay of 18 sec after the pipe rupture is assumed to initiate the quenching, although cooling during the blowdown period occurs for only 9.5 sec. In this case a heat-transfer coefficient of 15 Btu/(hr) (sq ft) (°F) provides sufficient cooling to control the temperature rise. A somewhat higher value is needed, however, to reverse the rise.

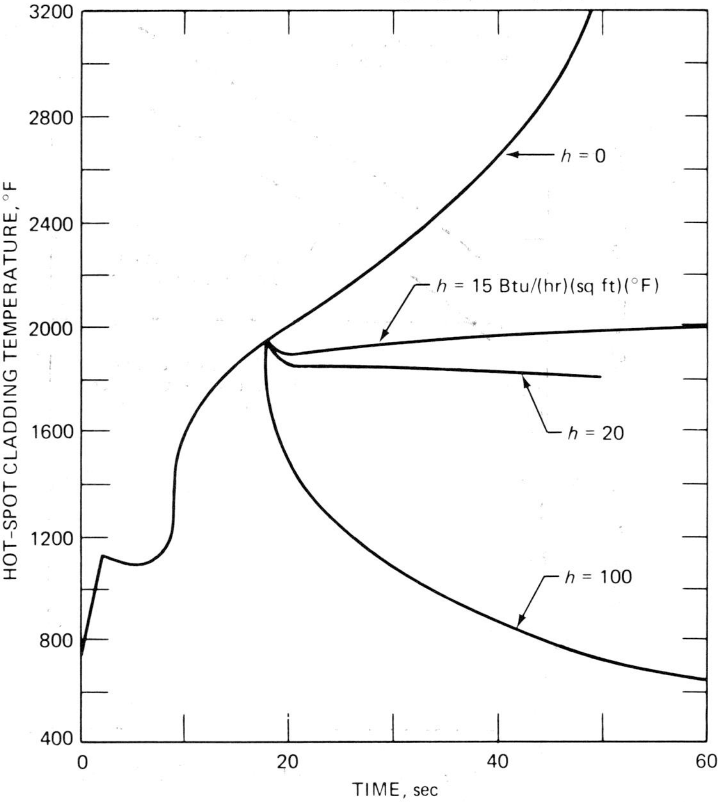

Fig. 4.21 Hot-spot cladding temperature vs. time for a 36-in.-ID double-end hot-leg pipe rupture and variable quench coefficient.

4.138 Another type of analysis, usually part of the core design, is to determine the maximum number of coolant-flow channels that could experience partial flow blockage without jeopardizing fuel-cladding integrity. For example, in many pressurized-water reactor designs, the flow to several of the innermost fuel assemblies can be reduced by as much as 85% before critical heat flux is approached. This type of consideration is often part of the safety evaluation required for reactor licensing (§6.245).

COMPUTER PROGRAMS FOR THERMAL–HYDRAULIC DESIGN AND ANALYSIS

4.139 Computer codes have been mentioned throughout this chapter as useful for the analysis of reactor thermal systems. This function is consistent with the vital role being assumed by computer methods in all phases of engineering design where calculations of some complexity are called for. Many codes have therefore been developed by reactor designers to meet specific needs. Some of these are proprietary and are not generally available. The remainder of the chapter discusses several of the published codes, indicating the types of approaches that can be used. Some design applications are given in Chap. 8.

4.140 Core-design calculations are normally carried out by computer methods that can be classified into two categories. Some programs include a solution to the neutron-diffusion equations and perhaps the kinetics equations (§5.23) in addition to an analysis of fuel-element thermal-transport and channel hydraulics. Other programs are limited to a detailed steady-state thermal–hydraulic calculation. Each type has its use. The nuclear–thermal–hydraulic approach, which generally includes some simplifying assumptions, is very useful for the type of transient analysis required for a safety evaluation. In the detailed core design, however, a complete thermal–hydraulic analysis is necessary and can be carried out with only limited iteration with the nuclear design. The second approach is useful for this purpose.

4.141 An example of the first type is SPLOSH-II, a two-group code that describes the kinetic behavior of the primary loop of water-cooled reactors.[42] The code is one-dimensional since the neutron diffusion, fuel element, and hydrodynamic equations are solved simultaneously for an average channel. Other channels can be related to the average channel by factors, although there is no provision for feedback. A number of subroutines are included to calculate initial conditions, heat flow, hydraulic flow, nuclear constants, and the nuclear kinetics. After the initial conditions are determined, the calculation proceeds by incremental time steps with neutron flux, fuel temperatures, etc., determined at each time step by an iterative routine. Computer time depends on the number of mesh points (§5.90) representing the system and the size of the time steps.

4.142 THEME 1 of the second type performs a steady-state thermal and hydraulic evaluation of a subchannel of a multirodded fuel element in a single-pass, organic-cooled reactor core.[43] Given the basic reactor parameters of power, exit temperature, temperature rise, fuel length, power distribution, and fuel-element geometry, the program will calculate the various thermal and hydraulic parameters for the limiting fuel channel. In addition, coolant, surface temperatures, local velocities, and critical-heat-flux ratios for various axial positions are calculated for the subchannel.

4.143 Another type of code, useful for hot-channel subfactor calculations, is for the detailed hydraulic analysis of a core with emphasis on flow distribution. COSMO makes use of a closed-channel model with adjustments in flow possible from channel to channel as local boiling occurs.[44] CAT-II, primarily for transient analysis, is based on an open-channel model in which there is no resistance to cross flow.[45]

4.144 REPP is another example of a comprehensive thermal–hydraulic code intended for core design.[46] It uses a one-dimensional representation to calculate heat removal from circular rods with a nuclear heat source. Two constraints, burnout heat flux (using the W-3 correlation) and fuel-center-line temperature, are used to determine reactor core size and fuel-pin diameters. The pressure drop, the coolant temperatures, surface heat fluxes, and fuel-center-line temperature are predicted from the input data and a thermal energy balance at various mesh points along the core. For a design problem the analysis is divided into two main divisions. The first division develops a mathematical model. The second division predicts the thermal and hydraulic conditions through the core using hot-channel analysis. An iterative process between the two subdivisions is used to design within specified fuel-center-line temperatures and heat-flux limits for a given reactor power level.

4.145 The code is particularly useful in conjunction with a nuclear burnup code (§5.100) for fuel-element and lattice design. For example, REPP can be used to determine a range of rod sizes and lattice spacings which optimizes heat removal within the core. Nuclear analysis can then be made over this range of potential cores to minimize fuel-cycle cost. Additional discussion is given in Chap. 8 (§8.75).

4.146 Other types of thermal-transport codes are designed primarily for studying the consequences of accidents considered as part of the safety analysis. For example, FLASH is a digital program that calculates flows, inventories, pressures, and temperatures in a pressurized-water-reactor primary system during a loss-of-coolant accident.[47] Starting with calculations of the rate of coolant efflux through assumed leaks, it calculates inflow from the fill system, inventory of water above and below the core, pressure drop across the core, flow through the core and loops, neutron and gamma heating in the fuel, and heat transfer from the fuel to the coolant. It calculates the fuel temperatures in both hot and average channels of the core. Thus it provides a direct measure of the effectiveness of the system design and operating procedures in preventing fuel melting. Since various assumptions regarding the behavior of materials under extreme conditions are necessary in accident models, codes such as FLASH are frequently updated as additional information becomes available.

REFERENCES

1. L. S. Tong and J. Weisman, *Thermal Analysis of Pressurized Water Reactors,* American Nuclear Society, Hinsdale, Ill., 1970.
2. S. Glasstone and A. Sesonske, *Nuclear Reactor Engineering,* D. Van Nostrand Company, Inc., Princeton, N.J., 1963.
3. M. M. El-Wakil, *Nuclear Heat Transport,* International Textbook Company, Scranton, Pa., 1971.
4. R. C. Daniel and I. Cohen, In-Pile Effective Thermal Conductivity of Oxide Fuel Elements to High Fission Depletions, USAEC Report WAPD-246, Westinghouse Electric Corporation, April 1964.
5. International Atomic Energy Agency, Thermal Conductivity of Uranium Dioxide, Technical Report Series, No. 59, Vienna, 1966 (STI/DOC/10/59).

6. J. L. Bates, High-Temperature Thermal Conductivity of Round Robin Uranium Dioxide, USAEC Report BNWL-1431, Battelle–Northwest, July 1970.
7. General Electric Company, General Electric Task III and V Report of 1000 MWe LMFBR Follow-On Work, USAEC Report GEAP-5710, February 1969.
8. H. B Meieran, Light Water Reactor Fuel Element Design, in Symposium on Sol-Gel Processes and Reactor Fuel Cycles, USAEC Report CONF-700502, pp. 534–550, Oak Ridge National Laboratory, May 1970.
9. C. M. Cox and F. J. Homan, Performance Analysis of a Mixed-Oxide LMFBR Fuel Pin, *Nucl. Appl. Technol.*, **9**: 317 (1970).
10. R. M. Carroll and O. Sisman, Fission-Gas Release During Fissioning in UO_2, *Nucl. Appl.*, **2**: 1942 (1966).
11. W. B. Lewis, Engineering for the Fission Gas in UO_2 Fuel, *Nucl. Appl.*, **2**: 171 (1966).
12. M. J. F. Notley, Calculation of Fission-Product Gas Pressures in Operating UO_2 Fuel Elements, *Nucl. Appl.*. **3**: 334 (1967).
13. C. W. Sayles, A Three-Region Analytical Model for the Description of the Thermal Behavior of Low Density Oxide Fuel Rods in a Fast Reactor Environment, *Trans. Amer. Nucl. Soc.*, **10**: 458 (1967).
14. A. Boltax, P. Murray, and A. Biancheria, Fast Reactor Fuel Performance Model Development, *Nucl. Appl. Technol.*, **9**: 326 (1970).
15. T. R. Bump, Mechanical and Thermal Analysis of Cylindrical Fuel Elements During Off-Normal Conditions After Extended Burnup, *Nucl. Appl. Technol.* **9**: 301 (1970).
16. L. S. Tong, *Boiling Heat Transfer and Two-Phase Flow,* John Wiley and Sons, Inc., New York, 1965.
17. L. S. Tong, Heat Transfer in Water-Cooled Nuclear Reactors, *Nucl. Engr. Des.*, **6**: 301 (1967).
18. R. P. Stein and P. A. Lottes, Boiling Burnout for Reactor Design, in Reactor Technology, Selected Reviews – 1964, USAEC Report TID-8540, pp. 131–176, 1964.
19. W. M. Rohsenow (Ed.), *Developments in Heat Transfer,* Chap. 8, The M.I.T. Press, Cambridge, Mass., 1964.
20. S. S. Kutateladze, Heat Transfer in Condensation and Boiling, 2nd ed., USAEC Report AEC-tr-3770, 1952.
21. O. Baker, Simultaneous Flow of Oil and Gas, *Oil Gas J.*, **53**: 185 (1954).
22. D. C. Groeneveld, An Investigation of Heat Transfer in the Liquid Deficient Regime, Canadian Report AECL-3281, 1969.
23. S. Levy, J. Batch, and J. Casterline, Critical Heat Flux Considerations in the Thermal and Hydraulic Design of Water-Cooled Nuclear Reactors, in *Proceedings of the Third International Conference on the Peaceful Uses of Atomic Energy, Geneva, 1964,* Vol. 8, pp. 114–127, United Nations, New York, 1965.
24. W. H. Sens and P. A. Lottes, Analysis of Heat Transfer, Burnout, Pressure Drop and Density Data for High-Pressure Water, USAEC Report ANL-4627, Argonne National Laboratory, May 1, 1951.
25. B. Thompson and R. V. Macbeth, Boiling Water Heat Transfer, Burnout in Uniformly-Heated Round Tubes: A Compilation of World Data with Accurate Correlations, British Report AEEW-R356, 1964.
26. L. S. Tong, Prediction of Departure from Nucleate Boiling for an Axially Non-Uniform Heat Flux Distribution, *J. Nucl. Energy,* **21**: 241 (1967).
27. J. M. Healzer, J. E. Hench, E. Janssen, and S. Levy, Design Basis for Critical Heat Flux Condition in Boiling Water Reactors, USAEC Report APED-5286, General Electric Company, September 1966.
28. L. E. Scriven, On the Dynamics of Phase Growth, *Chem. Eng. Sci.*, **10**(1): 1 (1959).
29. R. P. Morgan, A Review and Discussion of Literature Concerning Transient Heat Transfer and Steam Formation, USAEC Report IDO-17226, Phillips Petroleum Company, March 1967.

30. Commissariat a l'Energie Atomique, Hydrodynamic Instabilities Limiting the Power of Boiling Water Reactors, Vols. I and II, USAEC Report EURAEC-1464, June 1965.
31. D. H. W. Allan, *Statistical Quality Control,* Reinhold Publishing Corporation, New York, 1959.
32. F. J. Moody, Probability Theory and Reactor Core Design, USAEC Report GEAP-3819, General Electric Company, January 1962.
33. A. C. Nelson, Jr., and W. S. Minkler, A General Method for Evaluating the Effects of Uncertainties in Design Variables on Core Thermal Performance, *Nucl. Sci. Eng.,* **17**: 101 (1963).
34. E. G. Stevens et al., FFTF Fuel Pin and Subassembly Conceptual Design Methods and Data, USAEC Report BNWL-1064, Pacific Northwest Laboratory, June 1970.
35. H. Chelemer and L. S. Tong, Engineering Hot Channel Factors for Open-Lattice Cores, *Nucleonics,* **20–9**: 68 (1962).
36. L. S. Tong, H. Chelemer, and E. A. McCabe, Jr., Hot Channel Factors for Flow Distribution and Mixing in Core Thermal Design, USAEC Report WCAP-2211, Westinghouse Electric Corporation, February 1963.
37. D. R. MacFarlane, An Analytic Study of the Transient Boiling of Sodium in Reactor Coolant Channels, USAEC Report ANL-7222, Argonne National Laboratory, June 1966.
38. H. K. Fauske, Coolant Ejection Mechanisms for Use in Analyzing Fast Reactor Safety, *Reactor Fuel-Process. Technol.,* **11**: 180 (1968).
39. D. R. MacFarlane, N. McNeel, D. A. Meneley, and C. K. Sanathanan, Theoretical Studies of the Response of Fast Reactors During Sodium Boiling Accidents, *Trans. Amer., Nucl. Soc.,* **10**: 694 (1967).
40. S. E. Hensen, Parametric Studies of Fuel Pin Temperature Response, USAEC Report IDO-17295, Phillips Petroleum Company, February 1969.
41. Sacramento Municipal Utilities District, Rancho Seco Nuclear Generating Station Unit 1., Preliminary Safety Analysis Report, Vol. III, 1967.
42. D. Moxon, SPLOSH-II, A Dynamics Programme for Nuclear-Thermal-Hydrodynamic Behavior of Water-Cooled Reactors, British Report AEEW-R-441, 1966.
43. J. P. Saunders, ORNL Version of THEME I Code, A Thermal–Hydraulic Code for Analysis of the Hot Channel in a Fuel Bundle, USAEC Report ORNL-4015, Oak Ridge National Laboratory, November 1966.
44. R. H. Young, COSMO, A Thermal and Hydraulic Steady State Design Code for Water-Cooled Reactor, Report CEND-196, Combustion Engineering Corporation, 1963.
45. R. O. Sanberg, CAT-II – An IBM-7090 Code for Predicting Thermal and Hydraulic Transients in an Open Lattice Core, USAEC Report WCAP-2059, Westinghouse Electric Corporation, August 1962.
46. R. M. Hiatt and C. Bromley, Jr., REPP: A Thermal–Hydraulic Design Code for Water-Cooled Reactors, USAEC Report BNWL-1013, Pacific Northwest Laboratory, March 1969.
47. S. G. Margolis and J. A. Redfield, FLASH: A Program for Digital Simulation of the Loss of Coolant Accident, USAEC Report WAPD-TM-534, Bettis Atomic Power Laboratory, May 1966.

Nuclear-Analysis Methods

5

INTRODUCTION

5.1 In the engineering design process (§1.27), a "solution" to problems must be selected from among alternates on the basis of various criteria. For a nuclear power system, an essential part of the design is the specification of the core, which depends on its nuclear characteristics. These nuclear characteristics affect the neutron behavior and, in turn, the neutron flux and power distribution. However, the power distribution markedly affects the thermal hydraulics, fuel design, and economics. Interplays with other design parameters are also effective.

5.2 Selecting specifications of the core therefore involves iteration with other areas to contribute to an integrated reactor-system design. Some aspects of these interplays are considered in Chap. 8. To contribute to the design, we must describe the core neutron behavior. In this chapter methods needed for such description and some of the parameters affecting them are examined.

5.3 An analytical description of core neutron behavior in terms of materials, temperatures, dimensions, time, etc., can be considered a separate discipline known by various names, such as reactor physics, reactor theory, and reactor analysis. Part of this subject is certainly concerned with theoretical nuclear physics and various ways of describing the nature and behavior of matter. On the other hand, the designer is primarily interested in applying calculation methods developed from the theory. Here the term "nuclear analysis" will describe the design-oriented part of the subject. The theoretical basis of a calculation method must of course be understood for the most intelligent use of the method.

5.4 Since nuclear-analysis calculations are normally carried out by specialists, the core designer can focus his attention on understanding why the core behaves as it does and take steps to correct deficiencies. He is likely to work closely with another member of the design team who is responsible for the calculations. However, the nuclear-engineering graduate student who is developing design background will want to carry out some core calculations for himself. This chapter provides the student with a brief survey of calculational approaches with occasional detail to improve understanding. To develop proficiency in a specific method, he should consult other sources.

CORE DESIGN FEATURES

5.5 Considerations affecting the core design are mentioned throughout this book, and some requiring a description of the core neutron behavior are given here.

Power Generation

5.6 Core energy production is primarily the result of neutrons interacting with fissile atoms. Since the power distribution as a function of both space and time is of major engineering importance, the designer must know the neutron distribution also as a function of space, energy, and time. Iterations are involved, of course, in fixing the core specifications that contribute to the neutron balance. These include the core enrichment, moderation, geometry, and planned changes with time as required for fuel management.

5.7 As emphasized in Chap. 4, nuclear power-peaking factors contribute strongly to the hot-channel analysis from which limits on the thermal design are derived. Nuclear peaking factors for various operating modes must also be considered part of the nuclear analysis. For example, Fig. 5.1 shows the axial distribution of power expressed as the maximum-to-average ratio, $P/\bar{P}$, for two conditions of operation for a large pressurized-water reactor.[1] The first condition, a group of control elements partially inserted for transient control after a power-level change, yields a power ratio of 1.70. The second, a somewhat idealized situation when all rods, including those for xenon overrride, are withdrawn, yields a cosine power distribution with a 1.50 power ratio in the center of the active core. Although the local linear heat rate is less in the second case, the approach to DNB (Departure from Nucleate Boiling) is closer since the peak occurs farther up the channel.

5.8 Another type of analysis for a different pressurized-water reactor[2] is illustrated in Fig. 5.2, where the shift in axial power pattern is considered at

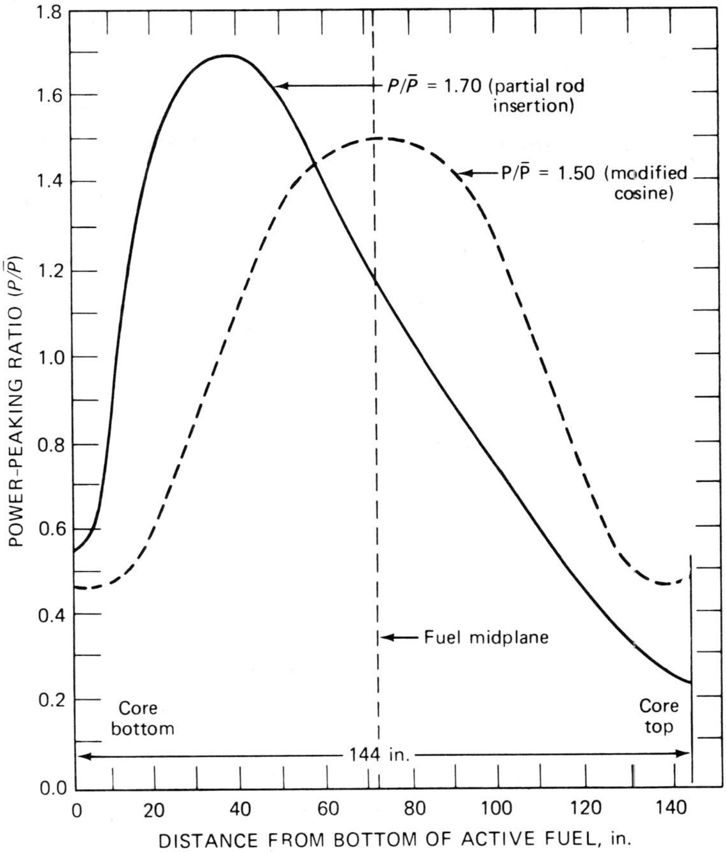

Fig. 5.1 Axial power patterns.

various overpower levels. The changing pattern is caused primarily by a decrease in moderator density near the top of the core as its temperature increases (§5.12).

5.9 Since the time-integrated spatial power distribution determines fission-product buildup and isotopic changes in the fuel, which, in turn, affect the instantaneous power distribution, a second major analytical requirement is to describe such time-dependent changes as burnup proceeds.

5.10 In general, the designer would like to choose parameters that will produce a radial and axial power distribution which is as flat as possible throughout the core lifetime. At the same time, it is necessary to provide sufficient reactivity,* as contributed by fuel enrichment, to yield optimum fuel exposure and yet maintain adequate reactor-control requirements.

---

*Reactivity is used here as a qualitative measure of the capability of the fuel to maintain a chain reaction.

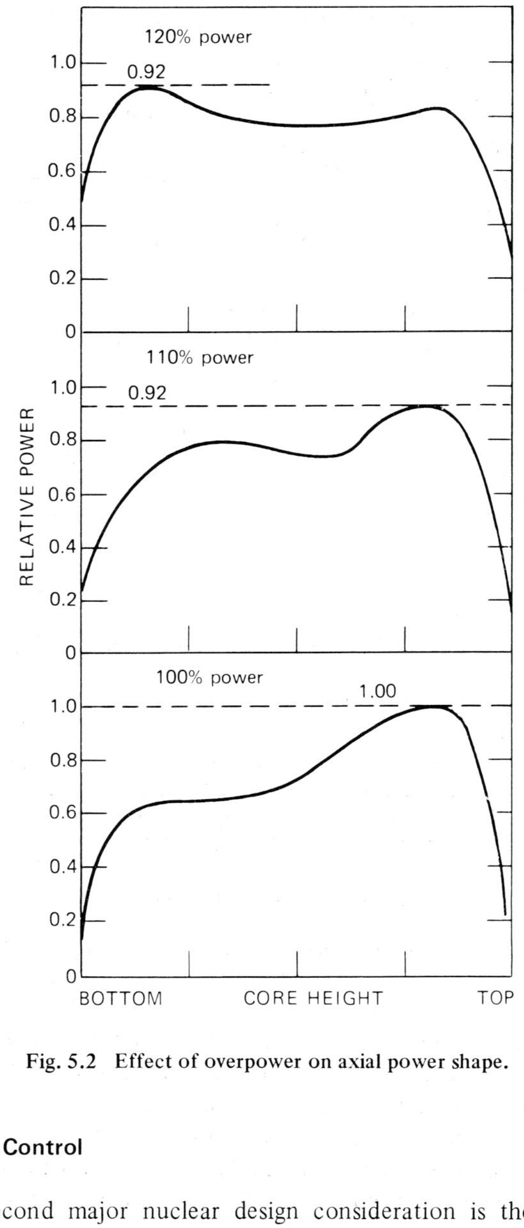

Fig. 5.2 Effect of overpower on axial power shape.

Reactivity and Control

5.11 A second major nuclear design consideration is the calculation of reactivity under all projected operating conditions. Control elements and their behavior also require special attention. Design approaches and requirements are treated in Chap. 6. For the nuclear description of the core, however, both the contribution of control features to the neutron balance and the control spatial arrangement specified by the designer are important. For example, the details of the neutron-flux distribution will be affected by the decision to use well-distributed small poison rods for control elements or large cruciform blades, as well as by the amount of "worth" assigned to burnable poison fixed in fuel locations and absorbers such as boron that may be dissolved in the water coolant.

5.12 Safety-related considerations are an important feature of the core design. As discussed in Chap. 6, the response of the reactor to both abnormal operating conditions and possible accidents depends on reactivity feedbacks based on changes in the neutron balance. For example, in a water-moderated and -cooled reactor, expansion of the coolant–moderator by heating reduces its density, which has several effects including enhanced neutron leakage and spectral hardening, with a resulting decrease in reactivity. The core neutron behavior must be described for such considerations.

Core Specifications

5.13 Many of the core specifications affect and are affected by the neutron behavior in an iterative way. In both thermal and fast reactors, the geometric layout, including length, diameter, and enrichment zones, is closely related to the neutron balance and hence is part of the nuclear design. Similarly, the fuel-element geometric array and the design of the fuel pin itself are related to the nuclear description. The composition of nonfuel materials, including the moderator, has a marked effect on the neutron balance and the nuclear design.

5.14 Since iteration plays such a role in the design, possible approaches differ, depending on where the iteration cycle is entered and what parameters are initially regarded as independent. It is sufficient here to emphasize that the nature of the interrelations justifies describing nuclear behavior as a design aid.

ROLE OF THE COMPUTER

5.15 The unique role played by the computer in reactor-core analysis can be appreciated by considering the parallel development of digital computers and reactor physics. During the early 1950s the first commercial stored-program computers were developed to meet the needs of AEC laboratories and contractors in weapons and reactor design.[3] As lengthy numerical computations became possible, it was no longer necessary to depend on difficult analytic solutions for very simple models, and more-general approaches were developed.

5.16 Since early computers were limited in speed, storage capacity, and other features, simplified computational methods giving inexact results were necessary. As the capabilities of computers increased, some of the computational restrictions were removed, and better approximations to theory became possible. The sophistication in analysis therefore paralleled advances in computer science and technology.[4] Since digital-computer-oriented methods are used almost exclusively for the nuclear analysis of practical cores, this discussion is primarily concerned with such approaches. Pertinent background material on reactor physics is available elsewhere.[5–8]

GENERAL COMPUTER APPROACHES

INTRODUCTION

5.17 Although the evolvement of calculation approaches as computers were developed is an interesting and useful story,[3,4] only a few comments are given here, primarily to provide perspective. Early criticality calculations were based on one- or two-group age-diffusion theory with the parameters derived from integral experiment correlations. As computer capability improved, a somewhat more sophisticated approach involved the use of three or four energy groups in one-dimensional diffusion-theory programs written for slab, cylindrical, and spherical geometry. A further sophistication as larger computers became available was the development of two-dimensional calculations using multigroup-diffusion theory. When cylindrical geometry is assumed, both composition and flux can be varied in both the radial and the axial direction. A representation very useful for design is then achieved which is widely used today.

5.18 An adequate description of neutron thermalization has been a challenge. An early method was to adjust cross sections obtained by crude spectrum-averaging procedures to agree with critical experimental measurements. This approach, indeed an art, was particularly difficult for a thermal-spectrum reactor, in which the neutron-absorption cross section is large relative to the moderating characteristics. Spectral shifts due to changes in reactor composition provided an additional complication.

5.19 A significant effort has therefore been devoted to the development of an accurate representation of neutron thermalization, taking into consideration complex neutron interactions with moderator atoms. Such treatments have been very useful for determining physical constants for criticality calculations (§5.34).

5.20 Another challenge was to develop transport-theory-based methods to permit calculations for reactors with small, fast cores, namely, problems involving regions of strong neutron absorption or large discontinuities (§5.68).

MULTIGROUP METHODS

5.21 Most design calculation methods involve a multigroup approach,[9–12] or a variation of it, wherein the reactor neutron-balance equation is separately stated for each of a number of neutron energy groups. Multigroup theory is covered in detail in the literature and hence will not be discussed here. For the representation of a reactor of practical design, some approximation to the Boltzmann equation that is convenient for machine computation is necessary.[13] Generally, the objective is to determine critical-enrichment requirements, with the shape and energy dependence of the neutron flux obtained as part of the calculation. Even for a one-group problem, materials, boundary conditions, and necessary nuclear constants must be specified. However, in a multigroup calculation the value of these constants also depends on the relative flux in each energy group. An iterative procedure that starts with an estimated flux and core size is therefore necessary. A typical procedure[14] for carrying out such a calculation, shown in Fig. 5.3, is discussed in the following sections.

5.22 The calculation first concerns the generation of cross sections and other nuclear constants for the subsequent criticality calculation. Computer programs (codes) specially designed for determining multigroup cross-sections and constants have been developed. Approaches for slightly enriched light-water-moderated reactor cores are discussed here to illustrate the principles involved. Except for specialized procedures to obtain constants, however, the multigroup approach is also applicable to epithermal and fast reactor systems.

5.23 Although transport theory is necessary for certain applications (§5.68), diffusion theory is very useful for many design requirements. As a basis for discussion, therefore, the familiar diffusion equation is considered in the following form for a lethargy group, j, and a single component:

$$\nabla \cdot [D_j(r)\, \nabla\phi_j(r,t)] - \Sigma_j(r)\, \phi_j(r,t) + S_j(r,t) = \frac{1}{v_j} \frac{\partial \phi_j(r,t)}{\partial t} \tag{5.1}$$

where j refers to the group, r the positional coordinate, and t the time variable.* In the second term, $\Sigma_j(r)$ is the absorption cross section for the j group neutrons, plus the cross section for scattering neutrons out of the j group. The source term consists of the net neutrons scattered into the group plus the number of fission neutrons in the j group from fissions in all other groups,

$$S_j(r,t) = \sum_{k=1}^{j-1} \Sigma_{k\to j}(r)\, \phi_k(r,t) + \chi_j \sum_{k=1}^{J} (\nu\Sigma_f)_k(r)\, \phi_k(r,t) \tag{5.2}$$

where $\Sigma_{k\to j}$ is the scattering (inelastic and elastic) cross section from group k to group j, χ is the fission spectrum normalized to $\sum_1^J \chi_j = 1$, and J groups exist.

5.24 The diffusion equation, particularly the source term, is sometimes expressed in different ways with various notation. Remembering the neutron-balance basis for the various terms used will minimize confusion, however.

*Unless specific exceptions are indicated, notation follows that used in Reactor Physics Constants, USAEC Report ANL-5800, 2nd ed., Argonne National Laboratory, July 1963.

5.25 The steady-state or time-independent solution of Eq. 5.1 is important to the reactor designer. We write this by introducing a term, λ, so that the equation will have a steady-state solution when λ is equal to unity. In this eigenvalue problem we can therefore study the effect of changes in parameters when λ is close to unity, and ϕ_j will change very little in shape as such parameters are varied.

$$-\nabla\cdot[D_j(r)\,\nabla\phi_j(r)] + \Sigma_j(r) - \sum_{k=1}^{j-1} \Sigma_{k\to j}(r)\,\phi_k(r) = \frac{1}{\lambda}\,\chi_j \sum_{k=1}^{J} (\nu\Sigma_f)_k(r)\,\phi_k(r) \quad (5.3)$$

Applicable boundary conditions include the satisfaction of flux and current continuity conditions at interfaces between regions.

NUMERICAL METHODS

5.26 Numerical methods are necessary for solving eigenvalue problems, except for those describing very simple reactor models. Generally, the required coupled differential equations are approximated by a set of coupled linear algebraic equations based on a "grid" or "mesh" being placed over the region of interest in the reactor. Finite-difference expressions then apply about each mesh point. Group fluxes and the source at each of the mesh points are then calculated by applying equations such as Eq. 5.3 in finite-difference form. Users of computer programs, to apply procedures properly and resolve difficulties, should be familiar with numerical solution methods described elsewhere.[12] A description of the specialized mathematics is not appropriate here, however. An outline to solution methods and computational strategy is also often found with code instructions.

5.27 Many matrix methods for solving the linear-difference (multigroup) equations use a *source iteration* procedure,[4] as indicated in Fig. 5.3. In general, the source term for each group equation can be estimated, starting with the fission-source distribution and an assumed flux and reactor size. The individual equations are then solved, starting with the highest energy group and proceeding to the thermal-energy group equation. From the flux distribution for the entire core obtained at the end of the cycle, a new source distribution is determined and the process continued until convergence within an acceptable range is obtained. As part of each cycle, each group diffusion equation is integrated numerically by iteration procedures to give the spatial distribution of the flux, a process known as *inner iteration.* The overall process of calculating a second-generation source distribution from a first-generation one is called an *outer,* or *source, iteration.*

5.28 The outer iteration is used to solve an eigenvalue problem. For the inner iteration the mathematical techniques are quite different since a set of inhomogeneous linear equations must be solved. Separate tests for convergence

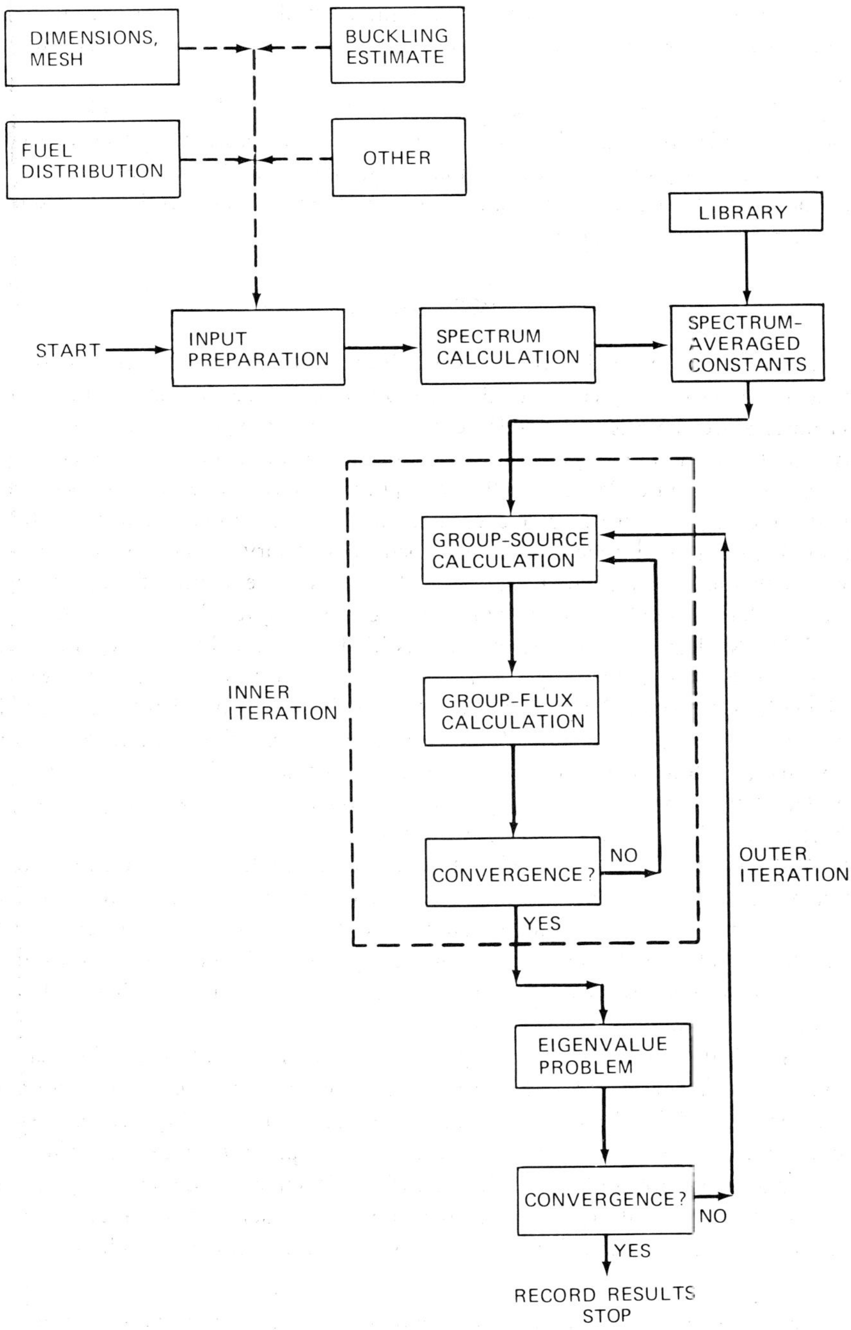

Fig. 5.3 Typical calculation flow.

are also used since it is often possible for the multiplication factor (eigenvalue) to converge satisfactorily before the flux shape has been adequately determined.[4]

5.29 Reactor-design procedures can include a somewhat different type of computer-code approach which also includes an outer iteration, concerned with fixing or searching for gross design parameters, such as reactor size (geometry) or composition. Adjustments are made until convergence criteria, perhaps based on a required eigenvalue,* are met.

TYPES OF CODES

5.30 Computer codes for nuclear design normally treat some areas of neutron behavior in greater detail than other areas, depending on the code purpose. Since there is a trade-off between computing expense and accuracy of representation, it is not practical to use a code that provides more detail than necessary or is intended for a different purpose. Often one variable, such as neutron energy or space, is treated in detail, and the other is approximated. Therefore many codes have been developed to meet varying needs. One purpose of this chapter is to indicate broad areas of code usefulness rather than provide a code catalog. Examples of some widely used codes are given for orientation.

5.31 Nuclear-design codes can be classified into three broad categories: (1) codes to develop energy-dependent cross sections for multigroup eigenvalue calculations, (2) static design codes to solve a wide class of problems that depend on the determination of the multiplication constant and the characterization of the flux distribution, and (3) time-dependent codes, which can be subdivided into depletion codes for which the time period is long and kinetics codes for short time periods as required in safety analysis.

5.32 Emphasis varies in each category. For example, neutron energy is the variable of major interest in cross-section development codes, whereas spatial representation is important in the static design codes. With time as the major variable for category 3 codes, a physical description less detailed than that for the static design codes is usually accepted so that computer running time will not become excessive.

5.33 Within each of the categories, a finer classification can be made. Several steps are required to transform experimental nuclear cross-section data to a form useful for design calculations, with each step requiring a code. Design codes may use diffusion or transport theory and represent a core with different degrees of dimensionality (1-D, 2-D, etc.). Time-dependent codes also vary in the extent of representation. For example, a point model may be adequate for calculating the changing fuel inventory in an economics code, whereas 2-D

*In the determination of conditions for criticality, an eigenvalue of unity for λ in Eq. 5.3 is desired. However, eigenvalues can be chosen to represent other specifications such as fuel concentration and dimensions (see §5.57).

representation is necessary to study changes in power-peaking factors. Approaches within these broad categories are surveyed in the following sections. Remember, however, that, to accomplish the overall core design, we mainly use codes that depend on one another and perform the calculation several times in an iterative manner as discussed in §5.144 et seq.

GENERATION OF NUCLEAR CONSTANTS

INTRODUCTION

5.34 The nuclear constants in the multigroup-diffusion or -transport equations are not at all constant but are energy dependent. Changes in material composition and temperature effects in practical core systems also cause variations from point to point. Furthermore, since the energy distribution is not determined until the equations are solved, some values must be initiated in the calculation for the constants and then iterated as the energy distribution is determined. Since neutron energy is the variable of interest rather than other reactor-core characteristics, special codes designed specifically for the generation of constants are preferable to codes intended primarily for core design (reactivity etc.).

5.35 The distinction between the two calculation approaches can be partially clarified by considering that design calculations are concerned with overall reactivity (criticality) variations, flux distributions, and material inventories whereas calculations for nuclear-constant development are concerned primarily with calculating reaction rates at a point or in a cell representative of a given portion of the core.

EVALUATION

5.36 Nuclear constants that are to be developed in a form useful for design calculations originate with experimental measurements and derived data. Before nuclear constants can be developed by special codes, the data must be put into usable form. The process of digesting the experimental data, combining it with the predictions of nuclear-model calculations, and attempting to extract the true value of a cross section is referred to as *evaluation*. The reactor designer wants evaluated data for all neutron-induced reactions covering the full range of incident neutron energies for each material used in a reactor. Evaluators generally do not supply the data in this form, however. Rather, they supply the bits and pieces, which, when put together, form a fully evaluated set of data for each material. A second requirement is therefore a data-processing system from which input can be retrieved to reactor design codes or to codes that generate multigroup sets.

5.37 The steps required to process measured nuclear data into a form suitable for input to a reactor design code are illustrated in Fig. 5.4, which is based on the Evaluated Nuclear Data File (ENDF) system.[15]

5.38 The Sigma Center Information Storage and Retrieval System (SCISRS) in Fig. 5.4 is an automated compilation of experimental nuclear data. The main items stored are the measured values of cross sections, but many secondary items of interest to evaluators (errors, method, references, etc.) are also stored. In this case the storage system ENDF/A accepts all evaluated data and ENDF/B stores complete sets of selected data for each material needed for a reactor calculation. The two boxes labeled SCORE and ENCORE represent the evaluation process carried out by a combination of hand operations and processing codes. FLANGE and ETOM are merely library-coupling programs. THERMOS and MUFT-GAM are examples of many codes that accept the library

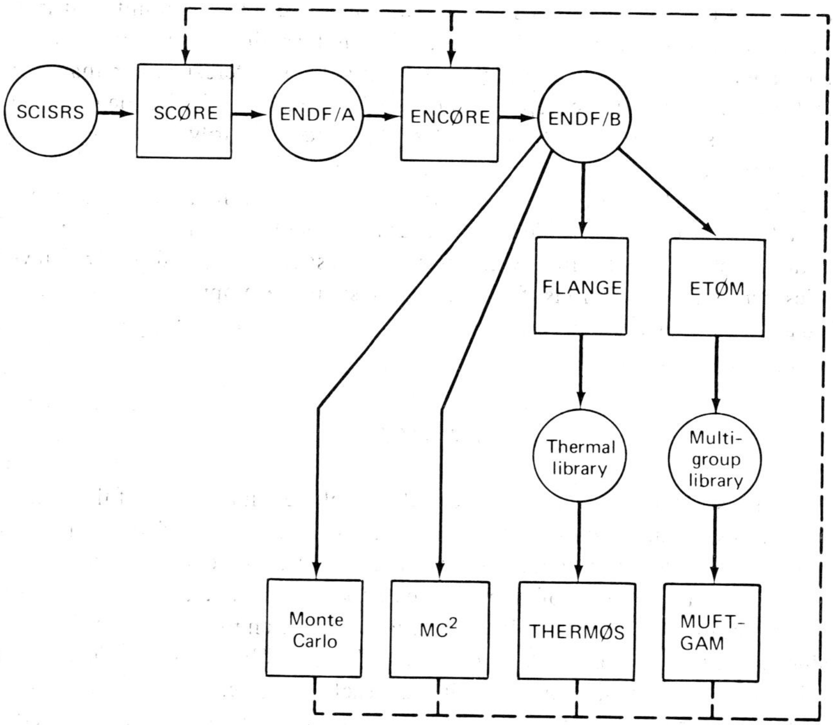

Fig. 5.4 Schematic of the flow of nuclear data from compilation to reactor calculation. Boxes: computer codes that perform some operation on the data. Circles: magnetic-tape (or card) files to store data. Solid lines: flow of data. Dashed lines: feedback to the evaluation phases from the testing of the data against the results of integral experiments.

data to develop thermal and epithermal reactor constants, respectively, as discussed in §§5.49, 5.52, and 5.53. MC^2 (Multigroup Constant Codes) was specifically developed to generate complete cross-section sets from ENDF data (§5.54).

GROUP-STRUCTURE REQUIREMENTS

5.39 Before we discuss some examples of codes for generating constants, readers may find helpful some general comments about the makeup of the groups needed for multigroup core calculations.

5.40 For slightly-enriched light-water-moderated systems with solid fuel elements, a "few-group" diffusion-theory approach using only three or four groups is normally adequate for approximating criticality requirements and power distributions. Although at least a two-dimensional treatment is normally needed, the lattice unit cell may be homogenized. The appropriate group structure depends upon predominant characteristics of various parts of the neutron energy spectrum. In a four-group treatment, for example, the energy range for each predominant effect might be as follows:

1. Fast fission and inelastic scattering (energy above 5.53 kev).
2. Elastic scattering (5.53* kev to 1.855 ev).
3. Resonance absorption (1.855 ev to 0.625 ev).
4. Thermalization (below 0.625 ev).

It is then possible to tailor the calculation method, which will normally require the use of a theoretical slowing-down model, so that it will be appropriate for the individual group in which one effect predominates. This particularly applies in the calculations for generating cross sections (§5.48 et seq.).

5.41 Selecting the energy ranges involves several considerations. In the high energy range, a homogenized approach is normally appropriate since the neutron mean free path is much longer than the dimensions of a lattice. A separate treatment of neutron motion is likely to be appropriate for each lower energy range. The thermal-group cutoff energy, for example, should be high enough so that upscattering of neutrons can be neglected. On the other hand, resonance peaks should be in the epithermal range. Although at resonance energies the mean free paths are smaller than lattice dimensions, a homogeneous transport equation, corrected by self-shielding factors (§5.43), may be used.

5.42 If a heterogeneous treatment of the lattice (§5.60) is used for the thermal range with homogeneous treatments for higher energy groups, some approximations are necessary in calculating the slowing-down source into the thermal range. The epithermal flux is assumed to follow a 1/E relation for several lethargy intervals above the thermal cutoff energy. The slowing-down source is then assumed to be isotropic as well as separable in space and energy. A

*Group cutoff energies are normally taken at even lethargy intervals which, when expressed in energy units, give the values listed.

choice of the thermal-group cutoff energy should therefore take into consideration the first resonance peaks listed in Table 5.1. In slightly enriched reactors the distortion in the 1/E flux caused by the lowest ^{233}U and ^{235}U resonances is not great. In plutonium-fueled systems, however, a significant distortion results from the 1.05-ev ^{240}Pu resonance.

TABLE 5.1

Peak Energies of the Low Energy Resonances of Thorium, Uranium, and Plutonium Isotopes

Isotope	First resonance (above 0.625 ev), ev
^{232}Th	21.9
^{233}U	1.45
^{234}U	5.20
^{235}U	1.14
^{236}U	5.49
^{238}U	6.68
^{239}Pu	7.90
^{240}Pu	1.05
^{241}Pu	4.31
^{242}Pu	2.65

5.43 Shielding on the effective cross section is another consideration. If the absorption mean free path of a material is small compared with its thickness, neutron absorption in the surface layers of the material will be high, and the flux will be depressed in the material away from the surface. The effective cross section therefore tends to decrease if the reaction rate is based on the surface flux. This is of particular concern in treatments for thermal reactors in the resonance region. Shielding factors for each energy group are defined as the ratio of the actual reaction rate and that for the same material exposed to the volume-averaged neutron flux.

5.44 Multigroup solutions of transport or diffusion equations are normally used for fast reactors. The number of groups required depends on the energy distribution of the reaction events. For a large dilute core, where considerable inelastic scattering occurs and a degraded spectrum results, 15 or more groups may be needed. Variations of cross sections with energy are large. Reasonably short lethargy intervals (about 0.5 Δu) are therefore needed to reduce errors caused by averaging the constant over the interval. This need for a many-group treatment is due to the wide energy spectrum for neutrons that cause fission. In contrast, in a thermal reactor, in which thermalized neutrons cause most fissions, a three- or four-group treatment is likely to be adequate.

5.45 The number of groups needed also depends on the *objective* of the calculation. A many-group calculation may be carried out, for example, for the sole purpose of developing a good picture of the energy dependence of the flux, which may be used as the basis for averaging effective few-group constants. Parametric design calculations can then be carried out with a few-group treatment at a considerable saving in computer time.

5.46 Desired energy-dependent nuclear constants are as follows:

D	Diffusion coefficient
Σ_a	Absorption cross section
$\Sigma_{j\to k}$	Scattering removal cross section from group j to group k
$\nu\Sigma_f$	(Number of neutrons emitted per fission) (macroscopic fission cross section)

Appropriate *group averages* of the constants are taken. Although the averaging procedures involve a number of considerations, depending on the pertinent reactor theory,[16] it is useful to list some of the usual definitions. Separability in space and lethargy are usually assumed for each group over which cross sections are averaged.* As a result the averages are spatially independent, provided the individual cross sections are initially independent of position. For a group, j, over a lethargy interval from u_1 to u_2, the averaging procedure can be as follows:

The flux $\phi_j(r)$ in the interval is defined as

$$\phi_j(r) = \int_{u_1}^{u_2} \phi(r,u)\, du \tag{5.4}$$

Similarly,

$$(\Sigma_a)_j = \frac{1}{\phi_j} \int_{u_1}^{u_2} \Sigma_a(u)\phi\, du \tag{5.5}$$

and

$$D_j = \frac{\int_{u_1}^{u_2} D(u)\, \nabla^2\, \phi(r,u)\, du}{\int_{u_1}^{u_2} \nabla^2\, \phi(r,u)\, du} \tag{5.6a}$$

or

$$D_j = \frac{1}{\phi_j} \int_{u_1}^{u_2} D(u)\, \phi(u)\, du \tag{5.6b}$$

*Although the assumption of separability is valid only for a uniform-composition large (asymptotic) core, it is appropriate as a first approximation for the subsequent numerical-solution iteration procedure from which a spectrum is obtained.[17]

Eq. 5.6b is a simplification of Eq. 5.6a as a result of the space and lethargy separability. It is important to keep in mind that the actual averaging process, particularly in regions of high resonance peaks, can be more sophisticated than we have indicated.

SOME EXAMPLES OF GROUP-CONSTANT CODES

5.47 Some typical codes used to obtain group constants are considered in this section. The physics approaches used are briefly discussed, with a level of technical sophistication somewhat higher than elsewhere in the chapter. The general reader, therefore, will receive the flavor of the approach. Others, if they wish, can refer to reference material.[7,9,10,12] This discussion can be omitted without loss of continuity.

5.48 For slightly enriched water-moderated reactors, usually having complex geometry and core structure, transport approaches are useful to determine the slowing-down distribution of neutrons, which, in turn, is needed to calculate nuclear constants. Different treatments are appropriate for the various energy groups and different moderating media (§5.49). The need to apply special techniques for the homogenization of "cells" containing strong absorbers, such as control rods, should also be recognized (§5.61).

5.49 Fast- and intermediate-group constants can be calculated by using a slowing-down code*[18] such as MUFT-4 or various modifications[20] of it, such as GAM-1. The approach used is based on a multigroup one-dimensional P_1 or, as an option, a B_1 approximation. The P_n treatment involves the spherical harmonics expansion of both the neutron distribution and scattering functions in the Boltzmann equation [7,12] with the resulting coupled set of differential equations in terms of Legendre polynomials truncated to a desired order. However, the B_1 approximation, which involves a slightly different representation of the flux and a similar representation of the scattering function by a finite number of terms, is said to give better estimates of the higher moments of the spatial neutron distribution.

5.50 Within the P_1 or the B_1 treatment, various slowing-down models can be chosen. Hydrogen moderation can be treated exactly or by various methods, and an age approach or Greuling–Goertzel approximation could be used for nonhydrogenous slowing down. An initial value of the buckling is also necessary. Since the calculated group constants are not very sensitive to the buckling, little iteration is required.

5.51 A "library" containing microscopic cross sections and other constants for various materials is included with the code. Code input information therefore

*Code-information material is available from the Argonne National Laboratory Code Center and similar sources.[19]

also includes the element number densities* and desired group structure. It is also necessary to account for spatial effects in the homogenized calculation, such as resonance self-shielding in the epithermal range and the Dancoff effect. So-called "L factors" are used in MUFT. These are equal to the ratio of the heterogeneous resonance integral to the integral calculated by the program on a homogeneous basis. Such "self-shielding numbers" are also required as input to the code.

5.52 For thermalized-neutron constants a code[21] such as SOFOCATE is useful. The energy spectrum of neutrons in equilibrium with a hydrogen-moderated homogeneous medium is determined by solving the Wigner–Wilkins equation.[22] The medium is assumed to be infinite in this equation and the moderator to have an absorption cross section following the 1/v law and a constant scattering cross section. Leakage effects can be accommodated by a function depending on the ratio of neutron absorption to scattering. Input information includes temperature, cross sections, atomic densities, buckling, etc. TEMPEST is a similar program.[23] It is important to remember that these codes have been improved from time to time and that many other codes are available for thermalized constant development.

5.53 More-sophisticated approaches than the preceding ones are available which use integral transport theory for describing lattice cells. The THERMOS code,[24] designed for water-moderated reactors, computes the scalar thermal-neutron spectrum as a function of position in a lattice. THERMOS makes use of a subsidiary code, GAKER, based on the Nelkin kernel,[7] which represents energy exchange between the neutron and the vibrational and rotational levels of a free molecule of mass 18. All other scattering transfers within the cell are based on a gas-model scattering kernel.

5.54 Group constants for fast reactor calculations are generated by procedures somewhat different from those for moderated systems. In fact, early "old-fashioned" approaches attempted to define a single set of averaged cross sections for an "average" spectrum appropriate for a given type of reactor (oxide or metal fueled) and then use these cross sections for all calculations for the given type. A number of computer programs have now been developed,[25] however, which convert basic data to multigroup constants for use with diffusion-theory, transport-theory, and/or Monte Carlo neutronic codes. These programs can produce both region-dependent and region-independent constants. In addition, techniques are available for treating resonance data in the resolved as well as the statistical energy regions, and for generating elastic-, inelastic-, and total-transfer matrices, which account for anisotropy in both elastic and inelastic differential scattering. GRISM and MC^2 are computational systems[25,26] that produce multigroup cross sections averaged over any given spectrum for either diffusion- or transport-theory calculations.

*Atom densities.

5.55 The value of sophisticated averaging techniques is limited, of course, by the accuracy of available microscopic-cross-section data. Both cross-section-measurement and critical-experiment[26,27] programs therefore play a major role in the development of nuclear design methods.

5.56 The preceding examples are intended merely to provide a very brief introduction to the many computer approaches available. Because modifications and improvements of programs are also continually being made, the designer must review current practice to be certain that he is using the best available methods to meet his requirements.

STATIC-ANALYSIS METHODS

CRITICALITY SEARCH AND REACTIVITY CALCULATIONS

5.57 Calculations in this category are intended primarily to relate variations in design parameters to criticality. Multigroup diffusion equations, such as Eq. 5.3, can be written in terms of λ as a parameter that corresponds mathematically to the eigenvalue and physically to k_{eff}. Here the desired design parameters are related to the coefficient matrices developed from the difference-equation set needed for the numerical solution. For orientation, however, primary attention is devoted to the use of the method and not to the mathematical detail. Criticality searches may be made in terms of parameters, such as buckling, material concentrations, and dimensions or positions. The determination of changes with fuel exposure is a related problem but will be considered separately.

5.58 An important computation challenge is to simplify the three-dimensional reactor-core model, which requires very complex code treatment, to a practical balance between adequate representation and reasonable computer effort. A common simplification is to consider the core cross section as a cylinder with a consequent reduction of the calculations to a two- or one-dimensional treatment. A two-dimensional treatment, for example, would be based on an r–z model, and a one-dimensional calculation would consider only radial variations. A zero-dimensional treatment, in which the reactor is considered a point, is of course the ultimate simplification and is useful for survey work.

5.59 A typical reactor core consists of discrete local regions of fuel, moderator, coolant, control rods, and structural materials, all arranged in some type of "semiregular" geometry. Whether or not the designer is attempting to use a simple zero-dimensional treatment or a complex three-dimensional code, he must work out the problem of homogenization. This is true both in the "local-region" situation concerned with developing nuclear constants representative of a lattice cell and in the "whole-core" calculations required for criticality parameters. In each case some degree of homogenization is necessary to simplify the physical representation so that a computer code of practical size can describe the nuclear behavior.

5.60 A variety of methods, many quite complicated, have been developed to account for the heterogeneity in a reactor lattice cell. Such homogenization techniques are used for calculating physics parameters for a unit cell that may be defined as a single fuel rod together with its associated cladding, gap between fuel and cladding, ferrules, and moderator. Fuel, water, and structural material are assumed to be homogenized over the volume of the unit cell, and the characteristics of the cell (e.g., densities of nuclei, thermal and nonthermal energy spectra, and resonance absorption probabilities) are computed for the homogeneous mixture.[28] It must be emphasized, however, that this *cell-theory* method of obtaining homogenized constants is approximate but useful for design. Separability of space and energy is assumed as well as an independence of the cell region from nuclear events in the remainder of the core.

5.61 An important problem associated with homogenization is the difference in flux levels in different materials within the cell. Consider the homogenized area shown in a typical water-reactor lattice shown (full size) in Fig. 5.5. A typical thermal-flux distribution[29] is shown in Fig. 5.6. Of greatest

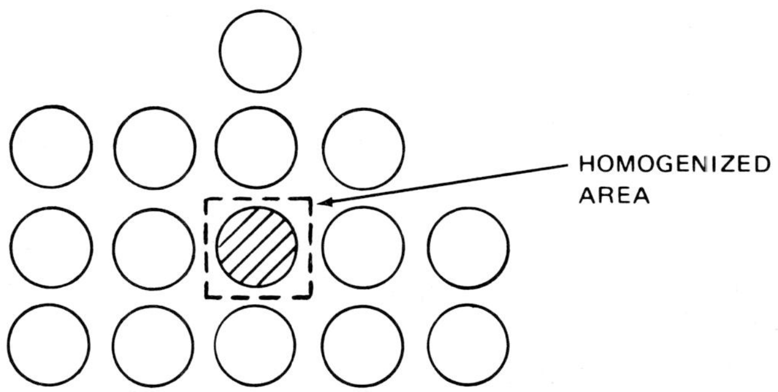

Fig. 5.5 Typical water-reactor lattice.

importance are the effects in the resonance and thermal energy ranges. In some codes factors are introduced to account for the changes in spectrum and resulting reaction rates in different cell regions. For example, in LASER, a depletion code,[30] the thermal-reaction rates for the various cell elements are obtained by multiplying the cross sections averaged over regional spectra by a corresponding disadvantage factor. A so-called "g" factor is used in other codes such as CANDLE.[31] The "g" factor, different for each element of the cell, is based on the local spectrum variation and, in turn, the energy dependence of the individual cross sections.

5.62 The effect of surrounding rods in the lattice should be accounted for in computing resonance absorption in a closely packed lattice. The so-called Dancoff–Ginsburg correction factors[32] are generally used to account for this "shadowing." Extensive tables of the factors are available in the literature.[33]

5.63 The preceding approach is in a sense oversimplified and represents merely one example of a calculation method, particularly for epithermal groups. For close-packed water lattices, the "thermal" group can be treated by using an appropriate spectrum-averaged cross section in a cylindrical P_3 calculation[7] for

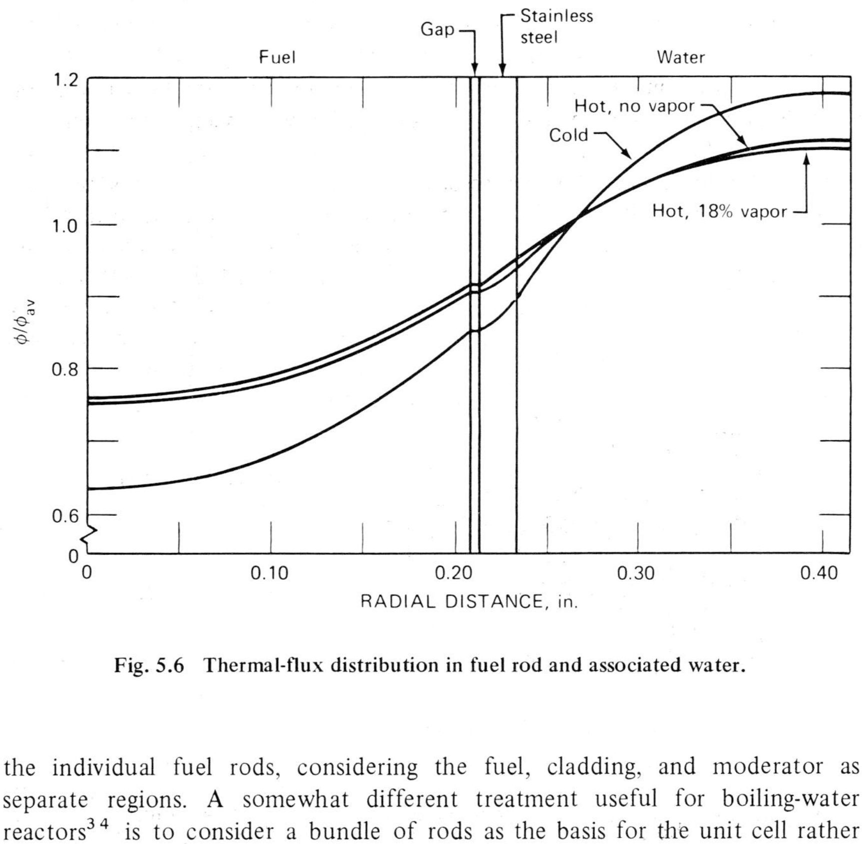

Fig. 5.6 Thermal-flux distribution in fuel rod and associated water.

the individual fuel rods, considering the fuel, cladding, and moderator as separate regions. A somewhat different treatment useful for boiling-water reactors[34] is to consider a bundle of rods as the basis for the unit cell rather than a single fuel rod. The unit lattice cell is then formed by a number of fuel rods, the surrounding associated water gaps, the flow-channel partitions, and the control elements or portions thereof. For analysis, the unit cell may then be subdivided into a number of regions, each of which has a fairly uniform composition and environment. For example, for the Dresden I boiling-water-reactor cell, all of sixteen inner rods are treated as one region, the corner rods as another, etc.[34]

EXAMPLES OF CRITICALITY SEARCH CODES*

5.64 The WANDA codes[35] are representative of the "classical" one-dimensional diffusion-theory approach, which is very useful for preliminary design wherein computer-time expenditures are modest and an approximate picture can be tolerated. In this type of code, only a *few groups,* normally not more than four, are used; the computation scheme, compared with that for many energy groups, is consequently simplified. Many regions may be described, however, with the possibility of varying the mesh size in each region. The approach is similar to that previously described with parameters and properties provided as input for an eigenvalue search and various output options available.

5.65 A number of *multigroup,* one-dimensional diffusion-theory codes are available, of which AIM-6 and MACH-1 are representative.[36] These codes are particularly useful for fast systems needing a large number of groups. Very flexible, they permit a variety of calculation approaches within the one-dimensional limitation. MACH-1 is described in the following sections in relative detail. Various ancillary subroutines can be used, e.g., DEL, a *perturbation* code for calculating fractional changes in k due to a perturbation of microscopic cross sections, buckling, etc. Such possibilities are described in the code manuals.

5.66 The concept of *transverse buckling* is useful for a problem with less than three space variables since some assumption must be made with regard to the remaining space variables. In a cylindrical radial calculation, for example, an estimate may initially be made of the transverse-leakage term represented by DB_z^2, where

$$B_z^2 = \frac{\pi^2}{H^2}$$

and H is the axial height of the active fuel plus the reflector savings. In this case a radial-buckling term also applies, where

$$B_r^2 = B_t^2 - B_z^2$$

and the total buckling, B_t^2, is obtained using the eigenvalue of the radial problem. An axial-diffusion calculation may be carried out, if desired, using DB_r^2 for the transverse leakage.[37] A number of assumptions are necessary, including separation of r and z variables as well as equality of the transverse-buckling terms for each lethargy group.

5.67 The following summary of the MACH procedure illustrates the preceding principles.

---

*As in the group-constant code examples (§5.48), the physics approach discussions are brief. Some readers may wish to refer to general reference material.[7,9,10,12]

1. *Equation Set*

The code solves the coupled set of one-dimensional, multigroup, diffusion equations in slab, cylindrical, or spherical coordinates. One of these equations is

$$D_j \nabla^2 \phi_j - \Sigma_{t,j}\phi_j + S_{s,j} + S_{f,j} + S_{e,j} = 0 \tag{5.7}$$

where*

$$\begin{aligned}
D_j \text{ (diffusion coefficient)} &= \tfrac{1}{3}\left(\textstyle\sum_m \Sigma_{tr}^m\right)^{-1} \\
\Sigma_{t,j} \text{ (total removal)} &= D_j B_j^2 + \textstyle\sum_{k>j} \Sigma_{j\to k} + \Sigma_{c,j} + \Sigma_{f,j} \\
B_j^2 &= \text{transverse buckling} \\
S_{s,j} \text{ (scattering source)} &= \textstyle\sum_{k<j} \Sigma_{k\to j}\phi_k \\
S_{f,j} \text{ (fission source)} &= \chi_{f,j} \left(\textstyle\sum_k (\nu\Sigma_f)_k \,\phi_k\right) \\
S_{e,j} \text{ (external source)} &= \chi_{e,j}\, S_e
\end{aligned}$$

The finite-difference approximations to the diffusion term at space point i are

$$\left.\frac{d\phi_i}{d\xi}\right|_i = \frac{\phi_{i+1} - \phi_{i-1}}{2\Delta\xi} \tag{5.8}$$

where ξ is the spatial distance from the problem origin, and

$$\nabla^2\phi = \frac{1}{(\Delta\xi)^2}\left[\phi_{i+1} - 2\phi_i + \phi_{i-1} + \frac{(N-1)\,\Delta\xi}{2\xi}(\phi_{i+1} - \phi_{i-1})\right] \tag{5.9}$$

In spherical geometry the trapezoidal integration of functions of the form $\xi^2\phi(\xi)$ leads to large errors near the origin. All such integrals in MACH 1 are corrected by the first-order error term of the trapezoidal rule.

As a preliminary calculation step, the reactor is divided into regions in which D_j, B_j^2, $\chi_{f,j}$, $\chi_{e,j}$, and all $\Sigma_{x,j}$ are constant. The external source S_e, if present, is specified at each spatial mesh point. Input of group-dependent transverse bucklings B_j^2 is accepted by the code in accordance with the following procedure, which is not mathematically rigorous.

Considering cylindrical geometry with symmetry in the θ direction and assuming that the (r,z) variables are separable, we can write the flux as

$$\phi(r,z) = \phi(r)\,\psi(z) \qquad \text{with } \psi(0) = 1$$

*Notation used is the same as in the MACH description report[37] and may vary slightly from that used in previous sections.

The diffusion equation for one energy group can then be written

$$D_j \nabla^2_{r,j}\phi_j + \phi_j D_j \frac{\nabla^2_{z,j}\psi_j}{\psi_j} - \Sigma_a \phi_j + \sum_{k<j} \Sigma_{k\to j}\phi_k \frac{\psi_k}{\psi_j}$$

$$+ \chi_{f,j} \sum_k (\nu\Sigma_f)_k\, \phi_k \frac{\psi_k}{\psi_j} + \chi_{e,j} S_e \frac{\psi_e}{\psi_j} = 0 \qquad (5.10)$$

The usual approach, and the one used in MACH 1, is to set

$$\frac{\nabla^2_{z,j}\psi_j}{\psi_j} = -B_j^2 \qquad (5.11)$$

and to assume that all $\psi_k/\psi_j = 1.0$. This condition is met only over the asymptotic range of the z-direction fluxes and can be true only if all B_j^2 are equal. Alternatively, one could normalize ψ_k to 1.0 at some point z_1 and calculate B_j^2 over the range $z_1 \pm \Delta z$, where Δz is chosen small enough to ensure that ψ_k/ψ_j does not vary appreciably over the range. This is roughly equivalent to performing the radial calculation at z_1.

At interfaces between regions the usual flux- and current-continuity conditions are satisfied by using the gradient approximation given by Eq. 5.8. The arrangement of the difference scheme requires that each region contain at least two mesh intervals.

Four boundary conditions are available at each of the inner and outer boundaries. The general expression is

$$A\phi_j + B\frac{d\phi_j}{d\xi} = C \qquad (5.12)$$

The options are as follows:

a. $A = 0, B = 1, C = 0$ (zero gradient)

b. $A = 1, B = 0, C = 0$ (zero flux)

c. $A = 1, B = \omega_j, C = 0$ (homogeneously mixed)
where ω_j is the linear-extrapolation length; it is negative at inner boundaries and positive at outer boundaries.

d. $A = 1, B = 0, C = \phi_j$ (constant flux)
This is an inhomogeneous boundary condition, unless all ϕ_j are zero, and requires some special procedures.

2. *Procedure for Solution*

With the input conditions of fission source, reactor size, concentrations, etc., the calculation starts with the highest energy group and proceeds from the reactor outer boundary to the inner boundary. After the sweep through all groups is completed, a new fission source is calculated from the resultant fluxes. Except in special cases this source is then normalized to a total of one fission neutron in the reactor in the direction of the calculation; i.e., no integration is carried out over the transverse direction in slab or cylindrical geometry.

From the first two inner iterations, the new normalized source distribution is used to calculate the fluxes in the following iteration. In the third and subsequent iterations, the starting source at a point for the following iteration is obtained by linear extrapolation from the two previous iterations. The equation is

$$S'_{f,l+1} = S_{f,l} + \theta_l(S_{f,l-1}) \tag{5.13}$$

where $S_{f,l}$ = the normalized resultant source from iteration
$S'_{f,l+1}$ = the source guess for iteration $l + 1$
$\theta_l = \theta_0 \; [1 - e^{-0.5(l-2)}]$
θ_0 = the source extrapolation factor $0 < \theta_0 < 1$

The term $S'_{f,l+1}$ will not be extrapolated to a negative value. If a negative value is found, it is replaced by zero.

Inner iterations are continued until one of two inner-convergence criteria, as selected by the user, is met or until the inner-iteration limit is exceeded.

The eigenvalue criterion requires that

$$\left| \frac{k^l - k^{l-1}}{k^l} \right| < \epsilon_1 \tag{5.14}$$

The pointwise criterion is satisfied when

$$\frac{\lambda_{\max} - \lambda_{\min}}{\lambda_{\max}} < \epsilon_1 \tag{5.15}$$

where

$$\lambda_{\max} = \max_i \, (kS_{f,l}/S'_{f,l})$$

and

$$\lambda_{\min} = \min_i \, (kS_{f,l}/S'_{f,l})$$

This test is carried out only for mesh points with nonzero sources. If a very small concentration of fissile atoms ($\sim 10^{-30}$) is inserted in nonfissioning regions, the fluxes in these regions can be tightly converged in this option.

When the inner-iteration loop has converged, the code exits to one of several search or outer-loop options, as specified by the user.

3. *Normalization*

During the diffusion calculation the fission sources are normalized so that

$$\text{Fission-source integral} = \int_0^{\xi_{\text{outer}}} \sum_k (\nu\Sigma_f)_k \, \phi_k \, d\xi = 1.0 \qquad (5.16)$$

Output resulting from this fission-source normalization is specified as source normalized or unnormalized output.

In slab geometry this source integral has dimensions of neutrons/cm^2/sec; in cylindrical geometry, its dimensions are neutrons/cm/sec; in spherical geometry, they are neutrons/sec.

4. *Outer-Loop or Search Options*

Three options that do not involve the outer loop, inhomogeneous problems, one iteration, and k-effective, are available.

a. *Inhomogeneous Problems.* Normalization is determined by the magnitude of the external sources or boundary values.

b. *One Iteration.* A single sweep is made through the spatial mesh for all groups using input source values. The resultant sources are normalized before printing.

c. *k-effective.* The fission sources are converged within the inner-loop criterion; the sources are renormalized after each inner loop.

The following options involve successive trials of a search parameter to reach a specified value of k, which need not be equal to 1.0.

d. *Dimension Search, Reactor Size Not Fixed.* The thickness of one or more specified regions is varied, and the position of the inner boundary of the innermost search region and the thickness of nonsearch regions remain constant.

e. *Composition Search.* The atom density of a specified fuel (which may consist of a number of nuclides) is varied either alone or inversely with the density of a specified diluent. A ratio may be input such that equal volumes of fuel and diluent are varied. Generally, this ratio is

$$R = \frac{N_d^{\text{tot}} \sum_f (N_f A_f / \rho_f)}{N_f^{\text{tot}} \sum_d (N_d A_d / \rho_d)} = \frac{N_d^{\text{tot}} \sum_f (\text{V.F.})_f}{N_f^{\text{tot}} \sum_d (\text{V.F.})_d} \qquad (5.17)$$

where N^{tot} = the total initial fuel or diluent atom density (a/cm^3)
A = atomic weight of a fuel or diluent nuclide
ρ = full density (g/cm^3) of the fuel or diluent nuclide
V.F. = volume fraction of the fuel or diluent nuclide

If regions with different fuel concentrations are specified as search regions, the ratio of fuel-atom density in one search region to fuel-atom density in any other search region is maintained constant. The ratio R is specified for each material.

f. *Dimension Search, Reactor Size Fixed.* The thicknesses of specified search regions are varied, and the thickness of the first region outside the outermost search region is also varied so that its outer radius remains constant. If, for example, regions 1 and 3 of a four-region reactor were being varied, the outer radii of regions 2 and 4 would remain constant.

g. *Buckling Search.* The transverse buckling in specified regions is varied, under the assumption that the bucklings in the search regions are independent of region and group.

h. *Alpha Search.* Either this option may be used as a constant input value in conjunction with any other search or the value of alpha may be varied to achieve criticality.

5.68 Multigroup *transport methods* have been developed which require no greater computer time than many diffusion codes. Such approaches are particularly useful when the nature of the physical problem requires the greater accuracy inherent in a transport-method calculation. The relative applicability of diffusion-theory, P_1 approximations, and transport equations is discussed in texts on reactor theory.[5-8] In general, from a Fick's law approach a diffusion-theory solution is valid when the ratio of the divergence of the diffusion current to the capture rate is small, i.e.,

$$\nabla \frac{D\nabla\phi}{\Sigma_a\phi} \ll 1 \qquad \text{or} \qquad \frac{D\nabla^2\phi}{\Sigma_a\phi} \ll 1 \tag{5.18}$$

From a slightly different viewpoint, $\Sigma_a \ll \Sigma_s$ so that the flux will vary slowly with position. The region of interest should also be at least one diffusion length from a material interface or source discontinuity. Small-size fast systems are therefore normally described by transport methods.

5.69 A highly useful one-dimensional, multigroup transport code[38] is DTF-II, an adaptation of an earlier code, DTK, which includes a reprogramming into FORTRAN. The multigroup transport equation is solved for various geometries by using the discrete S_n approximation.[39] Various types of problems can be solved. These include:

1. Fixed-source calculations (the source consists of either a distributed neutron source or a shell source).

2. Reactivity calculations.

3. Time-absorption calculations.

4. Concentration searches.

5. Zone-width searches.

6. Outer-radius searches.

7. Transverse-buckling searches.

5.70 Two-dimensional few-group diffusion codes such as PDQ-4 are very useful[40] for lattice problems (§5.63). In boiling-water reactors, for example, the lattice may consist of a number of fuel rods in a bundle, associated water gaps, flow-channel partitions, and a section of a cruciform control element, as shown in Fig. 5.7. Also shown in Fig. 5.7 are the different representations required for a fuel assembly for a boiling-water reactor when the reactor is first started up and when an equilibrium fuel-management loading scheme is achieved (§7.65). One-dimensional representation of such a complex geometry would be quite crude. A two-dimensional x–y representation based on a model with the cell subdivided into a number of regions, each having approximately uniform composition for calculating the nuclear constants, is therefore useful.

5.71 In a lattice, such as that in Fig. 5.7, as many as 12 different regions[34] may be desirable to accommodate large differences in composition and temperature. Group constants can then be determined for the subcells by using codes such as MUFT and SOFOCATE. Such a procedure can be fairly complicated, however, with boundary conditions and various corrections tailored for the specific geometry and energy group. Flux-weighted cell averages of the constants can then be used for the PDQ eigenvalue problem. Whether the procedure for generating constants is simplified or complicated depends on the objectives of the calculation. A comparatively crude representation would be adequate, for example, for survey calculations, whereas a control-rod-worth evaluation requires careful analysis.

5.72 In a cell problem such as the preceding one, the designer may wish to supplement the two-dimensional diffusion-theory analysis with one-dimensional transport calculations,[41] particularly in the neighborhood of the control blades, where a slab geometry could be applied. Laboratory integral experiments are important in verifying such design procedures[42] (§5.155).

5.73 The PDQ codes[40] solve from one- to four-group diffusion equations over a rectangular region of the x–y or r–z plane. Usual boundary conditions are zero flux or the symmetry condition, $d\phi/dr = 0$. Setting up the rectangular mesh-line pattern is important in solving the problem. Internal and external boundaries should lie on mesh lines. Furthermore, a fine-mesh spacing is used where marked changes in flux are expected and a coarse spacing where moderate changes are expected. Other parameter-input problems include such items as boundary conditions, buckling, and compositions. In addition, the respective nuclear constants for each material and a flux approximation are needed.

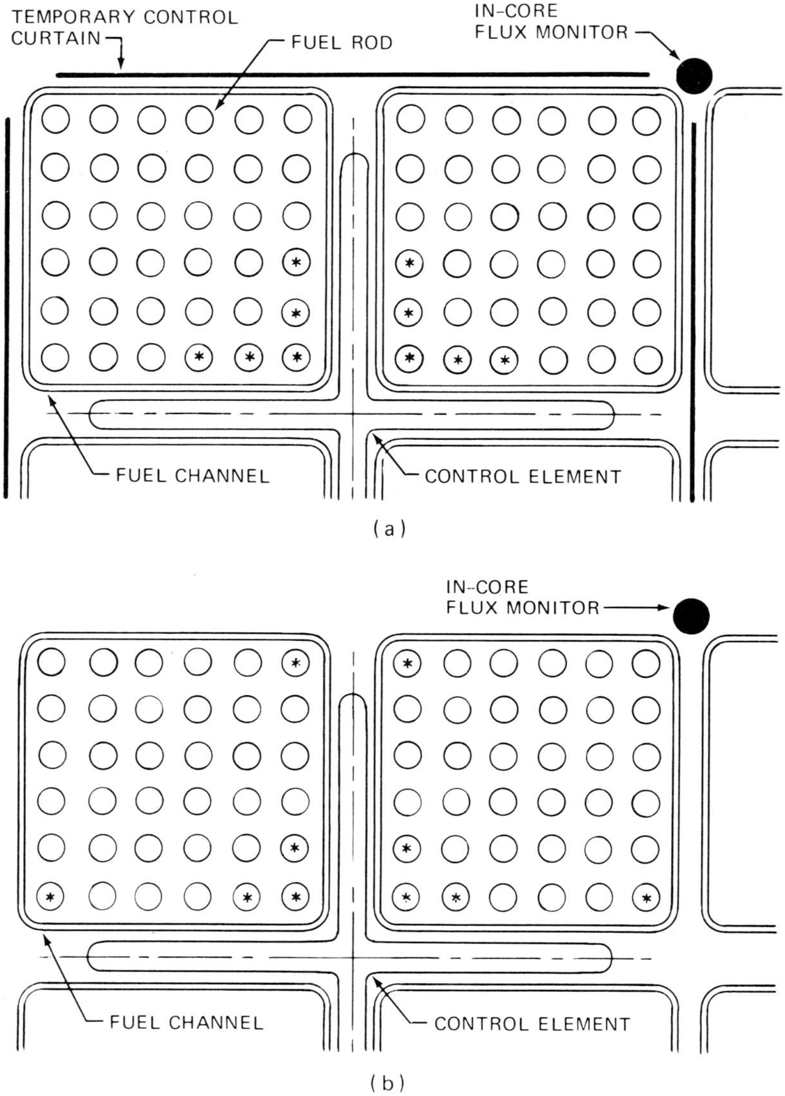

Fig. 5.7 Core lattices. (a) Initial. (b) Equilibrium. Each fuel bundle contains 36 fuel rods, as shown, or 64 in an 8 by 8 lattice. *, special corrected corner rods having enrichments different from those of standard rods in the bundle.

Various computer-output options, including the condensed results of each outer iteration, are available. The most desired results, of course, are the flux and source values.

5.74 Two-dimensional many-group diffusion codes are available for the detailed energy representation needed for fast reactor systems. One such code,[43] CRAM, uses a unique numerical solution wherein the multigroup equations are solved within a geometric "channel" rather than on the basis of energy group by energy group. A two-dimensional *transport-theory* code,[44] TDC, can be used for difficult cylindrical-geometry problems requiring an S_n calculation. Such codes generally require a substantial amount of computer time and therefore are generally used for special applications in conjunction with simpler diffusion codes.

FLUX AND POWER DISTRIBUTION, SYNTHESIS METHODS

5.75 In addition to developing an assessment of criticality and contributing parameters, the reactor-core designer has a major interest in the power distribution. The thermal design of the core is very dependent on the ratio of the peak to average power. A two-dimensional "map" of the power distribution can easily be developed from the fission-source output options available in the computer codes previously described. A reactor core is three dimensional, however, not merely a flat plane. Methods to develop a full picture of the power distribution throughout the core are therefore needed which consider the effects of varying control-rod insertions, boiling voids, etc., in the axial direction. Although three-dimensional diffusion-theory codes have been developed, their use for design parametric studies is normally impractical since the computing-time expense is excessive. For many preliminary calculations, it is satisfactory to use two-dimensional codes with space variations in the radial and axial directions of a cylinder, assuming azimuthal symmetry. On the other hand, the detailed representation possible with a radial traverse is frequently important. As a result *synthesis methods* that combine a number of plane calculations into a three-dimensional pattern have been developed.[45] The term is also sometimes applied to comparatively simple methods used to synthesize two- and three-dimensional flux distributions from one-dimensional calculations.

5.76 Synthesis methods vary in complexity. A simple approach, based on the transverse-buckling concept (§5.66), uses the buckling terms to express y and z leakage in the one-dimensional multigroup diffusion equations,

$$D_j \frac{d^2 \phi_j}{dx^2} - [D_j(B_{yj}^2 + B_{zj}^2) + \Sigma_{aj}]\,\phi_j + S_j = 0 \tag{5.19}$$

where an equivalent transverse loss, $[D_j(B_{yj}^2 + B_{zj}^2]\,\phi_j$, has been added to the absorption. Initially, if a radial problem is considered, the transverse buckling can be estimated from the axial fuel and reflector dimensions. Next, axial diffusion can be calculated with the previous results as input. Iteration can then continue until convergence is obtained. A cylindrical-core power distribution may be obtained with the *buckling iteration method*, for example, by using PDQ in the horizontal plane and a one-dimensional code, such as WANDA, for the axial calculations.

5.77 In a very similar technique, *the single-channel synthesis method*, the reactor is divided into a small number of horizontal layers, and a radial calculation is applied to each layer. The layers are then connected axially by a one-dimensional calculation in which each layer is homogenized. The dividing planes are normally chosen so that the pattern of inserted control rods is uniform throughout the height of each layer. Actually, the calculation details are more complicated than implied by this summary since considerable attention must be given to the proper averaging of regional properties. Flux separability of axial and radial components is a basic assumption of the method. The consequent assumption that leakages are constant within a region, causing flux discontinuities at interfaces, limits the method to some extent.

5.78 An extension of the single-channel approach makes use of several simultaneous axial calculations to couple the radial planes together. A challenge in the *multichannel synthesis method* is to relate the various transverse leakages to one another. Various methods using special computer programs for so-called coupling coefficients have been proposed.

5.79 Variational principles are the basis of a third synthesis method. The radial-flux distribution is assumed to be dependent on the axial flux and, indeed, to vary continuously with the axial position. It is necessary initially to make reasonably good guesses of trial radial-flux distributions and then to use variational methods for optimizing the mixing of such distributions. In *nodal methods* (§5.122), the core is divided into many large volumes, or nodes, in which materials and flux do not vary. Flux and power distribution are calculated by matching leakage from one node to another.

DEPLETION CALCULATIONS

INTRODUCTION

5.80 Depletion calculations* provide a picture of the changes that take place in core composition, power distribution, and reactivity as a function of time. Calculations for reactor dynamics (kinetics), as needed for control and safety analysis, apply over a much shorter time interval during which the composition does not change significantly. Such calculations are considered in

*Some of the material in this section was contributed by R. L. Stover.

§5.131 and in Chap. 6. The various causes and effects in depletion calculations are interrelated since the variation of fissile-atom and fission-product concentration with time and position depends on the time-integrated power, which, in turn, depends upon the local enrichment. To the core designer the lifetime behavior is extremely important to know since he must ensure that the various thermal-design limitations are not exceeded as the fuel burns and is partially replaced according to some fuel-management scheme (§7.53), with accompanying shifts in control rods and changes in the power-density pattern. Also necessary is sufficient excess reactivity to achieve the desired fuel exposure but remain within safety criteria. Isotopic changes in the fuel must be accounted for to optimize the exposure and plan the fuel management primarily from economic considerations. Fuel-depletion calculations are also important in assessing the isotopic content to determine the value of discharged fuel as needed for corporate accounting purposes.

5.81 Depletion information is also needed as an aid to operation. The operator, faced with such decisions as those involving the effects of control-rod placement on power distribution, a changing reactivity account, burnable-poison changes, or power-coefficient changes, needs a model to help him. Such a model may also be helpful in the development of adaptive control methods where a computer is used for on-line control.

5.82 Once a reactor design becomes final and construction begins, most of the design variables become fixed. Much flexibility can remain in the design of the fuel, however, which will need to be replaced throughout the lifetime of the plant. Over a 30-year period, the investment in fuel for a 1000-Mw(e) reactor may be of the order of $300 million. Optimum design of the initial and replacement fuel, which depends largely on accurate depletion calculations, therefore offers considerable economic incentive.

5.83 The need to follow a time-dependent process involving reaction rates that depend on the product of fluxes and isotopic compositions makes the nonlinear, noneigenvalue depletion calculation inherently complex with many approximations needed for methods to be practical. Furthermore, burnup behavior does not lend itself to simulation by critical experiments with the corresponding opportunity for comparing predictions and measurements as needed to develop semiempirical techniques. Therefore analysis must be used to obtain the desired description of reactor performance.

CALCULATION METHODS

Introduction

5.84 Practical depletion calculations must be carried out with a computer.[46] An introduction to the general nature of such calculations can best be obtained by considering the steps usually needed, as shown in the flow sheet of Fig. 5.8. Initial macroscopic cross sections and other constants can be prepared

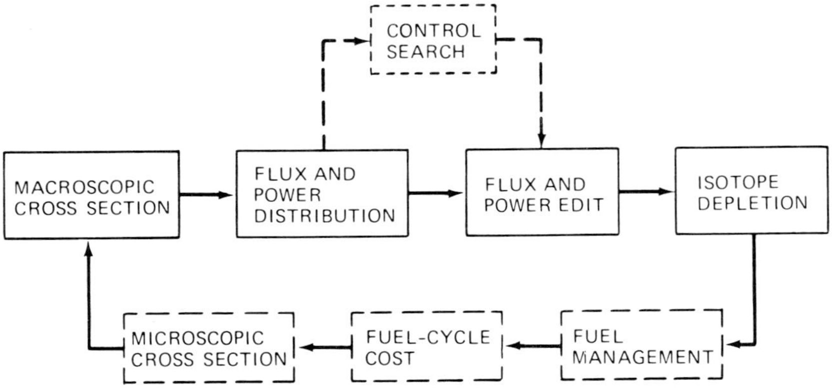

Fig. 5.8 Basic depletion cycle.

in a manner similar to those previously described (§5.34). A number of methods, varying in complexity from a one-group zero-dimensional approach to a three-dimensional synthesis method, can be used to determine flux and power distribution. A three-dimensional picture is normally needed to meet engineering design requirements, however.

5.85 Before isotopic-depletion changes are actually calculated, the flux and power calculations are edited in light of the mesh pattern to obtain input values appropriate for desired core volumes. The heart of the overall procedure is the calculation of time-dependent number densities.* Either a constant flux or a constant power is used to solve the differential equations over a finite-time step (§5.90). Finally, the cycle can be repeated for additional time steps, as desired.

5.86 In addition to the basic procedural steps, several optional calculations are shown in Fig. 5.8. If the control-search option is used, an iterative calculation determines an isotope density or buckling so that the core reactivity is some specified value. Fuel-management schemes can be included by allowing isotope concentration to be changed after the depletion calculations. If the initial and final isotope densities for the cycle are known, the fuel cost can also be calculated. Since spectrum-averaged cross sections change as the spectrum changes, new microscopic cross sections can be calculated at the end of the cycle or even during the cycle if the spectrum changes too rapidly within the cycle time.† Most depletion codes, however, assume constant cross sections over all of the depletion cycles.

---

*Densities of individual isotopic nuclei.

†LEOPARD and LASER recompute spectra and group cross sections at the end of each depletion step.

Cross Sections

5.87 Microscopic cross sections may be dependent on isotope concentrations, either because the cross sections are spectrum averaged over a given energy group or because they are also flux weighted as a function of spatial location in the reactor. Both the neutron spectrum and the spatial-flux distribution will change as isotope depletion takes place. An example[47] of typical thermal cross-section changes with burnup is shown in Fig. 5.9.

5.88 It is useful to sort out the cross sections used and classify them as either slowly varying or rapidly varying with depletion. For slowly varying cross sections, recalculation during an analysis can be kept to a minimum. Scattering and transport cross sections are frequently of this type. Sometimes strong energy dependence may require the use of many groups for the analysis. With a large

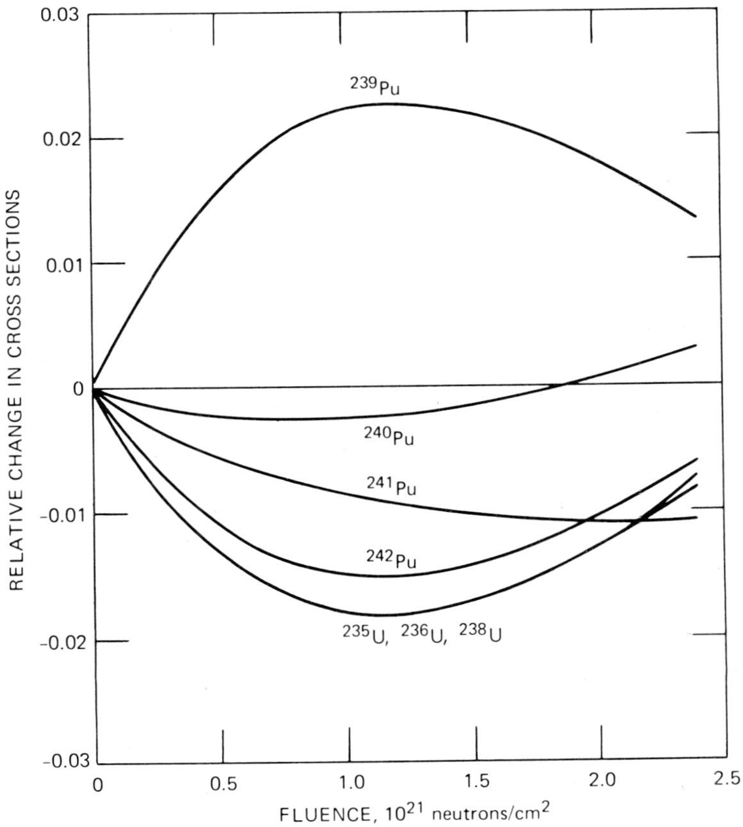

Fig. 5.9 Relative change in the thermal microscopic absorption cross sections of fuel nuclides with burnup.

number of groups, however, the spectral dependence of the *average* cross section in each group tends to be small.

5.89 Rapidly varying cross sections arise for several reasons, including strong depletion dependence of self-shielding factors or strong dependence of resonance absorbers on neutron spectrum and number density. With strong dependency it is necessary to account for the variation during each computer time interval or to employ short time intervals and make appropriate adjustments. When lumped poisons are treated by using self-shielding factors expanded into polynomials that are functions of the poison concentration, the expression can be incorporated into the depletion calculation. Similarly, concentration-dependent polynomials can be used to determine new cross sections for heavy resonance absorbers.

Time-Step Calculation

5.90 Depletion equations describe the isotopic-concentration changes with time at a point or a region in a reactor. The point or region is a unit of volume which either is homogeneous in composition or has been properly homogenized by volume and flux weighting. Fuel-burnup or depletion equations at a point have the general form

$$\frac{dN_A}{dt} = -N_A \left(\lambda_A + \sum_i \sigma_{a_i}\bar{\phi}_i\right) + N_B\lambda_B + N_c \left(\sum_i \sigma_{c_i}\bar{\phi}_i\right) \qquad (5.20)$$

where N_A = number density (nuclei/cm^3) of the isotope of interest
B = precursor of isotope by decay
c = precursor of isotope by neutron capture
λ_A, λ_B = decay constants
σ_{a_i} = microscopic total absorption cross section for energy group i
σ_{c_i} = capture cross section for energy group i
ϕ_i = average, power-normalized flux for energy group i

N, ϕ, and σ are all functions of time, but λ, the decay constant, is invariant with time. Since the equations are nonlinear in time, the usual method of solving them is to linearize the equations by assuming that flux and cross sections are constant over a short time interval. Using the *constant-flux assumption* (§5.92), the resulting linear equations can be solved exactly by analytical solutions or by approximate numerical methods using finite-difference equations.

5.91 The depletion equation for a single *fuel* isotope can also be expressed in general form as

$$\frac{dN_F(r,t)}{dt} = -N_F(r,t) \int_0^\infty \phi(r,E,t)\, \sigma_a(E)\, dE \qquad (5.21)$$

where N_F = fuel-isotope number density
σ_a = fuel-absorption cross section
r = position dependency
E = energy dependency

If the *constant-power assumption* (§5.94) is used, the complete right side of Eq. 5.21 is independent of time, and the equation can be integrated directly over a time interval. In this case the exact and "approximate" solutions are identical. In summary, coupled depletion equations can be solved exactly or approximately for several fissile isotopes by using the constant-flux assumption or can be solved exactly for a single fissile isotope by using the constant-power assumption. Since these approaches are important but are described in only a few sources, they are discussed here (§§5.92 to 5.97). (These sections can be omitted without loss of continuity.)

Constant-Flux Assumption

5.92 If the flux spectrum is assumed invariant over a time interval, then

$$\phi(r,E,t) = \phi(r,t)\ \psi(E) \tag{5.22}$$

where $\psi(E)$ is normalized so that

$$\int_0^\infty \psi(E)\, dE = 1 \tag{5.23}$$

If a few-group energy model is used, the integral can be replaced by a summation, and Eq. 5.21 becomes

$$\frac{dN_F(r,t)}{dt} = -N_F(r,t) \sum_{i=1}^{G} \phi_i(r,t)\ \sigma_{a_i} \tag{5.24}$$

where G = number of energy groups

$$\phi_i(r,t) = \phi(r,t) \int_{i\text{th group}} \psi(E)\, dE \tag{5.25}$$

$$\sigma_i = \frac{\int_{i\text{th group}} \psi(E)\ \sigma_i(E)\, dE}{\int_{i\text{th group}} \psi(E)\, dE} \tag{5.26}$$

Now, if the flux is assumed constant in time, Eq. 5.24 can be written

$$\frac{dN_F(r,t)}{N_F(r,t)} = -\sum_{i=1}^{G} \phi_i(r)\,\sigma_{a_i}\,dt \tag{5.27}$$

which has the exact analytical solution

$$N_F(r,t+\Delta t) = N_F(r,t)\,e^{-\alpha\Delta t} \tag{5.28}$$

where

$$\alpha = -\sum_{i=1}^{G} \phi_i(r)\,\sigma_{a_i} \tag{5.29}$$

5.93 The approximate numerical solution is found by expanding the exponent in a Taylor series and dropping all high-order terms. The finite difference equation is then

$$N_F(r,t+\Delta t) = N_F(r,t)\left[1 - \sum_{i=1}^{G} \phi_i(r)\,\sigma_{a_i}\,\Delta t\right] = N_F(r,t) - \beta\,\Delta t \tag{5.30}$$

where

$$\beta = N_F(r,t)\sum_{i=1}^{G} \phi_i(r)\,\sigma_{a_i} \tag{5.31}$$

It can be seen from Eqs. 5.23 and 5.24 that, if the flux is assumed constant over the same time interval in both cases, the finite-difference solution requires smaller time steps than the exact solution.

Constant-Power Assumption

5.94 If the power is assumed constant over a time step, Eq. 5.24 can be written

$$\frac{dN_F(r,t)}{dt} = -\beta \tag{5.32}$$

which has the solution

$$N_F(r,t+\Delta t) = N_F(r,t) - \beta\,\Delta t \tag{5.33}$$

From the definition of β, the solution for constant power can be written

$$N_F(r,t+\Delta t) = N_F(r,t)\left[1 - \sum_{i=1}^{G} \phi_i(r)\,\sigma_{a_i}\,\Delta t\right] \qquad (5.34)$$

which is the same as the finite-difference solution at constant flux. If a first-order difference solution of a *single* fuel-isotope equation is used with the assumption of the constant flux, the solution therefore has the same form as that obtained with a constant-power assumption. Indeed, as the time interval decreases, the constant-flux and constant-power solutions approach each other.

Comparison of Constant-Flux and -Power Assumptions

5.95 The constant-power assumption has the advantage of being consistent with normal reactor operating practice, i.e., constant power, with large time intervals permissible in the calculation. Another advantage is that the exact energy extracted is obtained from the fraction of fuel depleted. The main disadvantage of using the constant-power assumption is that the equations do not apply for coupled fuel chains with more than one fissionable isotope since the term β (Eq. 5.31) is composed of all power-producing nuclides. Another disadvantage is the difficulty in depleting nonfuel nuclides.

5.96 The advantage of the constant-flux solution is that poison depletion and coupled fuel chains are easily handled. Variations of cross sections during a burnup interval can also be accounted for. The disadvantage of power variation during a time interval can be minimized by performing the depletion calculation in small steps and renormalizing the flux amplitude (but not the shape) at each small step.

5.97 If the solutions of the exact-constant-flux, approximate-constant-flux, and constant-power calculations are compared for a single fuel isotope over a small-enough time interval, the solution paths would take the shapes shown in Fig. 5.10. It is stressed that the approximate-constant-flux solution and the constant-power solution are identical only for a single fissile isotope. In a reactor with more than one fissile isotope, the constant-flux solutions must be used since the absorption rates of the individual isotopes (β_i) change with time, although the total absorption rate, $\beta = \sum \beta_i$, remains constant.

SPATIAL SUBDIVISION

5.98 The depletion equations must be applied to some core-volume unit. Since the power and material composition are assumed constant over an interval of time throughout the volume unit, a high degree of subdivision is desirable for an adequate description of the flux pattern. On the other hand, the finite difference in time solution is applied to each mesh point, and thus computer

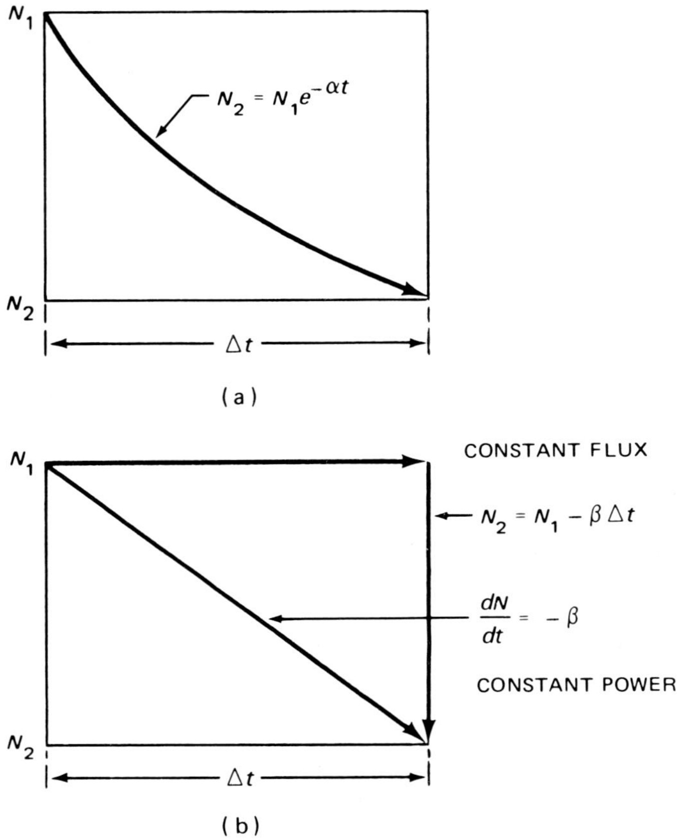

Fig. 5.10 Comparison of depletion-calculation incremental-time-step behavior. (a) Exact solution to constant-flux calculation. (b) One-step constant-flux calculation and exact solution to constant-power calculation.

time can quickly become excessive as many mesh points are accommodated. As a result a compromise is needed.

5.99 Two spatial subdivision approaches are commonly used. In one the depletion equations are solved for each spatial mesh element used in developing the flux- and power-density representation. Many calculations are required for this, the so-called *point-subdivision method,* although a continuous pattern of concentration variations is obtained. In the second method, called *regionwise depletion,* the composition is assumed constant over a group of mesh elements. Since the number of compositions to be considered is now less than the number of mesh points, spectra and appropriate cross sections can be calculated for each composition. This method produces composition and power discontinuities at region interfaces, however, since the power density averaged over the region must be used in the stepwise depletion calculation. Therefore region size should be adjusted to give small power variation, perhaps less than 5%. Large regions can be used, however, where the power is flat or low. The varying state of subdivision in the core is therefore determined by the balance between accuracy desired and computing expense justified.

DEPLETION CALCULATION CODES

5.100 The depletion codes[48] that have been developed vary considerably in complexity because of the possibility of having either zero-, one-, two-, or three-dimensional representation and a single, few-group, or multigroup energy treatment. Generally, the programs search for a multiplication constant of unity during a given time step with control-material neutron absorption a variable. The flux distribution is then calculated, the fuel depleted during the time step, and the cycle repeated. Zero- or one-dimensional codes with four or fewer neutron energy groups are used most often. Such codes are normally used for determining isotopic content of a batch of discharge fuel, and the more elaborate two- and three-dimensional codes are used to check the power distribution as well as such operational features as shim position, control-rod movement, and fuel management. Although it is generally desirable to represent the core in as much detail as possible consistent with the computing expense appropriate, some simplifications are helpful in obtaining the desired representation. Cylindricization of the core, for example, permits one- or two-dimensional codes to be used for many depletion calculations for which fuel assemblies can be grouped into "batches" represented for this purpose by annular rings. Although some of the features of multidimensional and multigroup eigenvalue calculations have been treated previously (§5.70), the following discussion also considers the complications introduced by the time-step calculation. Some readers whose interests do not extend to examples of specific code approaches may wish to omit the discussion.

Zero-Dimensional Codes

5.101 Zero-dimensional calculations can be used to generate curves of isotopic concentration as a function of fuel exposure (Mwd/tonne) for the average core neutron spectra. Although such codes have many limitations, they are comparatively simple and hence inexpensive to run. Since the reactor is considered as a single point, however, the spatial effects of burnup are neglected. Furthermore, the point model is unable to treat the neutron leakage out of the reactor core accurately since the leakage strongly depends on the existing flux shape. Neutron-energy variations can be accounted for, of course, by using a multigroup approach, and, if needed, spectrum-variation effects can be introduced by recalculating the group constants after each time step.

5.102 The zero-dimensional codes follow the basic depletion calculation described. Since all materials in the core, including the control rods, are homogeneously mixed, self-shielding factors must be used to account for inhomogeneities that actually exist in the core. Core leakage is accounted for by using a value for buckling which is kept constant during the calculations or varied to adjust for criticality.

5.103 Most zero-dimensional codes provide a criticality search for either the control-poison concentration or the neutron leakage. One or the other is adjusted within each time step until the multiplication factor attains a predetermined value. This is usually done before the depletion portion of the time step. Isotopic compositions are determined at the end of a time step during which the flux is assumed constant. Normally, uranium and plutonium chains are programmed into the code. Burnable poisons can usually be handled by methods similar to those used for the fuel chains.

5.104 In fast reactor calculations fission products can be lumped into an aggregate, and little error will result. For thermal calculations, however, the high-cross-section fission products must be treated separately. The two most important nonsaturating fission products, ^{135}Xe and ^{149}Sm, are always treated separately. The saturating fission products whose concentrations do not build up in proportion to the production rate are usually grouped into a saturating fission-product aggregate.

5.105 Many of the zero-dimensional codes can follow burnup in each of several fuel regions simultaneously.[49] The flux levels are the same in each region, but the isotopic composition can be changed after each time step to simulate refueling. Fuel cycles and fuel-cycle costs can be studied in codes having the subroutine to perform this calculation.

5.106 Interesting examples of zero-dimensional codes[49,50] for depletion are GAD and LEOPARD.[50] In GAD, a multigroup code, up to eight fuel zones can be treated, but the flux must be uniform. Irradiated fuel can be removed and fresh fuel reloaded in one or more zones as often as desired. A provision for fuel-cost calculation is included. LEOPARD recomputes the spectra after each burnup step and adjusts the constants.

One-Dimensional Codes

5.107 One-dimensional codes are best suited for investigating fuel depletion in zonal batches of fuel assemblies. For this purpose the reactor core is cylindrical and the computations are carried out along the radial dimension. Axial variations in the power distribution are neglected, but the code input must include a transverse buckling (§5.67) to account for axial leakage. The cylindrical core may consist of concentric circular regions, each being appropriately homogenized. Although an individual fuel assembly cannot be accounted for in one-dimensional codes, the zones can be subdivided to the width of one fuel assembly so that all fuel assemblies with the same radial location are in one annular ring.

5.108 In one-dimensional codes the isotopic concentrations used in each region can also be axially averaged. This is particularly important for boiling-water reactors where the water density varies axially because of steam voids. Homogenized poison concentrations in each zone are used to represent control-rod insertion in that region. Alternatively, the control rods can be "banked" into individual annular control rings, where the poison concentration can be varied to represent rod movement in the axial direction.

5.109 Also, with a one-dimensional code axial calculations in a cylindrical core are possible. With the slab-geometry option, the calculation can proceed with radially averaged isotopic compositions and a transverse buckling to represent radial leakage.

5.110 Radial variations in the neutron spectrum are considered by using different few-group constants in the different regions. One code,[51] CNCR, also provides an option to calculate changes in the neutron density due to variations in the coolant temperature. The neutron density changes with the coolant density, which, in turn, depends on the axial coolant-temperature rise. If the variation in coolant density with temperature rise is fed into the computer, an iteration loop results, since the coolant-temperature rise depends on the neutron density.

5.111 Several criticality-control searches are available in one-dimensional codes. For instance, the poison-concentration search can be varied from zone to zone in a predescribed manner so as to represent preferential movement of control rods. In a radial calculation where axial leakage is approximated by a transverse buckling, the critical core height can also be found from a transverse buckling search. In an axial calculation a poison-boundary search that represents the degree of control-rod insertion is possible.

5.112 CANDLE is an example[31] of a classic one-dimensional few-group diffusion-theory depletion code that will calculate flux shape and deplete 13 isotopes at up to 500 mesh points in 50 regions. A maximum of four energy groups can be used; a library of fast cross sections is available in the code, but the thermal cross sections must be input. Thermal self-shielding factors and region-dependent resonance escape probabilities for ^{238}U must also be supplied as input.

5.113 The isotope-depletion equations are solved by the finite-difference method at each mesh point. Provision is not made, however, for ^{232}Th, ^{233}U, ^{234}U, and ^{242}Pu. Xenon and samarium are treated separately, and the rest of the fission products are grouped into two fission-product aggregates. Three types of control searches are available in CANDLE. Criticality can be maintained by varying a poison concentration, transverse buckling, or the interface between a poisoned and a nonpoisoned region.

5.114 FEVER is a second example[52] of a one-dimensional few-group diffusion-theory depletion code. A maximum of 25 isotopes can be depleted at up to 150 mesh points in 20 regions. A maximum of four energy groups can be used, for which all cross sections must be provided as input. Slowing down is allowed only to the next adjacent group, and the fission source may be only in the highest group.

5.115 The isotope-depletion equations are solved by the finite-difference method for the ^{232}Th and ^{238}U chains. The equations are solved by region instead of at mesh points. One nonsaturating fission-product aggregate and two rapidly saturating fission products (^{135}Xe and ^{149}Sm) are included. Self-shielded burnable poisons can be depleted by using poison density-dependent shielding factors. Besides a poison-search option, the code includes a xenon override calculation, a shutdown–start-up option, and a cold shutdown calculation of k_{eff}. The running time is ~½ min per time step (IBM 7094 basis).

5.116 SIZZLE is a one-dimensional diffusion code[53] for calculating burnup in intermediate or fast reactor cores. A maximum of 18 energy groups can be used to calculate isotope densities at a maximum of 100 points in up to 20 regions. The diffusion calculations are based on the AIM-6 code in which down-scattering may occur to any of five energy groups. A built-in microscopic cross-section library is available in the code.

5.117 Both the ^{232}Th and the ^{238}U fuel chains are included in the code. These isotope equations are solved exactly by the analytical solution. In the original version of the code, all fission products were lumped together and depleted. A modified version allows individual depletion of xenon and samarium. The only criticality search permitted is a poison-concentration search. Running time is ~½ min per depletion step (IBM 7094 basis).

Two-Dimensional Codes

5.118 If the neutron flux is assumed to be separable into an axial and a radial component, a two-dimensional code can calculate the power distribution and isotope concentrations as a function of time in either x–y, r–z or r–θ geometry. For x–y calculations in a radial plane, the isotope concentrations are axially averaged, and this introduces again the same uncertainties in water density as those introduced by the one-dimensional codes. However, the x–y geometry allows representation of individual control rods and fuel assemblies, which is important for refueling studies, especially scatter reloading (§7.69). If isotope composition in zonal batches of fuel asssemblies is of interest, the r–z geometry is appropriate. Although the r–θ geometry gives a three-dimensional composition mapping, initial isotope concentrations in each r–θ zone are difficult to calculate. Regardless of the geometry two-dimensional depletion codes are expensive to use because they take ~⅓ hr for each time step. Since the neutron-flux calculations take up most of this time running time can be decreased by computing a fraction of the core and using symmetry to map the remainder of the core. Also, relatively large time steps are usually chosen to reduce expense. Criticality searches are quite limited in two-dimensional codes owing to the expensive neutron-flux calculations. However, region-dependent poison searches are normally available.

5.119 TURBO is an example[51] of a two-dimensional code which, in a sense, represents a combination of the PDQ-type diffusion calculation (§5.73) and the CANDLE-type (§5.112) depletion approach. Input includes the dimensions and compositions, on a rectangular-region basis, of the core and reflector radial cross section. After the diffusion calculation, flux determination, and eigenvalue solution, changes in fuel and poison composition are determined over the time step for each mesh rectangle. Owing to the complexity of the calculation, however, the program differs from CANDLE in that only one time step at a time is carried out automatically. In this way adjustments can be made of transverse buckling, power level, etc. A criticality search is not provided.

Three-Dimensional Calculation Methods

5.120 The most accurate method of determining isotope concentrations and control-rod configurations as a function of time is the three-dimensional depletion code. But a few-group code with 50,000 mesh points takes about one day of computer time for just one flux calculation. When time dependence for fuel depletion is added, the running time becomes almost prohibitive. Two different code types have therefore been developed for reducing computer time. One type,[54] FLARE, simply uses fewer mesh points and only one-group diffusion theory. Since the representation is crude, experimental measurements are used to fix adjustable parameters. The other type, e.g., the TURBO-ZIP system,[55] retains the large number of mesh points and several energy groups by combining one- and two-dimensional calculations into a synthesized solution. Some versions of PDQ also allow restricted three-dimensional solutions.

5.121 A synthesis code provides an approximate three-dimensional picture by combining solutions of the one- and two-dimensional problems. If separability of the flux into two components is assumed, the solutions are combined by multiplying the axial-flux variation by the two-dimensional flux shapes (TURBO-ZIP System). In another method[56] of synthesis (TNT02), the axial-flux solution is replaced by an axially dependent mixing function $F(z)$. The three-dimensional flux is then represented by a linear combination of the x–y flux components of each value of z:

$$\phi(x,y,z) = \sum_{n=1}^{N} F_n(z)\, \phi_n(x,y) \tag{5.35}$$

Synthesis techniques have several disadvantages (§5.75). Discontinuities develop at the interfaces between axial zones; moreover, separability of the flux is not a good assumption for highly heterogeneous cores. Also, in the second synthesis method, the functions $\phi_n(x,y)$ are difficult to develop.

5.122 FLARE permits a three-dimensional representation by providing a maximum array of 14 by 14 by 12 nodes. The core is therefore normally divided into up to 12 axial zones with symmetry used to permit a one-half or one-quarter core radial representation. The calculation consists of a source iteration and a fuel-burnup iteration. Actually, there is no depletion calculation. Only fuel exposure is calculated for each node with an independent calculation necessary to determine isotopic compositions.

5.123 A modified diffusion-theory model is used which involves only two parameters, k_∞ and M^2 (migration area), which are inputted as analytical fits to a fuel-exposure function based on experimental data or a more precise calculation. The calculation is further simplified by using an albedo to represent the reflector.

5.124 FLARE, developed for boiling-water reactors, includes an iteration calculation for finding the coolant-void fraction, as well as power and fuel-exposure iteration. The fuel-exposure iteration can be used to simulate control-rod programming specified in the FLARE input for each step. Typical running time for a converged-power and void solution is ~2 min.

5.125 The TURBO-ZIP system is a three-dimensional synthesis depletion program that combines the two-dimensional TURBO code and the one-dimensional ZIP code. This method is based on the assumption that the solution to the three-dimensional diffusion can be represented by the product of the two-dimensional x–y flux from TURBO and a one-dimensional z-flux function from ZIP. Several radial zones of a core are depleted over the expected core life by TURBO. Each of these radial zones corresponds to an axial point in the ZIP-calculation. For each axial point, flux-averaged diffusion coefficients, macroscopic cross sections, bucklings, and isotope densities are known from the TURBO output data. These data are input to ZIP either in tabular form or as a least-squares fit to a polynomial with fuel fraction as the independent variable.

5.126 The diffusion-equation coefficients at each axial point known, ZIP is used to calculate the flux and power distribution in slab geometry. A depletion calculation then determines the fraction of fuel remaining at each axial point. The specific isotopic compositions can then be found from the TURBO data known as functions of fuel fraction. New diffusion coefficients are then calculated, and the calculation proceeds to the next depletion step. After all depletion steps are finished, the flux, power, and fuel fractions are known for each axial point at each time step. From these data and the TURBO data, the complete three-dimensional time-dependent map of flux, power, and isotope concentrations can be found.

5.127 A depletion code[29] of special interest, particularly for water reactors, is LASER. Fuel burnup in a single heterogeneous lattice cell is obtained by transport theory for a one-dimensional representation. MUFT (§5.49) and THERMOS (§5.53) are included to develop constants as burnup proceeds. The model therefore accounts for depletion spatial and energy effects on a lattice *microscopic* scale (lattice cell).

5.128 In using depletion codes, the core designer is interested not only in reactivity and concentration from a given set of input specifications but also in how changes in the input design parameters affect the results. Such changes are necessary as part of the design iteration process (§1.27) in the search for the most satisfactory specifications. One approach is to vary the input specifications systematically and develop a set of parametric plots. The many parameters and the relative complexity of depletion calculations make this approach an expensive one. Another approach is to use *perturbation theory*[57] which is useful in criticality calculations to determine the sensitivity of the multiplication factor to input parameters in terms of the flux calculation and its adjoint. Application to depletion calculations is difficult, however, since the equations are nonlinear and do not represent an eigenvalue problem. ARIADNE-I, a burnup perturbation code[58] based on a two-group point model, represents an initial attempt to develop a useful design approach.

APPLICATIONS

5.129 Although applications have been indicated in the preceding discussion, a few additional comments are in order for light-water-cooled and -moderated cores. Studied extensively, these cores have had operating experience, and a design practice has therefore evolved. The approach for pressurized-water reactors uses a multigroup slowing-down treatment, such as MUFT; the development of group cross sections describing resonance capture and the fast effect by lattice-theory calculations; and a lattice thermalization method, such as SOFOCATE or THERMOS. Four groups, as collapsed from the multigroup representation, are usually adequate for the spatial calculation. Since flux variation is smooth, a coarse mesh is satisfactory. Similarly, large time steps can be used in the depletion calculation.

5.130 In boiling-water reactors local reactivity and flux change markedly at the axial level where boiling begins. As a result hydraulic parameters affecting the boiling point contribute to the criticality determination. A simplified neutronic model is therefore useful to determine the spectrum on a local basis as a function of water density. Neutron reaction rates can then be determined from cell calculations that can be performed at different water densities. The power shape is then determined, followed by a point-by-point depletion calculation.

FLARE (§5.122), a one-group three-dimensional program, gives good results for the representation.

DYNAMIC-ANALYSIS METHODS

INTRODUCTION

5.131 A major requirement of the nuclear design is to carry out sufficient analysis to ensure satisfactory dynamic behavior. Kinetics design calculations can be considered in three general categories. (1) Adequate control and reactivity margins must be provided for in accordance with the fuel-system design. (2) Reactor stability parameters, such as temperature- and power-feedback coefficients, must be determined. (3) Accident mechanisms and behavior must be analyzed to guide in establishing safety criteria. Although these areas are related to one another, the objectives from the viewpoint of the designer are quite different.

5.132 The nuclear design must provide sufficient excess reactivity in the fuel initially loaded in the core to compensate for the loss during operation as the fissile atoms are depleted and fission products are formed between fuel loadings. An additional margin must be provided, of course, for the reactivity changes associated with start-up and day-to-day operational requirements and the possibility of xenon buildup. Control rods, burnable poisons, and other schemes are in turn necessary to compensate for this built-in excess reactivity as well as to provide shutdown capability in a wide range of circumstances. Such considerations are considered in Chap. 6 (§6.37 et seq.).

5.133 Comparatively long term control and reactivity requirements of this type are included as part of the steady-state and depletion core calculation. The kinetics equations are not involved. Problems introduced in the analysis, however, can be formidable, particularly those in the heterogeneous cores of thermal reactors caused by strong absorbers and axial variations arising from control-rod movement. The need for fuel-management schemes (§7.53) wherein fuel of different enrichments may be scattered in the core further complicates the analysis.

5.134 A knowledge of the transient or "kinetic" behavior of the complete reactor system is an important part of the evaluation of a design in terms of safety criteria (§6.95). An analysis of such a system involves much more than a description of the nuclear behavior of the core, however, since heat transfer, fluid flow, and other aspects of the plant must be considered in the various reactivity-feedback paths that determine whether or not the system is stable.[59] Since stability analysis is treated at length in texts on control theory and kinetics,[59–61] this complicated subject is not discussed here. Reactor kinetics itself is a challenging discipline that cannot appropriately be treated here.

Pertinent to the preceding discussion of nuclear design calculations, however, are approaches used to determine kinetics parameters. This effort is only a small part of what might be done in the stability analysis.

REACTIVITY-CHANGE CALCULATIONS

5.135 The most familiar reactivity coefficient, or combination of coefficients, is based on temperature changes in the core. As explained in detail elsewhere,[62] reactivity changes can result from a combination of mechanisms, such as a change in the mean energy of neutrons in a thermal reactor, changes in density of materials, and changes in core size. Appropriate coefficients have been expressed as derivatives in terms of various reactor parameters after various algebraic manipulations to develop the desired functional dependencies. The temperature pattern in a core is far from uniform, however, since a gradient all the way from the center of a fuel pin to the coolant is required to accomplish the desired energy removal. Indeed, the gradients depend not only on the power generated but also on the rate of a "transient" change. Reactivity coefficients based on very crude models that neglect these effects therefore tend to be very approximate.

5.136 The *change in reactivity** caused by change in reactor power and a corresponding change in temperature pattern can be found by solving the eigenvalue problem for each condition. The multigroup diffusion-theory approaches previously described, for example, can be used.

5.137 *Perturbation theory,* another approach useful in determining reactivity coefficients, allows small changes of a system to be studied by using the characteristics of the original system. Studying such small effects by proceeding through a multigroup eigenvalue calculation is difficult since the effect of the change can be lost in the calculations owing to various approximations involved. Localized changes, not easily treated by other methods, can be treated by perturbation theory. Here, however, we alert the reader to the existence of perturbation-theory methods rather than describe them, since the mathematical background is sophisticated and space consuming.

5.138 The foregoing methods relate changes in reactivity (k_{eff}) to temperature or power changes in the core. Although such reactivity coefficients are important in the stability analysis of the reactor, they have limited usefulness in predicting kinetic behavior unless associated with *rate-of-change* information in a time-dependent problem. A solution of the reactor kinetics equation is the first step in providing such information. The normal approach is to use a one-point reactor model with only one neutron energy group. The familiar[63]

---

*In this case reactivity refers to k_{eff} in a steady-state system, not to ρ as discussed in §5.138.

coupled kinetics equations that serve as the basis for various computational procedures are as follows:

$$\frac{dn}{dt} = [k_{\text{eff}}(1-\beta) - 1]\frac{n}{l} + \sum_{i=1}^{I} \lambda_i C_i$$

$$\frac{dC_i}{dt} = \frac{k_{\text{eff}}\beta_i n}{l} - \lambda_i C_i \tag{5.36}$$

and

$$\frac{dn}{dt} = (\rho' - 1)\frac{\beta n}{l} + \sum_{i=1}^{I} \lambda_i C_i$$

$$\frac{dC_i}{dt} = \beta_i \frac{n}{l} - \lambda_i C_i \tag{5.37}$$

A frequently used modification of the first system is

$$\frac{dn}{dt} = \frac{k_{\text{ex}} n}{l} - \sum_{i=1}^{I} \frac{dC_i}{dt}$$

$$\frac{dC_i}{dt} = (1 + k_{\text{ex}})\beta_i \frac{n}{l} - \lambda_i C_i \tag{5.38}$$

where n = neutron density
k_{eff} = effective multiplication constant
$k_{\text{ex}} = k_{\text{eff}} - 1$
β = delayed neutron fraction
a_i = relative yield of the ith delayed neutron group
$\beta_i = \beta a_i$, delayed neutron fraction for ith delayed neutron group
λ_i = decay constant of the ith delayed neutron group
C_i = density of precursors for the ith delayed neutron group
I = total number of delayed neutron groups
l = prompt-neutron lifetime
$\rho = k_{\text{ex}}/k_{\text{eff}}$ = reactivity [$\rho' = k_{\text{ex}}/(k_{\text{eff}}\beta)$, reactivity in dollars]
t = time

A zero subscript to any of these quantities indicates an initial value. External source terms are often included in the neutron-density equation.

5.139 A classical computer code[64] for solving these equations is AIREK-3, which allows the reactivity to be described as a function of time and in terms of a feedback function. Outputs from the code include the neutron density, the inverse period, and various feedback variables, all as a function of time. The numerical methods for this type of computation are somewhat challenging. One is based on transforming the kinetic differential equations into integral equations and then applying numerical approximations to the integrals. With this method, called *collocation,*[65] it is assumed that the time variation of reactivity and neutron density can be matched by quadratic functions in a short time interval. Time steps as large as 1 sec can be used for many problems, and an analysis can thus be made with a moderate expenditure of computing time.

5.140 It is desirable to incorporate the neutron-behavior analysis provided by AIREK into a more general systems-analysis study as required for hazards analysis and design. The feedback functions provide such a possibility. In AIROS, feedback equations are available to represent multichannel heat transfer, coolant flow, and other parameters.[66] Although a point model [illegible]ed for the neutron behavior, spatial flexibility is possible in the representation of the heat transfer.

5.141 Since most commercially interesting reactor cores are large, spatial effects make an important contribution to the dynamic behavior.[67] Therefore spatially dependent kinetics and the development of analysis methods have received much attention. Various approximate methods have been proposed to treat the space–time behavior of such large systems. One approach is to expand the flux, source, and precursor densities in a set of orthogonal spatial eigenfunctions (*modal expansions*) and then use various techniques to obtain the description of system behavior.[68] In a somewhat different approach, the computer code[69] WIGLE, using diffusion theory with time dependence, can accommodate one dimension and two energy groups. Since this subject is in a state of vigorous development, the designer should carefully consult the current literature to determine what method can best meet his requirements.

5.142 The foregoing discussion on control parameters and kinetic design is intended only as an orientation to some of the computations. In applying such procedures, the designer must understand the kinetics and important parameters as applied to a specific concept. For example, a set of kinetics design principles for fast reactors can be quite different from those for large, water-moderated systems. In fast reactors coolant-void and Doppler effects are very important results of a reactivity increase from spectral hardening. Also, poison absorbers are difficult to use since cross sections are generally low. In fact, the whole subject of fast reactor control is a very special one that plays a major role in the design of such systems.[70] Knowledge of the specific system parameters and their behavior is therefore essential.

5.143 Reactor-kinetics analyses of the third category concern large transients possible after accidents. A study of the consequences of such improbable events may also be included. Such analyses, covering a range of subjects, are treated in Chap. 6 (Safety and Related Design Requirements) and in Chap. 8 (Design Considerations).

INTERPLAY OF CALCULATION METHODS AND CORE NUCLEAR DESIGN

5.144 Methods of *representing* the nuclear behavior of the core in terms of parameters that can be adjusted by the designer have been emphasized here. Calculation methods, therefore, provide a tool for the designer to relate cause and effect. So many parameters are interrelated, however, that an important secondary role of the calculation methods is to *identify* how a given change in a specification can lead to otherwise unforeseen effects. In many cases the use of various computer-code output options can provide the designer the needed insight. In other cases, where the approximations of the calculation, the mode assumed, or the constants used become important, a basic understanding of the calculation method is essential to the design process. Some of these general considerations are clarified by the following examples.

5.145 A primary nuclear-design requirement is to plan the core fuel within the limits of a satisfactory power-density distribution, control margin, and reactivity lifetime. Economic parameters such as processing cost could affect the optimum enrichment point, however. Thus the fuel-management program design may require a good deal of iterative parameter adjustment. For a heterogeneous core, however, such as that for boiling-water or pressurized-water reactors, the calculation is complicated.

5.146 Let us consider an example: The enrichment originally specified for a pressurized-water-reactor core needs to be increased to provide a longer fuel lifetime. The core at a given time will contain several different fuel batches, each of a different enrichment, depending on the fuel management strategy adopted. For the present, however, this factor can be neglected. Because of the new enrichment, important changes will occur in the lattice unit cell as the spectrum changes, and new constants will need to be generated. Even the code-group energy structure may have to be adjusted. An increase in enrichment will probably also require a corresponding increase in control poison to compensate for the increased initial reactivity. The resulting changed neutron absorption in the lattice cell might then possibly affect the validity of the cell-theory assumptions (§5.71) made in the original calculation for the initial enrichment conditions. Therefore, when changes are made as part of the design process, the calculation method itself often deserves as much attention as the design parameters.

5.147 The pin-by-pin power distribution in the core will also be affected by any parameter adjustment that causes a change in flux pattern. A check on pin temperatures to determine if there is a danger of overheating can indeed be a calculation challenge requiring that fineness-of-lattice representations and approximations be evaluated.

5.148 In each of the preceding cases, an apparently simple change can lead to a number of interrelated effects, some of which are identified in the calculation. Such effects should be considered in parameter adjustment as carried out in the design process. Furthermore, the applicability of the calculation method to the problem may be affected. The designer must therefore be careful not to blindly accept a computer readout without assuring himself that it is valid for his purpose. In addition, he must be alert to side effects resulting from what may appear at first glance to be a straightforward cause-and-effect calculation.

ROLE OF EXPERIMENTAL METHODS

INTRODUCTION

5.149 Experimental measurements of various kinds are an important part of the overall reactor-core design activity. Since design calculation *methods* are not exact, it is necessary to relate them to known physical behavior. In addition, uncertainties are involved in some of the basic physical constants used in the calculations. Consequently the range of reliability of calculation models and methods must frequently be determined by using the results of experiments systematically planned for comparison.

5.150 The designer, in attempting an analytical description of the desired reactor core, is faced with the need to use more and more complicated models as he tries to represent an actual core more realistically. The complexity is caused by lattice details, such as control-rod effects, flux variations, and isotopic changes during depletion. In addition, various parameters, such as temperature coefficients and rod worths, are important in the design. A comparatively simple model has the advantage of permitting relatively inexpensive calculations, provided the results are adequate for the design purpose. Experimental measurements can provide diagnostic information helpful in deciding whether a given model is indeed satisfactory.

5.151 Most experiments can be placed in the following categories:[71]

1. Basic parameter studies. Measurements may be made of the basic physical parameters inherent in the model, such as absorption and scattering cross sections and resonance effects of material.

2. Neutron behavior. Measurements of neutron thermalization in various media and the determination of *reactor physics* constants represent a type of

basic experiment that can be made separately or in a lattice. It is usually *differential* in nature.

3. Critical experiments. Included are two areas, so-called *clean critical* experiments in which simple geometry is used and core *mock-up* experiments. These are of the *integral* type.

4. Subcritical experiments.

BASIC PHYSICAL CONSTANTS

5.152 The reliability of nuclear-design methods depends on the nature of approximations in the calculational models used, and also to some extent on the accuracy of cross sections and other nuclear constants. Various approaches are used in the required experimental physics effort. Since the only objective here is to draw attention to the activity, descriptions are not appropriate. Accurate cross sections as a function of energy for all reactor isotopes continue to be an important need, however, and an experimental challenge. Fission properties, resonance integrals, decay constants, and other basic parameters can be included in this category. The experimental facilities required for the measurements generally do not involve a reactor core except possibly as a neutron source.

NEUTRON-PROCESS DIFFERENTIAL EXPERIMENTS

5.153 Other experiments, usually differential in nature, are intended to measure neutron processes, such as thermalization, that occur in a lattice region. Such experiments, which verify the neutron behavior in a core lattice, aid the designer by providing a bridge between the basic information on physics constants and integral-experiment results. Such information not only provides a better description of the building blocks of the core analytical model but also improves the usefulness of such models. For example, predictions based on an analytical model might well agree with results merely because various errors tend to compensate with one another. Errors can be reduced by suitable differential experiments. The reactor model might then be used to predict with confidence parameters of interest throughout the reactor-fuel-depletion cycle for other cases where an integral-experiment confirmation may be difficult.

5.154 In general, these experiments are concerned with the thermal or epithermal neutron distribution in a core lattice as a function of space, time, and energy. This category may also include measurements of resonance escape, thermal utilization, fast fission ratios, etc. The techniques developed for such work emphasize the microscopic characteristics of the core rather than the overall behavior. Activation foils, for example, have been very useful as spectral indicators.

ZERO-POWER CRITICAL EXPERIMENTS

5.155 Integral experiments that test the design model are very useful. Operation at extremely low power levels has the advantage of producing little heat to be removed, as well as ensuring that composition will not change with time. Since fission products are produced in negligible amounts, the core is also accessible after an experiment, and necessary geometric adjustments can easily be made. Two different approaches are used. In one, referred to as *clean critical experiments,* a simplified geometry is used. "Clean" refers to freedom from such complications as the partial insertion of control rods. The objective is to uncover major problem areas in the analytical model by systematically varying parameters and determining the effect on criticality. Fairly simple experiments of this type are often very useful during the early development stages of a concept for evaluating its feasibility. Once the feasibility is proven, however, a more accurate representation is needed to supplement the detailed design.

5.156 In *mock-up experiments* a reasonably accurate representation of the core is attempted. One use is to check the validity of heterogeneous-lattice design techniques such as those necessary for the boiling-water-reactor core shown in Fig. 5.7. Here the presence of large water gaps and cruciform control rods represents a substantial analytical challenge.[72] Core mock-ups can also provide the designer with such information as approximate critical-rod positions, rod worths, shutdown margins, and, to some extent, temperature coefficients. The designer should remember, however, that mock-ups are to partially confirm analytical methods and are not as satisfactory as an actual reactor core for predicting characteristics.

5.157 The need for separate zero-power experiments varies with the stage of development of the concept. In well-developed pressurized-water- and boiling-water-reactor systems, for example, critical experiments for design purposes are generally no longer needed; experiments and measurements begun after the reactor is built and carried out during the start-up activities are emphasized. These confirm design calculations and establish characteristics necessary to meet licensing requirements (§6.205).

SUBCRITICAL EXPERIMENTS

5.158 Various types of experiments on subcritical systems are also useful. *Pulsed-neutron-source techniques* are one example. A repetitive pulse of high-energy neutrons is injected into a subcritical assembly, and the time rate of decay of the prompt neutrons generated by each burst is measured. The degree of subcriticality can then be obtained from the decay rate. In this case, comparison of the experimental results with analytical predictions can test the model used to describe the leakage. The technique can also be used to determine the worth of control elements.

5.159 Neutrons in subcritical assemblies can be studied by using the slow-chopper technique. Neutrons of different velocities are separated by the different times of arrival at a detector after having been chopped into pulsed bursts by a rotating shutter. Characteristics of the multiplying medium can then be compared with those predicted by calculations.

5.160 Fast subcritical assemblies[73] can be designed to provide a neutron spectrum at the core–blanket interface, and into the blanket, that simulates the spectral behavior in these regions of large fast breeder reactors. Blanket design parameters and related fuel-cycle questions can be experimentally studied in such a facility more conveniently than in a critical assembly.

COORDINATION OF EXPERIMENT AND DESIGN

5.161 Basic physical-constant data from neutron-behavior experiments generally apply to a variety of designs. Although coordination with specific concepts is not necessary, an awareness of design needs for specific materials or specific parametric ranges should guide the experimentalist, however. For lattice parameters, close coordination is important between the designer and the experimentalist during the various stages of a new project. Mock-up critical experiments can be performed if needed to verify design decisions before manufacturing is approved.

5.162 Experimental methods are helpful to the designer even after the reactor has been built. A comparison of measurements made in an operating power reactor with those made during start-up, and possibly with a zero-power critical experiment can provide a basis for the extrapolation that may be needed to design additional reactors of the same general type. In large power reactors, small variations in parameters may represent a considerable sum of money over the lifetime of the reactor. An optimization of the reactor design through all possible methods, including experimental verification of calculation models, is therefore economically highly justified. Such effort is particularly appropriate for the design of replacement cores.

REFERENCES

1. Sacramento Municipal Utilities District, Preliminary Safety Analysis Report, Rancho Seco, Nuclear Generating Station Unit No. 1, 1967.
2. Portland General Electric Company, Preliminary Safety Analysis Report, Trojan Nuclear Power Plant, 1969.
3. W. J. Worlton and E. A. Voorhees, Recent Developments in Computers and Their Implications for Reactor Calculations, in Proceedings of the Conference on the Application of Computing Methods to Reactor Problems, USAEC Report ANL-7050, pp. 15–32, Argonne National Laboratory, August 1965.
4. M. K. Butler and J. M. Cook, One-Dimensional Diffusion Theory, in *Computing Methods in Reactor Physics,* H. Greenspan, C. N. Kelber, and D. Okrent (Eds.), Gordon and Breach, Science Publishers, Inc., New York, 1968.

5. S. Glasstone and M. C. Edlund, *The Elements of Nuclear Reactor Theory*, D. Van Nostrand, Inc., Princeton, N.J., 1952.
6. J. R. Lamarsh, *Introduction to Nuclear Reactor Theory*, Addison-Wesley Publishing Company, Inc., Reading, Mass., 1966.
7. G. I. Bell and S. Glasstone, *Nuclear Reactor Theory*, Van Nostrand Reinhold Company, New York, 1970.
8. H. S. Isbin, *Introductory Nuclear Reactor Theory*, Reinhold Publishing Corporation, New York, 1963.
9. A. Radkowsky (Ed.), *Naval Reactors Physics Handbook*, Vol. I, Selected Basic Techniques, USAEC Report TID-7030, Division of Reactor Development, 1964.
10. W. C. Sangren, *Digital Computers and Nuclear Reactor Calculations*, John Wiley & Sons, Inc., New York, 1960.
11. B. W. Roos and W. C. Sangren, Some Aspects of the Use of Digital Computers in Nuclear Reactor Design, in *Advances in Nuclear Science and Technology*, E. J. Henley and H. Kouts (Eds.), Vol. 2, p. 303, 1964.
12. M. Clark, Jr., and K. F. Hansen, *Numerical Methods of Reactor Analysis*, Academic Press, Inc., New York, 1964.
13. Ref. 12, pp. 183 et seq.
14. Ref. 12, p. 175.
15. H. C. Honeck, ENDF/B: Specifications for an Evaluated Nuclear Data File for Reactor Applications, USAEC Report BNL-50066, Brookhaven National Laboratory, May 1966. Revised by S. Pearlstein, July 1967.
16. Ref. 9, p. 197.
17. J. H. Ferziger and P. F. Zweifel, *The Theory of Neutron Slowing Down in Nuclear Reactors*, p. 225, M.I.T. Press, Cambridge, Mass., 1966.
18. H. Bohl, Jr., E. M. Gelbard, and G. H. Ryan, MUFT-4 – Fast Neutron Spectrum Code for the IBM-704, USAEC Report WAPD-TM-72, Westinghouse Electric Corporation, July 1957.
19. M. Butler, B. Maskewitz, and J. Rosen, Activities of the ANL–ENEA Computer Program Libraries and the ORNL Radiation Shielding Information Center in Proceedings of the Conference on the Application of Computing Methods to Reactor Problems, USAEC Report ANL-7050, pp. 125–139, Argonne National Laboratory, 1965.
20. G. D. Joanou and J. S. Dudek, GAM-I: A Consistent P_1 Multigroup Code for the Calculation of Fast Neutron Spectra and Multigroup Constants, USAEC Report GA-1850, General Atomic Division, June 1961.
21. H. J. Amster and R. Suarez, The Calculation of Thermal Constants Averaged over a Wigner–Wilkins Flux Spectrum: Description of the SOFOCATE Code, USAEC Report WAPD-TM-39, Westinghouse Electric Corporation, January 1957.
22. Ref. 17, p. 244.
23. Reactor Physics Constants, USAEC Report ANL-5800, p. 735, Argonne National Laboratory, July 1963.
24. B. J. Toppel and I. Baksys, The Argonne-Revised THERMOS Code, USAEC Report ANL-7023, Argonne National Laboratory, March 1965.
25. H. Alter, The Atomics International Evaluated Nuclear Data Files and Associated Computer Programs for the Automated Preparation of Multigroup Constants, USAEC Report ANL-7320, pp. 33–46, Argonne National Laboratory, October 1966.
26. B. J. Toppel, A. L. Rago, and D. M. O'Shea, MC^2, A Code to Calculate Multigroup Cross Sections, USAEC Report ANL-7318, Argonne National Laboratory, June 1967.
27. G. T. Peterson, P. G. Aline, and R. J. McWhorter, The Role of Experimental Reactor Physics in G. E. Core Design Philosophy, Report APED-4329, General Electric Company, 1963.
28. Ref. 9, pp. 620 et seq.

29. PL Final Design Report, USAEC Report IDO-19030, Combustion Engineering, Inc., October 1961.
30. C. G. Poncelet, Burnup Physics of Heterogeneous Reactor Lattices, USAEC Report WCAP-6069, Westinghouse Electric Corporation, June 1965.
31. O. J. Marlowe, Nuclear Reactor Depletion Programs for the Philco-2000 Computer, USAEC Report WAPD-TM-221, Westinghouse Electric Corporation, January 1961.
32. H. S. Isbin, Introductory Nuclear Reactor Theory, p. 524, Reinhold Publishing Corporation, New York, 1963.
33. Ref. 23, p. 280.
34. T. M. Snyder, Boiling Light Water Reactor Physics, USAEC Report APED-4038, General Electric Company, 1962.
35. Ref. 9, p. 1461.
36. D. A. Meneley, L. C. Kvitek, and D. M. O'Shea, MACH 1, A One-Dimensional Diffusion-Theory Package, USAEC Report ANL-7223, Argonne National Laboratory, June 1966.
37. Ref. 9, p. 169.
38. W. W. Engle, M. A. Boling, and B. W. Colston, DTF-II – A One-Dimensional, Multigroup Neutron Transport Program, USAEC Report NAA-SR-10951, Atomics International, March 1966.
39. Ref. 32, p. 326.
40. W. R. Cadwell, PDQ-4 – A Program for the Solution of the Neutron-Diffusion Equations in Two Dimensions on the Philco-2000, USAEC Report WAPD-TM-230, Westinghouse Electric Corporation, June 1961.
41. J. I. Hapen and R. W. Goin, Borax-V Neutronics, USAEC Report ANL-6964, Argonne National Laboratory, December 1965.
42. B. J. Toppel and R. Avery, Some Physics Aspects of the Plutonium Experiment in EBWR, USAEC Report ANL-7019, Argonne National Laboratory, April 1965.
43. A. Hassitt, A Computer Program to Solve the Multigroup Diffusion Equations, British Report TRG-229 (R), 1962.
44. B. Carlson, C. Lee, and J. Worlton, The DSN and TDC Neutron Transport Codes, Appendix II: The TDC Code, USAEC Report LAMS-2346, Los Alamos Scientific Laboratory, 1960.
45. Ref. 9, pp. 656 et seq.
46. Ref. 9, pp. 751 et seq.
47. R. T. Shanstrom, M. Benedict, and C. T. McDaniel, Fuel Cycles in Nuclear Reactors, USAEC Report NYO-2131, Massachusetts Institute of Technology, August 1959.
48. M. F. Valerino and Z. R. Rosztoczy, Analytical and Experimental Methods of Determining Heavy-Isotope Content of Operating Fuel Elements, USAEC Report CEND-540, Combustion Engineering, Inc., September 1965.
49. F. W. Todt, GAD, An Infinite Medium Multigroup Burnup Code for Fuel Cycle Analysis, Report GA-4150, General Atomic Div., General Dynamics Corp., 1963.
50. R. F. Barry, LEOPARD – A Spectrum-Dependent Non-Spatial Depletion Code for the IBM-7094, USAEC Report WCAP-3269-26, Westinghouse Electric Corporation, September 1963.
51. S. M. Hendley and R. A. Mangan, CNCR-2 – A One-Dimensional Few-Group Depletion Code for the IBM-7094, Which Includes the Effects of Water Density Variations on Neutron Moderation, USAEC Report WCAP-6058, Westinghouse Electric Corporation, January 1964.
52. F. Todt, FEVER – A One-Dimensional Few-Group Depletion Program for Reactor Analysis, USAEC Report GA-2749, General Atomic Div., General Dynamics Corp., November 1962.

53. D. P. Satkus and H. P. Flatt, SIZZLE; A Fast or Intermediate Reactor Burn-up Code, Report AE 64-Memo-183, Atomics International, 1964.
54. D. L. Delp, D. L. Fischer, J. M. Harriman, and M. J. Stedwell, FLARE, A Three-Dimensional Boiling Water Simulator, General Electric Company, USAEC Report GEAP-4598, July 1964.
55. Ref. 9, pp. 789 et seq.
56. C. A. Flanagan, R. J. Breen, O. J. Marlowe, and A. J. Vigilotti, TNTO2 – A Three-Dimensional Neutron Flux Synthesis and Depletion Calculation Computer Program, USAEC Report WAPD-TM-468, Westinghouse Electric Corporation, August 1964.
57. J. R. Lamarsh, *Introduction to Nuclear Reactor Theory,* pp. 523 et seq., Addison-Wesley Publishing Company, Inc., Reading, Mass., 1966.
58. E. U. Vaughan and R. F. Berland, ARIADNE-I, A Burnup Perturbation Code, USAEC Report NAA-SR-12376, Atomics International, June 1967.
59. M. A. Schultz, *Control of Nuclear Reactors and Power Plants,* 2nd ed., McGraw-Hill Book Company, Inc., New York, 1961.
60. M. Ash, *Nuclear Reactor Kinetics,* McGraw-Hill Book Company, Inc., New York, 1965.
61. G. R. Keepin, *Physics of Nuclear Kinetics,* Addison Wesley Publishing Company, Inc., Reading, Mass., 1965.
62. Ref. 60, p. 79.
63. Ref. 23, p. 812.
64. L. R. Blue, and M. Hoffman, NAA Program Description, AIREK-3 – Generalized Program for the Numerical Solution of Space Independent Reactor Kinetics Equations, USAEC Report NAA-SR-Memo-9197, Atomics International, November 1963.
65. H. Flatt, *IBM Nucl. Computing Tech. Bull. No. 5,* 1962.
66. R. A. Blaine and R. F. Berland, AIROS – A Digital Simulator for Power Reactor Dynamics, USAEC Report NAA-SR-9943, Atomics International, August 1964.
67. J. B. Yasinsky and A. F. Henry, Some Numerical Experiments Concerning Space–Time Reactor Kinetics Behavior, *Nucl. Sci. Eng.,* **22:** 171 (1965).
68. Ref. 61, p. 166.
69. W. R. Cadwell, A. F. Henry, and A. J. Vigilatti, WIGLE – A Program for the Solution of the Two-Group Space–Time Diffusion Equations in Slab Geometry, USAEC Report WAPD-TM-416, Westinghouse Electric Corporation, January 1964.
70. Ref. 61, pp. 298 et seq.
71. Ref. 9, pp. 1143 et seq.
72. G. T. Petersen, P. G. Aline, and R. J. McWhorter, The Role of Experimental Reactor Physics in G. E. Core Design Philosophy, Report APED-4329, General Electric Company, 1963.
73. K. O. Ott and C. W. Terrell, Subcritical Fast Reactor as a University Research Facility, *Trans. Am. Nucl. Soc.,* **13:** 456 (1970).

6 Safety and Related Design Requirements

INTRODUCTION

6.1 Public health and safety as well as safe operation of a plant are important responsibilities of the design engineer whatever the nature of the industrial process involved. For nuclear energy systems, however, the designer has some special responsibilities. The fission-product inventory of a large nuclear power plant, if somehow completely released, would indeed be catastrophic. Although the probability of such a release, or even a small fraction of it, is normally extremely low even under abnormal conditions, the designer must recognize that the public, being generally aware of weapons effects, fallout, etc., is very sensitive to the possibility of nuclear hazards.[1] Furthermore, these considerations and a growing general awareness of the effects of the environment on the quality of life have brought nuclear power plants to the focus of public concern.

6.2 In the United States a proposed plant design is subjected to a searching analysis as part of a licensing procedure intended to protect the public. Various accident possibilities are analyzed, primarily to evaluate the safeguards provided in the design against such accidents. Since the project cannot become a reality unless the licensing requirements are fulfilled, meeting safety criteria is vital to the design process right from the beginning. Environmental considerations associated with the reactor site and not directly concerned with safety enhance the challenge. For example, the effects of discharging waste heat to rivers or lakes are likely to be a vital design consideration.

6.3 The broad area of design for safety includes several subjects. Questions of reactor control and kinetics are certainly part of the story although they are likely to be included in the nuclear design. Coolant thermal–hydraulics limitations, particularly during transients, relate to safety criteria. Accidents that could lead to fission-product release and the engineering design to avoid such accidents are included in yet another subject category. A major part of safety design is the containment and its relation to the site.

6.4 Each subject can have considerable depth, depending on the needs of the designer. For example, an analysis of the release and transport of fission products could involve detailed study of various diffusion transport processes as well as isotopic changes. Site considerations could involve complex social and economic as well as engineering factors. In addition, a knowledge of the requirements for power reactor licensing, which involve legal considerations, is essential background for the designer.

6.5 Since the subject of reactor safety is so broad and involves so many disciplines, only the role of specialists as they might interrelate with the designer and the availability of information in book form are considered in this chapter. For example, the principles of reactor kinetics and control are well described. In addition, this subject tends to be the responsibility of a specialist. Therefore the discussion here is for the nonspecialist designer or student who is concerned primarily with orientation and interplay with other phases of the design. Thus conceptual relations rather than technical detail are emphasized.

6.6 In this conceptual approach the design features needed to ensure reliable and stable reactor operation are given high priority. The probability of abnormal operating conditions and accidents should be made as low as practical. Although this probability can be made extremely low, it cannot be made equal to zero. Hence the *control* of fission products becomes a prime reactor-safety consideration.

6.7 Fission products build up in the fuel and are kept there by the first barrier, namely, the fuel material and the fuel cladding. If for any reason this barrier is violated, some fraction of the fission-product buildup will be released to the primary coolant system, the second barrier. If, then, the primary system is breached, fission products encounter the third barrier, the containment structure; finally, if this fails, fission products can move into the environment. What must be considered, therefore, are the factors that enter into the possible violation of the integrities of the fuel cladding, primary system, and containment structure and the relation of these factors to each other and to plant design.

6.8 Several aspects of reactor control which tend to prevent abnormal behavior are therefore initially considered in this chapter. Background material on fission products is next presented as a basis for the subsequent discussion on types of abnormal reactor behavior. Next, engineered safety features incorporated in the design to control the fission products, should they be released, are considered. Since all the preceding subjects can affect the safety of the public, licensing requirements also deserve attention.

KINETICS AND CONTROL REQUIREMENTS

6.9 Reactor safety design logically begins with an analysis of the kinetic behavior, particularly in relation to other design parameters. Of course, the control system must be considered together with the kinetics to determine the transient behavior of the plant, including questions of stability. Transients result from both routine operation and unexpected nonroutine situations. In the establishment of limiting conditions that affect component design, the nonroutine situation is generally more important but is more difficult to describe analytically. This is particularly true of a large transient associated with an accident. Since it is also difficult to separate operational and safety matters, we shall first discuss simplified operational transients.

KINETIC BEHAVIOR

6.10 A starting point for studying the time behavior of reactor power as a function of core reactivity is the conventional form of reactor kinetics equations, such as

$$\frac{1}{P}\frac{dP}{dt} = \frac{1}{n}\frac{dn}{dt} = \frac{\rho(t) - \beta}{l^*} + \sum_{i=1}^{6} \lambda_i \frac{C_i}{n} \tag{6.1}$$

$$\frac{d}{dt}\left(\frac{C_i}{n}\right) = \frac{\beta_i}{l} - \lambda_i \frac{C_i}{n} \tag{6.2}$$

where P is the reactor power and the other symbols have the usual meaning (§5.138). Solutions to these equations and their application are well described elsewhere.[2-4] However, Eqs. 6.1 and 6.2 are substantially simplified from the basic transport-theory description dependent on neutron energy, direction of motion, and position. The applicability of the approximations used should therefore not be overlooked.

6.11 Although various methods treat spatial dependency as well as energy effects, more important here is the time dependency of the neutron chain reaction in relation to various reactivity-feedback effects. The space-independent or point-reactor-model relations are therefore considered adequate for identifying important parameters. Spatial effects are considered separately (§6.31).

6.12 The core reactivity can be affected by a wide number of so-called disturbances that tend to change the neutron balance. Since many of these disturbances can be caused, or at least be affected, by the reactor power, the behavior can be described by a typical control feedback loop (Fig. 6.1). In the diagram an externally applied change in reactivity operates through the neutron kinetics equations to yield an incremental change in reactor power. Various effects from the change in energy production affect, in turn, the reactivity (comprising the power coefficient) and therefore feed back and combine with the original externally applied reactivity, either reinforcing or diminishing it, depending on the sign of the coefficient. Remember that the ultimate effects depend largely on the time required for the chain of events in the feedback loop.

6.13 Such feedback-controlled reactor behavior can be described by three systems of equations:

1. The neutron-kinetics equations relating multiplication factor (reactivity) and neutron density, normally assumed proportional to the reactor power.

2. A family of thermal-transport and hydrodynamic equations that relate the reactor power to temperature distributions within the reactor core.

3. Equations relating the changes in reactor temperature to reactivity, providing the feedback effect.

6.14 It is useful to express the feedback effect in terms of a so-called power coefficient of reactivity, which relates power-dependent changes, such as a shift in temperature pattern, a change in the relative position of one component to another, or other effects, all to the reactivity. Actually, the power coefficient of reactivity is not a single coefficient but a composite of a number of individual contributions, in the sense of partial coefficients.

6.15 Such partial coefficients merely serve to simplify the analysis by permitting separate consideration of temperature effects on the properties of different core components and, in turn, the effects of such changes on the

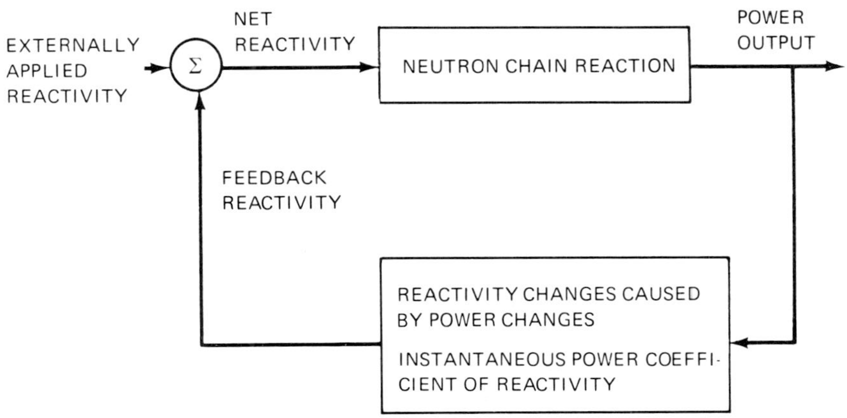

Fig. 6.1 Concept of reactivity feedback through power coefficient of reactivity.

neutron balance. Temperature-induced feedback mechanisms range from the Doppler effect, primarily a nuclear interaction phenomenon, to the results of the expansion of various reactor components, such as the fuel, coolant, or moderator. Such expansion leads to secondary nuclear changes affecting the reactivity.

6.16 Since the terms associated with these various coefficients are not standard, we shall clarify their meaning. First, consider separately the effects that do not depend on the steady-state operating power at the time of the transient. For example, an *isothermal-temperature coefficient* can be used to describe the reactivity change caused by a unit change in the uniform temperature of the core on a zero-power basis. In other words, temperature *gradients* such as those which would occur in the fuel at power are not considered. In some cases it is useful to consider the coolant and the moderator separately from the fuel, still on a zero-power basis, or at least with each region at its own uniform temperature. Isothermal-temperature coefficients, determined experimentally by measuring the reactivity at several uniform core temperatures, provide a basis for calculations applicable to other conditions involving temperature gradients.

6.17 Reactivity changes due to power transients are caused by temperature changes of core components, which, in turn, are caused by shifts in the heat-generation–heat-removal balance. Bear in mind that the temperature of a component at any instant depends on the relative rates of thermal input and heat removal, as well as its heat capacity. Temperature changes occur after the time delay required for the necessary heat transfer and hence are coupled to the initiating power transients by time-dependent dynamic relations. Therefore the power coefficient of reactivity in Fig. 6.1 is really the composite of a number of different feedback effects, each with a separate time constant.

6.18 Rather than use the power-coefficient concept, it is frequently preferable to consider feedback effects individually to develop a clearer picture of the relative contributions. For example, the dynamic performance of a boiling-water reactor, which depends on the interaction of fuel power, heat transfer to coolant, void formation, channel flow, and void reactivity feedbacks, can be described by a diagram[5] such as Fig. 6.2. Although temperature gradients exist, average temperatures that take the gradient into consideration are often used for the moderator, fuel, etc. Use of such effective temperatures of various regions in developing a feedback picture is sometimes termed a "lumped-parameter" approach.

6.19 Fortunately temperature-dependent effects, such as the expansion of moderator or the formation of voids in a boiling-water reactor, tend to dampen reactivity increases instead of reinforcing them. It is important to keep exceptions in mind, however, such as when a significant neutron absorber expands and a lower density of absorbing nuclei becomes effective in the neutron balance. A dampening effect dependent on a long thermal-transport path and hence slow acting when compared with the rate of a possible reactivity increase may also not come into effect until the temperature of the fuel becomes excessive. Fast-acting nuclear effects, such as the Doppler coefficient, which are dependent on the fuel temperature itself are therefore important to safety design, particularly for fast reactors.

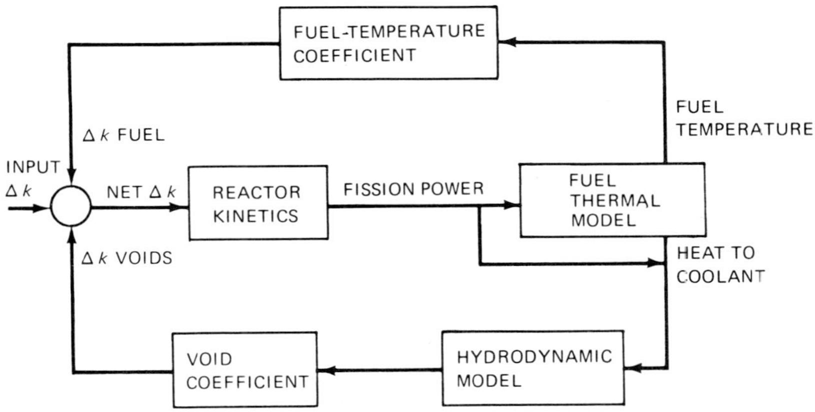

Fig. 6.2 Simplified block diagram of nuclear–thermal–hydraulic feedback loop.

THE DOPPLER COEFFICIENT

6.20 An important prompt-acting coefficient is the result of the Doppler effect. For the *resonance* absorption of neutrons by heavy fissile and fertile elements, the thermal agitation of the target nuclei tends to affect the relative velocity of the incident neutron and the nuclide and hence the cross section. As the temperature of the material is raised, the increased thermal agitation broadens an initially narrow resonance energy cross-section band.

6.21 This broadening increases the neutron absorption since the flux now within the resonance energy band is enhanced. The actual integral of the energy-dependent absorption cross-section over the energy region remains unchanged, however. As the resonance peak broadens, the *self-shielding* effect for heterogeneous fuel pins decreases; this tends to increase the absorption further. For a very narrow resonance band with an extremely high cross section, a large self-shielding effect occurs since the surface nuclides can deplete the neutrons in the incident flux having energies within the band. Inner nuclides are therefore not effective. With broadening, however, a greater fraction of incident neutrons will have energies within the band. Although the maximum absorption cross section per nuclide is now reduced somewhat, the total absorption is increased since the underlying nuclides can now become effective absorbers.

6.22 For pure absorbing materials, such as ^{238}U and ^{232}Th, the Doppler effect causes the reactivity to decrease as the temperature increases. Since, for pure fissile materials, such as ^{239}Pu and ^{235}U, capture competes with fission, increased Doppler-effect absorption could cause the reactivity to increase as their temperature is raised. On the other hand, virtually every practical reactor system has enough pure absorbing (fertile) material in the core to make the net effect negative. The relative importance of the Doppler effect depends greatly on the spectral distribution of the fission events in a given core, however. In a well-thermalized system, for example, the fraction of fissions in the resonance region is likely to be comparatively small. Doppler broadening is therefore likely to have only a minor effect on the reactivity. Considering the high end of the spectrum, a small fast reactor fueled with highly enriched metal that will have a rather hard spectrum will likewise not have a large reactivity change from the Doppler effect. In a large fast reactor fueled with oxide containing dilute plutonium and substantial ^{238}U, however, the Doppler effect tends to be very important.

STABILITY ANALYSIS

6.23 The stability of the reactor from the servomechanism viewpoint is, of course, a question of safety that the designer must consider. However, since this is a sophisticated subject that is normally handled by specialists in kinetics and control and is well described elsewhere, it will be only briefly mentioned here.

6.24 A reactor is considered stable if the neutron density will not grow without bound after a reactivity perturbation. Because of the possibility of oscillations due to feedback, however, an unsatisfactory, and hence unstable, situation could exist in a bounded system. Feedback effects include not only those associated with such core characteristics as temperature, moderator voids, and poisons but also many external reactor-plant effects, such as those associated with coolant transport, thermal transport, and control-rod mechanisms. The effects of these reactivity-feedback contributions are commonly analyzed by using transfer functions, which involves transformation to the complex plane and the frequency domain with linearity assumed.

6.25 The transfer function, defined as the quotient of the Laplace transform of the output signal and the Laplace transform of the input signal, has the desirable property of permitting the analysis of a complex feedback system by using block diagrams and the simple algebraic manipulation of individual functions.

6.26 Transfer-function methods for handling feedbacks and determining stability have evolved directly from servomechanism theory. As a result, a large body of literature describing these methods is available. However, since such methods are no better than the accuracy of the description of the various feedback coefficients, these feedback effects merit a good deal of design attention. It is useful to consider internal feedbacks, which include core-temperature coefficients and void coefficients, separately from external plant feedbacks, which depend on such design features as coolant circulation, control-valve characteristics, and rod-servomechanism behavior. Internal feedback effects, considered part of the study of the kinetic behavior of the core, are emphasized in books on the subject,[3,4] and external system effects receive attention in books on reactor control engineering.[6]

6.27 For small disturbances for which the assumption of linearity is valid, the stability can be determined by a variety of both analytical and graphical methods,[6,7] some making use of transfer functions. An important advantage of these approaches is that experimental measurements of the system response at low and intermediate power levels can be used as input to predict the stability limitations at higher power levels.

6.28 As disturbances become larger, a point is reached at which certain terms in the solutions of the kinetics equations can no longer be neglected and the behavior becomes nonlinear. In safety analysis nonlinear theory that can be applied to these large perturbations is of particular interest. Nonlinear partial differential equations also are introduced when the reactor system is described, taking into account the coupling of heat transfer, hydrodynamics, and time-dependent neutron diffusion. The prediction of conditions required for the stability of nonlinear systems is a challenge that has been only partially solved. One promising class of stability-analysis techniques for nonlinear systems is based on the "second method" of Liapunov,[8] in which various criteria are applied to a real, positive, definite scalar function of the differential-equation system being considered. A number of other analytical methods, some equivalent to the Liapunov method, have also been developed for nonlinear systems.

6.29 In connection with stability analysis, the role of the delayed neutrons should be mentioned. Delayed neutrons are often neglected to reduce the complexity of the problem. Although it is argued that a system shown to be stable on prompt neutrons is usually more stable when the delayed neutrons are considered, such reasoning can produce an overly conservative design. Therefore, if the analysis accounts for delayed neutrons, a higher stable reactor power than might otherwise be considered acceptable might result.

ANALYSIS MODELS

6.30 In large commercial reactor cores, spatial effects can become important, and representation by a point kinetics model may not be accurate. On the other hand, a model for design analysis should be no more complicated than required since the calculation expense increases with the sophistication of the spatial dependency introduced. The calculation model and the corresponding computer code must therefore be tailored to the specific need, with work indicating the expected accuracy considered. For example, for boiling-water reactors one-dimensional representation appears adequate for the coupled thermal–hydraulics kinetics.[5] We shall not detail the various possibilities here, however.

6.31 Various computational simplifications may also be applicable. In considering the power coefficient, for example, when changes are not rapid, we may frequently assume that the temperature distribution at any time is close to the steady-state temperature distribution corresponding to the power level at that time. In this, the so-called *quasi-static analysis,* coolant flow rates and inlet temperatures are also held constant.[9] Such separation of driving force and feedback effects tends to be more appropriate for thermal reactors than for fast reactors, where the time constants are shorter. Both quasi-static and nodal techniques, in which the reactor is divided into a number of regions, are also very useful in spatially dependent kinetics.

6.32 The model used also depends on the extent of the transient considered. An analysis of an accident in which the fuel is melted, for example, with primary interest in subsequent events, requires a treatment much different from that of an operational transient. In the analysis of an excursion, it is sometimes useful to break down the evaluation into three phases: (1) the initiating events, (2) the ensuing nuclear excursion, and (3) the resulting physical and chemical effects. Computations can be simplified for kinetic effects in bursts (phase 2). For example, a steady-state calculation can be used at several discrete time steps to determine a time-dependent power distribution from which feedback effects and spatial changes can be developed.

6.33 Spatial effects generally become important in large, loosely coupled cores in considering both modest transients and large excursions. For example, the sudden local insertion of cold water from one of the cooling loops of a pressurized-water reactor can produce not only an overall power excursion but also a “tilt” in the power pattern. Such tilting of the neutron flux combined with reactivity feedback effects is likely to lead to oscillatory transients across the core.

6.34 A slower type of oscillation can result from the poisoning effect of ^{135}Xe. Should there be a localized perturbation leading to an increase in the neutron flux, the rate of ^{135}Xe consumption, or burnout, will increase, and its concentration will thus decrease since its formation depends on ^{135}I decay. The reduced neutron absorption will cause the flux to increase locally even further until the negative reactivity coefficient can become effective.

6.35 Weak coupling questions also apply to severe excursions. If one part of a system goes prompt critical and the remainder is only loosely coupled, space and time normally cannot be assumed as separable in the kinetics analysis. Of course, if material also becomes displaced, an analytical description of the transient becomes even more challenging.

6.36 Although simplified models can be used for practical reactor-safety calculations, the great complexity of the physical situation, particularly after fuel failure, must be recognized. Not only is the nuclear space–time problem difficult to describe but also associated thermal–hydraulics mechanisms are poorly understood, most liquid-coolant systems being likely to boil. Also, the properties of materials are likely to be uncertain under the extreme conditions of the transient.[10] The designer should therefore continually keep in mind that useful approximations developed by workers in the field are primarily a means to an end and should be used with caution.

CONTROL OF REACTIVITY

6.37 The control system and its design have an important bearing on the safety of the reactor plant. Since the kinetic behavior depends on reactivity changes, the *amount* of reactivity that may be inserted or removed, called the *reactivity inventory,* is one area of concern. This is particularly true because the inventory normally required is well above that needed for prompt criticality. The *methods* for inserting and removing the reactivity, or portions of it, as needed are the second area of concern. Particularly important are the methods for quickly shutting down the reactor in the event of an operational difficulty or equipment malfunction.

THE REACTIVITY INVENTORY

6.38 The fuel loading must provide sufficient positive reactivity to compensate for such typical reactivity losses as those caused by heating the core to operating temperature in water-moderated reactors and those caused by the depletion of fissile atoms during operation. This excess reactivity must, in turn, be compensated by control elements. In addition, sufficient *shutdown margin* must be available at all times to reduce the reactor power quickly. An accounting of reactivity needs and its apportionment among various compensating control elements is therefore a design requirement.

6.39 A few definitions are useful in discussions of reactivity inventory.[11] The term *excess reactivity,* ρ_{ex}, refers to the core (and blanket) reactivity when all control elements are at their maximum reactivity state. Another important reference state occurs during shutdown when all control elements are adjusted to a minimum reactivity state. The difference between criticality and the minimum reactivity condition is called the *shutdown margin,* and the range between the excess reactivity and the minimum reactivity is the *reactivity worth* of the control elements. However, in the specification of reactivity, ambiguity can be avoided only if the exact state of the reactor is described. For example, the cold, clean* state is quite different from hot and clean at zero power, hot and clean at full power, and states at different degrees of fuel burnup. Unfortunately the states used in published reactor-inventory specifications are not always uniform and must therefore be interpreted carefully.

6.40 The amount of excess reactivity required for a given reactor design is related to its safety only indirectly. Although a reactor with a large built-in excess reactivity requires more compensation than one with a small excess reactivity and therefore is not inherently less safe, the potential for prompt criticality exists in each case. Provisions for manipulating the needed excess reactivity and especially the corresponding compensation are therefore of primary importance to safety, with the amount of excess reactivity involved somewhat secondary. However, a small excess reactivity is economically preferable to reduce unproductive neutron absorption in control materials and to reduce the number of control rods with associated perturbations on the power distribution.

6.41 The shutdown margin, or the amount of negative reactivity available, affects the rate at which the reactor power level is reduced after the rods have been scrammed. The important parameter is the quantity $1/[1 - (\rho/\beta)]$, the fraction achieved quickly after scram of the initial reactor steady-state power.† This behavior is shown in Fig. 6.3. For water-cooled reactors the normal requirement for shutdown margin is based on the reactivity appropriate to the hot, "at power" core at some degree of fuel burnup. In the design of the control system, the so-called stuck-rod criterion must be considered; i.e., it must be possible to render the reactor subcritical with some specified number of control rods stuck in the fully withdrawn position.

6.42 Typical reactivity-inventory specifications for boiling-water and pressurized-water reactors are given in Table 6.1. Although the total worth of the control elements in each reactor is approximately the same, an important difference is the use of chemical shim in the pressurized-water system. With this concept the reactivity that must be controlled by movable control rods can be reduced significantly. A soluble absorber, such as boron, provides a uniformly distributed poison that avoids the local flux depressions produced by rods. Boron also has the characteristic of being depleted by reaction with the neutron

*Here "clean" refers to the initial value for the fuel concentration generally with no fission products present. In critical experiments "clean" refers to a simplified geometry in analytical representation.

†This relation corresponds to the stable-neutron-density level after a step reactivity change, which can be expressed as

$$n \approx n_0 \frac{\beta}{\beta - \rho} e^{t/T_p}$$

where n_0 is the initial level and T_p is the stable period.[12]

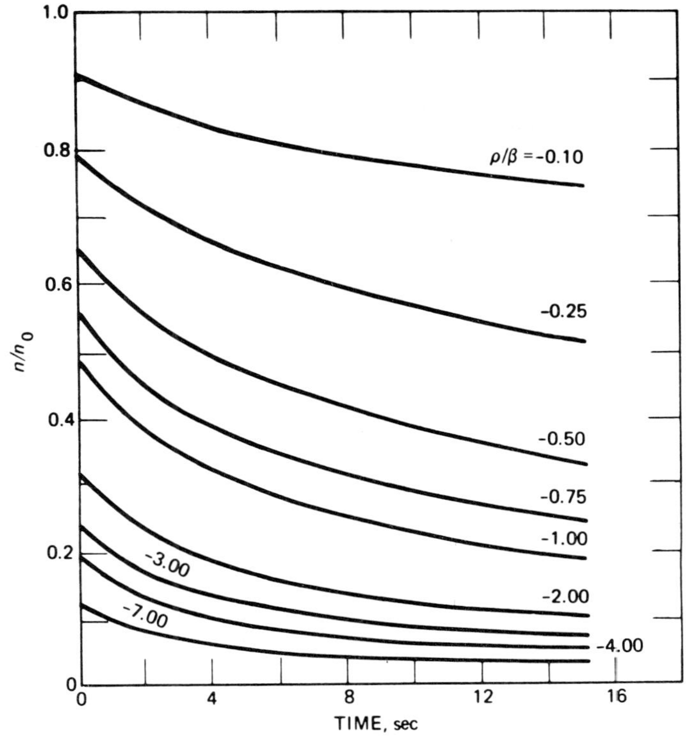

Fig. 6.3 Decay of reactor power from the steady state after an instantaneous reduction to reactivity ρ. The value n/n_0 is the ratio of the power to the steady-state power before shutdown. The initial power decrease is not instantaneous as implied by the figure but appears instantaneous because of the rather coarse time scale.

flux, similar to that of fuel. The resulting loss of poisoning effect, therefore, tends to compensate for the loss of fissile atoms during fuel burnup, as is desirable during shim control. In practice, however, this depletion tendency does not play a strong role in controlling the reactor, since the amount of water in the reactor core at one time is a small fraction of the total primary-system water with a resulting large inertia to change. The chemical-shim control system includes provisions for adjusting the boron concentration as needed for control purposes.

6.43 Soluble absorbers cannot be employed for rapid reactivity variations. Control-rod elements must therefore be used for the power variation caused by Doppler broadening, for changes in the core average coolant temperature, and for a shutdown margin at hot conditions.

TABLE 6.1

Typical Reactivity-Inventory Specifications

Control characteristics	Reactivity (ρ) BWR	PWR
Excess multiplication of clean core (uncontrolled):		
At 68°F	0.25	0.293
At operating condition (clean)		0.248
At operating condition (xenon and samarium equilibrium)		0.181
Worth of control rods	−0.17	−0.07
Worth of borated control curtains	−0.12	None
Total worth of soluble poison	None	−0.25
Total worth of control	−0.29	−0.32

6.44 In a boiling-water reactor, borated control curtains with about 12% reactivity worth are installed in the core until an equilibrium fuel-cycle condition is achieved. Under equilibrium conditions the shimming function is accomplished by the depleted-fuel fraction. Control elements, inserted from below, may also be used for some shimming and for axially shaping the flux in combination with the boiling-void distribution (§6.48). The effect of the control elements on the flux can be partially compensated by fuel pins of slightly greater than average enrichment used in the corner of the fuel assembly adjacent to the control blade. However, in planning such an approach, the designer must also consider the effect of the water gap when the control elements are withdrawn, which would lead to an *increase* in thermal flux at the outer region of the assembly from the improved moderation in the gap.

6.45 It is important to realize that dissolved boron, movable control rods, and the effects of fuel loading and burnup all contribute to a reactivity inventory which tends to change with time and which must be maintained to provide both an adequate shutdown margin and sufficient operational maneuverability.

6.46 Typical values for control requirements for plutonium-fueled fast reactor systems are given in Table 6.2.

This lower percentage $\Delta k/k$ requirement compared with the requirement for a thermal reactor is due primarily to the relatively high core conversion ratio and the large fissile-isotope inventory compared with the amount fissioned per exposure cycle. High conversion tends to replace the fissile atoms as they are depleted, and the relatively high enrichment tends to reduce the importance of the loss through depletion. The poisoning effect of the fission products produced is also less important than in a thermal system because of the high fissile inventory and the general tendency for cross sections to be less at higher neutron energies. Bear in mind, however, that, at least for plutonium-fueled systems, the delayed-neutron fraction is less than that for ^{235}U and hence the reactivity inventory of 7.4% is equivalent to about $21.

TABLE 6.2

Control Requirements for Typical Fast Reactor System

Control characteristics		$\Delta k/k$, %
Burnup (equilibrium cycle)		3.7
Cold to hot at zero power		0.5
Zero to full power		0.8
Shutdown margin		2.4
	Total	7.4

CONTROL METHODS

6.47 The principal movable control element is the so-called control rod, which has several different configurations. A typical boiling-water-reactor cruciform-shaped blade,[13] for example, is shown in Figs. 6.4 and 5.7. In a boiling-water system, the rods are normally inserted from the bottom of the pressure vessel. In contrast, in the rod-cluster control assembly for a pressurized-water reactor shown in Fig. 6.5, small rods are distributed in a number of fuel-element assemblies. The rod drives for a pressurized-water reactor normally enter through the top of the vessel.

6.48 The bottom insertion for the boiling-water system is made necessary by a liquid–vapor mixture in the upper portion of the core which reduces the local reactivity. Poisoning from below therefore produces a more even power distribution as well as some flexibility in shaping flux as the zone of voids and the partial insertion of the absorbing elements are independently adjusted. Also, the bottom insertion prevents interference with the steam-separator and dryer equipment above the core (Fig. 6.6). The steam-separator assembly is an array of standpipes containing vanes that impart a vortex motion to the steam–water mixture, with centrifugal forces accomplishing the separation. Remaining moisture is removed in the steam dryer assembly mounted above the separator.

6.49 The control-rod-drive system for the bottom-actuated boiling-water reactor elements is designed differently from that for the top-mounted pressurized-water reactor. Hydraulic drive using water under pressure is used for the boiling-water system, as shown in Fig. 6.7. Under emergency conditions the system hydraulically drives all control rods into the core to make the reactor subcritical. Normal specifications are for a scram velocity of 5 ft/sec within 30

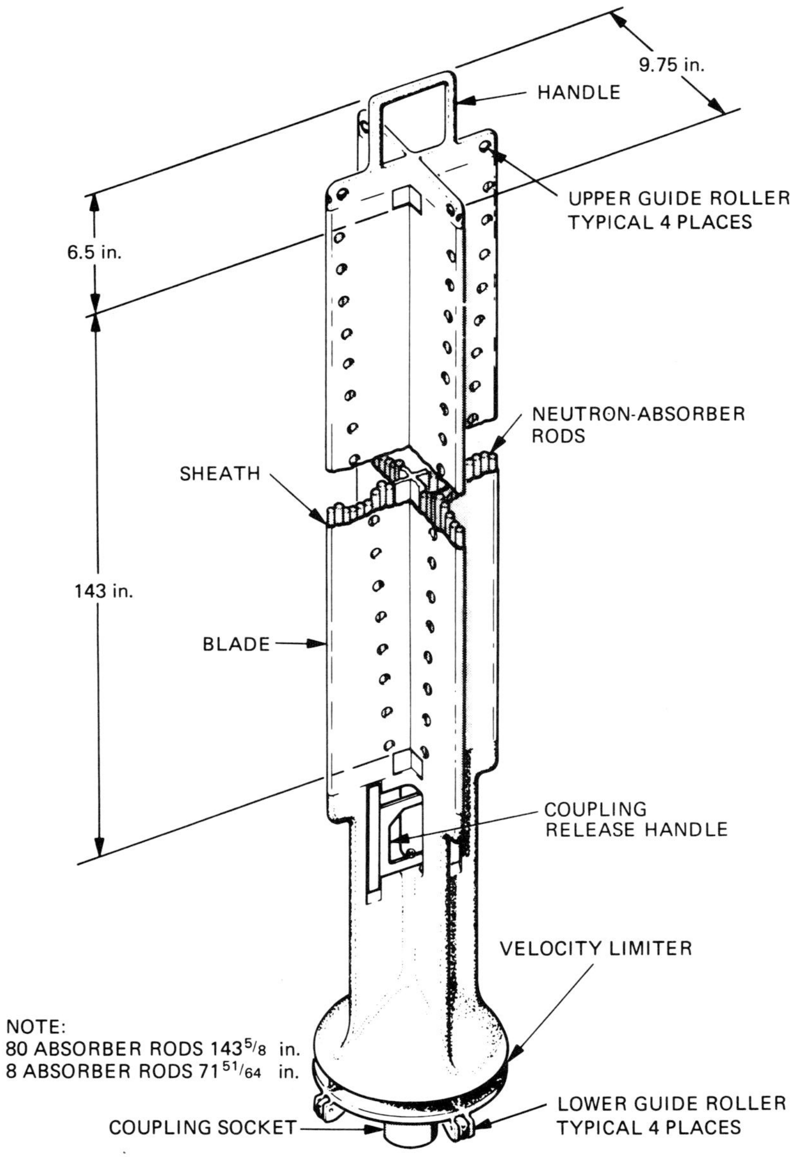

Fig. 6.4 Boiling-water-reactor control rod with cruciform-shaped blade.

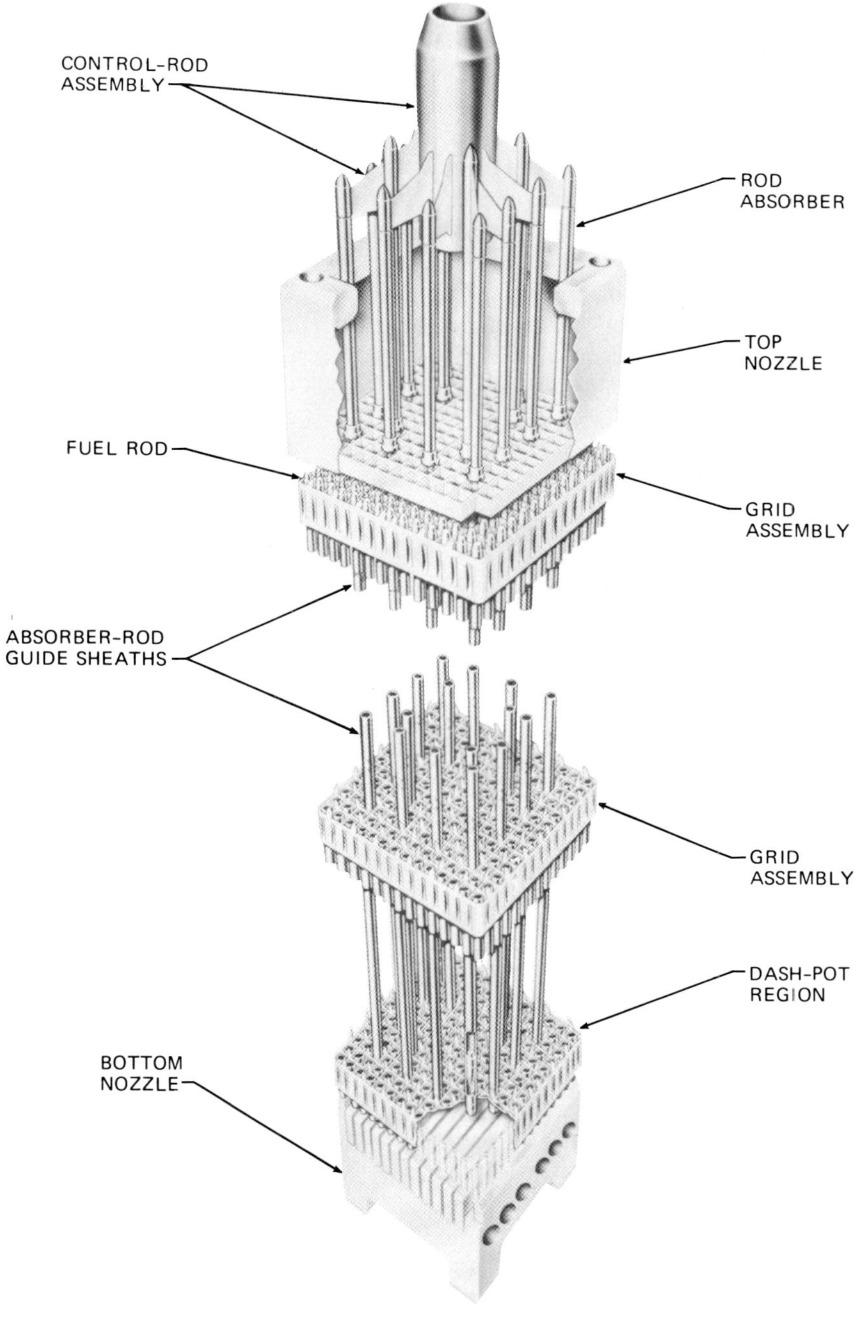

Fig. 6.5. Pressurized-water reactor rod-cluster control assembly.

msec after the start of motion. Each blade therefore performs the dual functions of shaping the power distribution during operation as well as being available for emergency scram. In a typical design the cruciform blades consist of an array of sheathed stainless-steel tubes filled with B_4C powder arranged in the core on a 12-in. pitch.

6.50 A control-rod-cluster assembly used in one type of pressurized-water reactor design is shown in Fig. 6.5. The control rods, distributed within the fuel assembly, are stainless-steel tubes containing a silver–cadmium alloy as the absorber material. This design tends to give more-even power distribution as well as more reactivity control per unit weight than cruciform rods because of a larger surface to volume ratio. In addition, the trend in pressurized-water core design has been toward rather large fuel assemblies compared with the needs for boiling-water systems. A cruciform system with such large assemblies would tend to perturb the flux excessively.

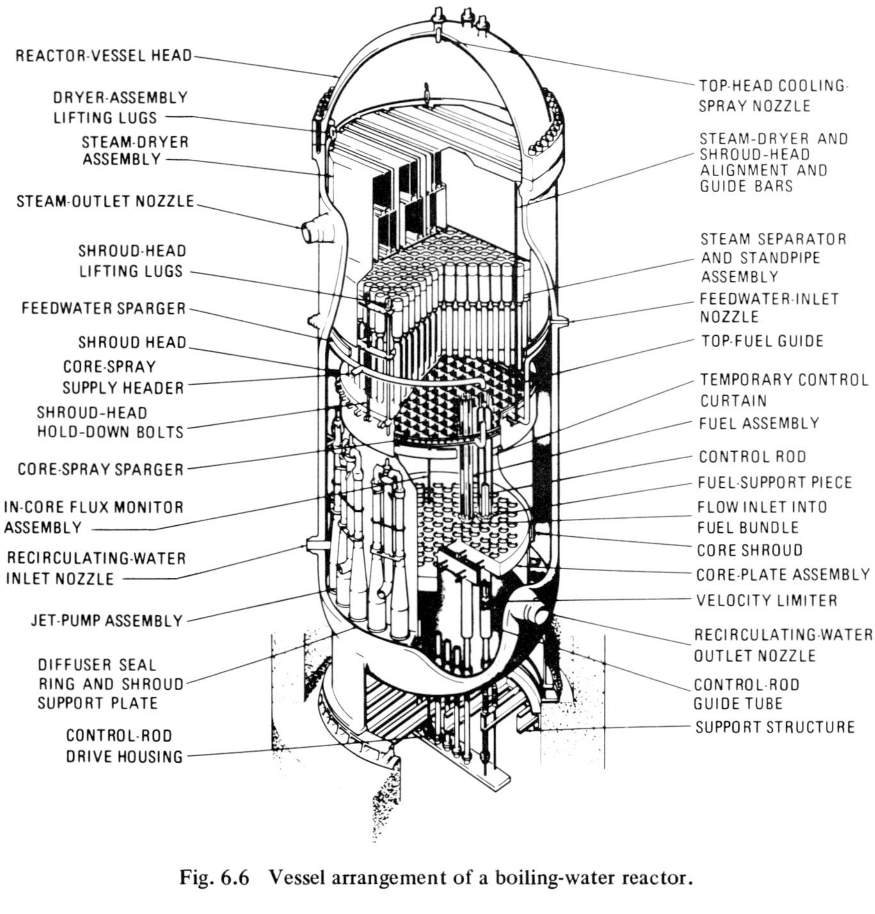

Fig. 6.6 Vessel arrangement of a boiling-water reactor.

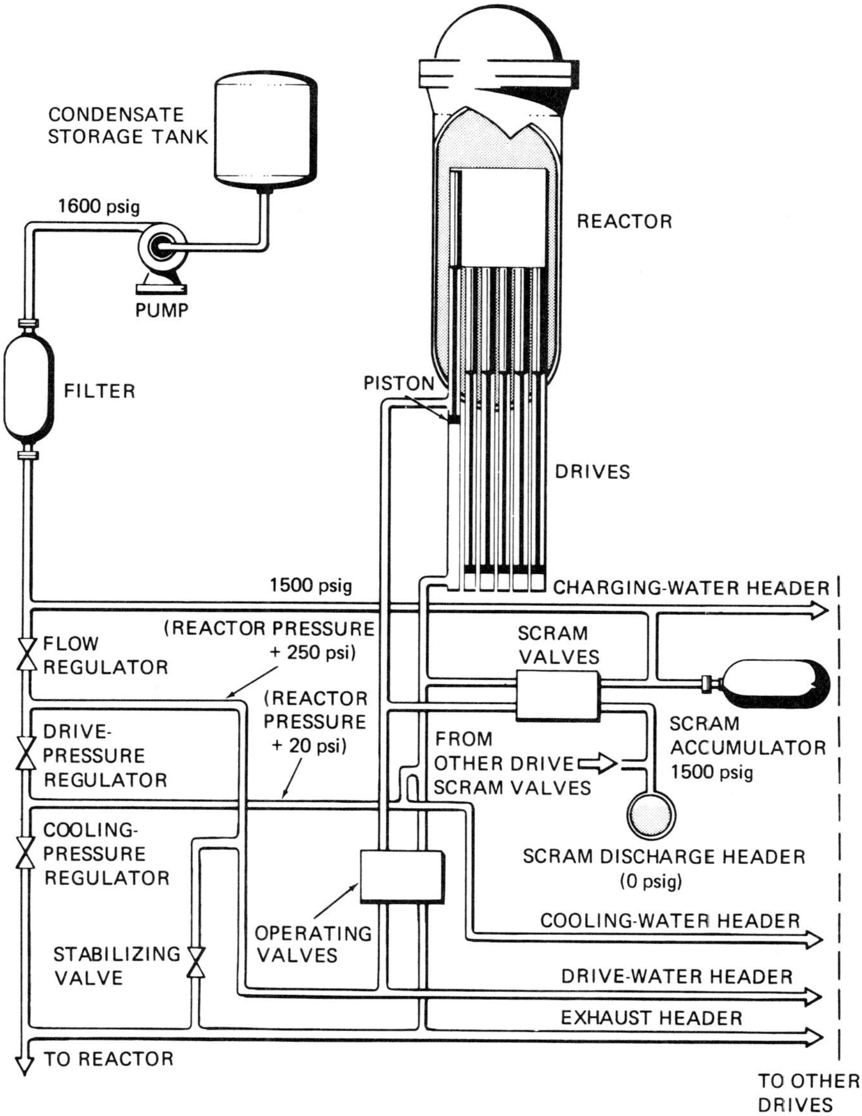

Fig. 6.7 Basic control-rod-drive system.

6.51 Control-rod-drive designs for pressurized-water reactors generally provide for gravity scram action. The drive mechanism may be of several types, such as the rack-and-pinion or the magnetic-jack drive. These concepts have been used in reactors for many years.[14] In the rack-and-pinion drive, the rod is moved vertically by a rack driven by a pinion, which, in turn, is driven by a motor through a gear box. The so-called magnetic-jack system permits the rod to move up or down in incremental steps by selectively energizing combinations of magnetic coils.

6.52 Control rods for high-temperature gas-cooled systems are somewhat different from those for water reactors. For example, in the Fort St. Vrain design,[15] the reactor is controlled by selective movement of 182 control rods, which operate in pairs. Each control rod consists of 11 cylindrical absorber sections containing a boron carbide-filled graphite compact and connected by a central spine to form an assembly approximately 16 ft long. The control rods are operated by winch-type devices located in the prestressed-concrete reactor-vessel (PCRV) (§6.179) top head. A pair of control rods is raised or lowered by the winding or unwinding of cables from a duplex drum driven by an electric motor. The scram velocity, which is only about 1.25 in./sec, is limited by operation of the motor as a generator. Relatively long scram times of about 150 sec are justified by the claim that very rapid consequences do not result from credible accidents, such as those involving loss of the coolant and depressurization.

PERFORMANCE OF CONTROL SYSTEMS

6.53 The expected performance of the control system is a very important design consideration, particularly from the safety viewpoint. For example, the need for redundancy in vital systems depends on the reliability expected. Furthermore, once the plant is designed, its inherent safety depends on the performance to be expected from safety-related components and engineered safety features (§6.166). A typical design iteration is therefore involved.

6.54 Although this discussion concerns the reliability of control mechanisms, the question is really part of a larger subject dealing with the quantitative assessment of the risks of operating nuclear reactors. Such an analysis of both the predicted frequency of failure and the consequences of such failures can guide the designer toward a desirable compromise between expensive conservatism giving maximum safety and unacceptable risks associated with a less costly system. As discussed in §6.101, a mathematical model for a suitable treatment is available, but the lack of sufficient failure data limits the application.

6.55 Since statistically significant experience with large water reactors is just being accumulated in the early 1970s, only limited data have been made available.[16] As frequency of failure experience is obtained and analyzed, however, a comprehensive treatment of control-system reliability should be possible which will provide valuable guidelines to the designer.

6.56 Published operating and safety experience with five power plants* provides some design guidance for control-rod systems. Break-in failures

*The plants examined were Dresden-1, Yankee, Indian Point-1, Humboldt Bay Unit No. 3, and the Shippingport Atomic Power Station.

continued at least 12 to 18 months after commercial operation began. Failure in control-rod systems was generally caused by:

1. Foreign objects or material in drive mechanisms or rod channels.
2. Material deficiencies.
3. Mechanical connector failures or interferences.

Although the nature of some of the malfunctions suggested the possibility of multiple rod failure, no such failure was actually found in the scram mode of operation.

6.57 Real and spurious scrams generally occurred about five times per year. A real scram can be defined as one resulting from a monitored parameter actually being at or above a design trip point; a spurious scram would include all other scram trips, such as those resulting from instrumentation-system malfunctions. Although failures have occurred in individual scram circuitry, no instances were reported wherein the overall system did not or could not respond.

6.58 Control systems for newer reactors similar to those previously described have appropriately been designed conservatively; not only improvements but also a redundancy of critical subsystems have resulted from experiences with the early plants. The designer must therefore treat data regarding failures and malfunctions very carefully, recognizing the systems nature of the problem and taking advantage of analytical techniques developed in the general area of quality assurance (§6.225).

ACCIDENTS AND FISSION-PRODUCT RELEASE

INTRODUCTION

6.59 In considering possible accidents and ways to prevent them, the reactor designer should appreciate the consequences of the accidents. Normally the release of fission products that might affect the surrounding population is a major consideration. Although detailed exposure calculations are quite complex[17,18] for a preliminary safety analysis, the exposure from the radioiodines is normally controlling. That is, if the dosage from the radioiodines can be kept within acceptable limits, other radioisotopes normally do not cause excessive exposure. However, bone-seeking isotopes such as ^{90}Sr which have accumulated over 2 years in cores or parts of cores should also be considered. Although the iodines reach a saturation value quickly, some of the bone seekers, particularly ^{90}Sr, continue to build up in concentration with exposure for long periods of time (Fig. 6.11).

6.60 A number of design considerations, such as siting, containment, and engineered safeguards, are very dependent on the parameters associated with fission-product release. We shall discuss these parameters, therefore, with emphasis on the radioiodines.

6.61 A wide variety of elements are formed in fission, the masses of which tend to fall in two broad bands, a light group with mass numbers from 80 to 110 and a heavy group with mass numbers from 125 to 155. Although there are some minor differences in the yields from different fissile isotopes, these may be neglected in an analysis of hazards. The fission products are initially unstable and decay through several stages by both beta decay and gamma emission. The primary concern here, however, is not a detailed representation of all of the isotopes, but merely an accounting of the radioactive sources of different types which contribute to the dosage that might be received from a cloud of released fission products. Methods of determining reactor inventory, fission-product transport, and matters related to biological characteristics are therefore considered, primarily from the simplified viewpoint generally adequate for conceptual design purposes.

REACTOR FISSION-PRODUCT INVENTORY

6.62 An inventory of specific fission products in the core fuel after a given period of reactor operation is the first step in considering the hazards from a possible release. The simultaneous rates of isotope production by fission, production by decay of a parent, loss by decay, and loss by neutron capture are described by differential equations that lend themselves to computer solution. For a fission-product nuclide, *i,* the familiar balance equation can be written as[19]

$$\frac{dN_i}{dt} = Y_i N_f \sigma_f \phi - N_i \sigma_i \phi + N_{i-1} \sigma_{i-1} \phi - \lambda_i N_i + \lambda_k N_k \tag{6.3}$$

where N_i = number of atoms per unit volume of fission product of species i
N_f = number of fuel atoms per unit volume
Y_i = direct yield from fission of the fission product ($\boldsymbol{\Sigma} Y_i = 2$)
λ_i = decay constant of fission product
λ_k = decay constant, where k indicates the precursor fission product
σ_f = effective microscopic fission cross section of the fuel
σ_i = effective microscopic absorption cross section of the fission product
ϕ = flux
$i - 1$ = fission product leading to i under neutron capture

For some approximate calculations fission products are assigned to two groups according to half-life, and all nuclei in the short-half-life group are assumed to decay at once and all nuclei in the long-half-life group are assumed to be stable. The last two terms in Eq. 6.3 then disappear. This permits the independent variable of Eq. 6.3 to be changed from the time, t, to the flux time $\int_0^t \phi(t)\,dt$, the variable $\int_0^t \sigma_f\,\phi(t)\,dt$, or the fuel burnup, β, where

$$\beta = 1 - \exp\left[-\sigma_a \int_0^t \phi(t)\,dt\right] \tag{6.4}$$

provided the microscopic cross sections can be assumed constant with time, despite changes that will occur as the neutron spectrum shifts with burnup.

6.63 If the operating time of the reactor has been sufficiently long to build up equilibrium concentrations of fission products, a nuclide concentration can be determined by setting dN_i/dt equal to 0. The equilibrium or saturation value therefore serves as a useful reference for charts, such as those of Blomeke and Todd, giving the results of rigorous computer calculations of fission-product concentrations.[20,21] Charted results present three ratios: N_S/N^0_{235}, the saturation values of the isotope in question expressed as a fraction of the initial number of atoms of ^{235}U originally present; N_T/N_S, the fraction of the saturation value at a time τ of irradiation at a constant flux; and N_t/N_T, the fraction of the shutdown value at cooling time t. Thus the fission-product inventory can be determined by referring to three charts. Saturation values for many isotopes are shown in Figs. 6.8 and 6.9, and a typical N_T/N_S chart is shown in Fig. 6.10. In this case ^{131}I approaches its saturation value in about one month. On the other hand, ^{90}Sr takes much longer, as shown in Fig. 6.11 and Table 6.3.

6.64 Tabulations of logarithmic constants are frequently used in calculating the fraction of the shutdown value that remains at various times after shutdown. For example, for ^{131}I, $\log N_t/N = -4.31 \times 10^{-7} t$ where t is in seconds after shutdown. Figure 6.12 is another type of graph giving the total fission-product activity as a function of irradiation time and thermal-neutron flux.[20] Some inventory information, similar to that given in Table 6.3, is also given with AEC siting criteria.[22]

6.65 Although fission-product inventory can be determined by hand calculations with charts or tabulations as mentioned, it is frequently more convenient to use computer codes, such as ISOGEN, which can handle up to 400 nuclide species and provide output in atoms, grams, and curies.[23] In a practical reactor system, the calculation of the inventory is complicated by the effect of a fuel-management scheme wherein the core consists of a number of different subcores, each having a different irradiation history. In addition, the isotopic spatial distribution in the core depends on the time-integrated power distribution. The integrated nuclear-reaction rates also depend on neutron-energy-spectrum effects. These complications can normally be accommodated within the accuracy required for safety calculations by the use of such computer codes as ISOGEN. Inventory calculations also provide a basis for determining the magnitude and distribution of decay power as part of a loss-of-coolant analysis (§6.120).

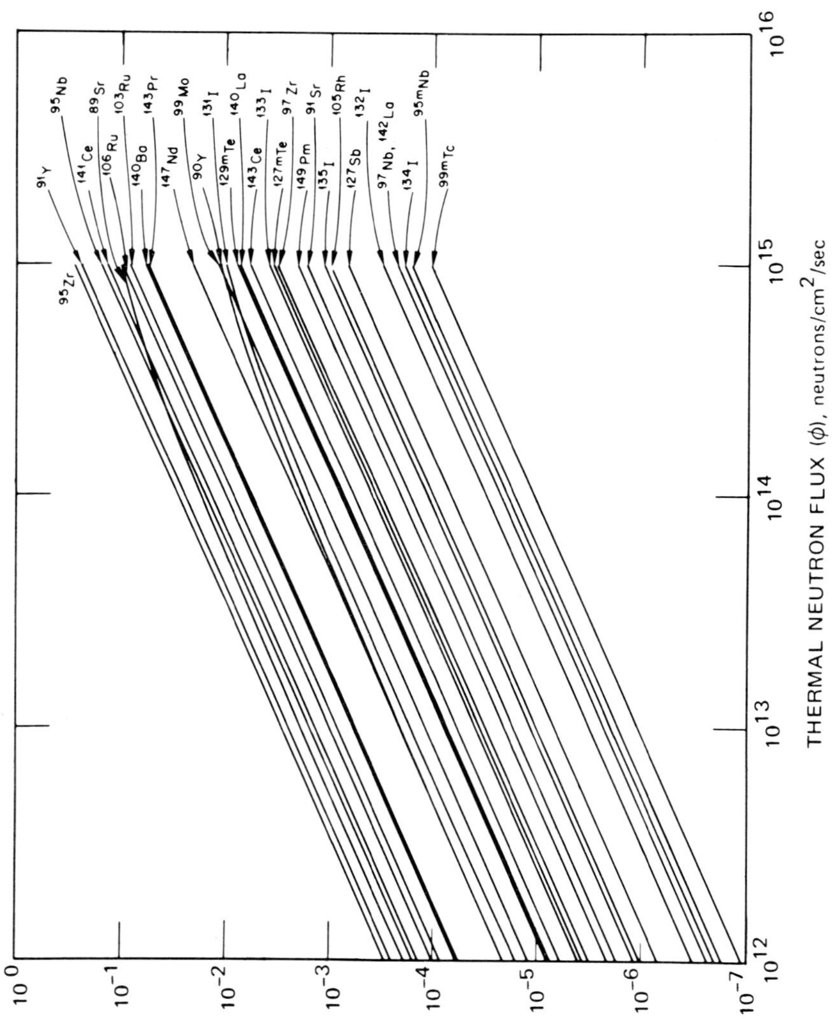

Fig. 6.8 Saturation values of radioactive fission products.

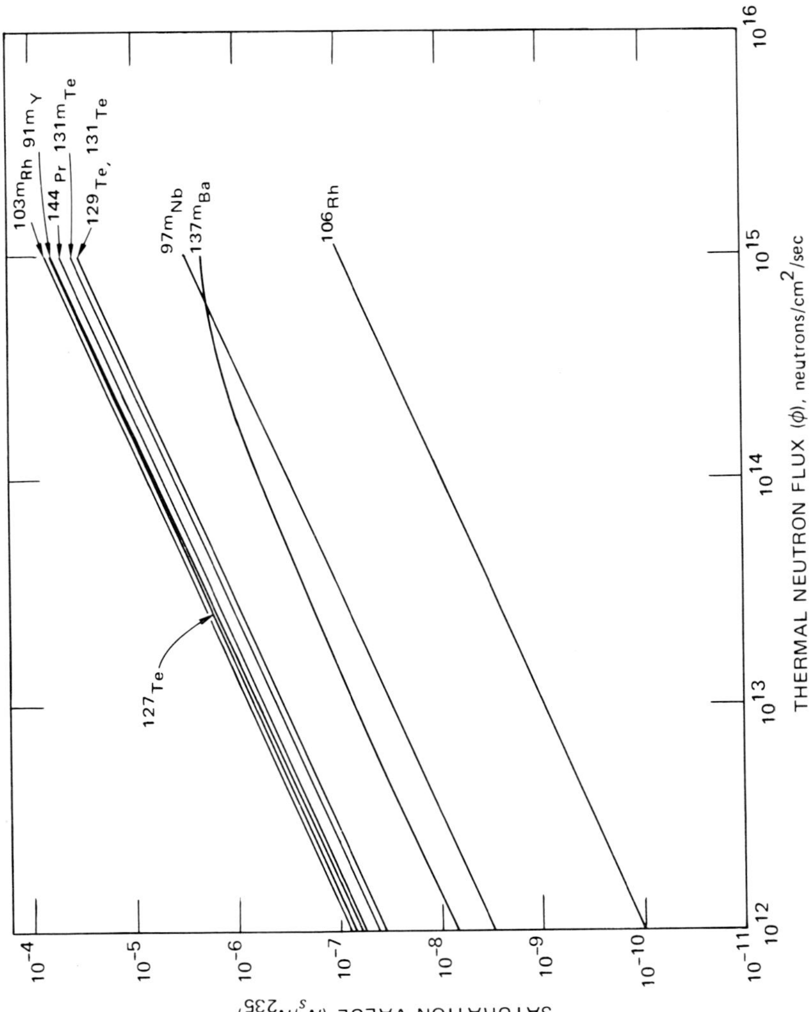

Fig. 6.9 Saturation values of radioactive fission products.

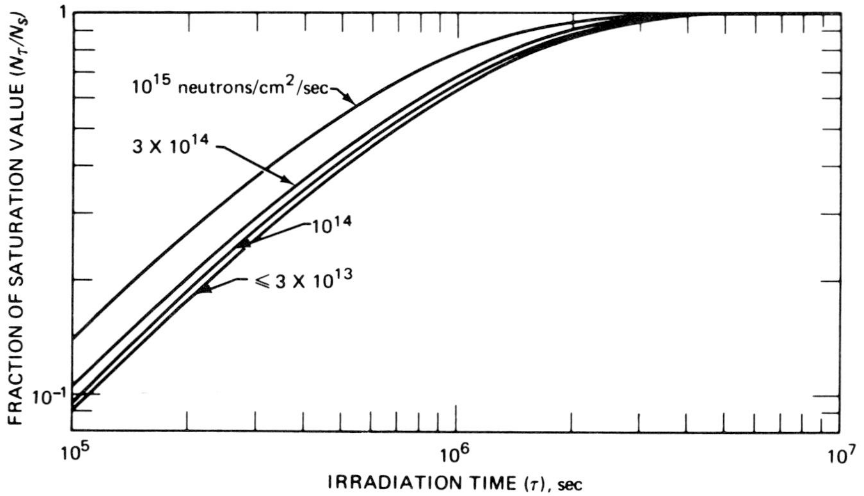

Fig. 6.10 Fraction of saturation value of 8.05-day ^{131}I as a function of irradiation time and flux.

NATURE OF FISSION-PRODUCT HAZARDS

6.66 The biological hazards of the fission products are of different types. In a release to the atmosphere, the most important hazards are the direct radiation from the "cloud" and the inhalation of radioiodine. Other internal effects, however, are worthy of consideration. Therefore we shall subdivide the fission products into those which affect the bones, thyroid, kidney, and muscles. Another important effect of a fission-product release is that deposited radioactive materials may directly contaminate foodstuffs or be incorporated into the nutrients of a food chain. Primary consideration here, however, is given to direct ingestion.

6.67 Small amounts of ingested material can cause considerable damage since the source of the radiation is very close to the exposed tissues and exposure can be for long periods of time. Alpha and beta emitters, which are not particularly hazardous when outside the body, are very dangerous when ingested since the emitted particles tend to dissipate all their energy within a small volume of tissue.

6.68 Important parameters are therefore the mass that enters the blood stream, the radioactive decay rate, the tendency of the isotope to concentrate in specific sites once it enters the blood stream, and the length of time the isotope is retained at the site. Such concentration is a result of the normal metabolic processes of the body followed by natural inactive isotopes. Iodine, for example, concentrates in the thyroid gland, and calcium-like elements are deposited in the bone.

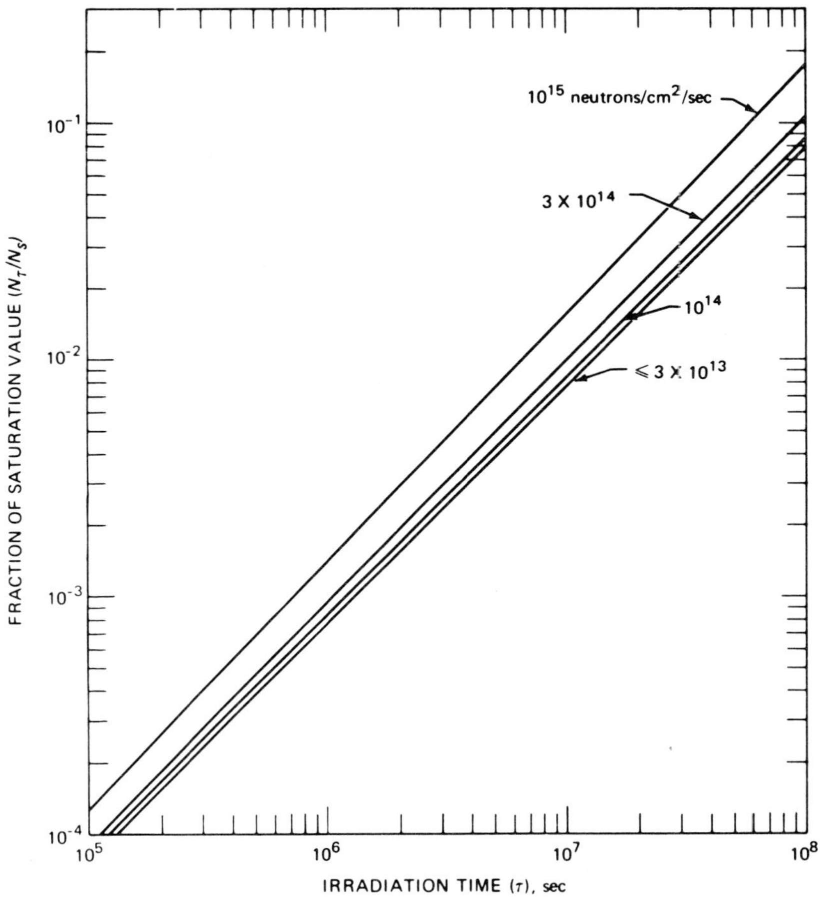

Fig. 6.11 Fraction of saturation value of 28-year ^{90}Sr as a function of irradiation time and flux.

6.69 One of the preceding parameters, the length of time an isotope remains in the body, is dependent on the isotope's "biological half-time," which is the time required for the isotope to decrease to half its initial value through elimination by natural processes. A combination of the radioactive half-life and biological half-time gives the "effective half-life," defined as the time required for the amount of a specified radioactive isotope in the body to fall to half its original value as a result of both radioactive decay and natural elimination (§6.75).

6.70 Isotopes representing the greatest potential internal hazard are therefore those with relatively short radioactive half-lives (high rate of emission) and comparatively long biological half-times. The iodine biological half-time

TABLE 6.3

Internal Dosage of Fission Products*

Isotope†	Radioactive half-life (T_r)	Yield, %	f_a	Effective half-life (T), days	D, mrems/μc	Reactor inventory, curies/kw: At 400 days	Reactor inventory, curies/kw: At equilibrium
Bone							
^{89}Sr	50.5 days	4.8	0.28	50	413	43.4	43.6
^{90}Sr-^{90}Y	27.7 years	5.9	0.12	6400	44,200	1.45	53.6
^{91}Y	57.5 days	5.9	0.19	58	337	53.2	53.6
^{144}Ce-^{144}Pr	282 days	6.1	0.075	243	1,210	34.7	55.4
Thyroid							
^{131}I	8.1 days	2.9	0.23	7.6	1,484	26.3	26.3
^{132}I	2.4 hr	4.4	0.23	0.097	54	40.0	40.0
^{133}I	20.5 hr	6.5	0.23	0.87	399	59.0	59.0
^{134}I	52.5 min	7.6	0.23	0.036	25	69.0	69.0
^{135}I	6.68 hr	5.9	0.23	0.28	124	53.6	53.6
Kidney							
^{103}Ru-^{103m}Rh	39.8 days	2.9	0.01	13.3	6.9	26.3	26.3
^{106}Ru-^{106}Rh	1.0 year	0.38	0.01	19	65	1.8	3.5
^{129m}Te-^{129}Te	33.5 days	1.0	0.02	10.4	46	9.1	9.1
Muscle							
^{137}Cs-^{137m}Ba	33 years	5.9	0.36	17	8.6	1.2	53.6

*Adapted from T. H. Burnett, *Nucl. Sci. Eng.*, **2**: 382-393 (1957).

†Mass number with *m* refers to *metastable* isomer.

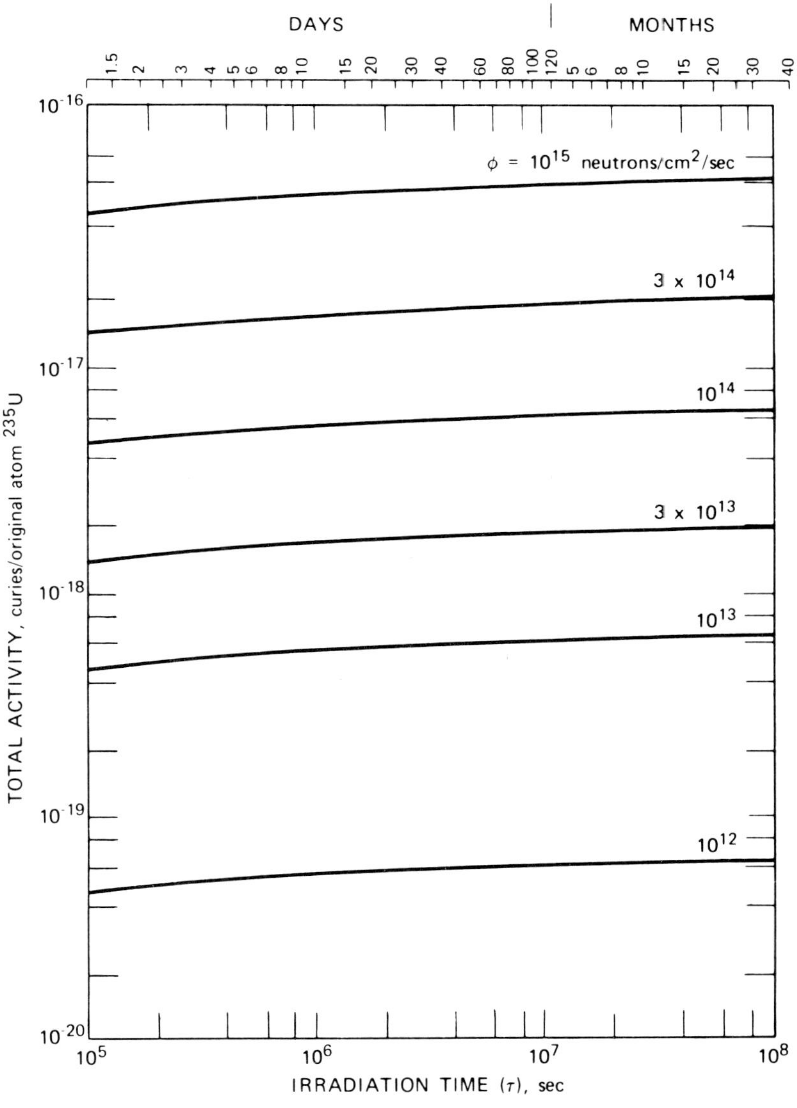

Fig. 6.12 Buildup of total fission-product activity with irradiation time and thermal-neutron flux.

varies considerably among individuals but averages about 90 days. Since ^{131}I has a radioactive half-life of 8 days, it remains in the thyroid gland during essentially all of its radioactive life.

6.71 Strontium and barium, fission products that are similar to calcium chemically, are deposited in bone tissue. Strontium-90 is a "bone seeker" of particular concern since it remains in the skeleton for a long time (Table 6.3), decays (0.54-Mev beta emission) to ^{90}Y, which, in turn, emits 2.27-Mev beta particles. Another dangerous bone seeker, though neither a fission product nor a short-half-life isotope, is plutonium. Plutonium-239, an alpha-emitting isotope, has a radioactive half-life of 24,000 years as well as a biological half-time of about 200 years. Consequently, once it is deposited in the body, mainly on certain surfaces of the bone, the amount of plutonium present and its activity decrease at a very slow rate. In spite of their short range in the body, the continued action of alpha particles over a period of years can cause significant injury.

6.72 Materials that do not gain access to the blood stream are normally not of great concern, at least to design. Particle size, which affects the ability of the nose to filter out the material, and solubility are therefore important parameters. On the other hand, evidence indicates that about 10% of inhaled plutonium remains in the lungs and bronchial lymph nodes.

6.73 The subject of biological effects of radiation is much more complicated than indicated by the preceding very simplified orientation treatment. Many isotopes, with their own decay schemes, particle energetics, and chemical, physical, and biological behavior, are involved. An analysis for a preliminary safety evaluation provides a picture of biological effects which is detailed only enough to establish design conditions.

INHALATION DOSAGE

6.74 Although the hazard downwind from a fission-product release consists of direct radiation from the cloud and inhalation of isotopes of such elements as iodine and strontium, the controlling factor is normally assumed to be radioiodine inhalation. We shall therefore consider a calculation of the inhalation dosage.[24]

6.75 Assume that the inhalation intake in microcuries can be expressed as

$$I = B\bar{Q}t \qquad (6.5)$$

where t = exposure time (sec)

$\bar{Q}$ = average concentration ($\mu c/m^3$)

B = breathing rate (m^3/sec)

If a breathing rate for moderate activity of 30 liters/min is assumed, $B = 5 \times 10^{-4}$ m^3/sec and the amount inhaled (in microcuries) would be

$$I = 5 \times 10^{-4}\ \bar{Q}t \qquad (6.6)$$

The total internal exposure (in rems) to a body organ can be expressed as

$$D = 73.8 f_a I \frac{NT}{m} \mathbf{\Sigma}\ E_i(RBE) \qquad (6.7)$$

where m = mass of body organ (g)

f_a = fraction of inhaled material residing in organ

I = amount of inhaled material (μc)

E_i = energy absorbed in the organ (Mev)

N = factor accounting for localized deposition

T = effective half-life (days)

RBE = relative biological effectiveness to convert dose units from rads to rems

The effective half-life in days is

$$T = \frac{T_f T_r}{T_f + T_r} \tag{6.8}$$

where T_f is the biological half-life and T_r is the radioactive half-life.

6.76 Exposure values for most fission products of interest have been calculated, and some are given in Table 6.3. The isotopes are grouped according to the critical organs they affect with energy values for the particular organ. A single inhalation of 1 μc of the respective isotope is assumed in the calculation of the D value. Table 6.3 also gives typical reactor-inventory values. The potential hazard in a release therefore depends on the product of the D value and the inventory. Radiation limits are discussed in §6.78.

6.77 Under accident conditions, however, the *relative* release and transport of the isotopes from the inventory may vary widely (§6.83). Solids, for example, depend on such matters as particle size, the possibility of chemical action or absorption by sprays in the containment, and, in fact, general physical and chemical properties. In addition, the more volatile fission products would be released to a greater extent than other equally hazardous but more stable elements, such as strontium. On the other hand, considering standard fraction-release assumptions of 1% bone seekers and 25% iodine,[22] the controlling hazard can shift from iodine to the bone seekers as the fuel exposure time is lengthened, as shown[25] in Fig. 6.13. The relative concentrations at a point of exposure, however, are very dependent on the nature of the release and possible removal patterns during transport.

Radiation Limits

6.78 As is true for handling certain industrial chemicals or other potentially harmful activities or agents that are part of our accepted industrial society, standards are necessary for the use of radiation so that harmful effects will be minimized. In the United States after World War II, the National Committee on Radiation Protection and Measurements (NCRP) was organized with representatives from professional societies and governmental agencies, as well as individual experts. The recommendations developed by this group have served as the basis for most radiation-protection programs and, later, for rules and codes adopted by the various regulatory agencies in the United States.

6.79 Exhaustive studies have produced the recommendations listed in Table 6.4. It is important to remember that these are intended as limits and not as design criteria. In fact, the recommendations call for exposures to be kept "at the lowest practicable level" for each particular set of circumstances. The role of radiation limits in reactor siting is considered in §6.94 et seq.

FISSION-PRODUCT TRANSPORT

6.80 Biological effects of released fission products depend on their being inhaled or acting as radiation sources at some location away from the reactor and near the exposed subjects. As a result, the *nature* and *rate* of the *release* and *transport* of the fission products from the point of release to the exposure location is an important link in the chain of events to be analyzed in considering the consequences of possible accidents, both as part of a safety evaluation and as a guide to design. We shall therefore discuss the parameters affecting

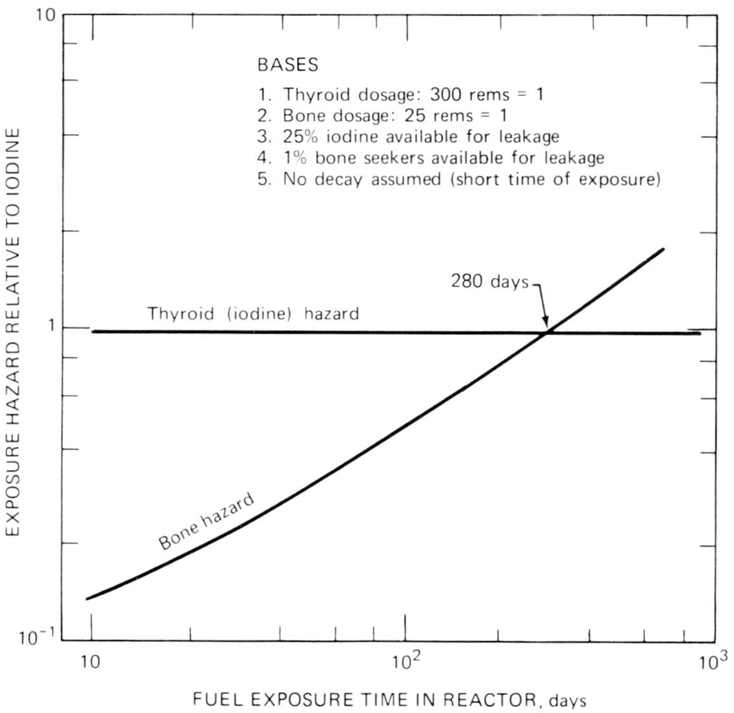

Fig. 6.13 Relative hazards of fission-product release.

TABLE 6.4

NCRP Radiation Dose Limits

Category	Value
Maximum permissible dose equivalent for occupational exposure	
Combined whole-body occupational exposure	5 rems in any one year
Retrospective annual limit	10 to 15 rems in any one year
Long-term accumulation	(Age − 18) × 5 rems
Skin	15 rems in any one year
Hands	75 rems in any one year (25/quarter)
Forearms	30 rems in any one year (10/quarter)
Other organs, tissues, and organ systems	15 rems in any one year (5/quarter)
Fertile women (with respect to fetus)	0.5 rem in gestation period
Dose limits for the public, or occasionally exposed individuals	
Individual or occasional	0.5 rem in any one year
Students	0.1 rem in any one year
Population-group dose limits* (*total* exposure from "man-made" radiation above and in addition to natural background which averages about 0.1 rem per year)	
Genetic	0.17 rem average per year
Somatic	0.17 rem average per year
Emergency dose limits (life saving)	
Individual (older than 45 years if possible)	100 rems
Hands and forearms	200 rems, additional (300 rems, total)
Emergency dose limits (less urgent)	
Individual	25 rems
Hands and forearms	100 rems, total
Family of radioactive patients	
Individual (under age 45)	0.5 rem in any one year
Individual (over age 45)	5 rems in any one year

*Average for population group is intended to ensure that no individual exposure exceeds 0.5 rem/year.

fission-product release and transport to gain background for treating accident possibilities.

6.81 The release and transport of fission products following an accident depend largely on the nature of the accident. For example, a simple rupture of the cladding in a ceramic-fueled system from a fuel-assembly-handling accident is likely to release only the accumulated fission-product gases. On the other hand, a hypothetical accident in which all coolant is lost and some of the engineered safety features do not operate might lead to melting and partial vaporization of the fuel with the release of essentially all the fission products. Even should this be the case, "back-up" design features prevent release to the public. In the analysis of the transport, the behavior of accident-released fission-product iodine is complex and quite dependent on the exact nature of the accident. The release is by no means instantaneous. Steps such as diffusion through the lattice of a ceramic fuel, mass transfer through a boundary-layer-type concentration gradient near a phase boundary in a molten fuel, and gas-phase diffusion all have definite transport rates that lend themselves to calculation, provided a reasonably descriptive model can be formulated. The development of such a model, which may be simplified or sophisticated, depending on the needs and the information available, is therefore the first step in an analysis. A simple approach is generally adequate for design purposes since the range of uncertainty can be limited because the maximum fission-product level to be considered is equal to that in the fuel inventory.

6.82 Even for a fairly simplified model, however, a number of process steps are involved, some with different transport paths. A simplified schematic-flow representation is shown in Fig. 6.14. The initial containment can be considered

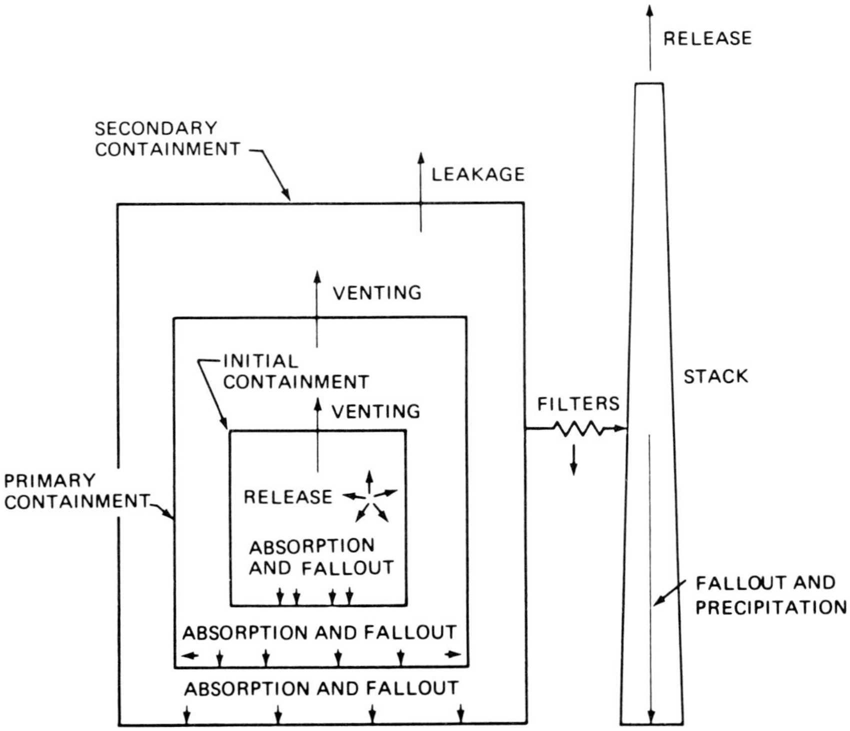

Fig. 6.14 Fission-product transport within containment.

as the cladding on a ceramic fuel pin, the primary containment as the coolant system, and the secondary containment as an outer pressure container. In each process the deposition and absorption of fission products is possible. Furthermore, the rates involved depend on numerous parameters. A controlled leakage or venting is indicated between initial and primary and between primary and secondary barriers. This means merely that the design provides for some leakage and includes cleanup devices to prevent release of radioactivity from the stack. The designer may provide methods for controlling the activity in the event of an accident causing failure of one or more of the containment shells; these are considered and described in *Engineered Safety Features and Siting* (§§6.166 et seq.). The process possibilities, including rate-dependent steps, remain essentially unchanged, however. An evaluation of the effects of the release on the surroundings should the outer containment fail, as desirable for a design iteration, also depends on the competing rate-dependent steps, including the effects of countermeasures.

6.83 Even if attention is centered on the behavior of the iodine isotopes as a simplification, a description of the transport processes occurring from the point of release from the fuel to leakage from the outermost containment still involves a large number of variables. For example, the fission products, which may include fine particulate matter, are transported by liquids, gases, or two-phase mixtures encountered along the path. The relative transport by such fluids and possible deposition on interior surfaces is very dependent on the physical and chemical form. Since iodine may have particularly varying characteristics that affect both adsorption and desorption rates on surfaces, an analytical description of transport and deposition becomes quite complicated. An analysis of iodine transport should include, of course, the effects of core and containment sprays (§6.186) that should activate during accident conditions.

6.84 Since fission products released from failed fuel elements but held within the primary containment constitute no hazard to the general public, accident analysis is normally confined to possible situations releasing fission products into the secondary containment, either directly or from failure of the primary containment system. Exposure of the public is therefore due to known allowable leakage rates for the secondary containment or the possible failure of such containment.

6.85 For such safety evaluation as that for licensing, the transport picture can be further simplified by making "conservative" assumptions that have been standardized.[22] It may be assumed, for example, that fission products are released from the core into the containment atmosphere with no credit taken for the preventive action of such engineered safety features as sprays and air cleaners. The containment concentration available for leakage to the environment may then be assumed as 100% of the noble gases, 50% of the halogens,* and

*An assumed release of 25% of the radioactive iodine inventory is suggested in several safety guides (§6.246).

1.0% of the solids. Such a release model is often called "a TID-14844 accident assumption." The containment may then be considered a fission-product source on the basis of the leakage specifications, which are normally 0.1 to 0.5% of the containment volume per day at standard conditions and at design pressure for light-water reactors.

6.86 As better models of fission-product transport within the containment become possible, more-sophisticated predictions will probably replace the preceding simplified release assumptions, with reduced concentrations likely through the action of engineered safety features. Computer codes,[26] such as EXDOSE, are available for calculating dosage from airborne fission products and for generating the fission-product inventory as a function of flux and time as well as providing for transport plate-out and filtration within the containment. Thus release and transport knowledge currently available[27] can be used to carry out a quick exposure evaluation.

ATMOSPHERIC DISPERSION

6.87 The next stage in the transport of released fission products is from a source outside of the reactor containment, such as from the top of a stack to a receptor at a so-called exposure point. Since the exposure depends on the concentration at this point, atmospheric-diffusion parameters affecting the transport and, in turn, the concentration must be considered. The meteorological principles that apply are described in a substantial body of literature.[28] The simple dispersion model described in the following paragraphs, however, is useful for many safety evaluations that permit order-of-magnitude estimates.

6.88 Attention is normally restricted to radioactivity released within several hundred feet of the ground and receptors within a few miles of the source. The behavior of the lowest layer of the atmosphere is therefore of primary interest. Let us consider a continuous point source which generates a smokelike plume that is carried along by the wind but is subject to dispersion by atmospheric turbulence as shown in Fig. 6.15.

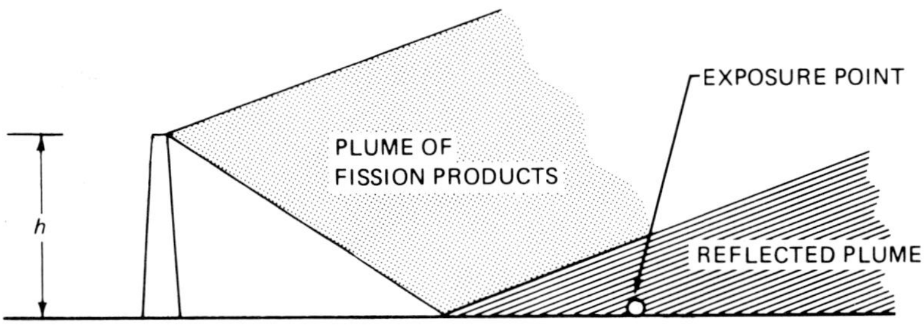

Fig. 6.15 Atmospheric dispersion model.

6.89 A basic equation for dispersion from continuous point sources is the so-called "generalized Gaussian plume formula," or the bivariate normal mode dispersion equation:*[29]

$$\chi = \frac{Q}{\pi \sigma_y \sigma_z \bar{u}} \exp\left[-\frac{1}{2}\left(\frac{y^2}{\sigma_y^2} + \frac{h^2}{\sigma_z^2}\right)\right] \tag{6.9}$$

where χ = ground-level air concentration (curies/m^3)
Q = source strength (curies/sec)
$\bar{u}$ = average wind speed along x-axis (m/sec)
h = height of the source above the ground (m)
y = lateral (crosswind) distance of the receptor from the plume axis (m)
σ_y = standard deviation of the cloud center-line concentration in the horizontal direction
σ_z = standard deviation of the cloud center-line concentration in the vertical direction

6.90 The extent of diffusion, expressed by σ_y and σ_z, can be determined experimentally. Values are given in Figs. 6.16 and 6.17 for six meteorological categories, described in Table 6.5, as suggested by Pasquill.[30]

6.91 Among the various conditions, category F is the most conservative for safety analysis since it yields the least diffusion and the highest concentration. Calculation results for a variety of conditions can be obtained from charts such as Fig. 6.18, where the quantity $\bar{u}$ (χ/Q) is shown as a function of distance from the source for a given source height. The concentration from a stack reaches a peak some distance downwind, depending on the weather condition, as might be expected from the plume behavior shown in Fig. 6.15.

Volume Source

6.92 Equation 6.9 should be modified for the effect of a volume source in determining atmospheric dispersion from the possible emission of airborne radioactive material through leaks in a reactor-containment structure. The source material is assumed to be distributed uniformly throughout the volume of the building enclosing the reactor with the enclosure designed to have, at most, some specified leakage rate under the postulated accident conditions. The source strength, Q, is therefore defined, but the leak can be considered as distributed.

6.93 Since a reactor building must have a turbulent wake in its lee, it has been generally assumed that any material escaping from the containment building would be dispersed rapidly into a volume proportional to the product of the building cross-sectional area and the wind speed. Although various treatments have been suggested for this effect involving modification of Eq. 6.9,

*This general formulation can be made equivalent to the Sutton and Pasquill relations with differences primarily in the use of constants and the application involved.

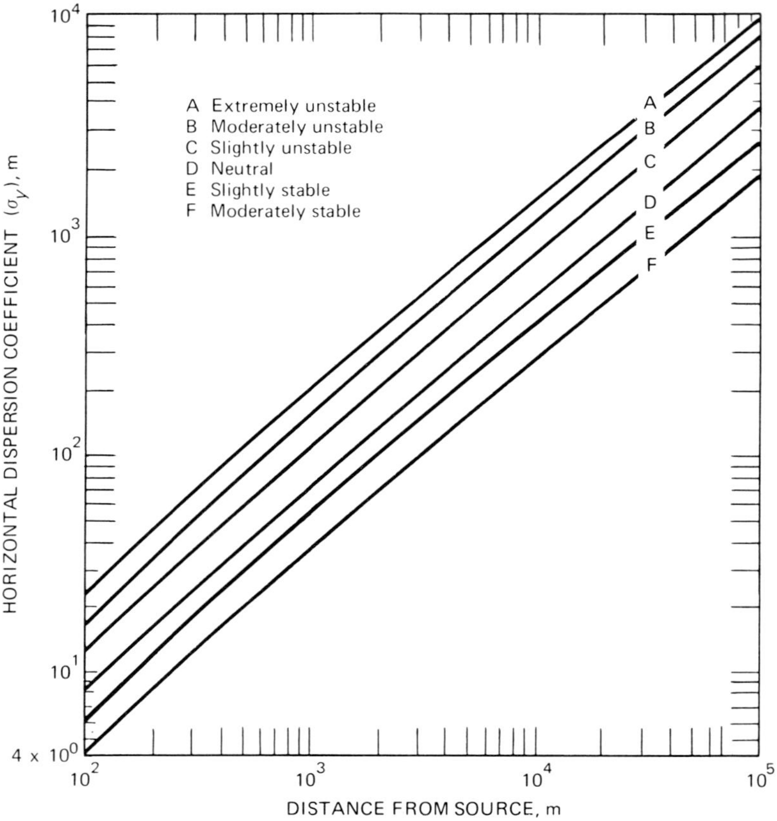

Fig. 6.16 Lateral diffusion, σ_y, vs. downwind distance from source for Pasquill's turbulence types.

a simplified approach involving a *virtual point source* is often adequate for preliminary safety studies. In this method an imaginary point source upwind of the containment building is defined so that the Gaussian plume produced would have a width equal to that of the building at its actual location.

Example*

Consider a 1060-Mw(e) pressurized-water reactor [3250 Mw(t)] with a containment vessel consisting of a 140-ft-diameter cylinder topped with a hemisphere dome. The top of the dome is 172 ft above grade, and the bottom of the

*This example is intended for instructional purposes as a simplified illustration of the use of Eq. 6.9. It is *not* representative of design practice.

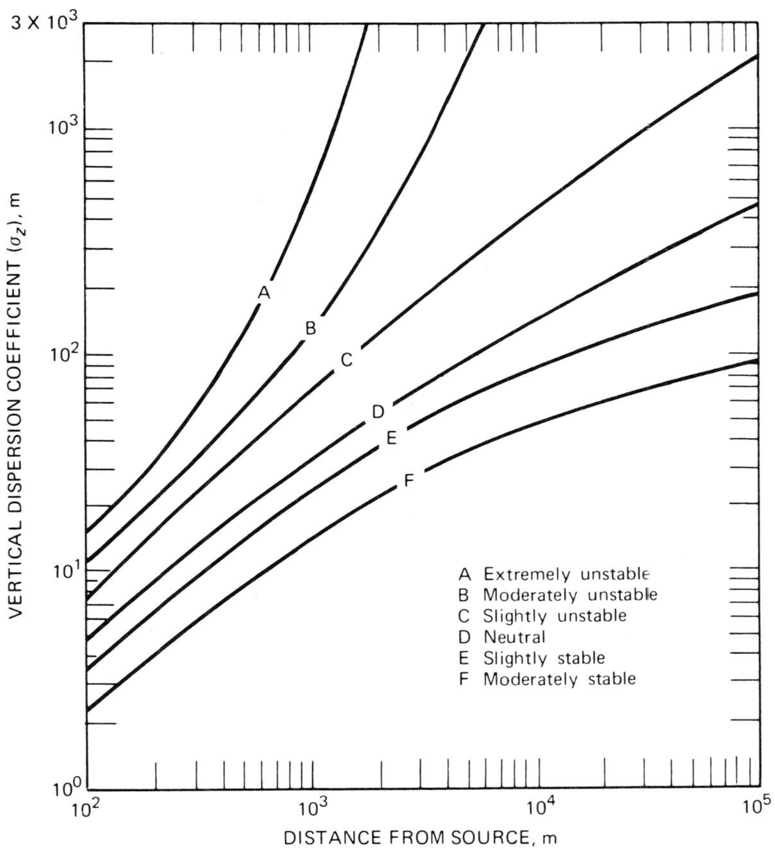

Fig. 6.17 Vertical diffusion, σ_z, vs. downwind distance from source for Pasquill's turbulence types.

containment vessel is 26 ft below grade. Assume that, owing to a hypothetical accident, 50% of the core dose equivalent* halogen inventory of 1.51×10^8 curies of ^{131}I is released immediately into the containment and that a containment leak rate of 0.1% per day applies. The closest approach to the plant site boundary is about 800 m. Estimate the 2-hr thyroid inhalation dose in rems at this exclusion distance.

Assume a virtual point source 680 m upwind of the center of the containment. The weather is moderately stable.

*Initial ^{131}I inventory is multiplied by factor of 1.88 to account for contribution of other iodine and tellurium isotopes.

TABLE 6.5

Meteorological Categories*

Surface wind speed, m/sec	Daytime insolation: Strong	Daytime insolation: Moderate	Daytime insolation: Slight	Thin overcast or ≥ 4/8 cloudiness	≤ 3/8 cloudiness
<2	A	A–B	B		
2	A–B	B	C	E	
4	B	B–C	C	D	E
6	C	C–D	D	D	D
>6	C	D	D	D	D

*A, extremely unstable conditions; B, moderately unstable conditions; C, slightly unstable conditions; D, neutral conditions (applicable to heavy overcast, day or night); E, slightly stable conditions; and F, moderately stable conditions.

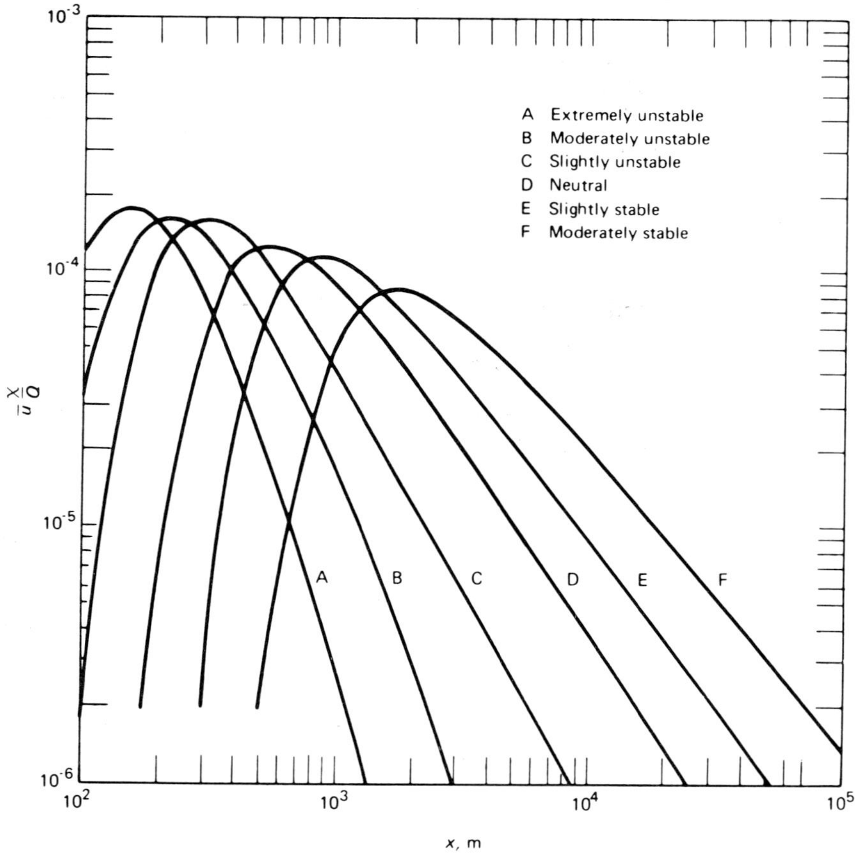

Fig. 6.18 Values of $u\,(\chi/Q)$ as a function of downwind distance, x, for various types of weather. Source at surface; h, 30 m.

The source strength is

$$(0.50)(0.001)(1.51 \times 10^8) = 7.55 \times 10^4 \text{ curies/day}$$

or

$$Q = \frac{7.55 \times 10^4}{(24 \times 3600)} = 0.875 \text{ curies/sec}$$

Other parameters are $\sigma_y = 55$ m (Fig. 6.16, $x = 1480$ m, moderately stable)
$\sigma_z = 19$ m (Fig. 6.17)
$\bar{u} = 2$ m/sec (Table 6.5)
$y = 0$
$h = 86$ ft, or 26.2 m

Applying Eq. 6.9 yields

$$\chi = \frac{Q}{\pi \sigma_y \sigma_z \bar{u}} \exp \left[-\frac{1}{2} \left(\frac{y^2}{\sigma_y^2} + \frac{h^2}{\sigma_z^2} \right) \right]$$

$$= \frac{0.875}{\pi(55)(19)(2)} \exp \left[-\frac{1}{2} \left(\frac{26.2}{19} \right)^2 \right] = 5.15 \times 10^{-5} \text{ curies/m}^3$$

$D = 1.484$ rems/μc for ^{131}I (Table 6.4)

Inhalation rate $= 5 \times 10^{-4}$ m^3/sec

$$\text{Two-hour dose} = (5.15 \times 10^{-5})(2)(3600)(5 \times 10^{-4})(1.484 \times 10^6)$$

$$= 274 \text{ rems}$$

Note that no source reduction through the operation of core sprays has been assumed. In a more realistic model, a time-dependent leakage source would be calculated considering the combined dynamic processes of spray cleanup, plate-out, conversion to methyl iodide, and desorption of methyl iodide. Typical results for this type problem would then be of the order of 20 rems.

EXPOSURE CRITERIA

6.94 Once the isotopic composition at a given point is established, the inhalation dose can be calculated. The consequences of such inhalation are then evaluated by comparing the dose with some reference criteria for conditions that would result in negligible injury. Since the question of what may constitute an acceptable risk in radiation injury is subject to some interpretation, guidelines established by the authorities are helpful to the designer.

6.95 The two principal guidelines published by the USAEC for evaluating design, safety, and site relations are 10CFR20 (Title 10, Code of Federal Regulations, Part 20) and 10CFR100. The former sets forth normal operating requirements, and the latter covers abnormal (accident) operations occurrences.* The guideline 10CFR100, when used with AEC publication[22] TID-14844, suggests dose–distance relations used to determine the safe distance at which a nuclear plant of a given size should be located from a population center and from a low population zone. It also establishes a circumferential distance called the exclusion area, within which the licensee has the authority to determine all activities. The area can be traversed by a highway and nonreactor activities permitted, provided arrangements are made to control the area in an emergency. Section 10CFR20 establishes guidelines that limit the maximum permissible releases of radioactive material to the environment from a nuclear facility. It also specifies allowable doses of radioactivity for operations, as well as nonoperations, personnel.

6.96 Different exposure criteria apply to each of the three zones considered. Within the exclusion area, consideration is given to the radiation worker who accepts a certain hazard connected with his job which would not be appropriate to apply to the general public. Current guidelines provide that an individual located on the *boundary* of the exclusion area for a period of 2 hr after an accidental fission-product release should not receive more than 25 rems whole-body radiation and 300 rems to the thyroid from iodines. In an example given, a breathing rate characteristic of the active portion of a working day is assumed equal to 3.47×10^{-4} m^3/sec. This contrasts with an average breathing rate of 2.32×10^{-4} m^3/sec for a person in the low population zone.

6.97 In the low population zone, "appropriate" measures could be taken on behalf of the people in the event of a serious accident. A person at the outer boundary is assumed to be exposed to no greater than 25 rems of whole-body and 300 rems of thyroid radiation from a radioactive cloud from a release *during the entire period of its passage.* For initial calculations of integrated thyroid dose, a 30-day exposure may be assumed.

6.98 As an alternate to accepting the standard 50% halogen release, the effect of various engineered safeguards in reducing the level may be considered. Sprays and filters, for example, could give dose-reduction factors as high as 100.[13] Offsite exposure could then be calculated by using the specified containment leakage as a point source, as before. Such considerations appear particularly desirable as sites for large reactors with large exclusion areas and low population areas surrounding them become harder and harder to find. Also, accidental-release criteria are quite different from *environmental*-dose considerations concerned with normal day-to-day exposures from radionuclides in the environment of operating facilities.[31]

---

*Further explanation of federal-regulation terminology is given in §6.246.

6.99 Environmental standards for all types of plant effluents are prescribed in 10CFR20. Standards that apply beyond the boundary of an exclusion area to the population at large as indicated in Table 6.4 are generally 10% of the maximum-permissible-concentration values that apply for the continuous exposure of the radiation worker. Since dilution normally occurs beyond the boundary, large population groups are not likely to be exposed to such values as the 170 mrem/year limit. Furthermore, most designs provide for effluent concentrations far below prescribed limits. However, the designer should keep in mind that environmental effects from nuclear power plants have produced some controversy, with doubts expressed about the adequacy of emission standards.[32] In fact, because of such controversy, some utilities have agreed to site-boundary exposure limits substantially below the standard maximum values given in Table 6.4. Recommended design guidelines for annual average boundary exposure rates are ⩽10 mrems for noble gases and ⩽5 mrems for iodines.

6.100 Accidental-exposure criteria for large population centers cannot be defined by simple man-dose guidelines since it is not reasonable to assume that a dosage level much higher than background can really be *acceptable* to the public. In an attempt to quantify the question, however, the trade-off between the probability of an accident that causes a release and the consequences of the release has been considered.[33] This entire matter is relevant to reactor-plant siting practices and is discussed in §§6.102, 6.103, and 6.193.

ACCIDENT MODELS AND FISSION PRODUCTS

6.101 The amount of fission products that can be accidentally released can be used as one measure of the severity of an accident. For example, Beattie,[34] relating a release of 10^4 curies of iodine activity to a reference 4×10^6 person population center within a 10-mile radius, obtained a total dosage of 2.2×10^6 man-rems, which would be likely to cause 33 cases of thyroid cancer compared with a natural incidence of 72 cases per year.

6.102 Farmer[33] assumed that a release of "a few thousand curies" might reasonably occur once during 10^3 reactor years of operation, or 1.5×10^4 reactor years for the 10^4-curie release. If there is one reactor on a site, the equivalent casualty rate is then 0.002 case per year. This approach was used for a release-criterion curve, such as that in Fig. 6.19.

6.103 A systems approach that considers accident sequences that may take several paths (§6.226) can be used to determine a probability of failure for a spectrum of accident possibilities. If an iodine release can be related to each accident possibility studied, the results can be plotted in probability–consequence space, as shown in Fig. 6.19. A boundary line representing points with units of curies per year is then established, separating an area of conservative design criteria at the lower left from an area of unacceptable hazards in the upper right. The line shown, a first attempt at such a standard, considers that a large release, even if rare, is probably less acceptable than a small

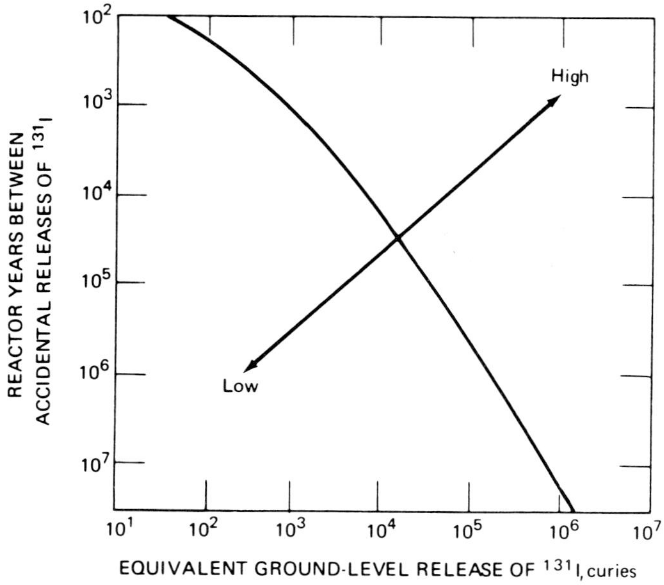

Fig. 6.19 Proposed release criterion.

release on an equal curie per year basis. This approach may therefore be used to provide guidelines for the engineered safety features that may be required in the design to control both the failure probability and the release. Experience data are needed, however, for such a probabilistic methodology design approach as that discussed under Engineered Safety Features and Siting (§§6.166 et seq.).

6.104 A different approach used in the United States and some other countries is to classify accidents as either credible or hypothetical. Credible accidents are considered within the realm of reasonable possibility and generally include the assumption that the safety and control devices in the design will indeed operate. In other words, a single initial failure is assumed. Hypothetical accidents are those considered primarily as a basis for design and presumed not to be within a reasonable level of probability. Safety systems are presumed not to operate in the sense that at least two simultaneous independent failures will occur.

6.105 The hypothetical accident normally produces a chain of events with such eventual grave consequences as core meltdown and release of fission products to the containment and, in turn, the possible subsequent release of fission products and perhaps fuel to the outside environment. Such accidents are termed design-basis accidents if they provide a reference for designing the containment and operating various installed engineered safety features. A determination of the adequacy of emergency core cooling, blast protection, and the specifications for the containment may therefore depend on the analysis of such a hypothetical accident, including calculated consequences such as the buildup of high temperatures and pressures from energy release. The design can then be tested by studying the fission-product release and transport under postulated accident conditions and determining the possible environmental exposure levels.

DESIGN ANALYSIS FOR REACTOR-SAFETY FEATURES

6.106 The reactor design must provide ways to avoid any serious consequences from a large number of relatively minor accidents, a variety of malfunctions, and sometimes serious failures. Each of these requires analysis and provision in the facility design for the appropriate safeguards. This subject includes loss-of-coolant incidents, which serve as design-basis accidents for containment for water reactors and for design of associated engineered safety features. Such design questions, however, are treated under Engineered Safety Features and Siting (§6.166 et seq.).

6.107 Although the type of incident and the corresponding counter-measures will vary from one reactor concept to another, there is a general similarity. A chemical-shimmed pressurized-water reactor will therefore be used here for illustration. For this analysis we shall classify abnormal situations that will require design provisions as being either operational in origin or due to mechanical malfunctions.

OPERATIONAL-FAULT REACTIVITY INSERTIONS

6.108 One type of incident results from various deviations from operational procedures. Although such incidents are not normally considered as serious problems, this is true only because of the various design provisions. Most of these provisions are standard design practice that has evolved as the relevant technology has developed. We mention here, however, some of the features of this area of safety design and analysis so that they will not be taken for granted.

Uncontrolled Rod Withdrawal from a Subcritical Condition

6.109 Although a continuous reactivity addition during start-up would cause an excursion, the possibility is minimized through strict administrative controls backed up by various instrumentation and control devices. Examples of such safeguards include:

1. An automatic rod-stop circuit prevents further rod withdrawal should the start-up rate measured by several independent channels exceed a predetermined value.

2. Automatic reactor trip is provided as backup should a start-up rate continue to be excessive.

3. An automatic-power-level reactor trip is provided to scram the reactor should a predetermined power level, normally set at a fraction of full power, be exceeded.

6.110 In the extremely unlikely event that these instrumented safety features become inoperative and an excursion takes place, the effect of the fuel Doppler coefficient will normally reduce the energy released to manageable levels before an overpressure condition can occur.

6.111 Analyses can be carried out to consider the effects of various combinations of possibilities.[35] For example, typical transient behavior for a start-up accident at hot, zero-power conditions is shown in Fig. 6.20 for the large pressurized-water reactor used as an example. Changes in fuel, cladding, and coolant temperatures are shown in response to a ramp reactivity insertion of 0.2×10^{-4} δk/sec when the initial reactivity is 0.99. Also shown is the bounding effect achieved by tripping the reactor when a 118% power level is reached. As a third possibility, a reactor trip actuated by a high start-up rate completely controls the transient.

6.112 In the analysis shown in Fig. 6.20, nuclear parameters were assumed according to the following rationale:

1. *Initial reactivity.* The initial value of reactivity (k_0 = 0.99) is selected from the range of expected values so that the greatest energy release will occur before any protective measures are taken.

2. *Doppler coefficient.* Since the magnitude of the power peak reached during the first part of the transient, for any given rate of reactivity insertion, strongly depends on the fuel-temperature reactivity coefficient, a "low" negative design value is used for the start-up accident ($\alpha_f = -1 \times 10^{-5}$ $\delta k/°F$).

3. *Moderator coefficient.* The contribution of the moderator reactivity coefficient is negligible during the initial part of the transient because the heat-transfer time constant between the fuel and the moderator is much longer than the nuclear-flux-response time constant. However, after the initial nuclear-flux peak, the succeeding rate of power increase is affected by the moderator reactivity coefficient. Accordingly, the most positive expected value is assumed for the moderator coefficient since this yields the maximum rate of power increase ($\alpha_w = +1 \times 10^{-4}$ $\delta k/°F$).

4. *Reactivity insertion rate.* For a given set of reactivity coefficients, the maximum energy release in the fuel corresponds to the maximum rate of reactivity insertion; thus the reactivity-insertion rate corresponding to withdrawal of the control-rod group at the maximum velocity in the region of maximum reactivity worth is assumed in the analysis (δk/sec = 2×10^{-4}).

5. *Reactor trip.* The behavior of the reactor is analyzed as though no protective trip were actuated until the overpower reactor-trip set point is reached. The most adverse combination of instrument and set-point errors, as well as delays for trip signal actuation and rod release, is taken into account.

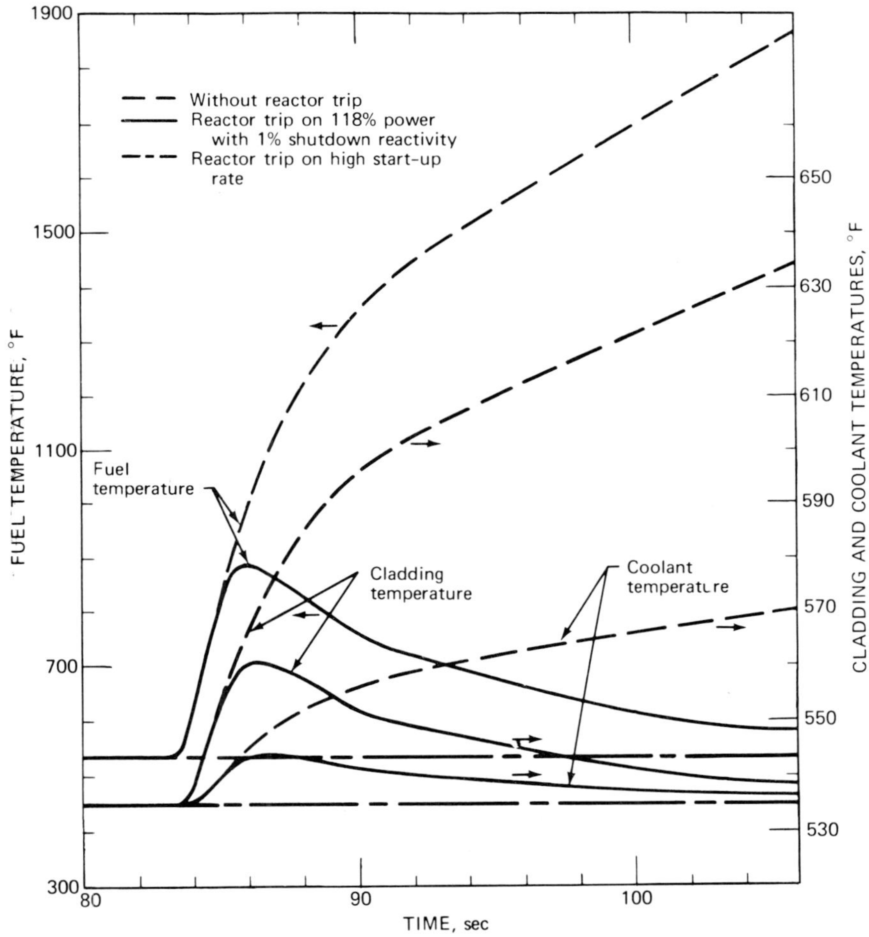

Fig. 6.20 Hot start-up temperature transients.

Also, the rate of negative reactivity insertion corresponding to the trip action is computed based on the assumption that the highest worth rod is stuck in its fully withdrawn position.

Rod Withdrawal at Power

6.113 A somewhat different type of incident occurs in water reactors if the control rods are withdrawn while the reactor is at power without a corresponding increase in the turbine-cycle load. Since the core power and in turn the heat flux increase but heat extraction increases very little, there is a net increase in coolant temperature. Unless terminated by instrumentation devices, this power mismatch would result in a critical heat flux or DNB (departure-from-nucleate-boiling) condition that would lead to fuel failure.

6.114 Various automatic features of the reactor instrumentation system reduce the probability of this type of incident. These include an overpower stop signal to trip the reactor if power ranges exceed a set point and other trips actuated by pressure, power, and temperature conditions. Redundancy is provided in each case. Analyses generally consider the effect of varying design parameters on the DNB ratio to determine trip ranges that will minimize DNB response.

Boron Dilution

6.115 In a chemical-shimmed pressurized-water reactor, it is possible to inadvertently increase the reactivity by changing the boron concentration. This type of change is very slow, however, and numerous alarms and indications of the impending condition provide very adequate opportunity for countermeasures.

Cold-Water Addition

6.116 The addition of cold water to the core results in an insertion of positive reactivity, the magnitude of which depends on the value of the negative moderator coefficient of reactivity. In a chemical-shimmed plant, dissolved boric acid reduces the negative moderator coefficient, thereby decreasing the potential reactivity insertion resulting from a cool-down. As a result, the most severe condition for a cold-water transient exists at the end of core life when the coolant system is essentially boron free and the most negative moderator coefficient exists.

6.117 Cold water may be inadvertently inserted into a pressurized-water-reactor core by the start-up of an inactive loop, the addition of excess cold feedwater to the steam generators, and a sudden increase in the steam discharge from the steam generators. Transients of this type can normally be accommodated by the reactor control system without a reactor trip resulting. In severe cases protection is obtained by reactor trips actuated by overpower or other deviations from normal operating conditions.

MECHANICAL MALFUNCTIONS

6.118 Possible safety problems caused by mechanical failures or malfunctions comprise the second design-analysis category. In addition to various loss-of-coolant conditions, fuel handling and other types of failures are included.

6.119 A coolant loss in the core may result from a "material failure" of a pipe or component in the reactor coolant system. Other causes are mechanical malfunctions, such as the improper opening of a valve or valves, allowing discharge from the high-pressure coolant system to a low-pressure system. Accidents are prevented, therefore, by ensuring a very high reliability of materials and components, rather than depending on instrumentation and control features. Should a failure occur, however, such control features play an important role in tripping the reactor and reducing the effects of the accident through the operation of "engineered safety features" (§6.166).

6.120 Loss-of-coolant accidents can be divided into three categories: minor losses that are compensated by various design features with no serious effect on operation; larger breaks that would result in reactor trip and for which provisions are available for cooling the core; and a design-basis-accident double-ended large pipe fracture that could release fission products to the containment.

6.121 For very small leaks or breaks, the coolant loss can be made up by the pressurizer level control (Fig. 6.21), which will add makeup coolant. In addition, in some systems the refueling water-storage tank can be used for supplementary coolant.

6.122 For larger breaks in which the discharge rate exceeds the ability of the charging pumps to replenish the loss, an initial decrease in pressure is sensed by instruments, which initiate the tripping (scramming) of the reactor, the closing of various isolation valves, and the safety injection of borated water at several points in the coolant system, as shown in Fig. 6.21. The possibility of core damage depends on the relative size of the break and the capability of the countermeasures in preventing an uncovering of the core with a resulting temperature transient. An analysis, normally using a digital-computer program and considering the various mechanisms involved, is needed to determine the extent of the incident. For example, expulsion of the coolant through the break is governed by the mechanisms of subcooled flow until the calculated system pressure falls to saturation at the mass-average temperature. After saturation is reached, two-phase flow through the break is assumed with no slip. System pressure remains at saturation level, with flow limited by sonic choking as long as the pressure ratio across the break exceeds the critical value for the mixture.

6.123 One such analysis for a 500-Mw(e) pressurized-water reactor[35] determined that neither would cladding burst nor fission products be released for a 10-in. break or smaller. In this case the so-called hypothetical design-basis accident consists of a double-ended break of a 27.5-in. main loop line. This type of accident has been extensively studied and analyzed, consistent with its importance in safety design. A brief discussion of the interplay of various parameters during the accident is useful here.

6.124 The loss-of-coolant accident can be considered in terms of a sequence representation,[36] as shown in Fig. 6.22. Here the accident is initiated either by a reactivity insertion or by failure of the primary system. A reactivity insertion causes fuel-cladding failure, pressure-pulse generation, and other large mechanical effects which induce primary-system failure and subsequent loss of coolant.

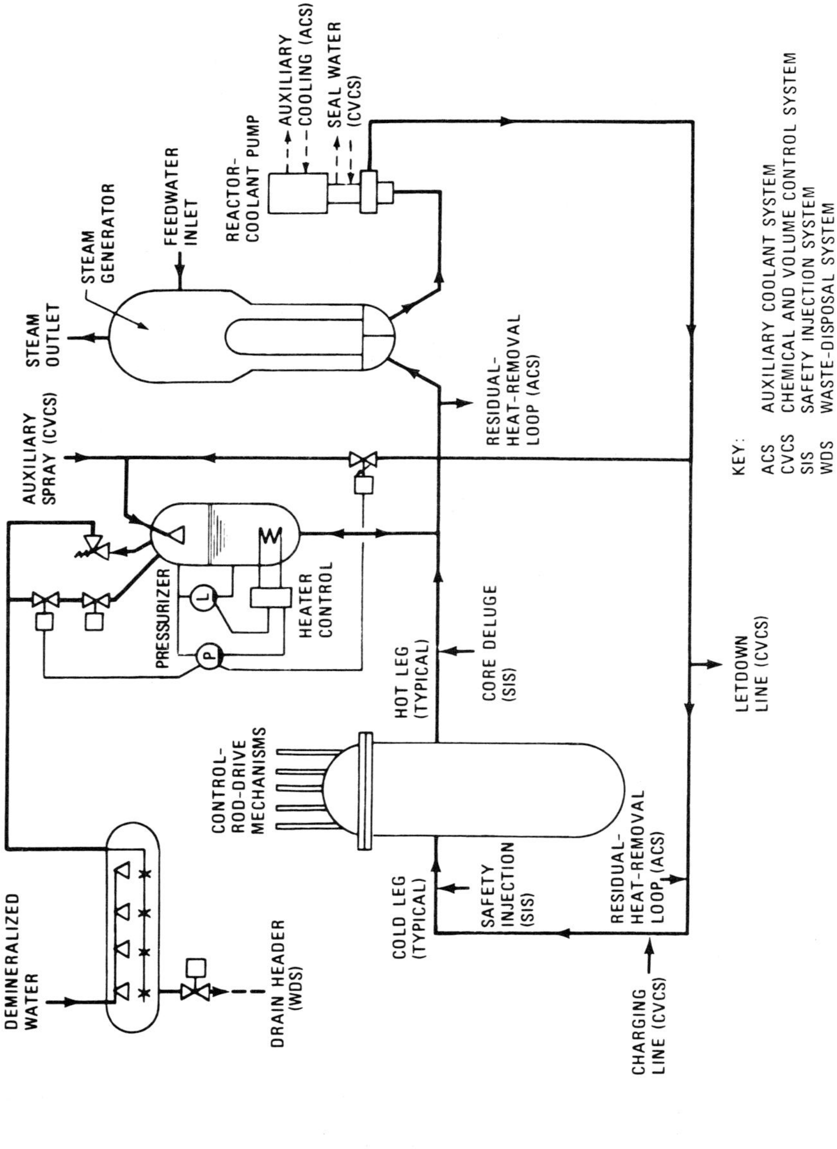

Fig. 6.21 Reactor coolant system.

Accidents are more directly initiated, however, when a crack develops and propagates to a critical-break size and causes primary-system failure, followed by rupture and loss of coolant. In the initial subcooled blowdown step, decompression may be accompanied by large transient pressure gradients that can impose damaging loads on the system. One reason is the different saturation pressures caused by the different temperatures in various parts of the coolant system.

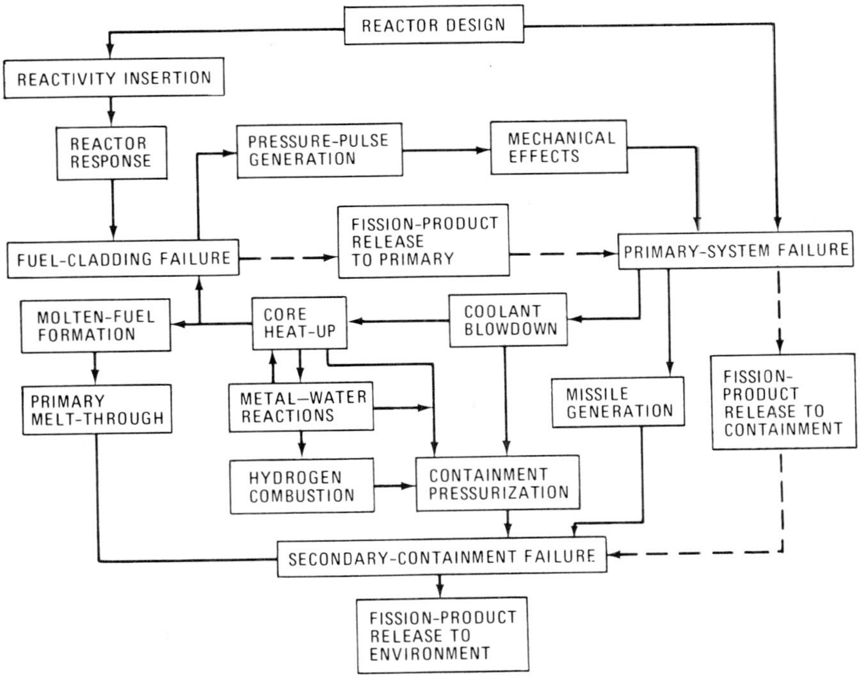

Fig. 6.22 Loss-of-coolant accident sequence.

6.125 In the subsequent saturated blowdown, flashing into vapor causes loss of core-cooling capability. The flow regimes are important in determining the emergency core-cooling requirement, and the output of mass and energy affect the containment response.

6.126 As the core begins to overheat, several processes come into play. First, with zirconium and stainless-steel cladding along with steam in the system, the potential exists for metal–water reactions, which add another source term to the core-heat-up process. Also, hydrogen, a by-product, can burn when released into the containment and thus affect the containment pressure. The core heatup can continue and eventually cause cladding melting followed by fuel meltdown. Since the melt migrates to the bottom of the reactor vessel, primary-system melt-through may occur, and the molten mass (about 800 tons of UO_2) can appear in the containment.* The potential for containment failure at this point is extremely high. Therefore the first requirement for the emergency core-cooling system is to prevent this chain of events from occurring – to stop core heatup before fuel melts and migrates into the containment.

6.127 Immediately after cladding failure, fission products are released to the primary system, from which they are expelled through the break and into the containment. The high containment pressure in turn expels these fission products into the environment. Engineered safety systems therefore include the emergency core-cooling system and the containment heat-removal systems, which limit the containment pressurization by removing energy from the containment atmosphere and hence reduce the dissipation of fission products. Safety systems, such as filters within the containment, also control fission-product release. These systems are discussed in greater detail in §6.186.

6.128 The accident sequence can also be divided, as shown in Fig. 6.23, into core response, primary-system response, and containment response. First, the system rupture occurs with the attendant blowdown. The potential then arises for large pressure drops and high fluid velocities which give rise to mechanical effects on the core and the remainder of the primary system. Fluid is dumped out of the primary system, and what remains flashes to steam. As shown, the subsequent chain of events begins with cladding and fuel failure and ends with failure of the containment.

6.129 If the interrelations due to operation of the safety system are introduced, the systems analysis becomes more complicated than that indicated by Figs. 6.22 and 6.23. The simplified diagrams are useful for designing safety systems, however, since they indicate the points in the accident sequence where engineered safety systems can (1) reduce the probability of initiation of an accident, (2) control the response of the system if the accident is initiated, and (3) limit the consequences of an accident if the accident runs its course.

6.130 In another phase of design, a number of analytical methods and computer programs have been developed to describe the effects of a loss-of-coolant accident under a variety of postulated situations.[37] Such studies can be compared with experimental work of AEC's LOFT (loss-of-fluid test) Program.[37] A typical blowdown pressure transient is illustrated in Fig. 6.24 for a large double-ended break. About 80% of the coolant mass may be lost in the first 10 sec after the break, and DNB is reached in about ¼ sec.

*The phrase *Chinese Syndrome* is sometimes used to describe the possibility that a molten core will break through the containment and slowly melt its way downward through the earth toward China.

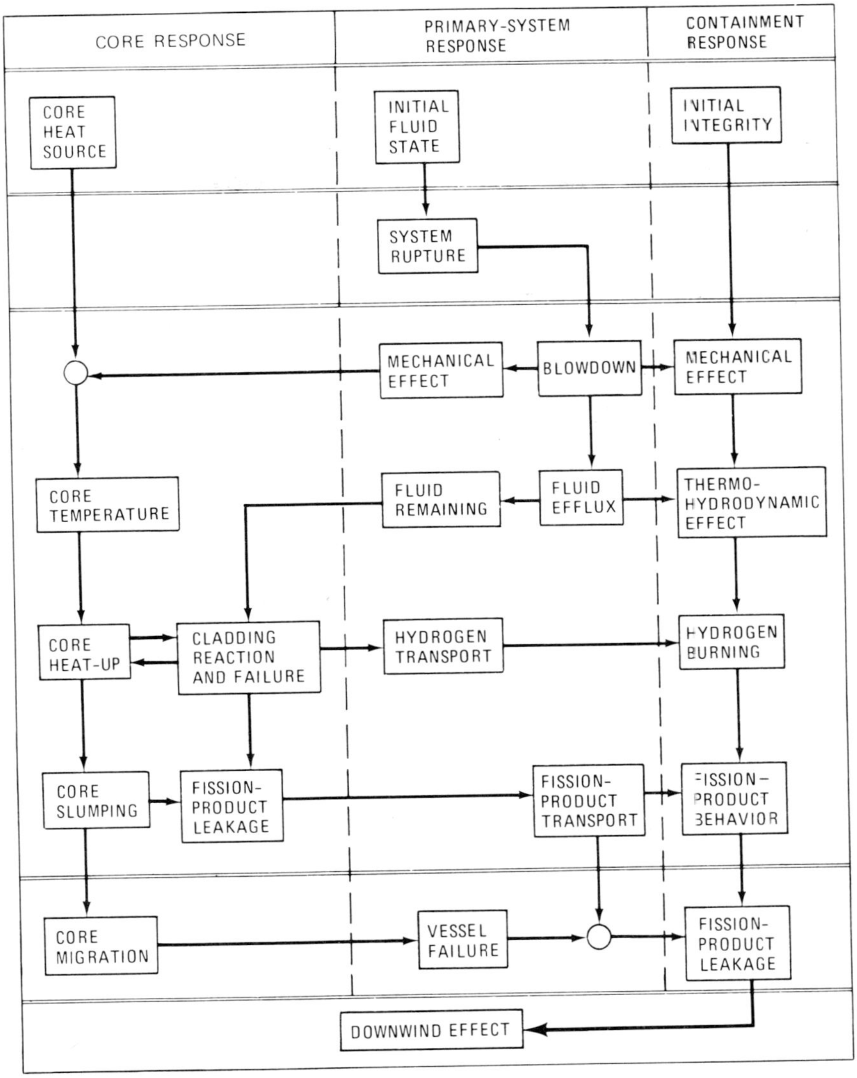

Fig. 6.23 Detailed division of the loss-of-coolant accident.

Loss of Coolant Flow

6.131 A related problem, normally not affecting public safety but important to design, is the possibility of a reduction in coolant flow while the reactor is at power. The resulting decrease in cooling ability could lead to an increase in coolant temperature with DNB exceeded and perhaps a plastic deformation of the cladding.

6.132 Design features include devices for tripping the reactor which are initiated by pump failure, coolant-temperature rise, or other deviations from nominal conditions. In addition, flywheel inertia can be provided on the coolant pumps to provide coast-down flow during the first 10 to 20 sec after a loss of power. Natural circulation of coolant through the core and reactor coolant loops should induce sufficient flow after the coast-down is completed to remove the residual heat of the core without damage to the fuel or cladding.

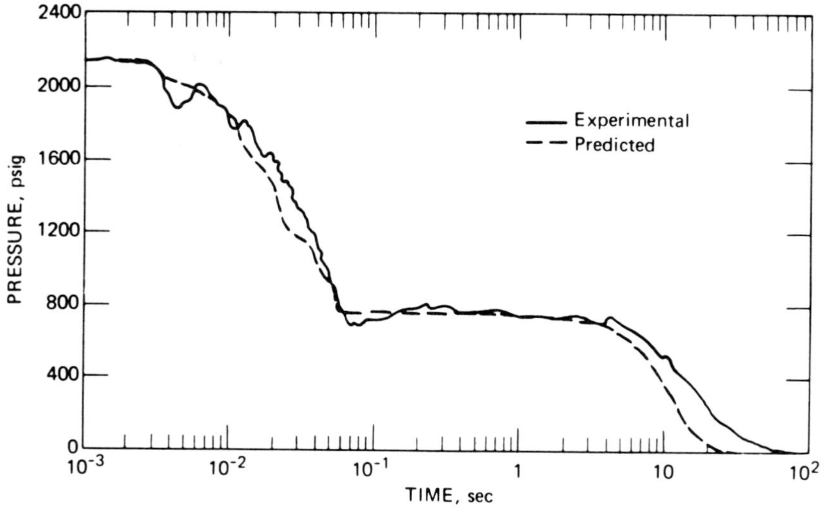

Fig. 6.24 Typical blowdown pressure transient for pressurized-water reactor.

Fuel-Handling Accidents

6.133 Since the reactor vessel is open to the containment during refueling operations, fission products released in the core because of an accident would quickly be transported into the containment volume. The designer must therefore be particularly concerned with potential hazards during this operation and provide adequate safeguards. In addition to mechanical damage to the cladding, possible hazards could be

1. Improper withdrawal of rod-cluster control assemblies.
2. Failure to maintain subcritical geometry.
3. Failure to maintain adequate shielding for personnel.
4. Failure to maintain adequate cooling of fuel assemblies.

6.134 Inadvertent control-rod withdrawals during head removal and control-rod drive-shaft-unlatching operations are prevented by a combination of detailed administrative procedures and alarm devices. In addition, the boron concentration of the water in the reactor is increased so that the core will remain subcritical even if all the rods are removed.

6.135 A critical geometry by spent fuel assemblies is avoided by proper rack spacing with the use of borated water as an additional safeguard. Hoists and manipulators are designed so that fuel assemblies cannot be lifted above a level at which water shielding would be inadequate. Adequate cooling of fuel during underwater handling can be maintained by natural convection.

6.136 Mechanical damage of fuel assemblies is prevented by various special gripping devices and handling fixtures, interlocks that restrict certain operations (e.g., the passing of the fuel shipping cask over the spent-fuel storage-rack locations), and careful procedures. On the other hand, mechanically initiated cladding failure with a subsequent release of fission products is normally one of the credible events considered in the preliminary safety analysis report (§6.248). In one such analysis the cladding of a row of 14 fuel rods is assumed to breach at approximately 90 hr after the reactor is shut down, a minimum period before fuel would be handled.

Control-Rod-Ejection Accidents

6.137 Though not considered a credible accident, the nuclear excursion from the failure of a control-rod-drive housing allowing the control element to be rapidly ejected from the core is sometimes analyzed. The reactivity insertion, a result of the full system pressure differential acting on the drive shaft, is limited by the Doppler reactivity effect and terminated by a reactor trip actuated by high-nuclear-power signals. The fuel damage likely from such an accident is governed mainly by the peak power attained in the transient, which, in turn, depends on the worth of the ejected rod and the power distribution attained with the remaining control-rod pattern.

6.138 The preceding incidents are primarily those anticipated for pressurized-water reactors. Somewhat different incidents are considered by the designer of other reactors. The design philosophy involving consideration of many types of credible events, careful analysis of the consequences, and necessary safeguards in the design is appropriate, however.

FAST REACTOR SAFETY CONSIDERATIONS

INTRODUCTION

6.139 For protecting the health and safety of the public and ensuring a very high level of system reliability, the same basic approach to safety is used for fast reactor design as for other reactor types.[38,39] Fast reactor safety features are evaluated by referring to analytically predicted consequences of many credible and hypothetical events with and without such features as is done for light-water reactors. Many of the features are different, of course, as would be expected for a different concept. For example, the primary-coolant system for sodium-cooled fast reactors is at relatively low pressure. Therefore the possibility of piping failures receives less emphasis than for light-water reactors. However, other phenomena, such as the possibility of very rapid sodium vaporization in the coolant channel, are emphasized.

6.140 A number of safety considerations in fast reactor design are sufficiently unique to warrant a separate discussion. Because fast reactor technology has not yet reached the same stage of commercial development as that for water-cooled reactors, we shall confine this discussion to rather broad aspects of the design, primarily for background purposes. Primary attention is given to sodium-cooled fast breeder reactors as a way of limiting the discussion. Attention is also focused on ceramic-fueled reactors of about 1000 Mw(e).

6.141 The design philosophy for nuclear safety in such large sodium-cooled fast reactors is still being discussed among designers. In consideration of the safety of fast reactors compared with current-generation thermal water reactors, the smaller neutron lifetime, the lower delayed-neutron fraction when plutonium is the fuel, and handling problems with sodium are frequently mentioned. Though relevant to inherent safety, such questions are not directly meaningful to the designer concerned with stability and engineering problems associated with accident possibilities and consequences. Fission-product release and transport as well as engineered safety features are extremely important, particularly since high-burnup fast reactors have a larger fission-product inventory than do thermal reactors.

REACTOR-PHYSICS CONSIDERATIONS

6.142 Some basic reactor physics considerations that are special for fast reactors are useful for orientation. Although accident analysis is considered separately, bear in mind that the analysis of the propagation of many accident-related events important to the design engineer depends on the neutronics of the system.

6.143 A key to many physics effects important to safety is the neutron spectrum and possible spectral changes caused by various events. The hardness of the spectrum depends on the neutron leakage and the materials used in the reactor. One of the most significant effects, particularly in sodium-cooled cores, is the coolant-voiding effect. In large systems tending to have small relative neutron leakage, a decrease in the coolant density due to voiding, or even in temperature, reduces the inelastic scattering and hence results in a hardened spectrum. As the spectrum hardens in dilute Pu–^{238}U systems, an increase in reactivity results primarily from the enhanced fissioning rate in ^{238}U and in ^{240}Pu. The spectral influence on reactivity is shown in Fig. 6.25, where the relative reactivity worth of a neutron (adjoint flux) increases as the neutron energy increases.[40] Remember that the resulting reactivity effects are spatially dependent since influences on the spectrum vary from point to point in the core. The detailed analysis of the resulting kinetic behavior required for a safety evaluation is therefore quite complicated.

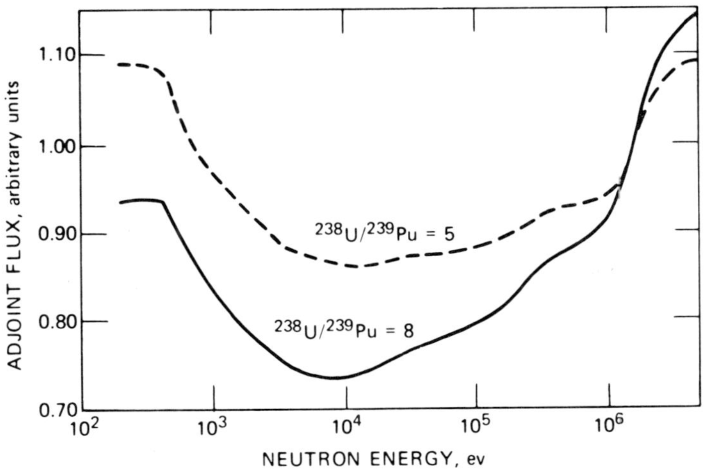

Fig. 6.25 Adjoint flux (neutron reactivity worth) for large $^{239}PuO_2$–$^{238}UO_2$-fueled reactors. (Reactors have 70 vol.% sodium. Effect of ^{240}Pu and fission products is not included.)

Doppler Effect

6.144 Since ceramic fuel does not provide an effective expansion reactivity effect, the primary negative feedback needed for stability is derived from the Doppler effect. The Doppler effect (§6.20) has an extremely short time constant; hence it becomes effective as soon as the temperature of the fuel changes. The effect is also larger for cores having a softer spectrum, generally true for large-size ceramic-fueled cores containing large quantities of ^{238}U. Studies of possible fast reactor accidents also show that a strong negative Doppler effect tends to rapidly quench the large reactivity insertion to a below-prompt-critical value before high pressures are likely to be produced. However, under certain conditions where there is a large positive reactivity being inserted the accident could still proceed to disassembly, although the Doppler effect tends to increase markedly the time scale of the accident. The consequences are thus likely to be less, and there is an opportunity for countermeasures to act.

Neutron Lifetime

6.145 Although the neutron lifetime is much shorter in fast reactors than in thermal reactors, for systems having a large negative Doppler coefficient, the shorter lifetime results primarily in a reduction in the energy release resulting from a prompt-critical severe excursion. A short neutron lifetime, for example, causes the power to increase at a greater rate for a given reactivity insertion with an effective early negative-feedback effect. On the other hand, a super-prompt-critical burst could take its course before mechanical control mechanisms had time to act. The picture is therefore not straightforward, particularly if oscillatory effects also enter the analysis.

6.146 The short neutron lifetime does affect many of the kinetics considerations and resulting transfer functions. Few control problems are introduced, however, as a result of the shorter lifetime. In general, the relations governing the behavior of fast reactors are essentially the same as those applying to thermal reactors. As long as the reactivity insertions do not approach prompt criticality, the short lifetime has only a negligible effect on the operational behavior.[41]

SPECIAL CONSIDERATIONS

6.147 In most fast reactor designs, the higher power density brings with it problems of providing shutdown cooling and emergency handling of the "afterglow." Similarly, some of the fission-product species that do not burn out can accumulate in rather large concentrations at the high burnup (100,000 Mwd/tonne) envisioned for the economic operation of fast breeder reactors. Special design approaches, such as the use of vented fuel, therefore, receive attention.

6.148 The vented-fuel design option is a result of the accumulation of considerable quantities of gaseous fission products in high-burnup high-power-density fuels required for low fuel-cycle costs in fast ceramic reactors. These gases must be either retained in the fuel element or released into the coolant. Since retention requires either thick cladding or low-density fuel, each leading to poor neutron economy, it is economically unattractive. Various devices have been proposed to permit controlled release of the gaseous fission products to the coolant.[42] Interestingly such venting might actually result in a burden lower in fission-product contamination in the coolant than that in a nonvented design if a lower rate of pin failure is obtained because of the pressure relief. The lower rate of pin failure could reduce the contamination produced when the coolant leaches out the solid fission products. However, since vented fuels require development, most designs are nonvented, with an extension of the fuel pin above the upper blanket region provided for a gas reservoir. Although this arrangement avoids the thick-cladding requirement to some extent, the approximate 5-ft increase in fuel-assembly length is a disadvantage.

6.149 The ability of the sodium to retain, or "getter," many of the fission products also needs special attention. Although this question is the subject of continuing research, there is evidence of significant retention, particularly of iodine.[42] As a result other isotopes may acquire controlling significance with respect to hazard determination.

6.150 Despite many of these special characteristics, the designer of fast reactor plants has essentially the same concerns he might have for a thermal reactor plant. It is important to provide for redundancy and fail-safe features of the reactor system as well as for component quality and operational surveillance. Various malfunctions or fault conditions and the resulting dynamic behavior are also considered in the safety analysis.

FAST REACTOR SAFETY ANALYSIS

6.151 As for thermal reactors, the safety analysis of a fast reactor design considers various accident situations and the response of the system to them.[43] Of particular interest for sodium-cooled reactors are various coolant-flow-reduction possibilities or reactivity additions which could cause fuel failure. If the heat-removal ability of the coolant is impaired without a corresponding reduction in core power density, the temperature of the fuel cladding or the sodium itself will exceed the sodium saturation temperature. The resulting production of sodium vapor and consequent two-phase-flow hydrodynamic regime tends to further restrict the coolant and provide a good possibility for overheating. The physical properties of sodium tend to accentuate such vapor problems when compared with water systems since the heat of vaporization is much less.

6.152 Various initiating situations for serious accidents and their interplay are shown in Fig. 6.26 and can be analyzed by such systems techniques as the "fault-tree" approach (§6.229). A relatively large number of initiating conditions must be considered.[44] Such systematic analyses are essential for the design of engineered safety features as well as suitable instrumentation to detect coolant changes and produce reactor scram. Analyses of the so-called core-disassembly or meltdown accident considering the effect of parameters on energy release and other consequences are also useful in the selection of various design specifications.[41] For example, the Doppler effect can be related with energy release and, in turn, containment specifications. Whether or not such an accident is credible and should serve as a basis for design is a question of safety-design philosophy, however.

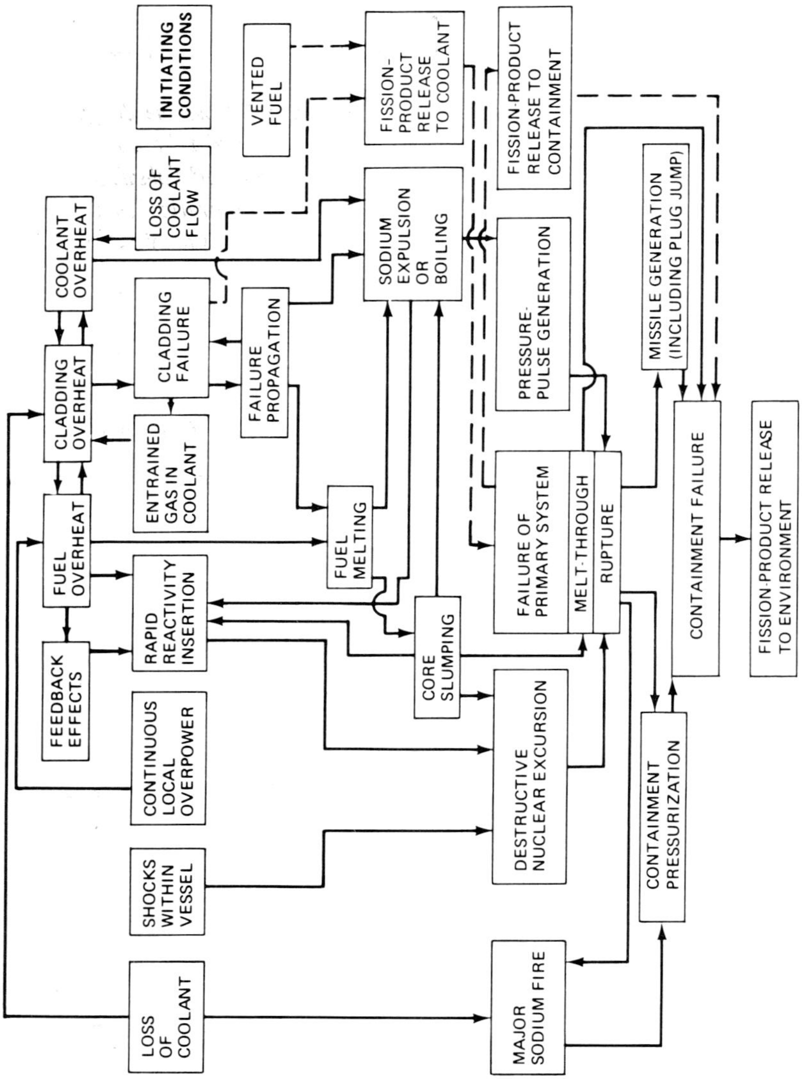

Fig. 6.26 Typical accident sequence diagram.

6.153 In fact, in fast reactor design, the severity of various possible accidents has evoked no definitive practices concerning the type of outer containment required. For example, studies for a high-leakage core having a barely negative sodium-void coefficient show results not very different from those for a core having an unspoiled geometry with lower leakage and with a somewhat positive void coefficient. Since safety-related research, analyses, and discussions are all being studied extensively, the reader should refer to the current literature for design guidelines.

Problem Areas

6.154 Several problem areas have been identified through attempts to describe postulated sodium-cooled fast reactor accidents in terms of basic mechanisms to guide the designer. For serious core-related accidents, generally the upper limits and, if possible, the lower limits of the damage caused by a sequence of events should be related to the design parameters of the system. For the desired analysis more information is needed in the following areas:

1. Sodium boiling.
2. Fuel-pin failure and failure propagation.
3. Motion of molten fuel and the possibility of vapor explosions.
4. Severe excursions.

The events associated with these areas tend to be interrelated, often in a complicated way. Some of the features of each category are discussed in the following sections.

6.155 The subject area of sodium boiling includes the events leading up to the expulsion of coolant. An analytical description of these events therefore requires an understanding of superheating, two-phase flow of the coolant after nucleation, and the effect of reactivity feedback on the fuel thermal input. Although several analytical models have been formulated (§4.128) to describe transient void-formation rates, the confidence level needs to be improved by more experimental information to better define the parameters.

6.156 A related safety problem in the liquid-metal fast breeder reactor is the behavior of coolant when flow channels become very hot and massive amounts of coolant are expelled. The expelled coolant may then condense in the upper plenum and flow back into the hot channel. On reentry, violent boiling may be produced by molten fuel interacting with coolant, the resulting vapor expanding so rapidly that the effect is similar to an explosion. Such an explosion could lead to a variety of consequences, including compaction of other regions of the core. Thus the mechanisms involved need to be better understood for accident modeling.

6.157 A vital question associated with fuel-pin failure is that of propagation. That is, under what conditions can the failure of one pin precipitate the failure of others and set up a chain reaction affecting so many of the pins in a subassembly that the subassembly has to be replaced? In addition, the possibility that the damage will spread to other subassemblies must be studied to understand the conditions under which a minor failure can lead to a serious accident.

6.158 An analysis of single-assembly accidents that could lead to failure propagation is illustrated in Fig. 6.27 together with a stepwise listing of key uncertainties in the analysis.[45] Five potential accident sources are indicated:

1. Partial-fuel-assembly blockage.
2. Fuel-pin-failure gas release.
3. Loading error of single pin or pellet in otherwise normal assembly.
4. Full-fuel-assembly blockage.
5. Loading error of fuel assembly in wrong location.

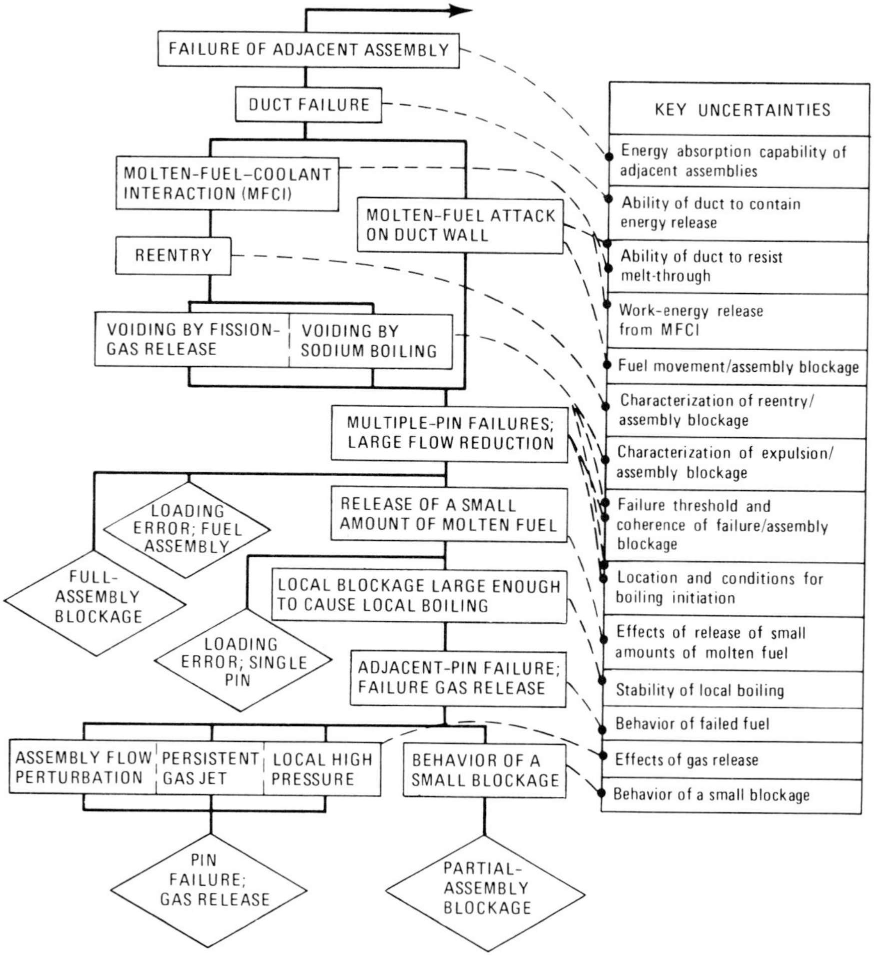

Fig. 6.27 Accident sequences and key uncertainties for single-assembly accidents.

6.159 Should fuel melt, more knowledge on fuel movement is needed to determine if reassembly is likely and a super-prompt-critical excursion possible or if the fuel is merely likely to be swept out of the core. Furthermore, in examining the consequences of a reassembly accident, we must know the rate of reactivity addition, which depends on the nature of the molten-fuel motion. Uncertainty in molten-fuel behavior therefore limits the confidence in predictions of accident consequences with analytical models.

6.160 The amount of energy released in a super-prompt-critical excursion should be estimated to aid in containment design and in selecting core-design specifications (§6.152). In a typical energy calculation, such as the Bethe–Tait analysis, energy is generated as a result of a prompt-critical condition until enough pressure develops to expand the core (or portion of the core) below the critical condition. Since material displacement is involved, the calculation model includes a coupling of the neutronics, hydrodynamics, and equation of state, each of which includes areas of uncertainty.

GAS-COOLED FAST REACTORS

6.161 Fast reactors using helium under pressure (~1200 psi) as coolant instead of sodium are receiving attention as an alternate approach to the breeder reactor.[46] Advantages include a high internal breeding ratio with small reactivity swing over the burnup cycle and only an insignificant coolant reactivity effect. Therefore the safety questions for coolant-void formation in sodium-cooled fast reactors do not apply for helium cooling. Also, the possible swelling of stainless-steel fuel cladding is more easily accommodated in the gas-cooled system since the pin spacing is not as close as that necessary for a sodium-cooled core, where the coolant volume fraction must be kept relatively low to avoid degrading the neutron spectrum.

6.162 The principal safety challenge in the gas-cooled fast reactor is to provide for the cooling of the core under such emergency conditions as loss of coolant flow or depressurization, which reduces the coolant density. Since natural circulation does not provide sufficient cooling capacity to remove the afterheat, the fuel would heat up rapidly after shutdown if cooling capability were not maintained. High integrity of the primary-coolant-system containment is provided by a prestressed-concrete reactor vessel that includes the circulators and steam generator in a self-contained steam supply system. The design ensures the operation of the coolant circulators under accident conditions. The use of steam turbine drives makes operation independent of an electrical power source. Auxiliary circulators also provide a backup capability for emergency cooling.

6.163 The safety analysis for a gas-cooled fast reactor follows the same general approach as that for light-water thermal reactors and sodium-cooled fast reactors. Potential accident situations are reviewed and the requirements for emergency systems examined. Since, like the liquid-metal fast breeder reactor, the gas-cooled fast reactor does not have an inherent shutdown mechanism, the Doppler coefficient is a very important power-limiting parameter. Although the entry of steam or water into the core would add reactivity,* complete core flooding is not considered credible, since the water inventory in the steam system is limited.

6.164 Considerable attention in the safety analysis must be given to the fuel-temperature behavior in response to various depressurization and coolant-flow-reduction possibilities. The rapid depressurization after a large break (area >90 sq in.) would seem to cause fuel damage, even with continuing circulator operation. A secondary containment of relatively small volume to maintain the coolant pressure at about 50 psi, despite a leak in the primary system, has therefore been suggested.

6.165 Some of the problem areas mentioned in §6.155 which relate to molten fuel but not to the sodium coolant apply also to the gas-cooled concept. Compared with the sodium-cooled concept, therefore, the gas-cooled fast reactor offers the important advantage of freedom from the safety concern resulting from the positive reactivity introduced by coolant voiding. On the other hand, adequate shutdown cooling represents a greater challenge. From the safety viewpoint each approach therefore has some advantages as well as some problems specific to the concept.

ENGINEERED SAFETY FEATURES AND SITING

6.166 In the broad sense engineered safety features include all provisions specified by the designer of a reactor plant to ensure that the system will operate without any unusual risk to the general public. Such features can therefore include special instrumentation and control devices, structural features, a large variety of such devices as cooling sprays intended to mitigate the consequences of an accident, and, finally, the containment.

6.167 In a somewhat narrower sense, various devices to lessen the consequences of an accident are emphasized, the instrumentation and containment being considered as separate items. Here, the term is applied to the total system, including the containment, intended to prevent and limit the consequences of a fission-product release.

6.168 Siting is related, of course, to the safeguards system. An exclusion area or siting of the plant away from large centers of population can limit the consequences of a release of fission products, just as design features can prevent fission products from reaching the atmosphere. Siting parameters can therefore be included during the design of the overall safety system since a site within

*Some designs provide for resonance absorbers in the fuel so that the introduction of steam into the core will cause a decrease in reactivity.

close proximity to a population center is acceptable only if fission-product release to the atmosphere and subsequent diffusion to the population center will be prevented by the design. In the United States, government requirements provide some guidelines based on exposure criteria.*[22]

6.169 As reactors increase in size, however, strict application of the TID-14844 guidelines results in quite large exclusion areas and distances to population centers. Since the guidelines were made with no assumption of engineered safety features, they may be modified on the basis of the safeguards provided by these engineered features. The extent of modification appropriate is generally evaluated on an individual site and reactor-design basis.

6.170 Siting close to urban areas in which load centers are located will avoid significant economic penalties of transmission costs for plants more remotely located. Thus extensive engineered safety features are usually economically justified if urban siting can thereby be made acceptable. The TID-14844 guidelines are compared with actual approved exclusion areas and distances to population centers in Table 6.6 for several large reactors.

TABLE 6.6

Siting Distances for Selected Plants

			Exclusion distance, miles		Population center distance, miles	
Nuclear plants	Size, Mw(t)	Type	Guideline	Actual	Guideline	Actual
San Onofre	1210	PWR	0.76	0.50	15.0	10.0
Oyster Creek	2000	BWR	1.07	0.25	22.0	10.0
Malibu	1700	PWR	0.95	0.16	19.8	10.5
Millstone Point	2030	BWR	1.07	0.50	22.0	3.5
Connecticut Yankee	1825	PWR	0.99	0.32	20.6	9.5
Dresden 2,3	2300	BWR	1.15	0.50	24.0	15.0
Indian Point 2	3218	PWR	1.44	0.3	30.0	3.0

6.171 If the site is considered part of the overall reactor safety system, designed engineered safety features can be emphasized when there is a strong incentive to reduce the site subsystem requirement for distance from population centers. Fission-product release to the environment can be prevented or reduced by three different approaches:

1. Prevent or reduce overheating of the fuel in the core through emergency cooling provisions.

*Guidelines are provided in various sections of the *Code of Federal Regulations*, such as Section 10 CFR100.

2. Provide for a reduction in the fission-product concentration by filtering, scrubbing, sprays, and other devices should fission products be released to the containment.

3. Provide various barriers around the primary system to reduce the probability of fission-product activity leakage. The outer containment system is part of this effort.

THE SAFEGUARDS – CONTAINMENT SYSTEM

6.172 The systems approach can be applied to the interplay between the engineered safety features and the containment since the respective design requirements are closely integrated. Questions of siting can be considered part of an even larger system that may comprise an entire geographic region.

6.173 The purpose of all containment is to provide a barrier between fission products that may somehow be released by the enclosed reactor system and the environment outside the containment.[47] Containment as a design requirement has evolved through several stages of development. Initial containers were primarily large static pressure envelopes with few penetrations. They were intended to control public exposure after a possible accident to a small experimental reactor not located at a remote site. In other words, the containment substituted for the remote location. Since one common design assumption is that all stored energy in the primary system is released into the containment structure, as water reactors became larger, a simple envelope became difficult to design at reasonable cost. Although a basis for design is still the containment of the stored energy in the coolant, which is hot and under pressure, ways to absorb this energy by means other than simple containment of the vapor in a pressure vessel have evolved.

6.174 As reactors have become larger, the total amount of fission products contained within a single vessel, as well as the stored energy, has increased. There is therefore greater concern over possible major failures of the primary reactor vessel and the generation of missiles that might breach the containment.[48]

6.175 Although the design-basis accident for water reactors is considered to be the rupture of a pipe in a main coolant loop in a manner that would permit full flow of the coolant from both ends of the ruptured pipe, it is assumed that the primary vessel remains ductile. If it becomes necessary in designing containment vessels to anticipate a sudden catastrophic failure of the primary reactor vessel, two effects not previously planned for in designs for containment must be considered. These are the containment of missiles formed by fragments of the vessel thrown apart at high energies and the containment of pressure shock waves that produce a concussion on the containment shell. The possibility of pressure-vessel loss of ductility is treated further in §6.244.

Pressure-Suppression Containment

6.176 Pressure-suppression containment is based on ducting the reactor coolant discharge from a hypothetical loss-of-coolant accident into a heat sink (usually a pool of water) and thus reducing the pressure and temperature inside the containment space by condensing the steam–water mixture. Some of the fission products entrained in the coolant would also be removed in the pool water. The system therefore consists of two main components, as shown in Fig. 6.28 for a boiling-water reactor: (1) a pressure vessel, or "dry well," enclosing at least the reactor and (2) a pressure-suppression chamber, or "wet well," partly filled with water. The dry well vents through a number of pipes whose discharge ends are submerged in the water within the suppression chamber. The primary-system lines penetrating the containment must be provided with isolation valves in duplicate as ensurance against the failure of a single valve or to account for breaks inside and outside the containment.[49]

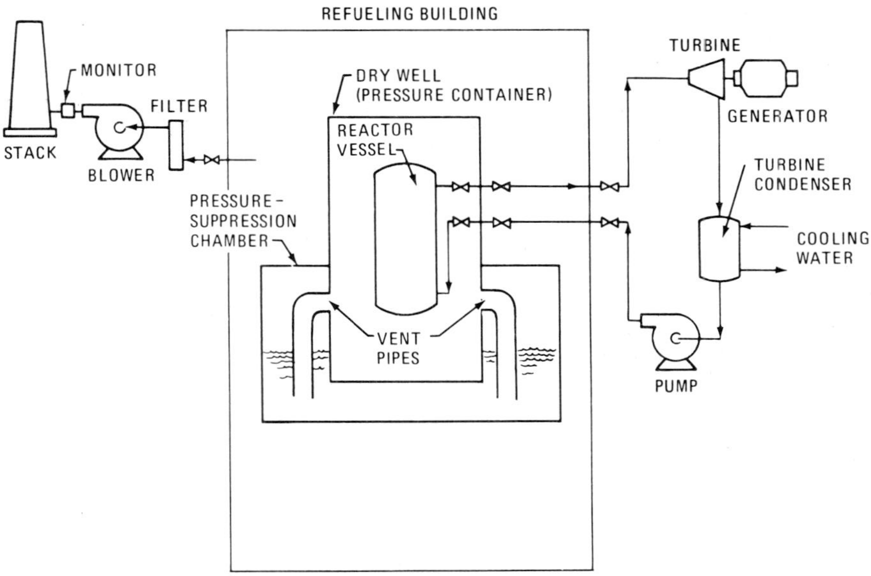

Fig. 6.28 Multiple containment shown with boiling-water reactor.

6.177 A type of pressure suppression containment (Mark III) proposed for large boiling-water-reactor plants is shown in Fig. 6.29. Here, a weir wall inside the dry well and the containment wall form the annular suppression pool. Horizontal vents in the dry well wall provide a path for condensing steam during blowdown conditions. A second water tank above the dry well provides

Fig. 6.29 Pressure-suppression containment for boiling-water reactor.

additional condensing capacity and serves as a reservoir for various emergency spray systems. Both the dry well and the suppression chamber are enclosed in a relatively low leakage building held at slightly negative pressure ($\sim\frac{1}{4}$ in. H_2O) by fans that exhaust through filters to a stack. This arrangement provides a form of multiple-barrier containment. In multiple-barrier concepts, leakage past the first containment barrier is collected within a reduced-pressure zone between the first and a second barrier and is either exhausted through a filter system and stack or pumped back inside the containment space. These concepts offer greater control of leakage than the single-shell containment vessel; further, leakage-rate testing may be more accurate, and continuous monitoring of the leakage rate may be easier.

Ice Condenser

6.178 A related method used for pressurized-water-reactor systems employs a bed of ice to absorb the energy that might be released from a break in the coolant system. The volume of the containment vessel may be about 50% smaller than that required for a simple pressure-containment envelope. In addition, the design pressure is reduced from about 50 psi to about 12 psi.[50] The ice compartment[51] rings the walls of the containment with access to the escaping vapor provided by a "trap door" activated by a slight differential pressure, as shown in Fig. 6.30.

Other Containment Systems

6.179 The preceding containments for water reactors are examples of several different possibilities, each appropriate for a different concept or, in some cases, a somewhat different safety philosophy. The coolant in gas-cooled reactors, for example, has much less stored energy than that in water-cooled systems. As a result the pressure–volume requirement is less, and one prestressed-concrete pressure vessel could contain the entire primary coolant system. In this approach, however, much design attention is given to other containment barriers and devices to reduce the effects of fission-product release. Some designs for large sodium-cooled fast reactors emphasize the integrity of the primary coolant system with an outer building containment that is not a pressure vessel but merely a secondary barrier.[52] A variety of containment designs therefore exists, each meeting individual requirements. The current design trend, consistent with this approach, is to consider the overall-safeguards-system requirements and individual contributions, the outer containment structure being one of the several subsystems.

SYSTEMS APPROACH

6.180 The preceding containment designs are primarily the result of a trend to consider the containment as only part of a total safety network with the performance requirements established only by looking at the overall system. This systems approach is illustrated here by considering a loss-of-coolant accident in a large boiling-water reactor.[53]

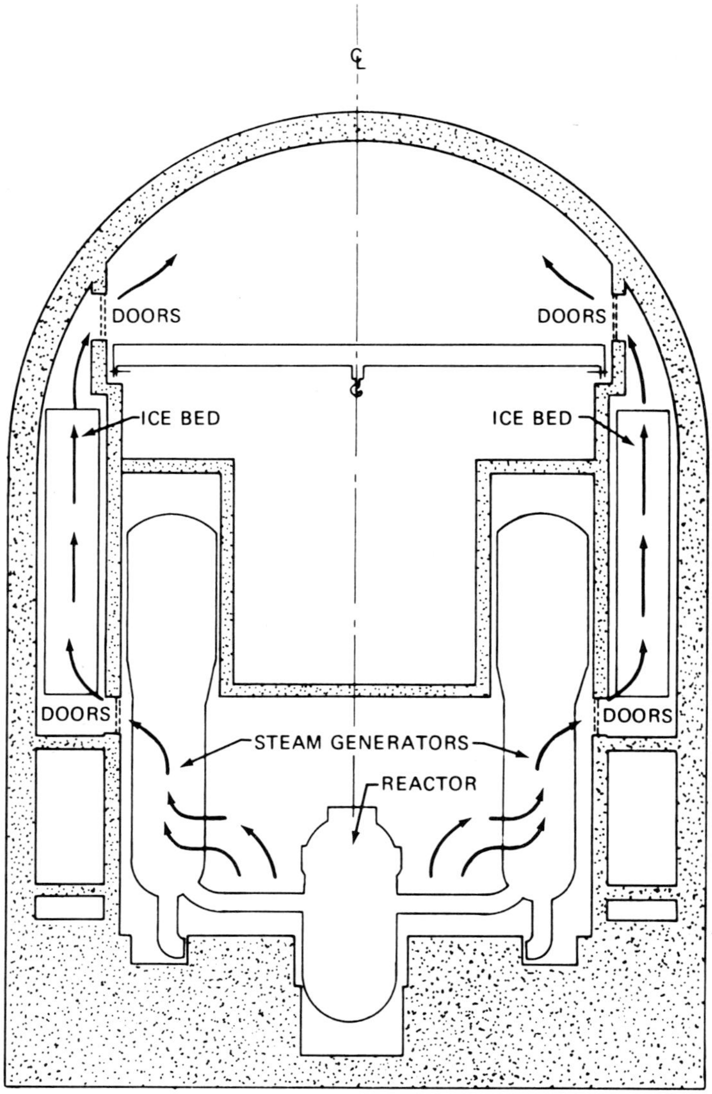

Fig. 6.30 Ice-condenser containment system.

6.181 The various interrelations resulting from an integrated-systems viewpoint of a loss-of-coolant accident are shown in Fig. 6.31. Three engineered safeguards subsystems – emergency core cooling, containment cooling, and containment fission-product removal – affect the failure mechanism. Containment failure can be achieved both by penetration and by overpressurization of the containment shell, each initiated by a breach in the primary system.

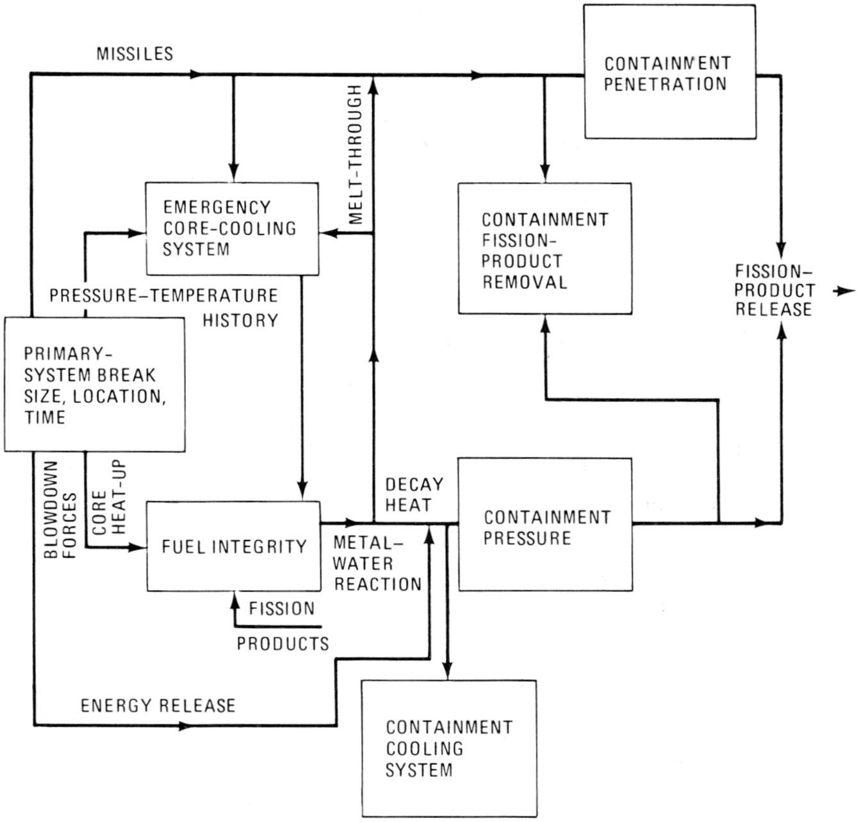

Fig. 6.31 Role of engineered safeguards in protecting barriers during a loss-of-coolant accident.[53]

6.182 Several interplays exist between the various subsystems. For example, the performance of the emergency core-cooling system could be influenced by the primary-system break through the formation of missiles, as well as by the forces and pressure–temperature history during blowdown. The blowdown forces and subsequent core heat-up also determine the early degree of fuel integrity, which, in turn, affects the operation of the emergency core-cooling system.

6.183 A systems analysis of the network should also consider changes in the role of subsystems as a function of time. For example, the emergency core-cooling system dominates the overall response during the early stages of the loss-of-coolant accident by controlling and limiting the increase in fuel temperatures. Later the containment cooling system begins its important role of removing the accumulated decay heat. Finally, much later, the fission-product removal system acts as a countermeasure by reducing the effect on the environment.

6.184 In the design, balance among the various elements of the overall network should be maintained. The performance of one element, for example, need not receive undue emphasis if it is coupled to a much weaker element in the system. For the boiling-water-reactor system, the primary-system integrity and the emergency core-cooling provisions emerge as the most important elements of the network. Both directly influence fuel integrity and the two possible modes of containment failure.

6.185 The network analysis permits several design approaches depending on where the emphasis on consequence prevention is placed. A preferred method is to emphasize a high-performance emergency core-cooling system and the integrity of the primary system. As a result the containment design requirements tend not to be too severe. Alternatively, the containment could be designed to handle missiles, melt-through, and metal–water reactions for all types of primary-system failures without depending on emergency core cooling. Such a design, which places the entire burden of protection upon the containment, is more difficult and more expensive than the first option, which provides for effective countermeasures during the early stages of the accident.

6.186 Since the emergency core-cooling system plays such a key role in the safeguards network for typical boiling-water reactor designs, some of its features are discussed here. Emergency cooling systems for 1000-Mw(e) water reactors can generally control afterheat generation rates greater than 100 Mw(t) for the first few minutes after shutdown and about 120 Mw-hr of stored energy and decay heat thereafter.[37] Actually, the system consists of a number of subsystems, as shown in Fig. 6.32, namely, a high-pressure injection, a semiautomatic blowdown, a core spray, and a low-pressure coolant-injection system. The high-pressure coolant-injection system pumps water into the reactor while it is fully pressurized and protects the reactor core should a small process line fail. Its flow rate maintains the reactor core covered until the reactor pressure decreases enough to make the core spray or the low-pressure coolant-injection system effective.

6.187 The semiautomatic blowdown system is a backup to the high-pressure coolant-injection system. It depressurizes the vessel by blowdown through automatic opening of the relief valves, which vent steam to the suppression pool. Two core spray systems protect the core by spraying suppression-pool water on top of the fuel from nozzles mounted on a sparger ring located above the core. The core spray system gives protection for breaks large enough to independently depressurize the reactor vessel as well as for smaller breaks after the action of the high-pressure coolant injection of the semiautomatic blowdown system. After the system pressure is released, the core is cooled further by the low-pressure coolant-injection system, which also uses suppression-pool water.

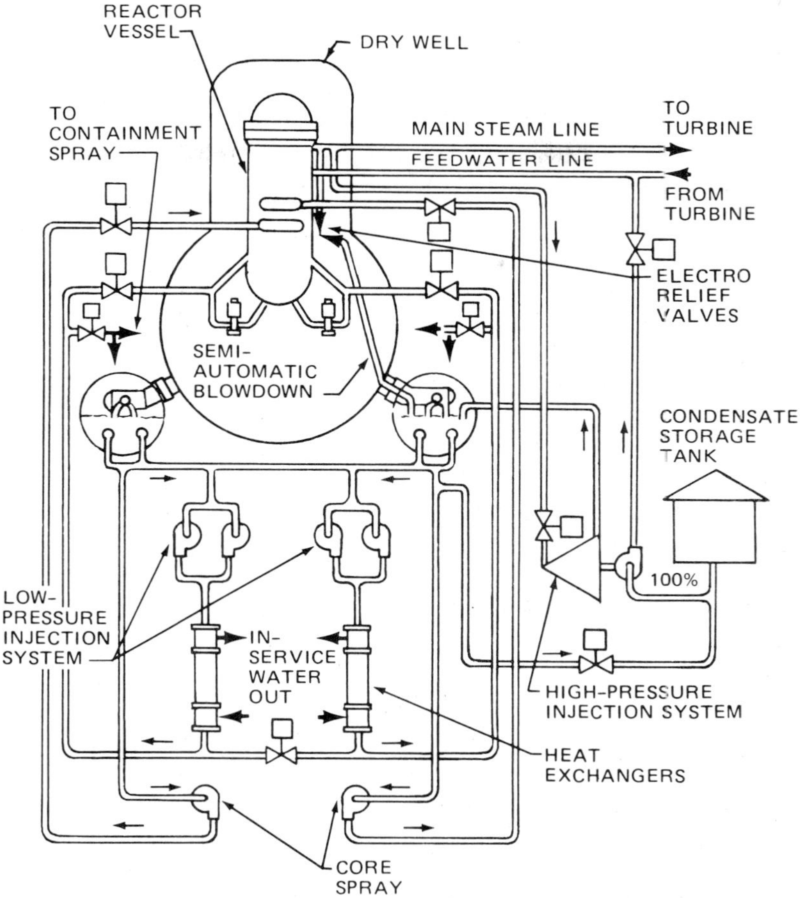

Fig. 6.32 Emergency core-cooling system.

6.188 Additional engineered safeguards are the containment coolant system and the fission-product removal system. In pressure-suppression boiling-water reactors, the containment coolant subsystem uses some of the components of the low-pressure coolant-injection system but sprays cooled suppression-pool water into the dry well and the suppression chamber. The fission-product removal system consists of a series of filters and other gas-treatment devices for processing the atmosphere of the reactor building and exhausting it to a stack.

6.189 Systems for pressurized-water reactors are similar in principle but different in detail. A major difference is reliance on reflooding of at least half the core shortly after a pipe break. This is accomplished with high-pressure, intermediate-pressure, and low-pressure emergency water-injection systems to protect the core for any break size up to a double-ended rupture of the largest pipe attached to the pressure vessel. Borated water is commonly used for such injection, normally into the coolant loops as shown in Fig. 6.21. However, in the early 1970s the injection effectiveness under certain circumstances was debated. Design changes, including the possibility of a core spray system, therefore received consideration.

CONTAINMENT AND SPECIFICATIONS FOR SAFETY FEATURES

6.190 Uniform design criteria are being developed by standards committees consisting of representatives from industry and government. In the past, containment structures have received the most attention since they can be tested more readily than the engineered safety features systems. On the other hand, a leakage rate during a controlled test could differ from that during an accident. Since the engineered safety features are vital in limiting the load on the containment structure, a shift in emphasis is likely.

6.191 Leakage rates of about 0.1% per day have been specified for containment structures for some large pressurized-water reactors under a pressure of about 60 psig. This is equivalent to leakage through a hole less than $^1/_{16}$ in. in diameter from a containment volume of 2×10^6 cu ft. Leakage specifications for boiling-water-reactor pressure-suppression systems are generally somewhat higher, ranging up to 0.5% per day. In general, leakage specifications for most systems fall within the range 0.1 to 0.5% of the volume per day. Design specifications for the various emergency core-cooling systems are determined by iteration with the accident analysis and will not be treated here. A measure of the size of the components, however, is given by a typical core spray requirement of 12,500 gpm for each of two pumps for a large boiling-water reactor.

SHIELDING

6.192 Since the purpose of the shielding system is to protect both personnel and components from the harmful effects of radiation, it is in a sense part of the engineered-safety-features system. This is particularly true if the design provides for both a shielding and a containment function for some of the structural components. As a result the design of the shielding system must be well integrated with the design of the containment and other safety features. Accident analysis provides the shield designer with input for the necessary iteration concerning radiation effects. On the other hand, shielding design tends to be a separate discipline with specialized principles which can best be treated elsewhere.[54] The role of shielding in the safety evaluation of the reactor design should not be overlooked, however.

SITING PARAMETERS

6.193 From the viewpoint of safety, the most important siting parameter concerns the possibility of exposure of the public to fission products in the event of an accident. Guidelines for the size of the exclusion area and suggested distances from population centers have therefore been established, as discussed in § 6.95. In addition to this primary requirement, however, several basic engineering factors can be very important in selecting a site for a nuclear power plant. Examples are the influence of electrical-energy-transmission charges and the availability of adequate cooling water.

6.194 Since the electrical load center is normally at the very same population center from which some separation is desirable for safety purposes, a direct trade-off is introduced between the need for such separation and the cost of transmission over the extra distance that may be required.

6.195 As a guideline, the cost of electrical power transmission* tends to average about 10% of the cost of energy delivered to the consumer. The transmission cost, in turn, does depend on a number of parameters including contributions to capital and operating expenses. For example, the capital cost of transmission lines per mile includes a fixed component, a component dependent on the voltage, and a component to allow for the cost of the right-of-way. Terminal equipment contributes an additional charge. Annual fixed charges are derived from the capital charges in the customary way. Lines losses (I^2R losses) represent the other important contribution to the annual operating cost.

6.196 Since the line losses may be reduced if a higher voltage is used, a design-optimization iteration applies in which there is a trade-off between the increased investment required and the savings in losses to be realized. Higher voltages are generally appropriate for longer lines. Estimating the cost of electrical power transmission is therefore fairly complicated. Computer programs have been developed for this purpose, however.[55] One example of such a calculation for a 155-mile 500-kv line with a 1000-Mw capability yielded a total capital cost of $18 million and a cost of power transmission of 2.65 mills/kw-hr.

6.197 When compared with a generating cost of about 6 mills/kw-hr, transmission over such a distance is seen to introduce a substantial cost increment, which utilities would prefer to avoid. Actually, the electrical distribution as it might affect the choice of a new generating plant site is much more complex. A new plant must be phased into an existing network that may supply several load centers and be interconnected to neighboring systems. Also, adequate system reliability may require alternate transmission-line routes. The economics of the transmission-cost factor of reactor siting are therefore far from simple. For some purposes, however, a rule-of-thumb estimate is useful which considers an arbitrary 20-mile shift of a plant away from the load center at a penalty of about $12/kw(e) in initial cost and capitalized losses.[56]

*Local-distribution cost excluded.

ENVIRONMENTAL FACTORS IN SITE SELECTION

Introduction

6.198 Growing national concern in the United States for preserving environmental values has focused much attention on the effects of nuclear power plants on their surroundings.[57,58] Such effects include low-level radioactive effluents during normal operation, thermal discharges from the thermodynamic power cycle, and releases that might occur during accident conditions. Controversy regarding the permissible levels of radiation and contaminants has introduced an additional factor in site selection which depends on public acceptance. In fact, the picture has been complicated, at least in the early 1970s, by debate on a variety of broad related subjects, including the desirability of using nuclear plants to meet pressing energy needs if unknown risks are involved. Since environmental dangers from fossil plants have also been pointed out, a dilemma has resulted. However, since much of the controversial material is journalistic in nature, we shall emphasize here the primary effluents from the reactor which interact with the environment and are of concern in site selection.

6.199 As part of the licensing procedure (§6.250), nuclear-power-plant applicants must provide information on the environmental impact of the proposed plant and assurances that all applicable quality standards will be met. In addition, certain reporting procedures are prescribed. These requirements, which were put into effect in 1971, implement the National Environmental Policy Act of 1969. The impact of the procedures is still evolving in 1972.

Sources of Radioactive Effluents

6.200 Under normal operation some fuel-cladding failures can be expected despite the very high level of integrity achieved through careful design and manufacturing.[59] Should a pinhole develop in the cladding, fission products escaping into the coolant would primarily be the gaseous or more easily vaporized constituents. Of these, the noble gases xenon and krypton are of special interest since they remain in the free form in the coolant. Other fission products are easily removed by the coolant purification system.

6.201 Other radioactive materials are formed in the coolant system by activation of corrosion products and other materials. Of particular interest is tritium, formed in small amounts as a fission product but also formed by fast-neutron reactions with the soluble boron usually present as a chemical shim in pressurized-water-reactor coolants. Secondary amounts are also formed in water-cooled reactors by thermal-neutron irradiation of the deuterium present. Since the tritium that is formed is converted to HTO, it behaves chemically the same as water and hence is not removed by the coolant purification system.

6.202 Radioactive-waste-treatment systems are an integral part of all reactor plant designs. Liquids from the reactor-coolant system and from other sources are usually concentrated by such processes as evaporation, ion exchange, and filtration to yield a material that can be shipped off site and suitably disposed. The effluent process liquid, which contains very little radioactivity, may be collected in holdup tanks, monitored, and then released under controlled conditions with the plant's condenser cooling-water discharge. This discharge has been debated in licensing proceedings since processing stages can be added so that the effluent can be reused in the plant and no radioactive liquid effluents discharged to the environment. Such methods, however, generally still result in a discharge of tritium in the form of gaseous effluent.

6.203 The requirements for gaseous-radioactive-waste management for a pressurized-water reactor differ significantly from those for a boiling-water reactor. Since a pressurized-water reactor uses a closed-loop reactor cooling system, the volume rate of gases containing radioactive isotopes from the cooling system and other sources is quite small. It is therefore a straightforward matter to compress the gases, provide for about a 30-day holdup in decay tanks, then release to the atmosphere the effluent, which now consists primarily of the long-lived ^{85}Kr. If desired, the krypton can be removed from the effluent gas stream by scrubbing with freon at low temperature, absorbing on charcoal, or by other methods.

6.204 Since the coolant that passes through the core in a boiling-water reactor is carried as steam directly to the turbine and then to the condenser, the cycle is not as closed as that for a pressurized-water reactor. The condenser vacuum required for high cycle efficiency is maintained by removing noncondensible gases by a steam-jet air ejector. Since the noncondensibles include air that has leaked into the system and water disassociation products as well as the radioactives gases, the discharge from a 1000-Mw(e) boiling-water reactor has a volume rate of about 140 cfm. In boiling-water-reactor plants, the air-ejector effluent gases are filtered to remove particulate material and provide at least a 30-min holdup for the gases to permit partial decay of the noble gases and other short-lived radioactive isotopes before release to the environment. As an alternative, hydrogen and oxygen can be catalytically recombined and condensed together with the original water vapor, yielding a process stream of about 20 cfm, which can then be treated by one of several possible recovery systems to remove the noble gases, e.g., that for the pressurized-water reactor.

6.205 It must be emphasized that the discharge of tritium (half-life, 12.26 years) and ^{85}Kr (half-life, 10.76 years) from reactor plants not provided with extensive secondary plants for their recovery has been shown not to lead to any significant effect on the environment, provided precautions are taken to ensure adequate dispersion. As larger fuel reprocessing plants are built, krypton will probably be removed from gaseous effluents, however. Because of extreme public sensitivity to radioactive discharges to the environment, secondary recovery systems may be installed in reactor plants to speed public acceptance and thereby avoid costly delays in licensing.

Waste Heat

6.206 Waste-heat-disposal requirements for power plants projected for the next several decades have caused some concern. A number of major streams in the United States on which new plants would logically be located appear not to have sufficient flow to accommodate the waste heat during the summer without a temperature rise above that permitted by quality standards.[60] Since this question has a significant bearing on siting, some of the important considerations are examined here.

6.207 Most thermal power plants in the United States use once-through cooling with fresh water to condense the steam leaving the turbine and then discharge the heated water to a river or other body of water. The amount of fresh water available in the United States is therefore of interest. The average rainfall in the United States is about 6,000,000 cfs. About two-thirds of this, which returns to the atmosphere by evaporation directly from water surfaces and by evapotranspiration of vegetation, is not recoverable by practical means. The remainder, about 2,000,000 cfs, contributes to ground storage and sustains the flow of streams, rivers, and lakes.[57]

6.208 In the mid-1960s only about 500,000 cfs of such fresh water was withdrawn from these supplies, and only 100,000 cfs was consumed. A considerable margin therefore appears to exist between demand and supply in both withdrawal and consumption. However, because of the very large seasonal variations in flow, the maximum possible withdrawal rate is estimated, by adding all practical reservoir capacity, to be only about 1,000,000 cfs, even with full reuse of the water. Water in the Mississippi and Ohio river basin is presently used on the average of seven times before reaching the river mouth. For example, for full use of U. S. rivers, reservoir capacity would have to be more than doubled, an unlikely prospect.

6.209 United States industry (including electric utilities) was withdrawing water at a rate of about 250,000 cfs in 1960 but was consuming only about 5000 cfs. Of this total industrial use, thermal electric power plants withdrew water at a rate of about 120,000 cfs and consumed about 1400 cfs. Power plants therefore accounted for about 50% of all industrial withdrawals and about 20% of all industrial consumption. Although average values can mislead, particularly when reuse is considered, they do give a preliminary picture. The 145,000 Mw(e) of nuclear capacity projected for 1980, for example, corresponds to a water withdrawal of about 220,000 cfs. New fossil power plants and industrial growth may account for an increment almost twice this value. Although their usefulness is indeed limited, such projections of average withdrawal rates do provide some strong indications that there are likely to be problems in specific areas.

6.210 Particular concern has been expressed on the effect of increased stream temperatures on aquatic life.[61] Furthermore, the proposed construction of a number of light-water nuclear plants, which may have a thermodynamic cycle efficiency* of about 32% compared with 40% for a fossil plant, has tended to draw attention to the general problem.

6.211 State water-quality standards typically specify both a maximum stream temperature and a maximum allowable temperature increase. The temperatures selected depend on the biological community living in the stream as well as the anticipated human use. If the stream supports a cold-water fishery, the regulations are generally quite restrictive, permitting maximum temperatures of the order of 65 to 70°F. If the stream has only warm-water fish, the temperatures permitted are of the order of 90°F from April through November. Various classes of streams range between these extremes. Since these maximum-permitted stream temperatures generally approach the stream temperatures already occurring for some portion of the year, it may not be possible to add greater heat loads to some streams during the summer months.[60] Allowable temperature increases of streams are generally 5°F or less. An additional complication is the tendency of the hot effluent to flow downriver as a separate plume for some distance and not mix well with the cooler water below or alongside. Restrictions of effluent temperature are therefore a possibility.

6.212 Because of various complications in using surface waters, cooling ponds or cooling towers have been considered as an alternate. Generally, if cooling ponds, which require large amounts of inexpensive land, are not feasible, natural-draft towers receive first preference. These are quite large, however, sometimes almost 450 ft high and over 300 ft in diameter for large power plants. In addition to the cost increment compared with that for using surface water, air pollution caused by fog and entrained slimicides and corrosion inhibitors may be introduced. Induced-draft towers are smaller and less likely to introduce fogging problems but may require as much as 0.4% of the output of the generating plant to drive the induced-draft fans. Since performance is quite dependent on humidity and ambient temperatures, the designer must consider a number of possibilities in seeking an optimum installation. Cost estimates vary considerably, capital costs generally being of the order of $10/kw(e) for a 1000-Mw(e)

---

*Some confusion in terminology exists since *thermal efficiency* is used both to express the fraction of *heat produced* that is converted into work and as a term having the same meaning as the *thermodynamic cycle efficiency* expressing the fraction of *heat adsorbed* in a cycle that is converted to work. If the former definition is applied to fossil power plants, the heat lost up the stack is included in the heat-rejection inventory with the heat adsorbed by condenser cooling water.

plant and the cost of energy, in turn, increasing by about 0.4 mill/kw-hr[62,63] for a cooling tower.

SEISMIC CONSIDERATIONS

6.213 Building codes include provisions for the seismic design of all types of structures; the criteria vary with the extent of earthquake accelerations to be expected where the site is located. Requirements evolved through licensing procedures for nuclear plants (§6.250), however, are much more stringent than for other types of public structures. In seismic areas such as California, the characteristics of the site itself must be carefully investigated as well as the design. In relatively nonseismic areas, the plant must be designed to withstand the effects of the greatest earthquake that can be reasonably anticipated at the site.

6.214 The initial design must ensure that no damage will occur under a specified horizontal ground acceleration condition. In addition, assurance must be provided that the reactor can be safely shut down at a specified higher horizontal ground acceleration although some damage might have occurred. Typical acceleration values for a site in a seismic-active region are 0.25g (25% that of gravity) and 0.50g, respectively. Where only minor earthquake damage is expected, lower accelerations may be specified.[57]

6.215 The design effort required to meet the preceding general criteria is by no means simple. Vital systems and components must be identified and analyzed to ensure that safe shutdown conditions can be maintained. Actual design approaches vary in complexity depending on the seismic activity anticipated. Needed "defensive construction" features are assumed to be incorporated in the design, of course. A "design-basis earthquake" for the site is a useful analytical reference. The details of the ground motion may be considered, including the response spectrum and time-dependent effects. Soil mechanics and foundation engineering are relevant in the force analysis. Although criteria may be primarily in terms of a horizontal-motion spectrum, vertical acceleration is also considered but is normally specified in terms of a fraction of the horizontal acceleration. Existing codes do provide considerable guidance in the stress analysis. Elastic-response characteristics are available, for example, for structural components, piping, and reinforced concrete.[64]

6.216 The possibility of differential displacement of the foundation material at a nuclear-power-station site, as might occur because of faulting, is a much more severe design challenge than uniform ground acceleration. Special designs are available, however, which tend to decouple the containment and foundations from the soil.[65] In one conceptual design study, the reactor building "floats" in a bentonitic slurry contained in a caisson.[66] For an extreme requirement such as this, a cost penalty of about 11% of the total capital cost appears to be involved.

METEOROLOGY

6.217 Two types of meteorological considerations are involved in site selection: (1) the effect of weather patterns on the diffusion of potential accident fission products to population centers and (2) ensurance that tornadoes or hurricanes will cause no safety problems.

6.218 Licensing procedures generally call for an evaluation of the meteorology of the site and adjacent areas to determine the pattern of possible fission-product dispersion. Normally, well designed gaseous-effluent holdup devices and stacks can provide adequate dispersion, even when the terrain and other factors are somewhat unfavorable. Usually, therefore, this aspect of the meteorology is more a design challenge than a siting factor. Also, meteorological evaluations generally show that the *least* favorable conditions on which dispersion estimates may be based occur at about the same frequency for most sites.[57] Of course, everything else being equal, locating a new plant "downwind" from a population center is preferable to locating it "upwind." Such a choice is seldom available, however, since site possibilities are normally rather limited and economic, cooling-water, and other considerations may affect the decision significantly.

6.219 The possibility of extreme conditions, such as tornadoes and hurricanes, is another design challenge. Provisions must be made for the integrity of the plant and its safe shutdown should such a storm pass directly over the plant. Critical structures must therefore be designed to withstand forces from about 300-mph tangential winds and tornado-generated missiles. Present practice is to consider average severe conditions as design criteria but not to consider probabilistic factors.

OTHER CONSIDERATIONS

6.220 The possibility of containment damage due to an aircraft accident has some effect on siting. Although the containment may be designed to withstand the impact of a crashing jet transport of perhaps the Boeing 707 size, the probability of such an accident should be kept as low as feasible by not locating the plant close to airport flight patterns. Probabilistic studies of the aircraft hazard to nuclear power plants located away from an airport indicate that the risk is acceptable by several orders of magnitude when compared with that of other accident possibilities.[67,68]

6.221 Other factors derived from normal engineering practice may also affect site selection. These include the cost of land and its preparation for construction, subsoil conditions and foundation requirements, accessibility for the delivery of components, local tax structure, and any other parameters associated with the site that might affect the plant operating cost.

SITING POLICIES IN OTHER COUNTRIES

6.222 Although the previous discussion applies primarily to practices in the United States, the basic principle of public-safety assurance by a combination of isolation and engineered safeguards generally applies. Interpretive policies and details of application may vary quite widely from country to country, however. In the United Kingdom and Japan, for example, the opportunity to use distance to control the hazards is quite limited.

6.223 The containment and the design features of the reactor system must therefore be emphasized. Criteria evolving from licensing practice may therefore depend on the judgments made in response to the situation in a particular country. An assessment of risks on a probabilistic basis as mentioned in §6.102 has met more favor in other countries than in the United States. Fission-product-release, diffusion, and dosage criteria are also not necessarily uniform. However, information is being exchanged and many standards are being evolved, and the possibility is strong that greater uniformity will be achieved in the future.

SAFETY ASSURANCE

6.224 A broad area that is part of the engineering of a reactor system concerns reliability, accident uncertainties, and general efforts to ensure that the system is within acceptable safety limits. Standards and other types of reference criteria are important as guides for objectively treating subjects which inherently involve some uncertainty.

6.225 The term *quality assurance* is used in several ways. In the broad sense it comprises all actions that give adequate confidence that the reactor system will operate satisfactorily in service. Performance and perhaps economic criteria are therefore involved in addition to questions of safety. In this sense, therefore, the term is more general than *safety assurance.* In the most common usage, component design and material selection are emphasized, with primary attention devoted to safety. Manufacturing procedures, testing techniques, and failure predictions are all likely to enter into the requirements. As a result the terms *quality assurance* and *safety assurance* are often used synonymously. Since quality assurances are required in the reactor-licensing procedure (§6.250), they are of considerable design importance.

PROBABILISTIC ANALYSIS

6.226 In the design and evaluation of many engineering systems, the reliability of components is closely related to estimates of safety. This is certainly true for nuclear reactor systems, where many of the accidents postulated in a safety analysis are initiated by a component failure. Therefore, for both design and analysis, reliability should be expressed on a quantitative basis. Fortunately, the aerospace industry and other manufacturers have progressed far in a discipline that can be called reliability analysis based on probabilistic approaches, with the principles applicable to the nuclear industry. The need to assess reactor safety on a quantitative basis, however, involves more than component reliability. As a basis for analysis, it is desirable to construct mathematical models that include various accident paths and consequences in addition to the initiating events. Such an approach, termed a probabilistic methodology for safety analysis, shows promise as an evaluation and design aid.[69,70]

6.227 Digital-computer techniques permit the practical combination of a number of reasonably sophisticated mathematical models describing parts of the reactor system into a larger overall model that can provide output in terms of dose levels. One such approach for a light-water-reactor system is expressed as a block diagram in Fig. 6.33. Most of the models designated on the diagram by a single block could, in turn, be described by a detailed computer flow chart.

6.228 The procedure is only one of a number of possible approaches. In each, however, the heart of the method is the *model representation* of a sequence of events with the opportunity to predict behavior under a given set of circumstances. Such modeling and simulation is a well-established operations research technique. The model can then be used for various types of analyses, such as the contribution of each component to the probability of system failure, the effectiveness of various safety trips and safeguards, and indeed a general assessment of risks. Keep in mind, however, that the representation is a simulated one and the analysis depends heavily on statistical data describing component reliability, which is generally lacking in the nuclear industry until much reactor operating experience has been accumulated.

Fault-Tree Analysis

6.229 Fault-tree analysis, a special type of graphic event representation, is particularly useful for safety evaluation. The approach provides a way of tracing back from an undesirable event to any of many possible key failures. The representation is generally in the form of a tree trunk for the event to be avoided and tree roots for the contributing events and actions. A simple example serves to clarify the approach.

6.230 Consider a typical gas-heated household water heater with a thermostatically controlled burner actuated by the temperature in the storage tank. A pressure-relief valve in the hot-water discharge line provides safety protection should the water overheat. Although the analysis of this system is straightforward, it illustrates the fault-tree approach. As shown in Fig. 6.34, the

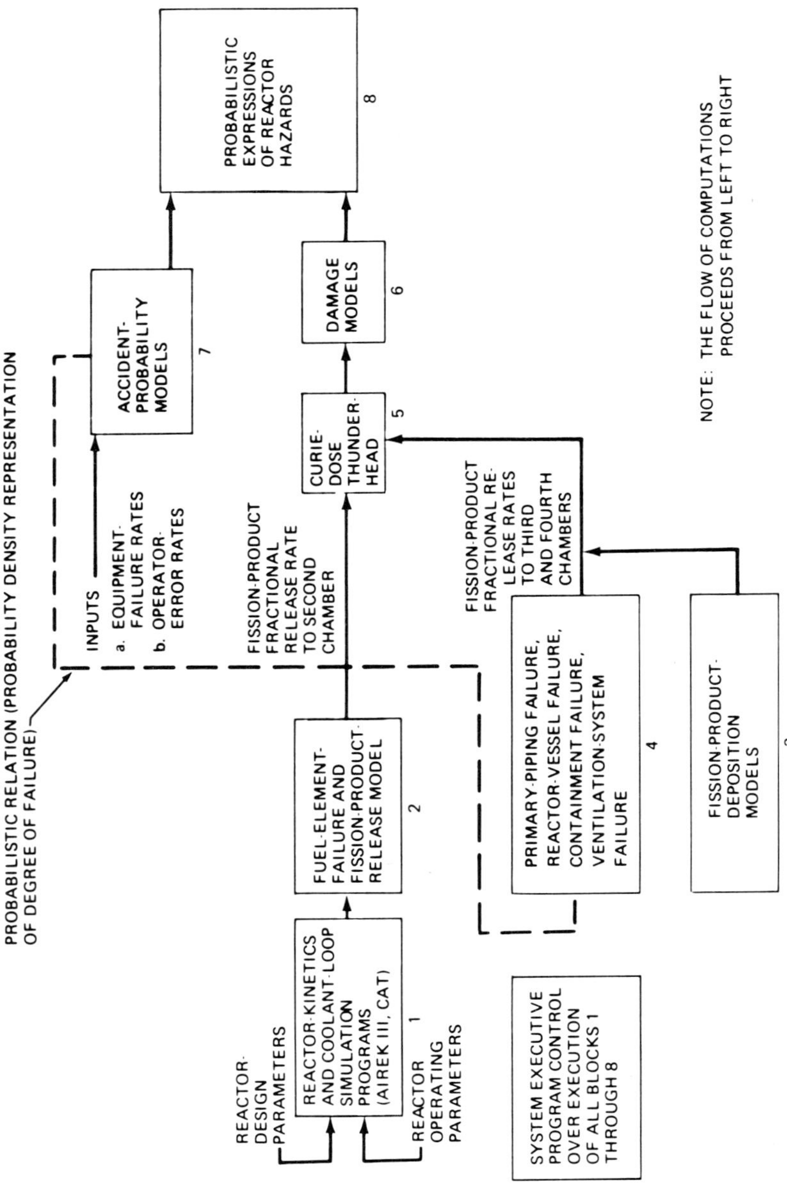

Fig. 6.33 Probabilistic model for light-water reactor.

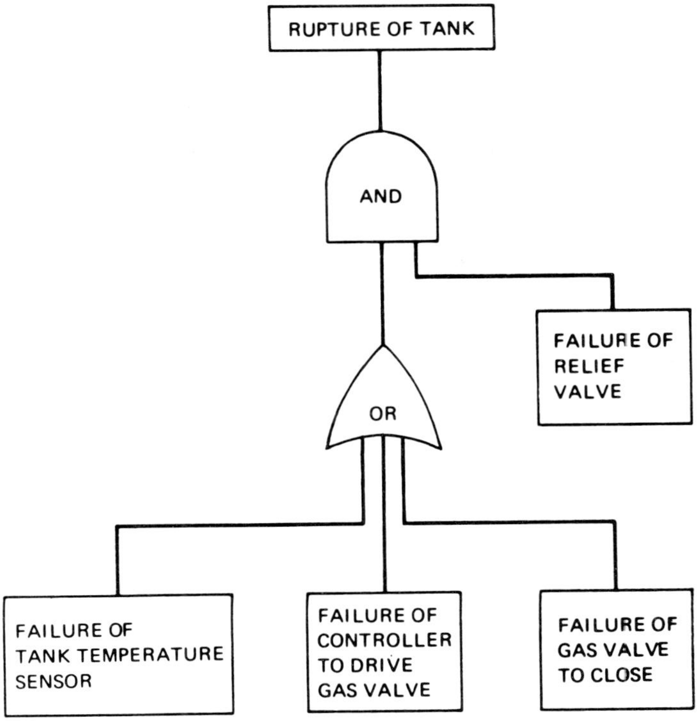

Fig. 6.34 Fault-tree representation.

undesirable event is the rupture of the tank, the "trunk" from which several "roots" flow. The roots are related through two types of logic "gates" shown on the diagram. The convex-bottomed OR gate is activated if any or all of the input events are present, and the flat-bottomed AND gate requires all inputs to exist. In Fig. 6.34, therefore, any of the failures leading to the OR gate would cause the tank to overheat, but it could not rupture without both overheating and relief-valve failure.

6.231 Figure 6.35 shows a fault-tree analysis for a loss-of-coolant-flow accident in a three-coolant-loop liquid-metal fast breeder reactor; the legend varies slightly from that in Fig. 6.34. Reduced coolant flow to the reactor could result from any one of the basic faults shown, which in turn would result in reduced coolant to the core. An incipient mismatch between reactor power generation and heat-removal rates would normally be corrected by the reactor control system. Core-coolant voiding could result only in the extremely unlikely event of failure of the two completely independent automatic control and trip systems as well as failure of the operator to respond to various alarms as indicated by the AND logic gate in the fault-tree diagram.

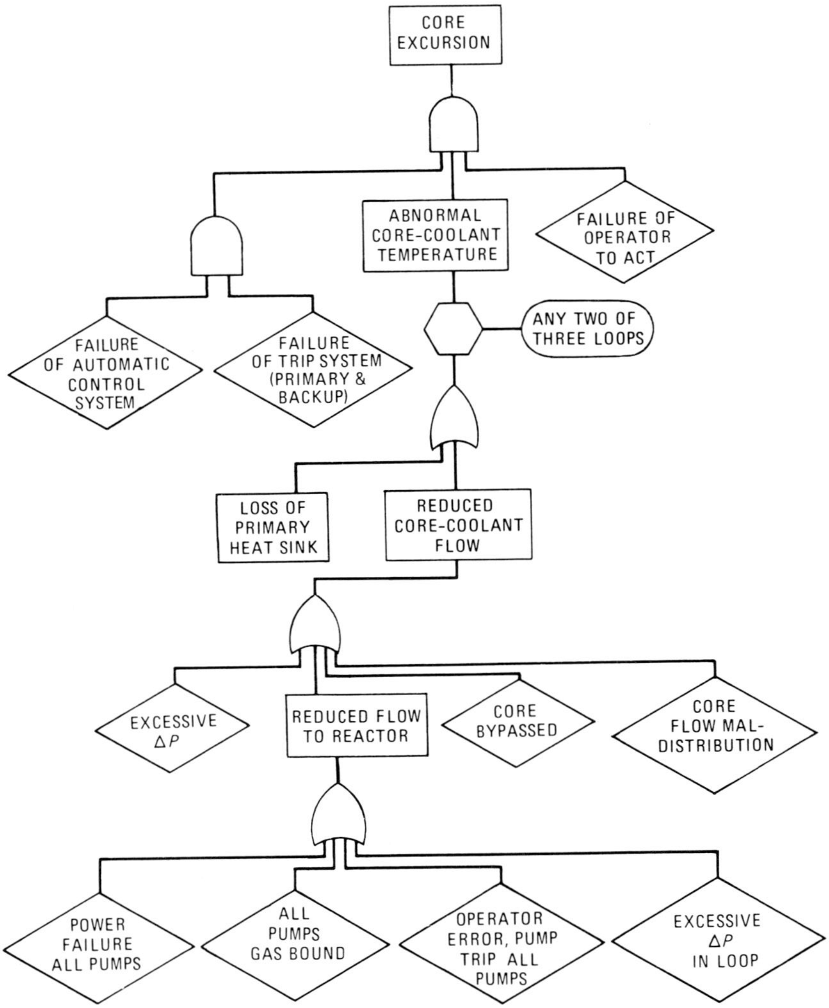

6.232 Additional roots of the fault tree may be developed downward to a detail as fine as desired, e.g., nuts, bolts, etc. Also possible are other gate options providing for priorities, restrictions, and time delays. Statistical parameters can also be included to yield a probabilistic analysis of the occurrence of the undesirable event or events.

Design Applications

6.233 As in most engineering design, analysis and evaluation iterate during the design of the parts of the reactor system related to safety. The probabilistic

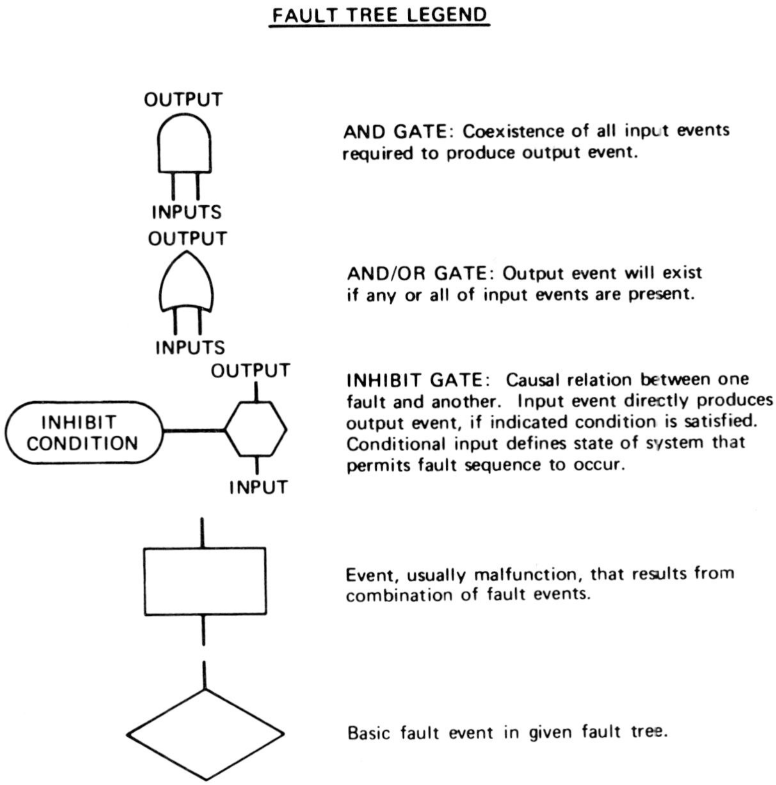

Fig. 6.35 Fault tree for loss of coolant flow.

approaches described are helpful in the evaluation. A difficulty, however, is to establish design goals in terms of acceptable failure probabilities. One approach is to assign such goals, considering the consequences of the postulated accident wherein an extremely low failure rate (1×10^{-7} per year) can be tolerated for a catastrophic result and a somewhat higher rate may be applied to a somewhat less serious result (§6.101). Although such a trend is straightforward, there is uncertainty as to what risk is appropriate.[33]

6.234 Fault-tree and other types of simulation analysis do provide a way of identifying critical subsystems or components whose failure could markedly influence the failure of the entire system (§6.229). Similarly, in a complex network, one event path may have a controlling effect on the total failure. Under such circumstances the designer may wish to introduce redundant elements to split the path and improve the reliability of the system.

6.235 Improved *component* reliability due to careful design and manufacturing quality control generally gives a higher level of system safety since the probability of failure leading to an accident is reduced. An economic trade-off therefore exists between the cost of such component reliability and the economic consequences of an accident. A different consideration is introduced, however, in the protective system required for the reactor. The protective system receives information from the reactor and its associated equipment, compares this information with established limits, and may shut down the reactor or initiate countermeasures. Generally, the more elaborate such a system is, the greater is the inherent safety of the reactor system. The more complex the protection system, however, the greater is the likelihood of unnecessary reactor shutdowns due to component failure within the protection system.* The operational reliability of the reactor *system* is therefore reduced although the safety is improved. Thus a design trade-off of this type between safety and system reliability must also be considered.

6.236 Reactor protection systems receive special design attention with industry-wide performance criteria a goal. Related to this is the periodic testing of protection systems according to standard guidelines (§6.192). As more reactors are put into operation, such testing as well as other accumulated experience helps develop a body of component-reliability data that can serve the designer of new plants.[72]

ROLE OF STANDARDS AND CODES

6.237 Engineering standards, codes, and specifications provide a basis for quality assurance which helps to quantify component reliability in the probabilistic safety analysis. In addition, the careful adherence to such standards by manufacturers as part of a quality assurance program generally improves component reliability and, in turn, the inherent safety of the reactor system.

---

*An indication of the reliability of a system consisting of many elements can be expressed by Lusser's product rule

$$R = \prod_{i=1}^{n} r_i$$

where R is the reliability (probability of nonfailure for a specific period of time), r_i is the reliability of the ith element, and n is the number of elements.[71] If there are 500 components in a system, therefore, each with a reliability of 0.99, the system reliability would only be about 0.01.

6.238 Standards have been, or are being, developed for many more features of the reactor system design than merely the component specifications. Hundreds of applicable codes and standards are being developed by technical society committees, trade associations, and other groups in an organized effort coordinated through the American National Standards Institute, which, in turn, represents one part of a worldwide effort.[73]

6.239 In addition to standards, criteria, and codes applying to the design and testing of the reactor plant itself, standards for various aspects of the fuel cycle and other related activities have been developed. Such documentation of proven engineering practices and procedures is therefore extremely helpful to the designer. However, the nuclear industry has developed so rapidly that the evolvement of these standards has not kept up with the need. This lag has become of considerable concern since quality assurance is vital for reliable operation of nuclear power plants.[74,75]

INCIPIENT FAILURE DIAGNOSIS

6.240 An inspection and testing program, begun after the reactor is placed into operation and carried out during the life of the plant, is also important to the safety assurance of the plant. Particular attention is given to monitoring the protection system to avoid a failure or loss of redundancy. Pressure vessels likewise receive special attention. The reactor design must therefore provide for the inspection of components and the installation of the devices planned for detecting incipient failures. Since techniques are rather specialized, we shall discuss here only a few of the many already developed.

6.241 *On-line* methods, effective while the reactor is at power, avoid loss of plant availability and therefore should be included in the design. This is particularly true of the safety-instrumentation system where on-line periodic testing is posible only if the appropriate circuitry and redundant components are provided in the design.[76]

6.242 One type of diagnostic approach concerns the detection of undesirable operating *conditions.* For example, the analysis of neutron-density fluctuations and other types of noise has been used to detect various incipient malfunctions. In some reactors, such as those cooled with sodium, it is desirable to detect traces of boiling.[77] In addition to neutron-flux noise, ultrasonic and acoustic detectors have been studied for this application. In fact, various types of transducers have been examined for detecting such conditions as cavitation, gas entrainment through the core, hydraulic transients, and the presence of mechanical vibrations.

6.243 Detecting flaws in *materials* by nondestructive-testing techniques is a second type of diagnostic approach.[78] Normally of most concern are the pressure vessel and the primary coolant system, for which acoustic methods show promise, although dye-penetrant and even visual methods are used for certain types of flaws. In fact, nonnuclear industries have used various types of nondestructive testing for many years as part of manufacturing procedures. Such approaches as those using ultrasonics, ultraviolet light, lasers, and penetrating radiation (X ray and gamma), therefore, have been highly developed and can be applied for reactor systems. Many of these, of course, may be useful only before initial operation, or at least during shutdown.

6.244 The surveillance of reactor pressure vessels, particularly for the NDT (nil-ductility transition) temperature, is worth special mention. Exposure to fast-neutron radiation causes a rise in the temperature below which failure will be brittle rather than ductile. Since operation at or below the transition temperature could be unsafe, the shift in NDT must be monitored. A useful guide is, of course, the cumulative exposure, which can be compared with predictions based on experimental results. Most surveillance programs for large water reactors feature a number of in-core specimens of the vessel material which can be removed according to a planned sequence and subjected to tensile and Charpy V-notch impact tests for ductility. The design must therefore provide for such specimen capsules. Nondestructive approaches, such as ultrasonic methods, are also used to detect radiation-induced embrittlement.[79]

LICENSING REQUIREMENTS

6.245 In the United States the Atomic Energy Commission is required by law to "regulate the peaceful uses of atomic energy in order to protect the health and safety of the public."[13,50] This charge has been implemented by the establishment of detailed licensing requirements for users of nuclear energy sources, including reactors. The regulatory functions of the AEC are quite separate from those concerned with operation and development. Organizations of the AEC concerned with the licensing of nuclear facilities are the Division of Reactor Licensing and the Division of Compliance, each of which is responsible to the Director of Regulation.

6.246 The Atomic Energy Act is very general in terms of licensing requirements and does not specify requirements or procedures in any detail. Such requirements are published as federal regulations, however. Since these regulations are frequently referenced by a standard notation, the terminology is reviewed here. Each Federal agency is assigned a chapter in the Federal Register. Chapter 10 (frequently called *Title 10*) refers to AEC regulations, and the designation *10CFR* means Chapter 10, Code of Federal Regulations. An additional number gives the category of the regulation. For example, 10CFR50 covers the licensing of nuclear power plants, research reactors, critical facilities, and reprocessing plants. Since the regulations are amended from time to time, the reader actually concerned with licensing should refer to the current version. In addition, *information guides* issed by the AEC from time to time should be consulted. These are not regulations, but they indicate the technical information desired in the license application. Similarly, *safety guides* describe design features that have been found acceptable from the regulatory viewpoint.

6.247 One section in 10CFR50 outlines in broad terms the standards that determine whether or not a license will be issued to an applicant. These standards require that reasonable assurance be provided that the facility and use of the facility will comply with the Title 10 regulations, that the applicant is technically and financially qualified to operate the proposed facility, and that the issuance of a license will not be "inimical" (harmful) to the common defense and security or the health and safety of the public.

6.248 The type of information to be included in an application is given first in general terms. Of particular interest is the statement, "The description of the process should be sufficiently detailed to permit evaluation of the radioactive hazards involved." The AEC has interpreted this statement to mean that a large amount of detailed information about all phases of the plant design and operation is required in the application. Instructions are therefore provided in some detail for the informational requirements in both the preliminary safety analysis report (PSAR) and the final safety analysis report (FSAR).

6.249 Information must be provided concerning the financial soundness of the utility and the competence of its staff.* In addition to site, plant design, operational, and safety-evaluation information, the preliminary safety analysis report must also describe and evaluate the quality-control program for the fabrication and construction of the structures, systems, and components of the facility. Research and development efforts must also be identified. Design criteria must be listed in detail, and the license applicant is expected to show how the criteria will be met in the final design. Since the criteria and the AEC regulatory interpretation of them change from time to time, this feature of the licensing requirements has been somewhat controversial.[80]

LICENSING PROCESS

6.250 The licensing process in the United States is somewhat complicated but is an important part of the safety-related engineering of the nuclear plant. A "flow sheet" of the process leading to a construction permit is shown[37] in Fig. 6.36. The construction permit allows the utility to build the facility with the understanding that safety problems defined during the review be solved before the reactor goes into operation. This review process is repeated when the reactor is ready to run and the utility submits a final safety-analysis report to obtain the operating license.

6.251 The "road" to the construction permit starts with the preparation of the preliminary safety analysis report by the utility, normally with the help of

*In certain cases information is also requested to permit an antitrust review by the Justice Department.

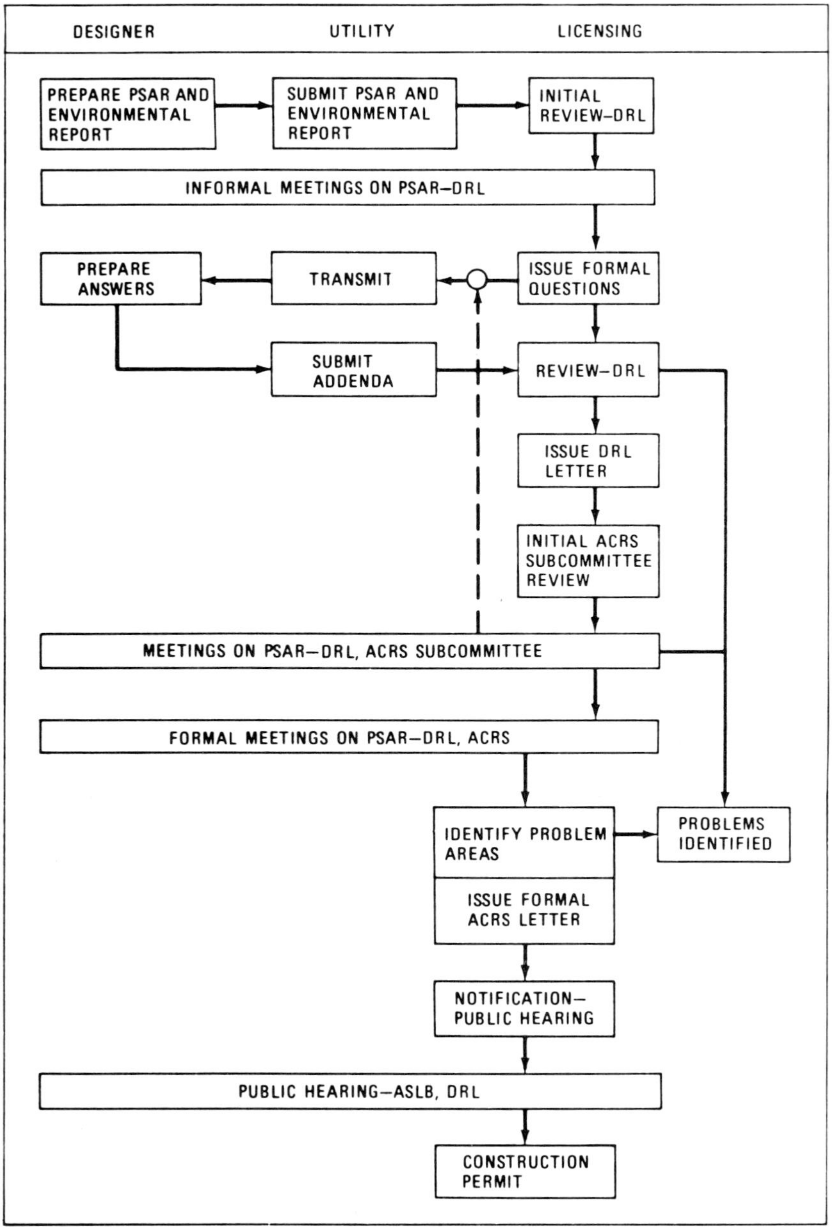

Fig. 6.36 Nuclear safety licensing process. PSAR, preliminary safety analysis report. DRL, Division of Reactor Licensing. ACRS, Advisory Committee on Reactor Safeguards. ASLB, Atomic Safety and Licensing Board.

the reactor manufacturer (vendor) and the architect–engineer (§1.7). The report is submitted to the Division of Reactor Licensing (DRL) for initial review, which may include several informal meetings with both the designer and the utility to clarify areas of uncertainty and obtain information that may still be required for legal licensing.

6.252 A report describing the impact of the proposed plant on the environment is also required (§6.198). Topics include thermal effects; radiological effects; chemical releases; effects on local economics, traffic, etc.; and effects on the area of construction activities. Such environmental reports may be comparable to the preliminary safety analysis report in scope and depth. Although normally the environmental report is not required until the application for a construction permit is filed, applicants are urged to schedule early meetings with the AEC regulatory staff concerning their siting plans. They also are encouraged to meet early in the planning stage with applicable federal, state, and local agencies authorized to develop and enforce environmental standards.

6.253 Next, the Division of Reactor Licensing prepares formal questions that are transmitted to the utility, which in turn may send some to the designer (vendor). Answers are submitted as an addendum to the original preliminary safety analysis report. The original report and the addendum all become part of the public record and are again reviewed. This chain may be repeated many times until the DRL is satisfied that the plant can be built without undue risk to the public.

6.254 The Advisory Committee on Reactor Safeguards (ACRS) was established by law to review safety studies and facility license applications and to advise the AEC. Since the ACRS is a semiautonomous group with no member employed by the AEC, an independent evaluation can be made. With many license applications in process in the early 1970s, however, the ACRS work load has become great. It has therefore been suggested that the ACRS charge be redefined to include the review of only novel concepts; applications for designs very similar to those previously reviewed would be processed without such review.

6.255 According to standard procedures, however, after completing its initial review, the Division of Reactor Licensing sends a letter to the Atomic Energy Commissioners, and the Advisory Committee on Reactor Safeguards assigns a subcommittee, which also reviews the preliminary report. Rather informal meetings, attended by the DRL, the designer, and the utility, are held on the preliminary safety analysis report. Here again, formal questions may be generated, answers prepared, and the chain repeated, if necessary. This same group then meets with the ACRS to identify and resolve problem areas. The ACRS subsequently issues a formal letter of approval for the plant. Next, the Atomic Safety and Licensing Board (ASLB) conducts a public hearing at or near the reactor site, where public "intervenors" may then raise questions about the proposed plant. Although such hearings have sometimes taken on the appearance of a court trial, the purpose is to help the ASLB decide if the plant will be safe or not. Once ASLB approval is obtained, the permit is issued and construction may start.

OPERATING LICENSE

6.256 The requirements for obtaining a facility operating license are similar to those for the construction permit except more extensive information is required. A final safety analysis report must provide complete technical specifications and high-calibre safety analyses. The follow-up nature of the report applies for both the quality-control program and the research and development effort. For example, only plans for the quality-control program could be described at the time of the preliminary safety analysis report, whereas the results of the program and an evaluation of performance are included in the FSAR. Similarly the results of research and development programs initiated to resolve safety questions are given.

6.257 Procedural requirements are, in general, similar to those prescribed for the construction permit. For "first-round" reactors of a given concept, the depth of inquiry is likely to be great and the time required long. As backlog of experience develops, the necessary reviews will probably be expedited, however.

LICENSING DEVELOPMENT

6.258 It must be emphasized that licensing requirements and procedures are subject to continual change. The preceding material is provided primarily for a picture of the care necessary in evaluating a new reactor rather than as an accurate description of procedures that are likely to remain current for a long period. In fact, studies of licensing trends show that the question and answer procedure has not always been carried out according to uniform standards.[81] Also, many of the detailed safeguard requirements involve combinations of mechanistic approaches and nonmechanistic assumptions* which are not always consistent.[82] The time required for licensing has also been criticized.[83] This has led to the suggestion that the procedures be modified to provide for an evaluation with public hearing of the acceptability of a site some time before a construction permit application is submitted. As a result, improvements will likely be made from time to time in the licensing approach without compromising public safety.

---

*For example, "nonmechanistic" assumptions of 50 to 100% fission-product release as given in TID-14844 are not consistent with "mechanistic" calculations considering various fission-product-retention possibilities which yield a much lower value.

NUCLEAR CRITICALITY SAFETY

INTRODUCTION

6.259 Though related to much of the previous discussion, nuclear criticality safety[84] deserves separate attention since it is important in many out-of-core operations involving fissile material in addition to those at the reactor plant site. Fuel fabrication and reprocessing are examples. In fact, several fatalities have occurred from criticality accidents during fuel reprocessing.

6.260 As large numbers of power reactors become operational, increasing amounts of fuel and fuel elements, both fresh and irradiated, will require handling. Shipping practices and the safety technology required in fabrication and reprocessing plants are not within the scope of this book. The principles involved and the general approaches followed are indicated, however.

PARAMETERS AFFECTING CRITICALITY

6.261 All the factors that influence the neutron balance affect criticality safety when a fissile isotope is being handled. Important considerations, most of which interplay with one another, are:

1. Mass. The mass of fissile isotope present is related to the neutrons produced. Of course, the amount of "critical mass" depends on the parameters that affect the balance between neutron loss and production.

2. Geometry. Geometry is important because it influences the neutron leakage. High-neutron-leakage containers in the form of long cylinders for fissile solutions are often used.

3. Moderation. An increase in moderation like that caused by accidental flooding of an array of fuel pins in water generally increases the fission rate and causes a reduction in the critical mass that may be required.

4. Reflection. Reflection has an effect similar to moderation. Other considerations include changes in poison concentration, concentration of the fissile isotope, and homogeneity.

APPROACHES TO CRITICALITY SAFETY

6.262 As for most engineering work, judgment is required to apply standard approaches to a given design problem.[85] Methods that may be considered are:

1. Geometry control. Where possible, sufficient separation and high-neutron-leakage geometries should be provided so that the system will be in an "always-safe" condition.

2. Neutron absorbers. Various poisons such as cadmium and boron, either in solution or in a geometric array, are useful.

3. Administrative control. Carefully planned operational practices that establish procedural limitations help to avoid accumulations of fissile materials that might cause criticality problems.

4. Double-contingency requirements. The process design should include sufficient factors of safety so that two unlikely, independent, and concurrent changes in conditions are required for an incident to occur.

6.263 Standard design practices for fissile-material operations outside reactors have been incorporated in standards and codes that are continually being revised to incorporate new experience (§6.237). Quality assurance considerations also apply. Many of the safeguards for the reactor plant may also apply for nuclear criticality safety. These include instrumentation and alarm systems, soluble poisons, and even the possibility of containment.

ANALYSIS

6.264 Design and analysis go hand in hand. The prediction of criticality depends on both calculation procedures and supporting experimentation. In a proposed operation with fissile material, the designer must consider normal and abnormal conditions, errors in analysis, equipment failures, and possible combinations of these. An acceptable design must therefore include some judgment on the probability of such circumstances and on the consequences of possible accidents which is similar to the evaluation of a reactor safety design.

6.265 In simple systems direct reference to experimental data from critical experiments may be adequate to establish the desired subcritical conditions. Such an approach may be useful in establishing minimum critical masses, diameters of infinitely long cylinders, and the critical thicknesses of infinite slabs or the effects of changes in moderator composition. Even if exact experimental information is not available, the desired subcritical condition can often be developed by interpolation. Machine computations may be necessary for more complex systems, however.

6.266 Machine computational methods are similar to those used in core design. Comparisons with experiment are useful since a measure of the reliability of the calculation methods is needed as a guide to the necessary design conservatism. The effective neutron multiplication constant (k_{eff}) frequently serves as a slowly varying parameter in normalization.

6.267 Since a nuclear criticality safety calculation should predict a noncritical condition with high reliability, the most conservative value should be selected when constants or other inputs are uncertain. The methods themselves can be relatively simple, with two-group approaches generally satisfactory. Special attention, however, must frequently be devoted to the interaction of units in which the neutrons escaping from a unit are replaced by a fraction of those escaping from another unit and vice versa. Angular distributions should therefore be recognized in the method.

REFERENCES

1. J. F. Hogerton, Background Information on Atomic Power Safety, Atomic Industrial Forum, New York, 1964.
2. T. J. Thompson and J. G. Beckerley, *The Technology of Nuclear Reactor Safety*, Vol. 1, The M.I.T. Press, Cambridge, Mass., 1964.
3. Milton Ash, *Nuclear Reactor Kinetics*, McGraw-Hill Book Company, Inc., New York, 1965.
4. G. R. Keepin, *Physics of Nuclear Kinetics*, Addison-Wesley Publishing Company, Inc., Reading, Mass., 1965.
5. R. L. Crowther, D. L. Fischer, R. O. Niemi, and J. E. Wood, Reactor Kinetics Effects of Significance to Large Boiling Water Reactor Design and Operation, Proceedings of the Brookhaven Conference on Industrial Needs and Academic Research in Reactor Kinetics, USAEC Report BNL-50117, pp. 1–36, Brookhaven National Laboratory, July 1968.
6. M. A. Schultz, *Control of Nuclear Reactors and Power Plants*, McGraw-Hill Book Company, Inc., New York, 1961.
7. G. I. Bell and S. Glasstone, *Nuclear Reactor Theory*, pp. 494–509, Van Nostrand Reinhold, New York, 1970.
8. Ref. 3, p. 140.
9. F. R. Nakache, Review and Evaluation of Temperature-Induced Feedback Mechanisms in Fast Power Reactors, USAEC Report UNC-5054, United Nuclear Corporation, 1963.
10. General Discussion Proceedings of the Brookhaven Conference on Industrial Needs and Academic Research in Reactor Kinetics, USAEC Report BNL-50117, pp. 449–459, Brookhaven National Laboratory, July 1968.
11. Ref. 2, p. 43.
12. S. Glasstone and A. Sesonske, *Nuclear Reactor Engineering*, p. 238, D. Van Nostrand, Inc., Princeton, N. J., 1963.
13. Current Status and Future Technical and Economic Potential of Light Water Reactors, USAEC Report WASH-1082, Jackson and Moreland, March 1968.
14. J. M. Harrer, *Nuclear Reactor Control Engineering*, pp. 194–210, D. Van Nostrand, Inc., Princeton, N. J., 1963.
15. Public Service Company of Colorado, Preliminary Safety Analysis Report, Fort St. Vrain Nuclear Generating Station, Docket 50–267, 1966.
16. B. J. Garrick, W. C. Gekler, and H. P. Pomrehn, An Analysis of Nuclear Power Plant Operating and Safety Experience (2 vols.), USAEC Report HN-185, Holmes and Narver, Inc., Dec. 15, 1966.
17. C. R. Russell, *Reactor Safeguards*, The Macmillan Company, New York, 1962.
18. S. Glasstone, The Effects of Nuclear Weapons, Superintendent of Documents, U. S. Government Printing Office, Washington, D.C., 1962.
19. Argonne National Laboratory, Reactor Physics Constants, USAEC Report ANL-5800 (2nd ed.), p. 400, July 1963.
20. J. O. Blomeke and M. F. Todd, Uranium-235 Fission-Product Production as a Function of Thermal Neutron Flux, Irradiation Time, and Decay Time, I. Atomic Concentrations and Gross Totals, USAEC Report ORNL-2127 [Pt. 1 (Vols. 1 and 2)], Oak Ridge National Laboratory, Aug. 19, 1957.
21. Wm. B. Cottrell and A. W. Savolainen, U. S. Reactor Containment Technology, USAEC Report ORNL-NSIC-5 (Vols. I and II), p. 4.36, Oak Ridge National Laboratory, August 1965.
22. J. J. Dinunno, F. D. Anderson, R. E. Baker, and R. L. Waterfield, Calculation of Distance Factors for Power and Test Reactor Sites, USAEC Report TID-14844, Division of Licensing and Regulation, AEC, 1962.

23. H. H. Van Tuyl, ISOGEN – A Computer Code for Radioisotope Generation Calculations, USAEC Report HW-83785, General Electric Company, September 1964.
24. Ref. 21, pp. 4.186–4.190.
25. R. W. Kupp and J. P. Cagnetta, Some Thoughts on the Fission Product Release Program, in International Symposium on Fission Product Release and Transport under Accident Conditions, Held at Oak Ridge, Tennessee, April 5–7, 1965, USAEC Report CONF-650407, p. 641, Oak Ridge National Laboratory, 1965.
26. M. M. Hendrickson, EXDOSE: A Computer Program for Calculating the External Gamma Dose from Airborne Fission Products, USAEC Report BNWL-811, Battelle Northwest, September 1968.
27. Idaho Nuclear Corporation, Water Reactor Safety Program Plan, USAEC Report WASH-1146, February 1970.
28. D. H. Slade (Ed.), Meteorology and Atomic Energy, 1968, USAEC Report TID-24190, Environmental Science and Services Administration, July 1968.
29. Ref. 28, p. 99.
30. Ref. 28, p. 101.
31. J. F. Honstead, A Survey of Environmental Dose Evaluations, *Nucl. Safety,* **9**: 383 (1968).
32. D. G. Jopling and S. J. Gage, The Pattern of Public Political Resistance, *Nucl. News,* **14**(3): 32 (1971).
33. F. R. Farmer, Reactor Safety and Siting: A Proposed Risk Criteria, *Nucl. Safety,* **8**(6): 539 (1967).
34. J. R. Beattie, Risks to the Population and the Individual from Iodine Releases, *Nucl. Safety,* **8**(6): 573 (1967).
35. S. N. Ehrenpreis and J. D. McAdoo, Reactor Safety Analysis, USAEC Report WCAP-3269-60, Westinghouse Electric Corporation, April 1966.
36. S. G. Forbes and T. R. Wilson, Loss-of-Coolant Accident Investigations, USAEC Report IDO-17272, Phillips Petroleum Company, June 1968.
37. C. G. Lawson, Emergency Core-Cooling Systems for Light-Water-Cooled Power Reactors, USAEC Report ORNL-NSIC-24, Oak Ridge National Laboratory, October 1968.
38. David Okrent, A Look at Fast Reactor Safety, *Nucl. Safety,* **6**: 317 (1965).
39. C. N. Kelber et al., Safety Problems of Liquid-Metal-Cooled Fast Breeder Reactors, USAEC Report ANL-7657, Argonne National Laboratory, February 1970.
40. W. J. McCarthy and David Okrent, Fast Reactor Kinetics in *The Technology of Nuclear Reactor Safety,* T. J. Thompson and J. G. Beckerly (Eds.), the M.I.T. Press, Cambridge, Mass., 1964.
41. R. A. Meyer and B. Wolfe, Fast Reactor Meltdown Accidents, in *Advances in Nuclear Science and Technology,* P. Greebler and E. J. Henley (Eds.), Academic Press, Inc., 1968.
42. G. W. Keilholtz and G. C. Battle, Jr., Fission Product Release and Transport in Liquid Metal Fast Breeder Systems, USAEC Report ORNL-NSIC-37, Oak Ridge National Laboratory, March 1969.
43. D. G. Sherer and W. G. Meinhardt, An Analysis of Fast Reactor Transient Response and Safety in Selected Accidents, USAEC Report GEAP-4787, General Electric Company, June 1966.
44. Argonne National Laboratory, Liquid Metal Fast Breeder Reactor (LMFBR) Program Plan, Vol. 10, Safety, USAEC Report WASH-1110, August 1968.
45. L. Baker, Jr., D. R. MacFarlane, and J. F. Marchaterre, Delineation of LMFBR Single Subassembly Accidents, *Trans. Amer. Nucl. Soc.,* **13**: 647 (1970).
46. Oak Ridge National Laboratory, An Evaluation of Gas-Cooled Fast Reactors, USAEC Report WASH-1089, April 1969.

47. Ref. 21, p. 1.46.
48. W. R. Gall, Design Trends in Systems for Containment, *Nuclear Safety,* 8(3): 245 (1967).
49. F. C. Zapp, Testing of Containment Systems Used with Light-Water-Cooled Power Reactors, USAEC Report ORNL-NSIC-26, Oak Ridge National Laboratory, August 1968.
50. *Licensing and Regulation of Nuclear Reactors,* Hearings Before the Joint Committee on Atomic Energy, 90th Congress, September 1967, Superintendent of Documents, U.S. Government Printing Office, Washington, D.C., 1968.
51. Indiana and Michigan Electric Company, Preliminary Safety Analysis Report, Donald C. Cook Nuclear Plant, Report DOCKET-50315-3, -4, and -5, 1967.
52. P. M. Murphy, D. B. Sherer, and A. B. Gibson, The Effect of Using Different Safety Criteria and Selected Safety Studies for a 1000 MWe Sodium-Cooled Fast Reactor in Proceedings of the International Conference on Sodium Technology and Large Fast Reactor Design, Part II: Sessions on Large Fast Reactor Design, USAEC Report ANL-7520, pp. 345–357, Argonne National Laboratory, Nov. 7–9, 1968.
53. S. Levy, A Systems Approach to Containment Design in Nuclear Power Plants in Containment and Siting of Nuclear Power Plants, Symposium Series, Vienna, Apr. 3–7, 1967, pp. 227–242, International Atomic Energy Agency, Vienna, 1967 (STI/PUB/154).
54. Ref. 12, Chap. 10.
55. L. D. Van Hees and E. L. Kelley, Jr., TRANCO: A Computer Program for Estimating the Cost of Electrical Power Transmission, USAEC Report BNWL-743, Battelle–Northwest, May 1968.
56. H. E. Vann, Influence of Safety Problems on the Development of Economic Nuclear Power, in Nuclear Power Reactor Siting, USAEC Report CONF-650201, pp. 110–124, American Nuclear Society, 1965.
57. W. K. Davis and J. E. Robb, Nuclear Plant Siting in the United States of America, in Containment and Siting of Nuclear Power Plants, Symposium Series, Vienna, Apr. 3–7, 1967, International Atomic Energy Agency, Vienna, 1967 (STI/PUB/154).
58. Office of Science and Technology, Electric Power and the Environment, Superintendent of Documents, U. S. Government Printing Office, Washington, D. C., 1970.
59. *Environmental Effects of Producing Electric Power,* Hearings Before the Joint Committee of Atomic Energy, 91st Congress, 1969–1970, Parts 1 and 2, 1970, pp. 175–194, 238–285, 1527–1544, Superintendent of Documents, U. S. Government Printing Office, Washington, D. C., 1970.
60. K. R. Wise and B. M. Cole, A Survey of Heat Sink Capacity of Major Streams Within the U. S., USAEC Report BNWL-951, Battelle–Northwest, January 1969.
61. J. R. Clark, Thermal Pollution and Aquatic Life, *Sci. Amer.,* **220**(3): 19 (1969).
62. Ref. 61, p. 1569.
63. *Nucl. Ind.* **17**(11–12): 67 (1970).
64. R. A. Williamson, Some Considerations in Earthquake Resistant Design of Nuclear Power Plants, USAEC Report HN-189 (Rev.), Holmes and Narver, Inc., March 1967.
65. J. M. Duncan et al., Antiseismic Design of a PWR Nuclear Power Station, USAEC Report HN-192, Holmes and Narver, Inc., April 1968.
66. E. R. Kummerle, Conceptual Design of a Boiling Water Reactor Plant to Resist Ground Fault Displacement, USAEC Report TID-24530, Burns and Roe, Inc., May 1968.
67. I. B. Wall and R. C. Augenstein, Probabilistic Assessment of Aircraft Hazard for Nuclear Power Plants – I, *Trans. Amer. Nucl. Soc.,* **13**: 217 (1970).
68. C. V. Chelapati and I. B. Wall, Probabilistic Assessment of Aircraft Hazard for Nuclear Power Plants – II, *Trans. Amer. Nucl. Soc.,* **13**: 218 (1970).

69. R. J. Mulvilhill, A Probabilistic Methodology for the Safety Analysis of Nuclear Power Reactors (2 vols.), USAEC Report SAN-570-2, Planning Research Corporation, Feb. 26, 1966.
70. L. M. Arnett, Quantitative Analysis of Reactor Safety, USAEC Report DP-1168, Savannah River Laboratory, November 1968.
71. B. J. Garrick and W. C. Gekler, Reliability Analysis of Engineered Safeguards, *Nucl. Safety,* **8**(5): 470 (1967).
72. L. G. Frederick, Reliability from the Supplier's Viewpoint, in Proceedings of the AECL Symposium on Reliability, Canadian Report AECL-3670, 1970.
73. Wm. B. Cottrell, National and International Nuclear Codes, Standards, and Regulations, *Nucl. Safety,* **7**(4): 415 (1966).
74. Milton Shaw, Standards and Reactor Safety, *Nucl. Safety,* **9**(5): 337 (1968).
75. Considerations Affecting Steam Power Plant Site Selection, Office of Science and Technology, Superintendent of Documents, U. S. Government Printing Office, Washington, D. C., 1968.
76. J. L. Anderson, On-Line Testing of Safety Instrumentation for the HFIR, in Incipient Failure Diagnosis for Assuring Safety and Availability of Nuclear Power Plants, USAEC Report CONF-671011, p. 16, Oak Ridge National Laboratory, 1968.
77. R. F. Saxe, Survey of Boiling Detection Methods in Reactors, in Incipient Failure Diagnosis for Assuring Safety and Availability of Nuclear Power Plants, USAEC Report CONF-671011, p. 41, Oak Ridge National Laboratory, 1968.
78. D. L. Parry, Nondestructive Flaw Detection in Nuclear Power Installations, in Incipient Failure Diagnosis for Assuring Safety and Availability of Nuclear Power Plants, USAEC Report CONF-671011, p. 107, Oak Ridge National Laboratory, 1968.
79. G. D. Whitman, G. C. Robinson, and A. W. Savolainen, Technology of Steel Pressure Vessels for Water-Cooled Nuclear Reactors, USAEC Report ORNL-NSIC-21, pp. 637, Oak Ridge National Laboratory, December 1967, 661.
80. *Nucl. Ind.,* **18**(3): 32 (March 1971).
81. K. P. Baskin and Alexander Sesonske, An Analysis of Reactor Licensing Supplementary Question Requirements, *Nucl. Safety,* **10**(3): 197 (1969).
82. *Nucl. Ind.,* **16**(4): 31 (1969).
83. F. C. Olds, Nuclear Plants – Late and Later, *Power Eng.* **73**(3): 55 (1969).
84. Proceedings of Nuclear Criticality Safety, USAEC Report SC-DC-67-1305, Sandia Corporation, 1967.
85. J. T. Long, *Engineering for Nuclear Fuel Reprocessing,* pp. 83–101, Gordon and Breach, Science Publishers, Inc., New York, 1967.

Fuel-System Analysis

7

INTRODUCTION

7.1 The design of the reactor core is one of the most challenging areas of reactor engineering because the interplay of many parameters must be considered. Parameters for heat removal and neutronics are discussed in Chaps. 4 and 5. Intimately related to these are a separate set of parameters concerned with the preparation and recovery of the fuel outside the reactor, the fuel-loading arrangement, and many of the mechanical design features of the fuel. It is therefore useful to designate this broad area of design attention as the *fuel system.*

7.2 The selection of a proper combination of system parameters to achieve a given result is the heart of the design process (§ 1.3). For a fuel system, however, choosing a criterion for the optimization poses some questions. The desire for a minimum energy cost does not necessarily lead to the same set of parameter values as an optimization to minimize the uranium mined for energy needs, although there are some economic factors associated with fuel conservation. Should fuel conservation be ignored and only a completely economic criterion be considered, the fuel design is likely to differ from one based on a compromise criterion. A different objective also applies to the fuel planned for a new reactor and that for a replacement core. When the reactor is being designed, both reactor and fuel parameters can be adjusted to achieve a minimum power cost. Once the reactor is built, however, the reactor design can no longer be changed. A replacement core is therefore designed to achieve a minimum fuel cost. The designer must also consider, however, the possible importance of fuel-conservation parameters, which are likely to have an economic input if the period studied is long.

7.3 An additional complication is introduced when a utility system has several nuclear plants and not all can be fully loaded. Loading strategy and fuel-system-design parameters are then based on the minimization of the discounted energy cost of the total system over the time span considered.[1]

7.4 The subject of this chapter is the role of the various fuel-system parameters in reactor design. Since much of the fuel cycle is pertinent, some discussion is included of such individual fuel-cycle steps as feed-material operations, enrichment, fabrication, and reprocessing. Fuel utilization is considered from the designer's viewpoint. Since fast reactors and advanced converters have some special fuel-system considerations, they are separately discussed. In many areas, however, a comprehensive treatment is not attempted since the overall subject is indeed a vast one.

FUEL-CYCLE OPERATIONS

7.5 A designer must be familiar with the various fuel-cycle operations before he can consider the effects of parameters on the fuel system in a meaningful way. Although descriptions of the various operations are generally available,[2] summary discussions are included here as a convenience and as a framework for subsequent treatments.

ORE AND FEED-MATERIAL OPERATIONS

7.6 Both uranium and thorium are found in various parts of the world in deposits which can be mined with varying degrees of difficulty and which yield ores of different grades. Therefore an ore price depends on both mining costs and demand. Since the ore price is a major fuel-system parameter, some of the factors that affect the cost of mining are considered here.

7.7 In the United States, uranium ore is obtained either by open-pit or underground mining, primarily in the Rocky Mountain area. If the ore deposit is within 400 ft of the surface, the decision to use one method or the other depends on an economic analysis. Generally, if there is no clear-cut advantage, open-pit mining is favored since lower grade ore and possible new deposits may be recovered during the stripping process. In many areas, however, stripping disfigures the land to an objectionable extent. Relative costs for different production rates are indicated in Table 7.1. "Development" refers to the activities required before production is possible at the normal rate.

TABLE 7.1

Relative Costs for Mining Uranium Ore*

Category	Open pit: 200 tons/day	Open pit: 1000 tons/day	Underground: 200 tons/day	Underground: 1000 tons/day
	Mining costs, $/ton			
Royalty	1.75	1.75	1.75	1.75
Primary development†	7.00	7.00	2.50	2.50
Other capital	0.50	0.25	1.20	1.10
Operating	2.75	2.50	9.00	7.75
Total	12.00	11.50	14.45	13.10

*Reference: Costs considered for mining 200 and 1000 tons of 0.25% U_3O_8 ore per day from an ore body having a seven-year life.

†Stripping cost is given for a 250-ft depth; underground mining is at a 500-ft depth.

7.8 Uranium ore after being received at the mill near the mine is normally crushed and ground. Physical beneficiation, in which ore is concentrated by flotation or other methods, is normally not practical for uranium ores since it is difficult to separate a residue low enough in grade to discard. Uranium is therefore extracted from the ore by leaching, usually with an acid. The so-called pregnant liquor, now containing the uranium, is then subjected to one or a combination of several different processes, such as precipitation, ion exchange, or solvent extraction, to recover the uranium. A "yellow cake" consisting of 75 to 85% U_3O_8 is produced. A typical mill process is shown in Fig. 7.1, and an indication of costs is given in Table 7.2. Finally, overall mining and milling production costs are summarized in Fig. 7.2 as a function of both throughput and ore grade.

7.9 Conversion is the term for purification of the ore concentrate leading to the production of pure uranium hexafluoride, the feed material for the gaseous-diffusion enrichment plants. In other cases, however, the term applies to operations yielding UO_3, UF_4, or even uranium metal. One objective of these operations is to reduce the impurities (materials with high absorption cross sections for thermal neutrons) found in the concentrate to a level acceptable for reactor use. One process for UF_6, used by the Allied Chemical Corp., is shown schematically in Fig. 7.3. Charges for the conversion to UF_6 range from $1.04 to $1.25 per pound of uranium.

URANIUM ENRICHMENT BY GASEOUS DIFFUSION

7.10 The gaseous-diffusion process[3,4] is normally used for the isotopic enrichment of uranium. Since fuel-system costs are very dependent on the price

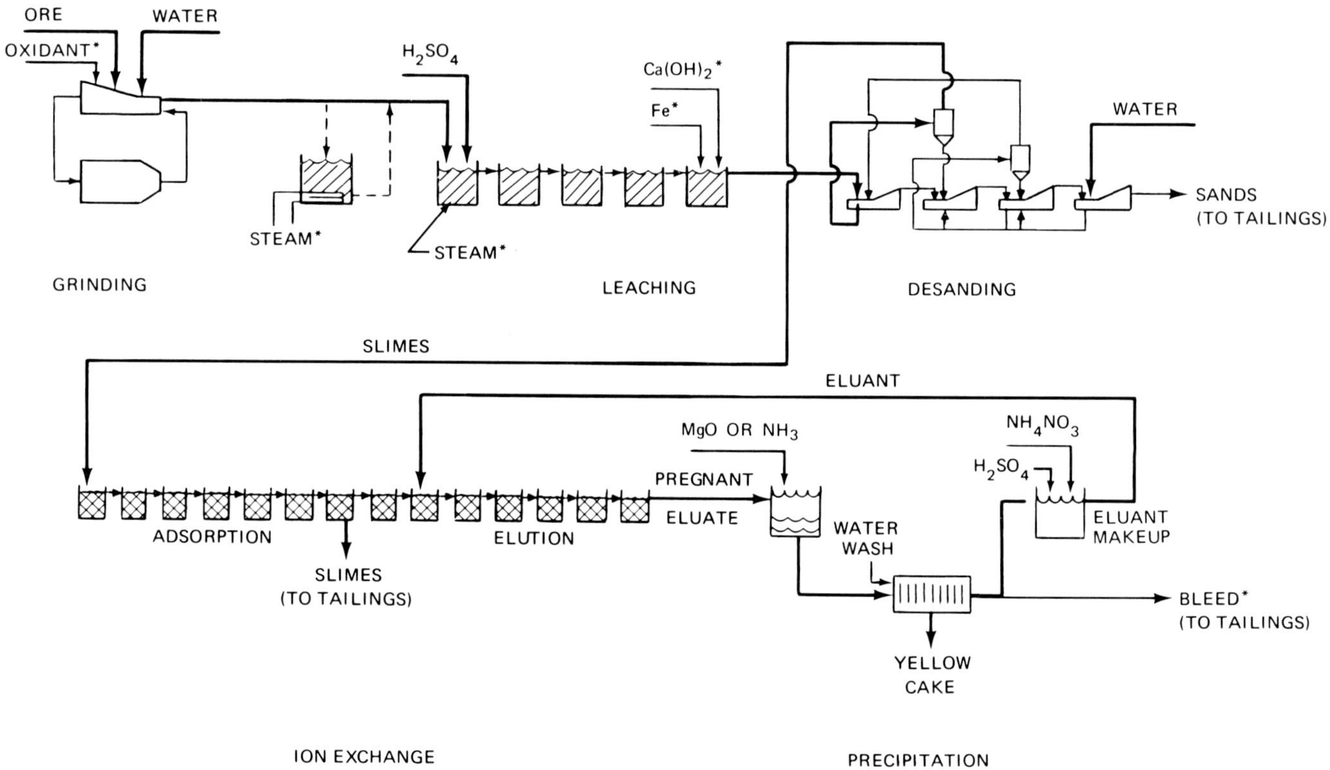

Fig. 7.1 Typical uranium mill process (acid leach–resin in pulp process). *, when required.

TABLE 7.2

Representative Overall Costs for Mining and Milling 200 and 1000 Tons of 0.25% U_3O_8 Ore Per Day

	Open-pit mining				Underground mining			
	200 tons/day		1000 tons/day		200 tons/day		1000 tons/day	
Operation	$/ton	$/lb	$/ton	$/lb	$/ton	$/lb	$/ton	$/lb
Mining	12.00	2.53	11.50	2.42	14.45	3.04	13.10	2.76
Ore hauling	2.00	0.42	2.00	0.42	2.00	0.42	2.00	0.42
Mill operating	9.75	2.05	5.72	1.20	9.75	2.05	5.72	1.20
Mill amortization*	2.78	0.59	1.64	0.35	2.78	0.59	1.64	0.35
Total	26.53	5.59	20.86	4.39	28.98	6.10	22.46	4.73

*Mill amortization is based on a 10-year period with a construction cost of $2 million at 200 tons per day and $5.9 million at 1000 tons per day.

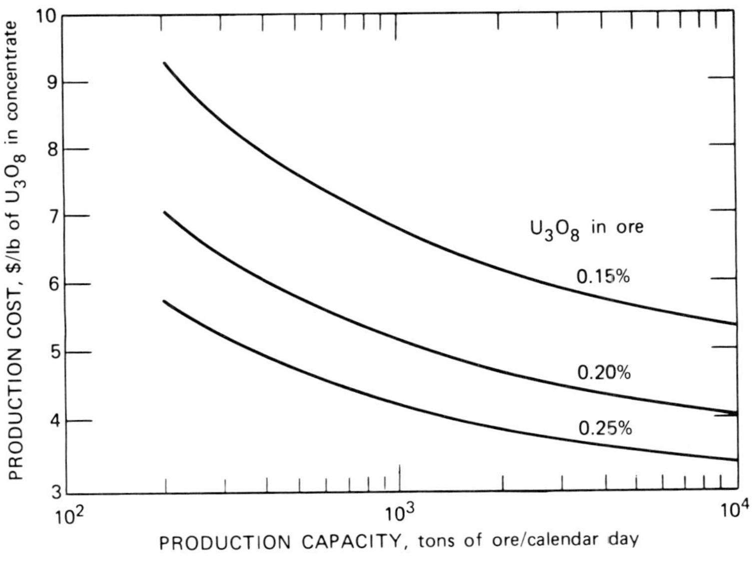

Fig. 7.2 Typical effect of production capacity and ore grade on production cost.

of the enriched product, which, in turn, depends on some of the process parameters, these are discussed here in some detail. The role of these costs in core management is considered in §7.55.

7.11 The separation principle is based on the difference in average velocity between a light and a heavy isotope in a gas with each isotope at identical kinetic energy at a uniform temperature. Molecules containing ^{235}U therefore strike the

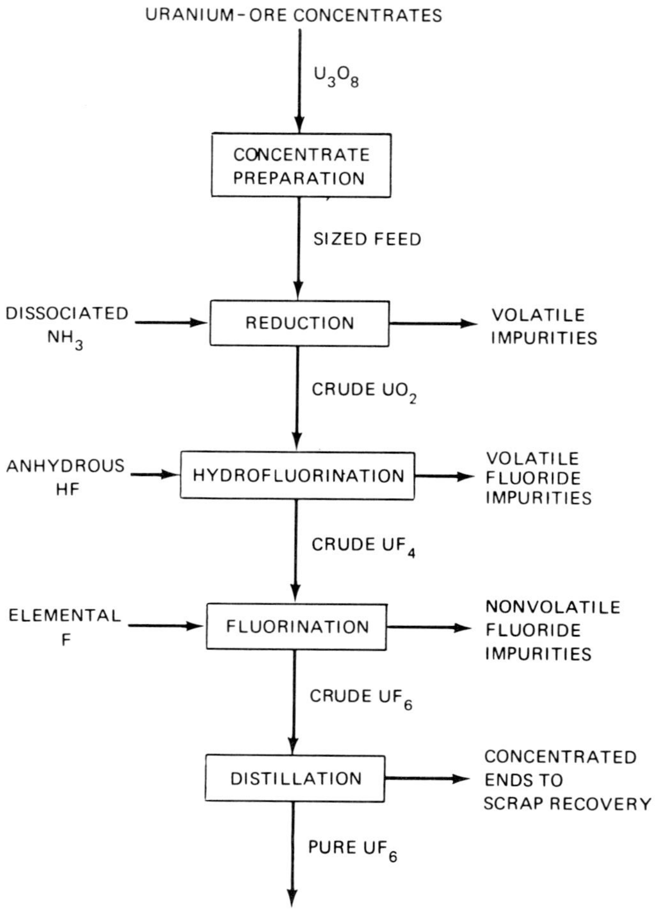

Fig. 7.3 Overall process flow diagram.

porous walls, which are in the form of tubes, more frequently than molecules containing ^{238}U. Since molecular transport through the porous barrier tube by "effusion" depends on the frequency of wall contact, the gas at lower pressure outside the tubes tends to be slightly enriched in ^{235}U. In fact, with UF_6 the separation factor (§7.15) is only 1.00429; the production of 90% ^{235}U from natural-uranium feed therefore requires about 3000 such separation stages, in series, arranged in a cascade.

7.12 For this process to work, a gaseous compound of uranium must be used, and the hexafluoride is the only known suitable compound. Since UF_6 is a solid at room temperature, the diffusion plants must be operated at temperatures and pressures necessary to maintain the UF_6 in gaseous form. Though a stable compound, UF_6 is extremely reactive with water, is very corrosive to most common metals, and is not compatible with organics, such as lubricating oils. This chemical activity dictates the use of such metals as nickel and aluminum and means that the entire cascade must be leaktight and clean. The corrosiveness of the process gas also imposes added difficulties in the fabrication of barrier material, which must maintain its separative quality over long periods of time.

7.13 The process concept is shown in Figs. 7.4 and 7.5. About half of the UF_6 gas flowing inside the barrier tube diffuses through the wall and is fed to the next higher stage; the remaining undiffused portion is recycled to the next lower stage. The diffused stream is slightly enriched with respect to ^{235}U, and the stream that has not been diffused is depleted to the same degree.

7.14 Figure 7.5 shows how the single stages are connected and the basic process equipment required. Motor-driven axial-flow compressors are used in the larger stages to compress the UF_6 gas to make it flow through the barrier. A gas cooler removes the heat of compression at each stage. Groups of stages are coupled to make up operating units, and such groups, in turn, make up the cascade.

7.15 The separation process, involving many individual stages, is similar in many ways to other stage-separation processes, such as fractional distillation, and makes use of many of the same principles. Our present concern, however, is primarily with the cost parameter, *separative work,* which can be considered initially in terms of a single stage.

Let $2L$ = the quantity of uranium fed to a single stage
z = the ^{235}U assay (weight fraction) of the stage feed stream
x = the ^{235}U assay of the depleted stream leaving the stage
y = the ^{235}U assay of the enriched stream leaving the stage

The stage-separation factor, α, is defined as

$$\alpha = \frac{y/(1-y)}{x/(1-x)} \tag{7.1}$$

For uranium isotopes in UF_6, $\alpha \approx \sqrt{352/349} = 1.00429$. Let

$$\psi = \alpha - 1 \tag{7.2}$$

Then, combining Eqs. 7.1 and 7.2 gives

$$y = \frac{x(1+\psi)}{1+\psi x} \tag{7.3}$$

Since $\psi \ll 1$, the right-hand side can be expanded in series form:

$$y = x(1+\psi)(1 - \psi x + \psi^2 x^2 + \ldots) \tag{7.4}$$

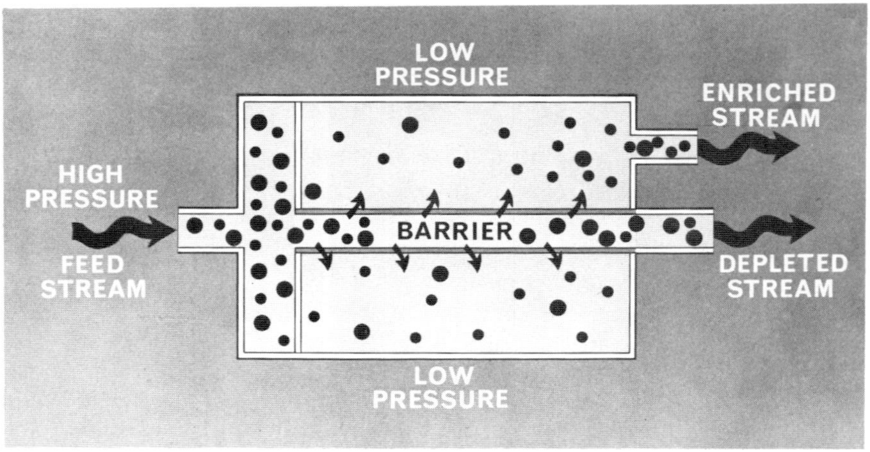

Fig. 7.4 Gaseous-diffusion stage.

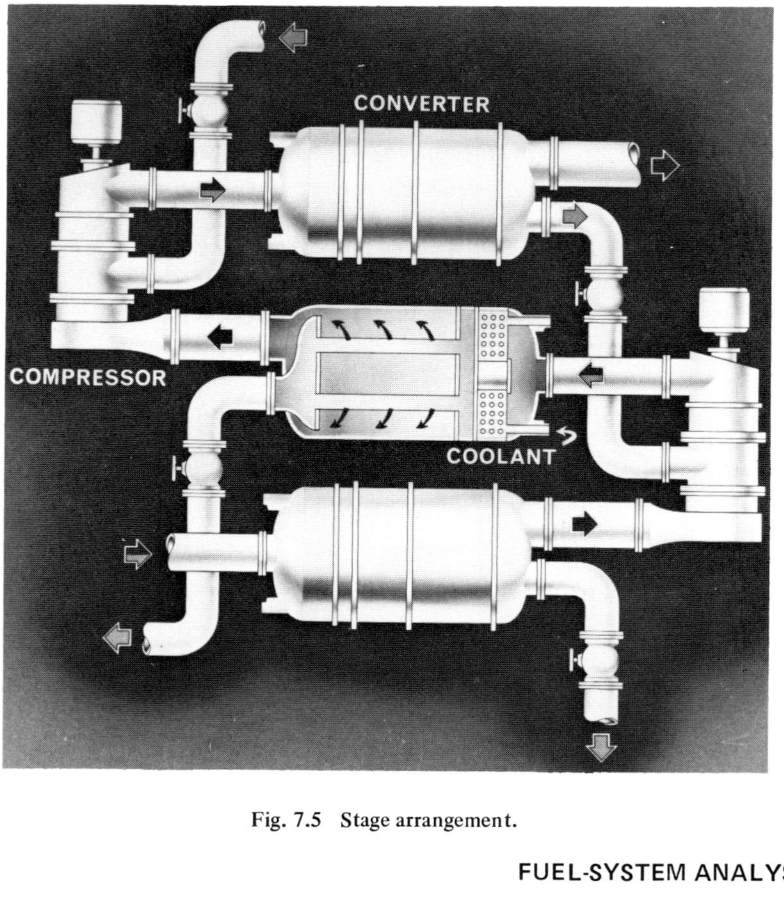

Fig. 7.5 Stage arrangement.

Neglecting higher powers of ψ and rearranging gives

$$y - x = \psi x(1 - x) \tag{7.5}$$

This gives the difference in ^{235}U assay between the enriched and depleted stream leaving the stage as a result of work done by the stage. Accordingly, the separated effluents from the stage should have more value than the feed to the stage. Assume the existence of a "value function," V, so that, for example, $V(x)$ represents the value of one unit of uranium at assay x. Then, making a "value" balance around a stage yields

$$\Delta = L\ V(x) + L\ V(y) - 2\,L\ V(z) \tag{7.6}$$

where Δ is the change of value effected by that stage. Expanding $V(y)$ and $V(z)$ about x in a Taylor's series gives

$$V(y) = V(x) + (y - x)\ V'(x) + \frac{(y - x)^2}{2}\ V''(x) + \ldots \tag{7.7}$$

$$V(z) = V(x) + (z - x)\ V'(x) + \frac{(z - x)^2}{2}\ V''(x) + \ldots \tag{7.8}$$

From a stage ^{235}U balance,

$$z - x = \frac{y - x}{2} \tag{7.9}$$

Substituting these expressions into Eq. 7.6 and neglecting terms containing the very small difference $(y - x)$ raised to powers greater than 2 gives

$$\Delta = L(y - x)^2\ \frac{V''(x)}{4} \tag{7.10}$$

Now, after substituting Eq. 7.5, we obtain

$$\Delta = \frac{L\psi^2}{4}\ [x(1 - x)]^2\ V''(x) \tag{7.11}$$

Since the stage work per unit of feed material is independent of assay, Δ is independent of assay; then

$$V''(x) = \frac{1}{[x(1 - x)]^2} \tag{7.12}$$

Equation 7.12 is an ordinary differential equation having the general solution

$$V(x) = C_0 + C_1 x + (2x - 1) \ln \frac{x}{1 - x} \tag{7.13}$$

where C_0 and C_1 are arbitrary constants. A particularly convenient form of $V(x)$ arises if we let $V'(0.5) = V(0.5) = 0$. Thus we completely define our "value function" as

$$V(x) = (2x - 1) \ln \frac{x}{1 - x} \tag{7.14}$$

It is to be emphasized that, though $V(x)$ is "value" per unit of material, it should never be confused with price or cost of material. In fact, it should be considered dimensionless. Equation 7.14 can also be developed from other considerations with the function $V(x)$ then called the *separation potential*.[2]

7.16 Now, for the overall plant, a material balance yields

$$F = P + W \tag{7.15}$$

where F = flow rate of the feed stream
P = flow rate of the product stream
W = flow rate of the tails stream

It is then possible to write

$$D = W\, V(x_w) + P\, V(x_p) - F\, V(x_f) \tag{7.16}$$

where D is known as the *separative duty*.

7.17 The separative duty is a measure of the expense of isotope separation. In an ideal gaseous-diffusion cascade, for example, the total flow rate, pump capacity, power demand, and total barrier area are all proportional to the separative duty. Having the same dimensions as the flow rates, the separative duty can also be expressed in terms of a unit quantity of material, normally in kilograms of uranium; it is then known as the *separative work,* which can now also be designated as Δ, or SWU (separative work unit).

7.18 Since the annual operating costs are proportional to the amount of separative work accomplished in a year, it is possible to determine the cost of a unit of separative work, C_s, normally expressed in dollars per kilogram. The cost of the product stream is then determined by summing the cost of the diffusion-plant operation (separative work cost), the cost of the feed, and the credit for the waste stream.

$$C_p P = C_s \Delta + C_f F - C_w W \tag{7.17}$$

where C_p = unit cost of the product
C_f = unit cost of the feed
C_w = unit cost of the tails

7.19 It is useful to develop a relation in terms of concentrations rather than flow rates by considering a ^{235}U balance.

$$x_f F = x_p P + x_w W$$

where the subscripts *f*, *p*, and *w* refer to feed, product, and tails, respectively. Then,

$$\frac{W}{P} = \frac{x_p - x_f}{x_f - x_w} \qquad \frac{F}{P} = \frac{x_p - x_w}{x_f - x_w} \tag{7.18}$$

Substitution of Eqs. 7.18 into Eq. 7.17 yields

$$C_p = \frac{C_s}{P} + \frac{x_p - x_w}{x_f - x_w} C_f - \frac{x_p - x_f}{x_f - x_w} C_w \tag{7.19}$$

Similarly, from Eq. 7.16, on a unit-product basis,

$$\frac{\Delta}{P} = \frac{x_p - x_f}{x_f - x_w} V(x_w) + V(x_p) - \frac{x_p - x_w}{x_f - x_w} V(x_f) \tag{7.20}$$

Combining Eqs. 7.19 and 7.20 gives the general uranium price schedule equation,

$$C_p = C_s \left[V(x_p) + \frac{x_p - x_f}{x_f - x_w} V(x_w) - \frac{x_p - x_w}{x_f - x_w} V(x_f) \right] + C_f \frac{x_p - x_w}{x_f - x_w} - C_w \frac{x_p - x_f}{x_f - x_w} \tag{7.21}$$

Note that the product cost consists of a separative-work component and a feed component. The waste credit is normally assumed equal to zero.

7.20 The optimum tails composition, the composition that minimizes the unit cost of the enriched product, is found by equating to zero the first derivative of Eq. 7.21 with respect to x_w. The optimum tails material of composition, x_0, can be shown to be that material which, if used as *feed* and assumed to have no value, will yield a product in an ideal cascade equal to the cost of the normal feed at the normal feed composition. Under such circumstances Eq. 7.21 reduces to

$$C_p = C_s \left[(2x_p - 1) \ln \frac{x_p(1 - x_0)}{x_0(1 - x_p)} + \frac{(x_p - x_0)(1 - 2x_0)}{x_0(1 - x_0)} \right] \tag{7.22}$$

where x_0 can be determined from the relation

$$C = C_s \left[(2x_f - 1) \ln \frac{x_f(1 - x_0)}{x_0(1 - x_f)} + \frac{(x_f - x_0)(1 - 2x_0)}{x_0(1 - x_0)} \right] \tag{7.23}$$

For a given feed concentration, the optimum composition therefore depends only on the ratio C_f/C_s. Some values of x_0 as a function of C_f/C_s for natural uranium feed ($x_f = 0.00711$) are given in Table 7.3.

TABLE 7.3

Optimum Tails Compositions

Ratio of feed cost to cost of separative work	Optimum tails composition, wt.%	Ratio of feed cost to cost of separative work	Optimum tails composition, wt.%
0.0	0.71150	1.3	0.19971
0.2	0.40248	1.4	0.19229
0.4	0.32750	1.5	0.18549
0.6	0.28240	2.0	0.15840
0.8	0.25078	2.5	0.13892
0.9	0.23806	3.0	0.12410
1.0	0.22684	4.0	0.10283
1.1	0.21684	5.0	0.08816
1.2	0.20783	∞	0.0

7.21 In the United States, enrichment services are provided by the U. S. Atomic Energy Commission according to a Schedule of Base Charges, in which both the separative-work component and feed-component coefficients (Δ/P and F/P) in Eq. 7.21 are listed as a function of product assay as shown in Table 7.4, an abridged version. An *off-optimum* tails assay of 0.2% has been officially adopted as a basis for the tabulation. As a result the table is not exactly represented by Eq. 7.22. Intermediate values are therefore determined by linear interpolation of the official, unabridged version. A toll-enrichment arrangement permits a customer to provide his own feed material and pay only for the enriching services provided. The various possibilities are clarified by the following example.

A customer desires 100,000 kg of 3% product. He has the option of furnishing (1) all the feed required as normal uranium; (2) 100,000 kg of 1%

assay uranium, the rest natural uranium; or (3) 100,000 kg of 0.6% assay uranium, the rest natural uranium. Compute the feed and separative-work requirements for each option. The results are summarized as follows:

	Factor from Table 7.4	Option 1	Option 2	Option 3
Feed required, kg U	F/P			
Normal U required	5.479	547,900 kg	547,900 kg	547,900 kg
Feed credit for 1% product	1.566		156,600	
Feed credit for 0.6% product	0.783			78,300
Normal U feed required		547,900 kg	391,300 kg	469,600 kg
Separative work, kg SWU	Δ/P			
All normal feed furnished	4.306	430,600 kg	430,600 kg	430,600 kg
Credit for 1% product	0.380		38,000	
Credit for 0.6% product	(0.107)			(10,700)
Total separative work required		430,600 kg	392,600 kg	441,300 kg
Charge for separative work at $32/kg SWU*		$13,779,200	$12,563,200	$14,121,600

*Rate subject to change from time to time.

FABRICATION OF FUEL

7.22 Fuel-fabrication operations and associated parameters are considered here primarily as a subsystem in a reactor system design. The identification of parameters having a noticeable effect on the reactor design is emphasized rather than a description of fabrication processes for their own sake.

7.23 For this purpose primary attention is given to uranium oxide fuel elements. Although carbides, alloys, and dispersions are also important, oxides are used in most present reactors and serve to illustrate the important parameters. Because mixed uranium oxide–plutonium oxide elements over a range of concentrations are also included, some of the design considerations for fast reactor and plutonium-recycle fuels are examined.

7.24 Once suitable materials are selected, fabrication-process parametric effects of interest to the designer are related primarily to the cost components. To clarify cost effects, we can consider the fabrication process to include all steps required for manufacturing the finished fuel element ready for insertion in the reactor, starting with uranyl nitrate solution from the chemical processing plant or with partially enriched UF_6. Such a designation has not been universally accepted, however, since frequently the fabrication-process label applies only to the steps required after the oxide powder or other solid fuel material has been prepared. The term *conversion* is then applied to the prior process yielding the

TABLE 7.4

Table of Enriching Services

Assay, wt.% ^{235}U	Standard table of enriching services*	
	Feed component (normal), kg U feed/kg U product	Separative work component, kg SWU/kg product
0.20	0	0
0.30	0.196	−0.158
0.40	0.391	−0.198
0.50	0.587	−0.173
0.60	0.783	−0.107
0.70	0.978	−0.012
0.711 (normal)	1.000	0.000
0.80	1.174	0.104
0.90	1.370	0.236
1.00	1.566	0.380
1.20	1.957	0.698
1.40	2.348	1.045
1.60	2.740	1.413
1.80	3.131	1.797
2.00	3.523	2.194
2.20	3.914	2.602
2.40	4.305	3.018
2.60	4.697	3.441
2.80	5.088	3.871
3.00	5.479	4.306
3.40	6.262	5.191
3.80	7.045	6.090
4.00	7.436	6.544
5.00	9.393	8.851
10.00	19.178	20.863
90.00	175.734	227.341
98.00	191.389	269.982

*The kilograms of feed and separative-work components for assays not shown can be determined by linear interpolation between the nearest assays listed.

solid from the chemical solution feed. Another term sometimes applied to the process yielding the fuel-fabrication material is *fuel preparation.*

7.25 Although fabrication operations are very important in the fuel cycle, we can consider here only the general nature of the steps involved and the directions possible to reduce costs. The extensive technology for these operations is described elsewhere.[5–8]

7.26 A number of routes are available for the fuel material from a chemical solution to a ceramic product suitable for loading into tubes. The nature of the product desired affects the process choice. Oxide fuel elements, for example, can be made by either of two general routes. Prepared oxide powder can be sintered into pellets in several ways. The initial powder is generally formed from ammonium diuranate (ADU) by a process that produces a high-density powder highly suitable for sintering. Most material is formed by a pressing process, although slip casting and extrusion are alternate methods.[6] The high-density pellets are then loaded into tubes. The other method involves vibrational compaction of the powder to form a high-density material in the tube. Although all U. S. light-water-reactor UO_2 fuel is made by some type of pellet process, vibratory compaction may have a future role with plutonium recycle fuels.

7.27 In the vibrational compaction process, if a mixed oxide is desired, impactable-grade UO_2 can be mechanically mixed with impactable-grade PuO_2, impacted in cans, crushed and sieved to the desired size, and finally loaded into fuel rods by automatic vibrator machines.

7.28 Another interesting method for preparing vibratory-compaction feed material is the sol-gel process. This process was developed primarily for the recycle of thorium reactor fuels, whose handling is complicated by gamma activity from decay products of ^{232}U and ^{228}Th. In addition to its adaptability to remote operation in a shielded facility, the process has the advantages of simplicity, flexibility, easy control of the size and shape of product particles, and a much lower calcination temperature required for densification. As a result this process may indeed become the preferred method of preparing ceramic-fuel materials.

7.29 It appears that oxides or carbides of thorium, uranium, or plutonium suitable for vibratory packing can also be prepared from nitrate solution by the sol-gel process, although some development is needed. The process consists of four simple steps: denitration to obtain reactive oxide, dispersion of this oxide in dilute nitric acid to form a hydrosol, gelation by evaporation or extraction of water from the sol, and calcination to produce the final powder.

7.30 After the powder or pellet is produced, a number of additional operations are required in the manufacture of a finished fuel-element assembly. A fabrication flow sheet for pelletized fuel is shown in Fig. 7.6 and a flow sheet for vibratory compaction is shown in Fig. 7.7. Studies of the processes show that for planning purposes cost differences are small.[9] Furthermore, parameters to be considered here, such as throughput, apply in about the same manner to each. We shall use, therefore, a mixed-oxide vibratory-compaction process for plutonium recycle fuel as a reference for which the *relative* effects of design parameters can be considered. The possible effect of changes in technology on such trends should be recognized, however.

7.31 One class of design variables concerns fuel-element design specifications, and the other, the manufacturing operations and the plant required for them. Examples of the first class are fuel diameter, length, cladding thickness, and dimensional tolerances. Production throughput, fixed-charge rates, and material prices are examples of the second type of fabrication parameter.

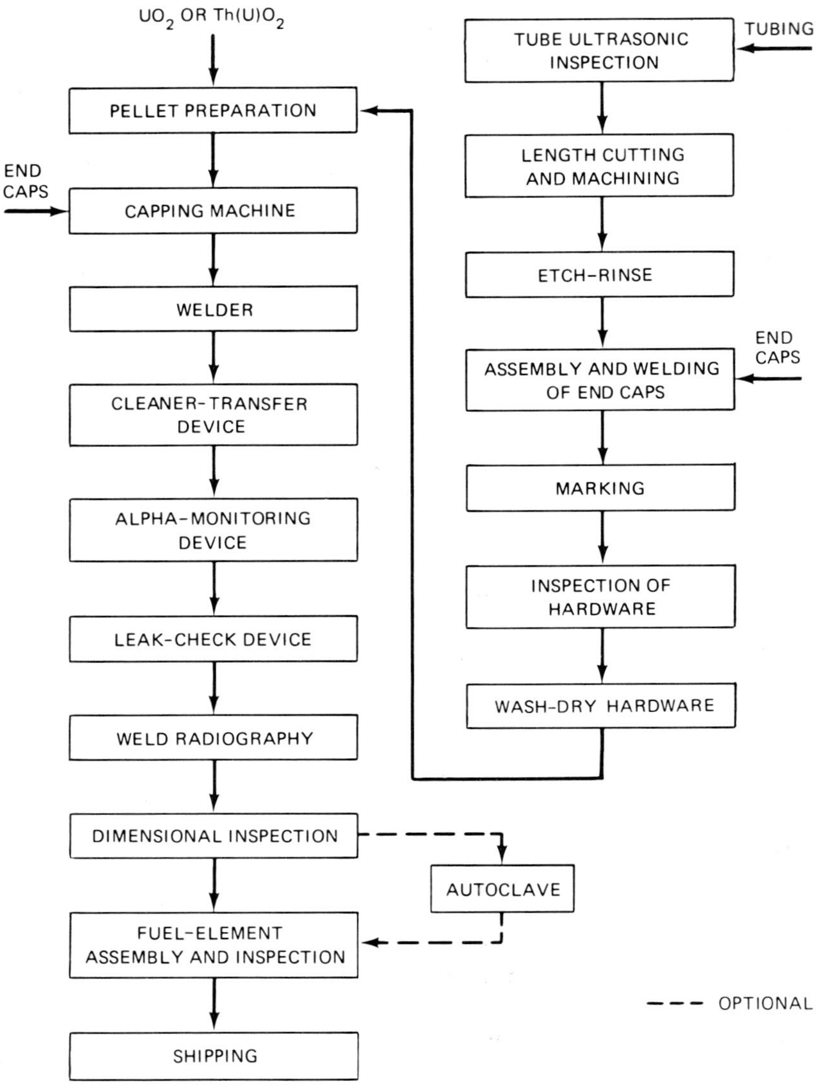

Fig. 7.6 Fabrication flow sheet for rod bundles containing pelletized fuel.

7.32 As an aid in analysis, it is useful to divide the fabrications charges into three components: (1) operating charges, (2) materials, including hardware, and (3) fixed charges on capital-investment items. The cost of the fissile- and fertile-fuel materials is normally considered separately, not as part of the fabrication charges proper. However, fixed charges associated with the investment in fissile and fertile material during the time required to perform the

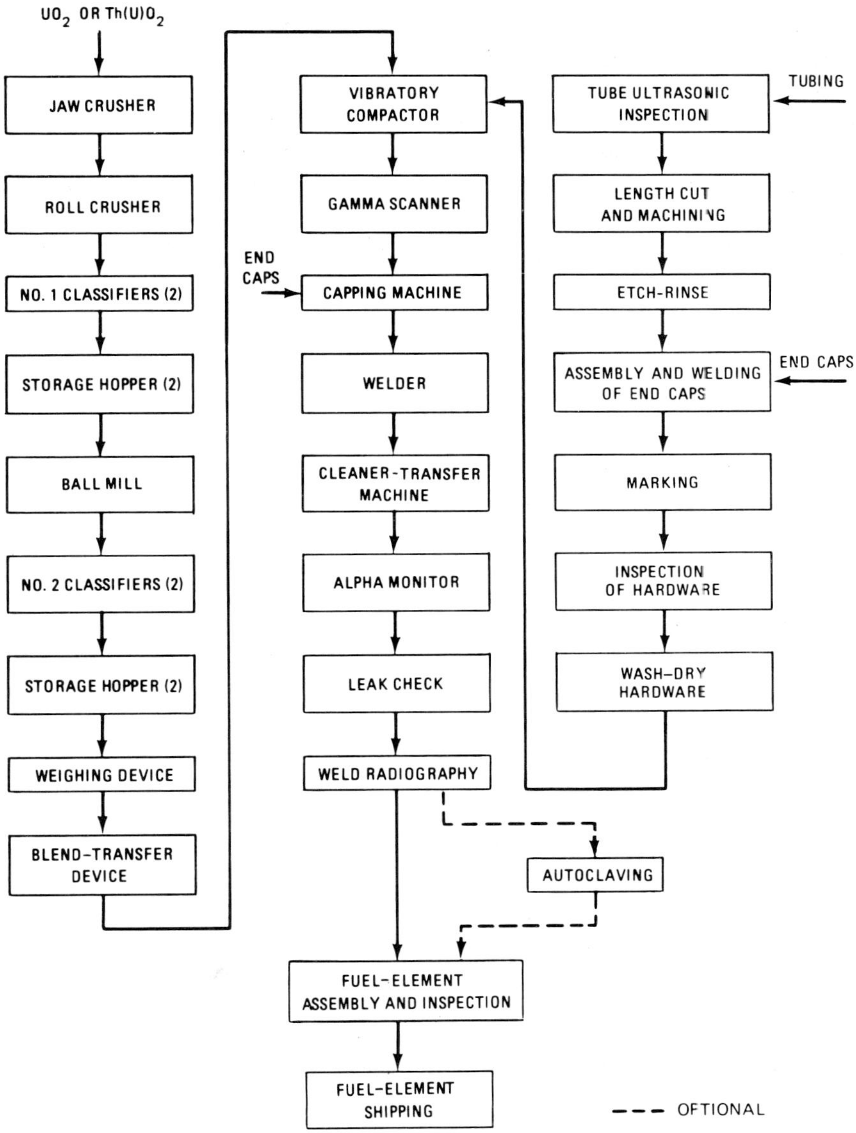

Fig. 7.7 Fabrication flow sheet for vibratory compaction of fuel in rod bundles.

fabrication operation are included. Hence the important parameters become the cost of operations, the charges for materials, and the time required for the processing. The fixed-charge rate is also an important secondary parameter since it affects the fixed-charge cost components.

7.33 Fabrication costs are commonly expressed in terms of the quantity of contained fuel on a metal basis, i.e., in dollars per kilogram of heavy metal atoms. Therefore design parameters affecting the amount of contained material associated with a cost contribution will, in turn, affect the fabrication cost when expressed on the unit-fuel basis. The fuel-pin diameter and length, for example, affect the amount of material that can be contained within the pin. Since many fabrication operations, such as handling, welding, and inspection, are applied to the pin as a unit, such cost contributions tend to vary inversely with the weight of fuel in the rod when expressed on a unit weight basis. A typical trend is shown in Fig. 7.8. A model was used to describe the various fabrication-cost components for a mixed-oxide case.[10] Total costs tend to be relatively constant on a unit weight of fuel basis provided the pin contains over approximately 1 kg of fuel.

7.34 Published estimates of fabrication costs are usually based on an analysis of individual operations in a hypothetical manufacturing plant. Such results indicate to the designer the relative importance of the individual cost contributions. The results of one such study completed in 1966 for a plant projected for about 1975 which will use 1.0 tonne (1000 kg) of fuel per day are given in Table 7.5. Relative costs are shown for fabricating a typical element (assembly) using ^{235}U as the fissile component and one using plutonium.

7.35 In general, the reactor fuel design affects only a few parameters that influence the cost of the fabrication-plant manufacturing operations. The sensitivity of fabrication cost to some fuel-design parameters is discussed in §7.137. Automation and process simplification can be applied, for example, as throughput increases and technology develops. A sensitivity to an inspection reject rate and nuclear losses is worthy of emphasis, however. The scrap rate is very important since the production of large numbers of unacceptable elements not only reduces the plant capacity but also introduces recovery expenses and increases the nonrecoverable loss of valuable fissile material.

7.36 The cost effect of changes in plant throughput depends on the design flexibility of the plant. Variations in the production rate of the order of 50% are likely to cause changes in cost of about 20%. Variations over a wider range, as determined by another study,[4] are shown in Fig. 7.9. A capacity of about 2 tonnes per day on a metal-atom basis is sufficient for a 15,000-Mw(e) power industry.

FUEL-REPROCESSING OPERATIONS

7.37 After the fuel assemblies are removed from the reactor and allowed to cool, they are shipped to a reprocessing facility where the valuable fissile material contained therein can be recovered. Commercial processes are based on the Purex solvent-extraction process with ion exchange and other processes sometimes used for product recovery (§7.39). The features of the Purex process

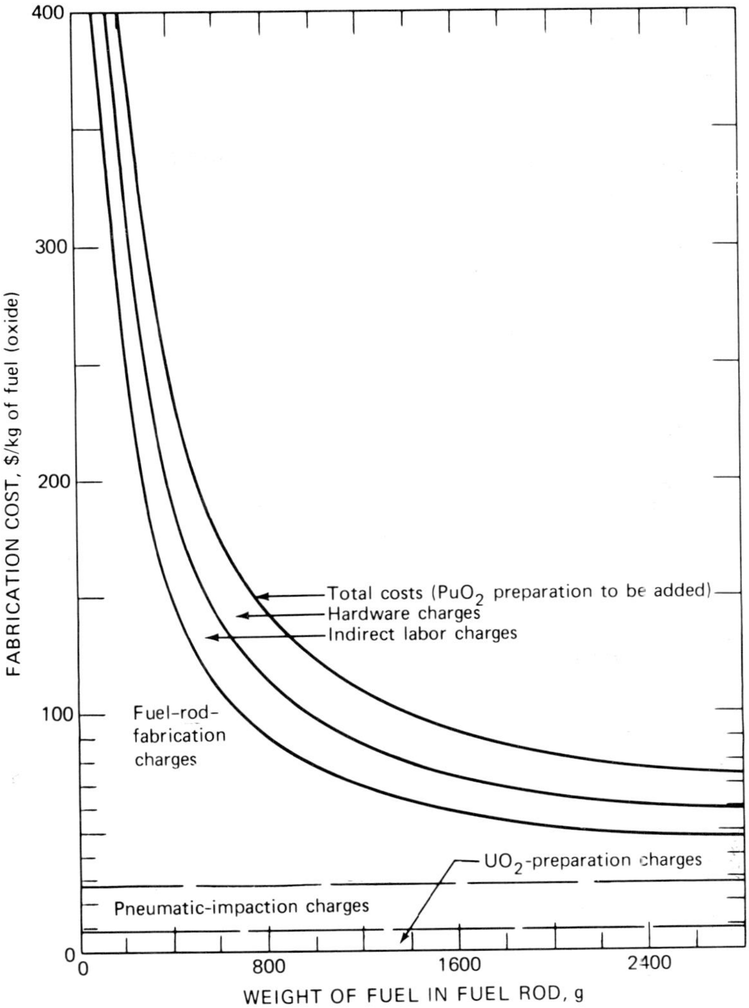

Fig. 7.8 Variation in fabrication costs with unit fuel weight per rod. (Individual cost components are represented by distance between curves.)

are shown schematically in Fig. 7.10 (see Ref. 2 or 5 for a detailed description of the process). Cost parameters, though significant, normally are not affected by the design of the fuel. Therefore only the general characteristics from the cost viewpoint are considered here.

7.38 The general layout for a nominal 1 tonne/day plant operated by Nuclear Fuel Services, Inc. (NFS) is shown in Fig. 7.11. Substantial effort is required to prepare the fuel for the solvent-extraction operations. In addition to the initial separation of fission products and uranium from plutonium, a number

TABLE 7.5

Fuel-Assembly Fabrication Costs[11]

	Cost per assembly,* $	
	3% UO_2	2.9% PuO_2-UO_2
Building and equipment	1,264	1,490
Operating labor	2,112	2,664
Rework labor	93	115
Direct materials	10,697	14,415
Indirect manufacturing expense	3,030	3,625
Nuclear losses	1,165	1,170
Ordinary losses	105	152
Working capital	254	336
Use charge	804	822
Fixed manufacturing cost	733	772
Fixed nonmanufacturing cost	368	368
Total ($/element)	20,627	25,929
Total ($/$kg_{U+Pu}$)	41.69	52.40

*Assembly contains 494 kg of oxide in 234 tubes.

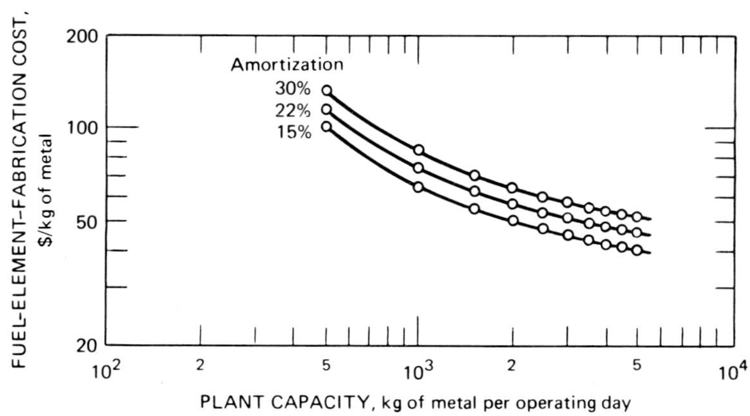

Fig. 7.9 Pressurized-water-reactor fuel-element costs for fabrication by vibratory compaction of the fuel in a contact plant.

of stages of extraction are required to obtain adequately pure product streams. For example, after the partitioning extraction cycle, the uranium stream is further treated by two additional, consecutive extraction cycles and then purified after concentration by passage through silica-gel beds. Similarly the plutonium-product stream from the partition cycle is reextracted once and then purified by ion exchange. Auxiliary operations are very important in the overall process and add to the complexity of the plant. These include organic solvent recovery, acid recovery, waste-stream recovery, and waste management.

7.39 A somewhat different approach, used by the General Electric Company in the Midwest Fuel Reprocessing Plant, the Aquafluor process, is shown schematically in Fig. 7.12. Mechanical feed preparation followed by leaching is similar to the Nuclear Fuel Services, Inc., operation; however, only a single cycle of solvent extraction is used to separate recoverable actinide elements from the fission products. Plutonium and neptunium are then recovered by ion exchange in the form of a nitrate solution. Next, uranyl nitrate hexahydrate stream from ion exchange is calcined in a fluidized-bed operation to form UO_3, which is then fluorinated in a second fluidized bed. Here the less volatile fission products that have been carried along up to this point are retained with the inert bed material. The UF_6 product is further purified by passing the gas through NaF and MgF_2 beds to remove traces of plutonium and fission products.

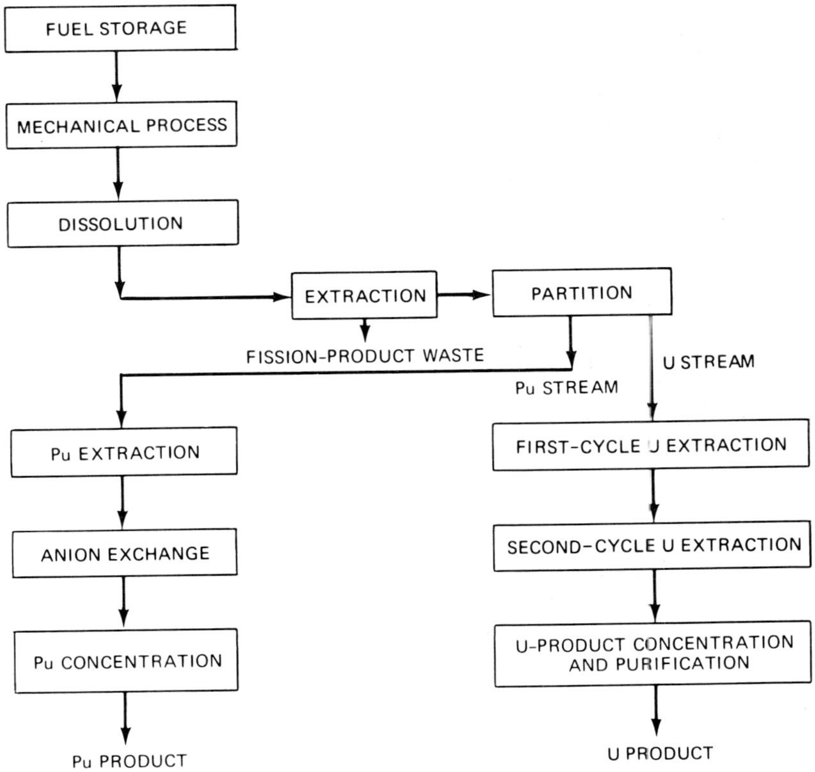

Fig. 7.10 Purex process operations.

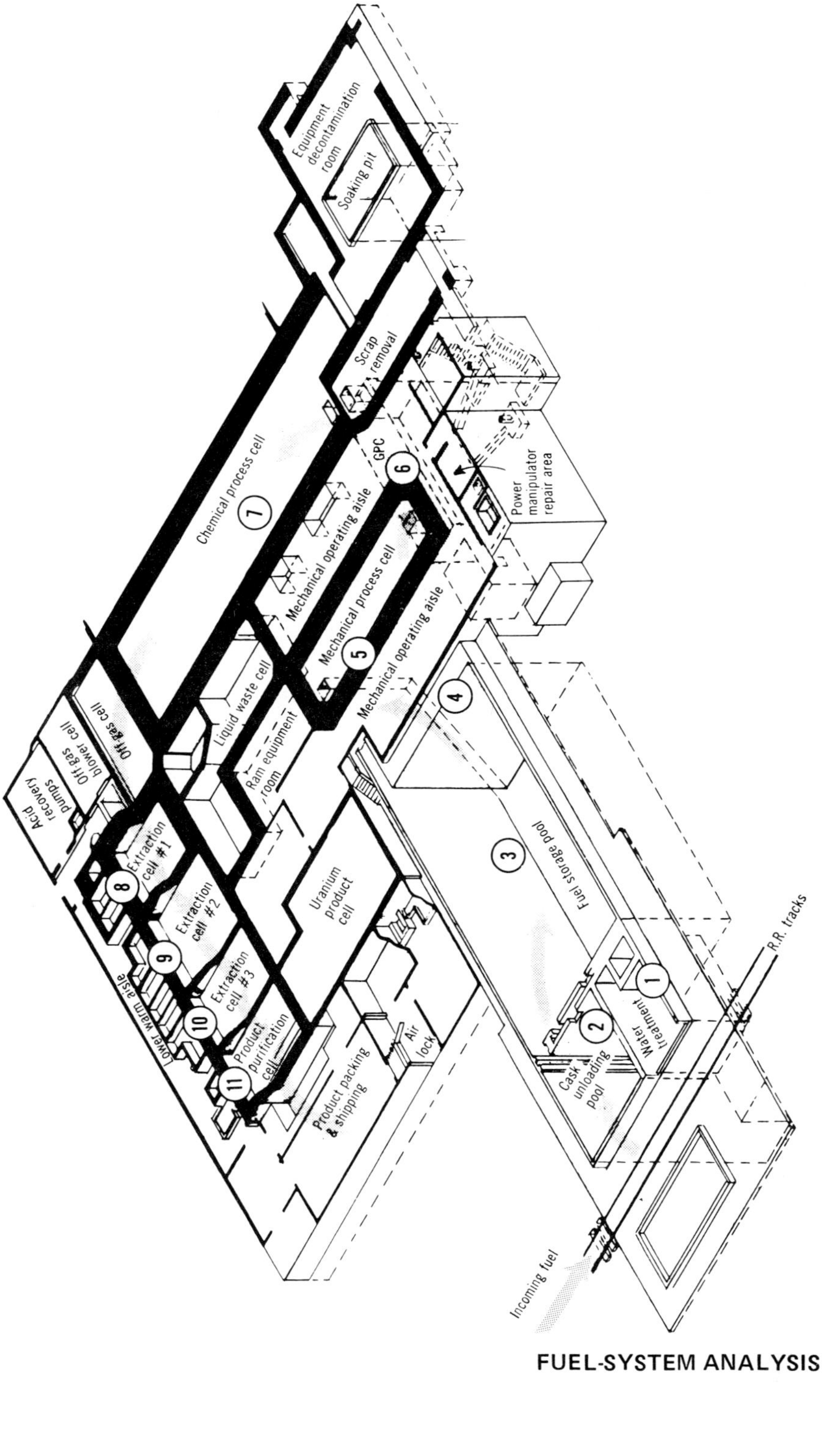
Equipment decontamination room
Soaking pit
Scrap removal
Chemical process cell
7
GPC
6
Power manipulator repair area
Mechanical operating aisle
Mechanical process cell
Mechanical operating aisle
5
4
Liquid waste cell
Ram equipment room
Off-gas cell
Off-gas blower cell
Off-gas pumps
Acid recovery
Extraction cell #1
8
Extraction cell #2
9
Extraction cell #3
10
Product purification cell
11
Lower warm aisle
Uranium product cell
Air lock
Product packing & shipping
3
Fuel storage pool
1
2
Water treatment
Cask unloading pool
R.R. tracks
Incoming fuel

Fig. 7.11 General plant layout for a nominal 1 tonne/day fuel-reprocessing facility. The primary areas of the process building are arranged in a "U"; fuel in shielded casks is brought by rail or truck to the unloading area at one end, and purified products are shipped from the other end.

The receiving and storage area, served by a 100-ton overhead crane, contains washdown facilities to clean external surfaces, cask decontamination pit, unloading pool, and fuel-storage pool. Fuel elements are taken from the storage pool to the mechanical-process cell by means of an underwater conveyor. Here, the fuel is sheared and then dropped through a chute into dissolver baskets in the general-purpose cell. Dissolver batches are transferred through a hatch into the chemical-process cell, where the fuel is leached and the cladding returned to the general-purpose cell for inspection and packaging. Except for the sheared fuel, all product transfers into and out of the chemical-process cell are as liquids. In the other leg of the "U" are three solvent-extraction cells and a cell for final product purification and concentration.

In the second building level (above that shown in the diagram) are chemical-process cell and extraction-cell operating aisles; in the third level, analytical facilities; and, in the fourth level, the control room, where flow adjustment and fluid transfer are controlled. Above the control room are cold-chemical makeup facilities and filters. [From T. L. Cramer, NFS: First Fuel Reprocessor, *Nucleonics*, **24**(12): 49 (1966).]

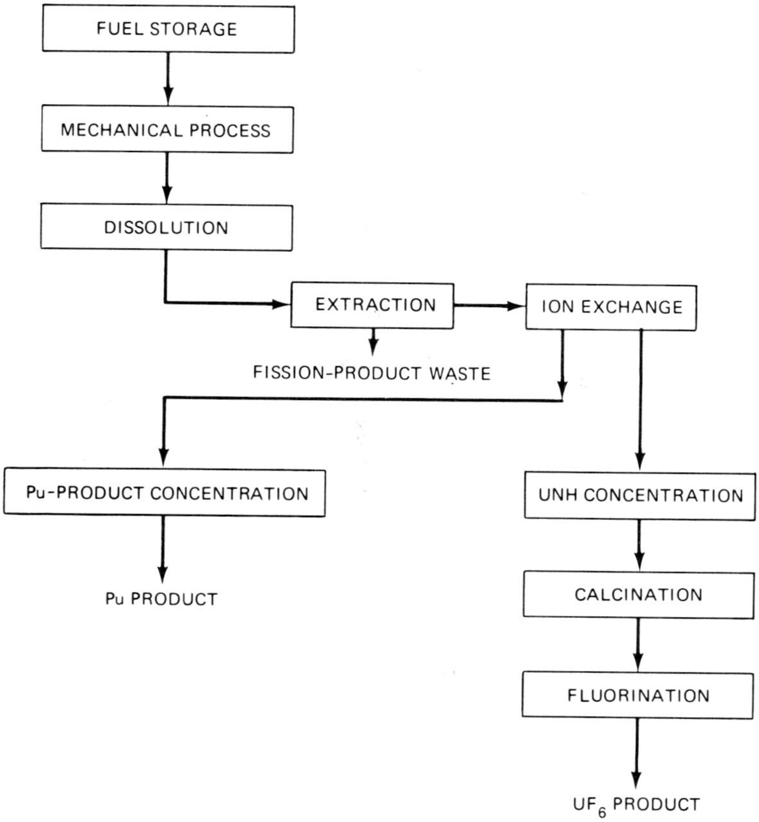

Fig. 7.12 Aquafluor process operations.

7.40 As for most kinds of chemical processing plants, unit costs tend to decrease as the capacity of the spent-fuel processing plant is increased. This trend, however, is limited by the necessity of maintaining nuclear-criticality control and, in aqueous processing, by possible limited ability to dispose of dilute wastes at a given site.[12]

7.41 For example, fission-product tritium, which initially appears in the dissolver solution, follows the dilute aqueous effluent streams that normally are not stored but are disposed as low-level wastes. Since a fuel having an exposure of 20,000 Mwd/tonne yields a tritium activity of 260 curies, a 1 tonne/day processing facility would require access to a river having a flow of 84 million gallons per day to meet an acceptable surface-water concentration of 10^{-3} μc/ml. Other radionuclides such as ^{90}Sr which have very low permissible concentration levels in waste streams must also be considered.

7.42 Nonaqueous methods of processing reactor fuels, such as fluid-bed fluoride volatility and pyrochemical processes, generally have smaller and more manageable volumes of waste than the aqueous methods have. These methods have not been developed to the same degree as the aqueous processes, however. For design purposes, therefore, it appears reasonable to assume reprocessing costs only slightly below the Nuclear Fuel Services, Inc., rates since the effect of process improvements and larger scale is likely to be compensated by inflationary economic trends. Cost contributions tend to be similar in nature to those in fabrication. The fixed charges associated with both the value of the fuel and the plant investment depend on the time required for processing. Day-to-day operating expenses, the cost of materials, and perhaps a separate category for waste disposal, complete the charges.

COST-ANALYSIS METHODS AND FUEL MANAGEMENT

INTRODUCTION

7.43 The fuel-cost approach described in Chap. 3 is useful in comparing reactor concepts and in highlighting the relative importance of the various contributions. In practice, however, an electric utility company would use a different system consistent with its accounting procedures. Furthermore, the utility must have cost information available for operational decisions and for the specification of replacement cores. Cost-analysis methods must therefore meet these requirements. Since there is considerable interplay between the operational and planning requirements, the overall effort can be called *nuclear fuel management.*

7.44 A cost analysis method generally should have the ability to report details for each accounting period, usually 1 month, as well as for longer term averages and totals. A given reactor may contain several different batches of fuel with different exposure histories. Other batches of fuel may also be at different stages of the fuel-cycle "pipeline" at a particular time. Since each batch has associated with it investment and processing-operation costs as well as a change in value due to the energy extracted from it while in the reactor, one of the engineering-economy concepts discussed in Chap. 2 must be applied to account for the time value of money. The various expenditures and credits that apply to each batch in the reactor can thus be reduced to a consistent basis.

7.45 The change in value of money with time could be handled by relating all changes in value of the fuel to the actual cash flow when payments are made for expenses and for credits received. More useful for the study of parametric effects, however, is the variation in fuel investment with time, as shown in Fig. 7.13. The present-worth concept can then be applied to the value on a monthly basis or according to an assumed payment schedule. In Fig. 7.13 the value of the fuel shown reflects the cost of operations performed on the fuel at each stage. Prior to reprocessing, for example, the fuel value is less than the salvage value

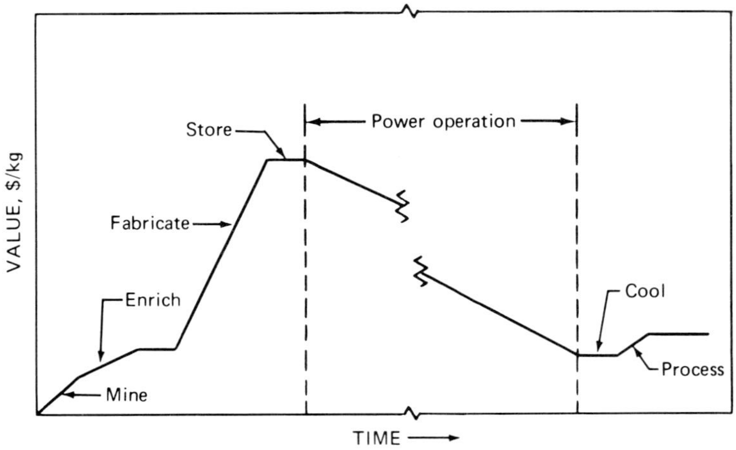

Fig. 7.13 Fuel-investment pattern (not to scale).

and increases during the reprocessing operation. Reprocessing would not be justified if this minimum value proved to be zero.

7.46 Cost analysis easily lends itself to computer calculation, and a number of codes have been developed for computing fuel costs.[13] Although most codes are written primarily for conceptual design, CINCAS-II is an example of a cost-analysis code[14] for use by the nuclear-power-station operator. A discussion of some of the features of this code brings out some of the general characteristics of cost analysis.

7.47 In CINCAS-II, fuel costs are calculated on an accrual basis for each month the fuel is in the reactor. The direct costs are allocated to each month in proportion to the fission heat produced during the month so that all accounts balance out when the batch is withdrawn from the reactor. Cost categories, or accounts, are set up for each of five "direct" costs: uranium depletion, spent-fuel shipping, chemical reprocessing and reclaimed-uranium conversion, fabrication, and plutonium credit. Debt charges, income taxes, and return on capital are treated by levying a composite-interest, or "inventory," charge each month based on the inventory value at the beginning of that month. These inventory charges are divided into four categories: uranium, plutonium, fabrication, and postirradiation. Preirradiation charges are treated as interest during construction and assigned to the beginning inventory values. Various engineering-economy parameters can be accommodated, results being available in terms of equivalent annual cash flows and other options.

7.48 The cost of fuel as a function of the energy generated in a given reactor is important knowledge in planning the operations for load changes. When several reactor plants are interconnected, for example, any increase in system load is assigned so that the incremental cost will be minimal. This will normally be true if the reactor having the lowest incremental fuel cost is loaded first and so on.

Computerized cost analysis is almost essential for such a system-wide operational approach.

7.49 The status of the fuel in the reactor core as a function of time and position must be known for an accurate assessment of the cost status. Thus a nuclear-depletion calculation giving an isotopic-composition distribution is important in the cost analysis. Once the isotopic composition is known, the reactivity inventory, fuel value, and energy-generation capability can be determined. Computer codes that can be applied to this phase of the analysis are described in Chap. 5.

7.50 A compromise is normally necessary in the spatial detail described. An individual fuel assembly, the smallest integral unit, is followed for accounting purposes from fabrication to reprocessing, since the number of assemblies in a batch may vary from one operation to the next. However, spatial averaging for the depletion calculation is normally required since the fine-detail representation necessary for individual fuel assemblies consumes excessive computer time (§5.98).

7.51 Questions also arise in the financial and accounting aspects of the analysis. Owing to possible differences in the rate-making basis and tax structure, the labeling of charges can affect costs. One question, for example, is whether nuclear fuel should be classified on the balance sheet as a current or a fixed asset and, if the latter, whether it should be shown separately or as part of the utility plant. Currently the fuel is treated as a separate plant item with its own write-off period.

7.52 Some cost codes (§3.61) provide for the capitalization of the first core and its depreciation over the life of the plant. This presents the possibility of a tax saving and suggests that all customers using the product of the plant should share a portion of the relatively high cost of the first core load. On the other hand, it is not considered sound policy to continue entering on the books sums relating to physical objects long since consumed and replaced. Core-management strategy schemes (§7.66) also tend to decrease the importance of the first core. Thus first-core capitalization no longer appears appropriate.

FUEL MANAGEMENT

7.53 In nuclear-fuel management, strategies and policies for the entire fuel cycle are formed to effect a minimum energy cost. Cost analysis, therefore, is an important tool. The problem lends itself to an operations research approach that requires an analytical model of the system and the use of optimization techniques to determine the best combination of parameters. Systematic search and analysis techniques applicable to this problem are being developed. Some examples are described in Chap. 9.

7.54 Fuel management has an impact on the fuel cost at various points in the fuel cycle where decisions are called for. These include:

1. Concentrate procurement.
2. Conversion.
3. Enrichment.
4. Fabrication.
5. Reactor use.
6. Spent-fuel shipment.
7. Reprocessing.
8. By-product disposition.
9. Uranium reconversion.
10. Plutonium disposition.

7.55 Since the fuel-cycle system is quite complex with extensive interplay between parameters, a sensitivity analysis that considers merely the effect of a change of a single parameter with all others remaining constant is not very meaningful. However, such a study does provide a qualitative guide to areas worthy of attention. The relative contributions of fuel-cycle cost parameters for a typical 1000-Mw(e) pressurized-water reactor[14] are indicated in Table 7.6. Here a 10% variation in each parameter is reflected in a corresponding change in fuel cost. Since substantial sums of money are associated with small changes in fuel-cycle costs, detailed analysis and the application of sophisticated operations-research methods appear justified as an aid in making management decisions at appropriate points in the fuel cycle.[15] The widely varying organizational approaches that can be used are not considered here, although some of the interplays involved are discussed.

TABLE 7.6

Fuel-Cycle Cost-Sensitivity Analysis

	Effect of 10% change in parameter	
Parameter	Change in fuel cost, %	Cost, $/year at 1000 Mw(e)
Enriching charge	3.7	415,000
Burnup	2.8	315,000
U_3O_8 price	2.8	315,000
Fabrication charge	2.5	280,000
Pu credit	1.2	135,000
Reprocessing charge	1.1	124,000
Shipping charge	0.2	22,000

POWER DISPATCHING AND OTHER OPERATING PARAMETERS

7.56 The utility company operator of a system consisting of a number of generating plants (units), which are normally at different locations, must decide how to adjust the power output of individual units as the power load on the entire system changes. This decision must effect minimum overall generating costs; therefore an increasing load is assigned first to units with lower incremental operating costs (§2.18).

7.57 For fossil fuels the incremental cost is easily determined once the cost of the fuel, its heating value, and the thermal efficiency of the unit as a function of capacity are known. However, for nuclear fuel, which remains in the reactor for several years, there are large financing charges that depend on how rapidly the fuel is used. Furthermore, since the reactor is loaded with fractions of the core at different times and the performance of one assembly affects that of its neighbors, the power-dispatching decision affects future decisions.[16] Parameters for subsequent batches include the batch size, enrichment, cycle times, in-core loading patterns, and design of the fuel. An economic model for the incremental costs for nuclear fuels is therefore quite complex and is undergoing development. Associated parameters are part of fuel management.

7.58 Although fuel-cycle optimization studies using simplified models provide a useful design reference, the possibility of operating-plan changes must also be considered. Sufficient flexibility in design of both the reactor and the fuel system should be retained to accommodate such changes. For example, unexpected outages or load changes might cause a change in refueling interval. Changes in economic parameters or advances in technology might also shift the optimum in a reference design. In addition to using fuel management to respond to many such changes, the reactor operator can call on the management of power, coolant, and neutron absorbers.[17]

7.59 The refueling cycle can be stretched out by *power management*, which could include shifting from base-load operation at rated power to load-following operation, perhaps on a partial basis. *Coolant management* provides additional flexibility, particularly for a boiling-water reactor. For example, the inlet temperature to the reactor core can be reduced by valving off some of the feedwater heaters, with a resulting shift in the axial power distribution due to the large negative coolant coefficient of reactivity. Similarly changes in coolant flow can shift the void fraction and hence the degree of moderation. This changes in turn the local reactivity and such characteristics as the power distribution and amount of plutonium production.

7.60 *Neutron-absorber management* includes control-rod positioning and, in boiling-water reactors, temporary absorber-removal strategies for optimizing available excess reactivity by shaping power distribution. The ability to use this flexibility depends on system requirements coupled with load-following strategy and the thermal margin (§4.54). For example, fuel-cycle costs can be decreased by operating the reactor, over most of the cycle, with the power peaked to the bottom of the core to increase steam voids and plutonium production and then, at the end of the cycle, with the power peaked toward the top of the reactor to gain reactivity. However, in practice the picture is more complex since the need to have bottom-entering control elements withdrawn at the end of cycle and the desirability of minimizing the power-peaking factor during operation favor a bottom-skewed power shape at end of life. Furthermore, unless the changes are completely programmed, the utility system interactions could offset the fuel-cycle cost gains since the increase in reactivity could move the refueling period into an interval wherein the cost of substitute energy could be large.

7.61 *In-core fuel management* is both an operating and a design parameter. To illustrate the type of challenge it presents, we shall treat the subject separately in the following section.

CORE MANAGEMENT

INTRODUCTION

7.62 Core management concerns the strategy of fuel loading in the core. Since the reactor core can be considered a subsystem of the larger fuel-cycle system, core management is part of the fuel-management challenge. Its primary aim is to optimize fuel-depletion parameters to minimize fuel costs while meeting a continuous power capability. To accomplish this, the strategy should yield the most uniform power distribution, consistent with other design requirements, at all times during operation. In addition, burnup among the discharged fuel assemblies should be as uniform as possible. For water reactors the desired uniform-burnup value is of the order of 30,000 Mwd(t)/tonne but is subject to optimization.

7.63 Various schemes have been proposed to meet these general objectives. In most cases the loading of the fuel is staggered so that at any one time there are in the core both fresh fuel containing few fission products and fuel subjected to substantial burnup. The reactivity associated with the insertion of fresh fuel extends the exposure possible with the depleted fuel; at the same time, the older fuel, serving somewhat like burnable poison, compensates for this reactivity without requiring a large reactivity inventory in the control rods.

7.64 As a general trend the enrichment of the fresh fuel must be increased as greater and greater burnups are sought. Local areas of high enrichment, however, can lead to undesirable local areas of higher power. Therefore the neutronic coupling between one area and another becomes important in planning the loading. A rather complicated interplay between a number of factors, such as enrichment, reactivity, neutron transport, and power, each dependent on both time and space, affects the problem. Compromises are therefore necessary in practice, particularly in attempting to meet requirements for burnup and power distribution. Conditions that would be off-optimum if each were considered alone may therefore be specified.

7.65 As an additional complication the initial core-loading pattern may differ markedly from the so-called equilibrium loading scheme, which may be repeated a number of times during the reactor lifetime. However, since a fuel batch remains in the core for about 3 years, the attainment of an equilibrium cycle may require a significant fraction of the reactor lifetime. Hence strategies that result in minimum costs over the reactor lifetime and not merely under equilibrium conditions are needed. Computer-based optimization procedures (§9.96) show promise for developing needed loading schemes.

FUEL-LOADING SCHEMES

7.66 There are a number of different fuel-loading schemes that can be evaluated by carrying out depletion calculations and determining the power distribution as a function of burnup. A study with a medium-size pressurized-water-reactor (PWR) core illustrates some of the problems that can be encountered with suggested schemes.[18] Some design specifications for a 1347-Mw(t) three-fuel-zone reference reactor used in the study are given in Table 7.7 Although the power is somewhat smaller than that in many current designs, the results are generally applicable.

7.67 In the simplest core-management approach, the entire core is loaded as a batch with fuel of a single enrichment. However, this arrangement suffers from both low burnup and a high peak-to-average power ratio. Power distribution can be improved by varying the enrichment of the fuel in the core. Generally, a

TABLE 7.7

Pressurized-Water Reference Reactor Characteristics for Core-Management Study[18]

Heat output, Mw		1347	
Core diameter, in.		110.9	
Core height, in.		120.0	
Loading technique		3-region: nonuniform	
Full-power lifetime, first cycle, hr		13,800	
Average first-cycle burnup, Mwd/tonne U		13,500	
Fuel-region values	**Central***	**Intermediate**	**Outer**
Initial weight of uranium per charge, kg	19,450	19,080	18,800
Discharge weight of uranium in first charge, kg	18,600		
Initial enrichment wt.% ^{235}U	3.15	3.4	3.85
Discharge enrichment wt.% ^{235}U	1.74		
Plutonium produced per charge, kg	134		
Initial conversion ratio		0.50	
Number of fuel assemblies		157	
UO_2 rods per assembly		180	

*Central region contains one more assembly than the other regions.

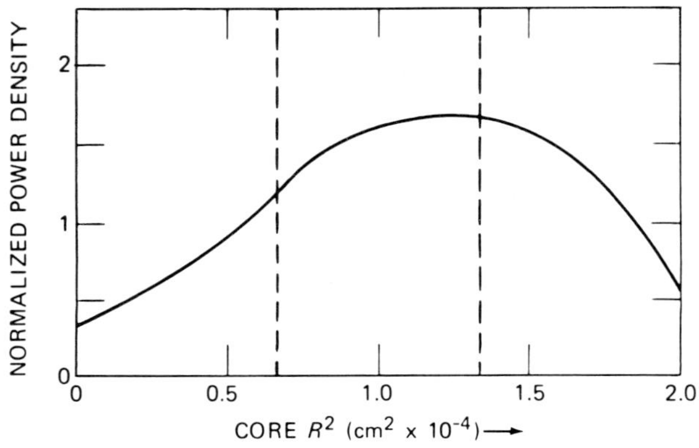

Fig. 7.14 Power-density pattern in a three-region core with out–in cycling.

cyclic scheme in which only part of the core is loaded at one time is necessary to achieve uniform burnup and yet provide the desired spatial variation of enrichment. Several such schemes are possible, however, and must be evaluated under a given set of conditions.

7.68 In an out–in procedure, fresh fuel is added to an outer region or zone of the core and then "shuffled" to an intermediate region when the next batch of new fuel is added at the next shutdown. Finally, in a three-zone core, the original batch of fuel is moved to the central region for its last period of exposure in the core. At first glance, this approach appears to meet most of the objectives listed since the depleted, poisoned fuel is in a region of high-neutron importance and tends to depress the central power density, which would otherwise be higher than the remainder of the core. Fresh fuel in the outside high-leakage area also tends to boost the power density in this region and to give an overall flat distribution. In a large core with high fuel burnup, however, a power-distribution graph at equilibrium is similar to that shown in Fig. 7.14. The square of the radius is used as the abscissa to show the equal-volume zones in the core better. Nominal peak-to-average calculated values that do not include any margin for uncertainties are shown. A power depression is found in the central region, a result of loose nuclear coupling characteristic of large cores and the high depletion due to the large burnup (25,000 Mwd/tonne U). In other words, the reactivity of the oldest fuel was not sufficient to give adequate power generation in the central region.

7.69 In a second approach, the *Roundelay* or *scatter* scheme (Fig. 7.15), the fuel is loaded in a checkerboard pattern where fuel elements of different degrees of burnup are evenly distributed. The fuel is loaded at staggered times but remains in one core position throughout its exposure. At the end of each loading cycle, the one-third of the fuel that has been in the reactor for three cycles, or burnup periods, is replaced with fresh fuel. The variation in reactivity

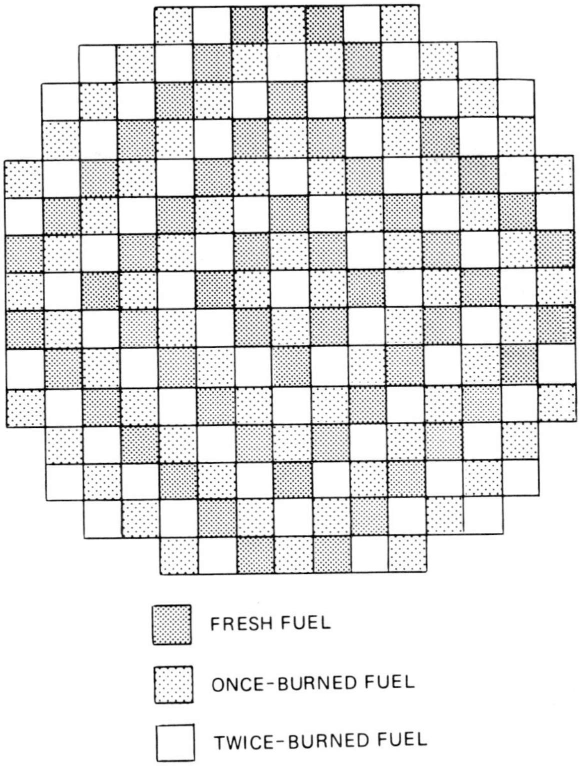

Fig. 7.15 Core layout for Roundelay or scatter refueling. (All fuel elements with the same pattern are discharged at the same time.)

between adjacent fuel assemblies of different previous exposures flattens the power-density distribution. The loading of fresh fuel adjacent to depleted fuel provides strong coupling between fuel assemblies, increases the power production from the burned assemblies, and at the same time reduces the peaking effect in the fresh fuel. An additional advantage is that only one-third of the elements is handled at each refueling.

7.70 Again power-distribution graphs provide a basis for evaluating the scheme. Figure 7.16 shows that fresh fuel loaded near the center causes excessive power peaking. The high-enrichment fuel introduced in a region of high neutron importance produces a high local fission rate despite the "damping" influence of adjacent depleted fuel. One reason for this is that the desire for high burnup requires the introduction of fuel having a higher enrichment than would otherwise be necessary. Therefore a logical improvement would be a scheme which avoids inserting fresh fuel in the center but which introduces enough reactivity in that region to prevent a flux depression.

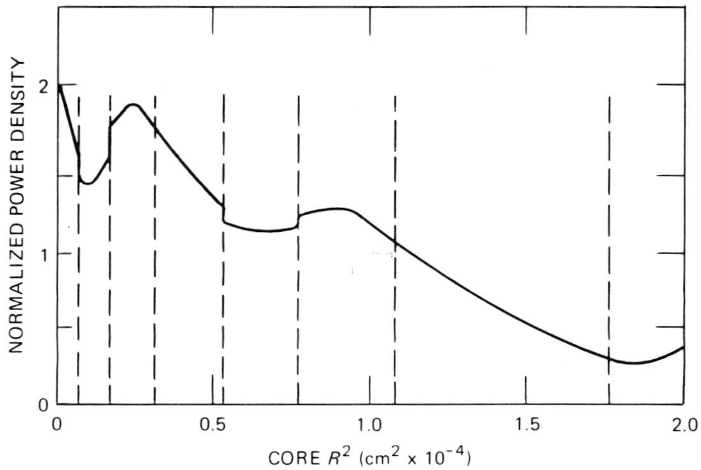

Fig. 7.16 Power distribution for Roundelay fuel cycling (equilibrium core).

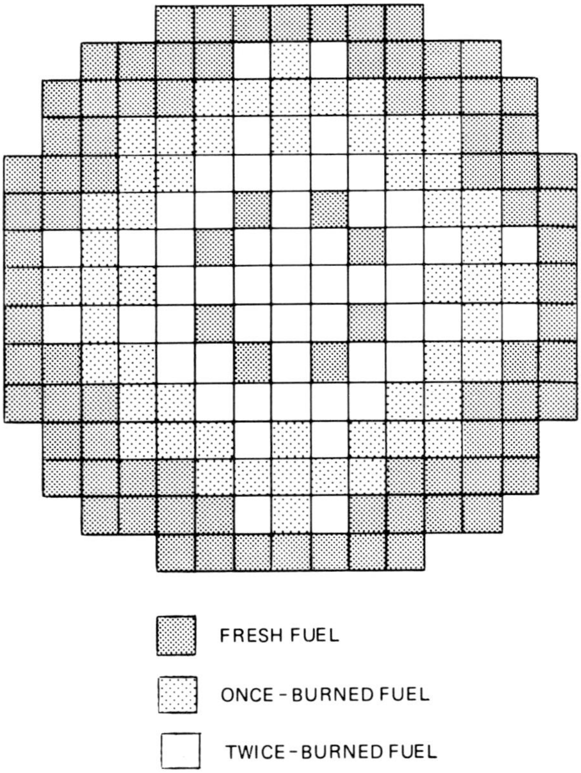

Fig. 7.17 Modified out–in equilibrium fuel-loading scheme.

7.71 One problem with scatter loading of PWR's is an excessive power-density ripple from one assembly to the next which could be reduced by reducing the *size* of the fuel assemblies. If rod-cluster control assemblies (§6.50) are used, however, such a reduction is not practical, since 16 small control rods are distributed throughout a number of the fuel assemblies.

7.72 Another approach to the problem, called the modified out–in fuel-cycling program, is shown in Fig. 7.17. In this method fresh-fuel assemblies are generally located on the perimeter of the core, and once- and twice-burned assemblies are placed mostly in the intermediate and inner regions, respectively. However, the more reactive fuel assemblies are placed in areas of power depression, and highly depleted assemblies, which contribute little reactivity, are placed in areas of high power density. In this way, coupling is improved among the fresh, once-burned, and twice-burned assemblies. Thus some fresh or once-burned fuel can be exchanged for some of the twice-burned fuel to eliminate the center region power "dishing" shown in Fig. 7.14. This technique, giving a favorable peak-to-average power-distribution ratio of 1.52, is undergoing more-detailed analysis. A similar scheme was actually used for the San Onofre Reactor initial loading.

7.73 The loading of initial cores and the arrangement for reload fuel batches for PWR's are planned on the basis of both operating experience and the physics analysis of many candidate arrangements. Since many parameters are involved, including the programming of control elements during operation, techniques vary somewhat from one reactor manufacturer to another, and standard approaches have not yet evolved. For example, Fig. 7.18 shows a recommended scheme[19] somewhat different from that in Fig. 7.17. A different type of representation for another PWR core-loading scheme is shown in Fig. 7.19, where numerals designate the number of cycles burned for each assembly in a ⅛-core arrangement. Also shown is the calculated power distribution expressed as the ratio of assembly to average power under equilibrium-cycle conditions, with the maximum of 1.545 indicated for the central assembly.[20]

7.74 Although the general objectives and approaches used for a boiling-water reactor (BWR) are similar to those for a pressurized-water system, there are some differences. The square BWR fuel assemblies tend to be smaller, of the order of 5.5 in. on a side, instead of 8.5 in., with perhaps 49 rods instead of 200 as in a PWR assembly. This design trend is consistent with the use of cruciform control elements in the BWR and rod-cluster elements in the PWR (§6.47). As a result the degree of homogenization is better in the BWR with less local peaking of power between adjacent fuel assemblies of different exposure. However, Fig. 7.20 shows that, within a given fuel assembly, fuel of different enrichments is used in the outer rods to reduce the power ripple, particularly when the cruciform element is not in position.

7.75 In a BWR, control-rod movement and adjustment of the void distribution by varying the recirculation rate can be used to flatten the power

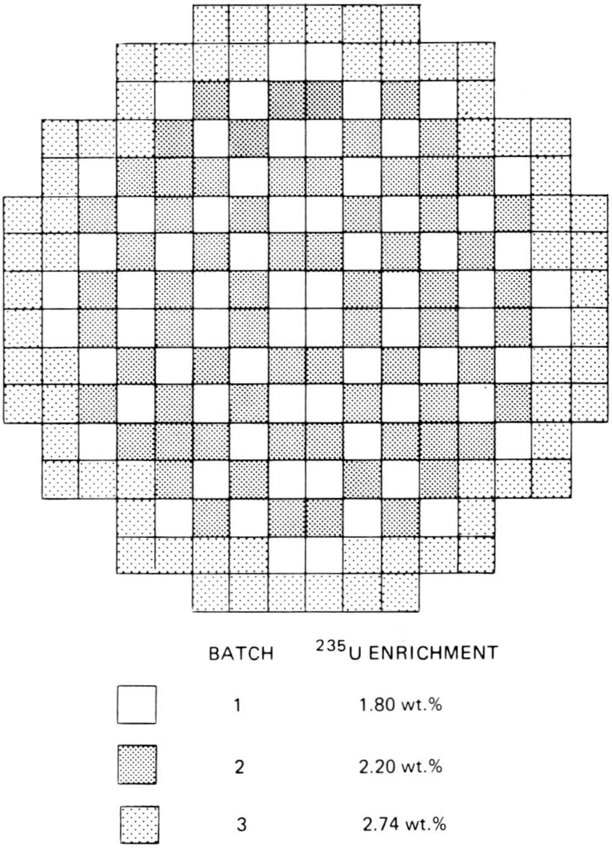

Fig. 7.18 Reactor fuel-loading scheme for an 800-Mw(e) pressurized-water reactor core.

distribution. In this way the spatial distribution of fuel burnup can also be controlled to counteract the effect of voids at the end of the fuel-exposure period when few control rods remain in the core. Chemical-shim control with burnable boron compounds in the coolant is normally used for this purpose in pressurized-water reactors. In this case flux distortion is avoided since the uniform distribution of poison permits normal operation with most of the control rods out of the core.

7.76 A typical large [1000-Mw(e)] BWR may have a modified-scatter, or modified-Roundelay, core-management scheme in which fresh fuel first is inserted in the outer 25% of the core and then, after one exposure period, is shuffled into a vacated position in a one-in-three scatter pattern in the central 75% of the core. The vacancy is produced when the depleted fuel is removed from the pattern in the central region. Once the fuel is moved into the scatter

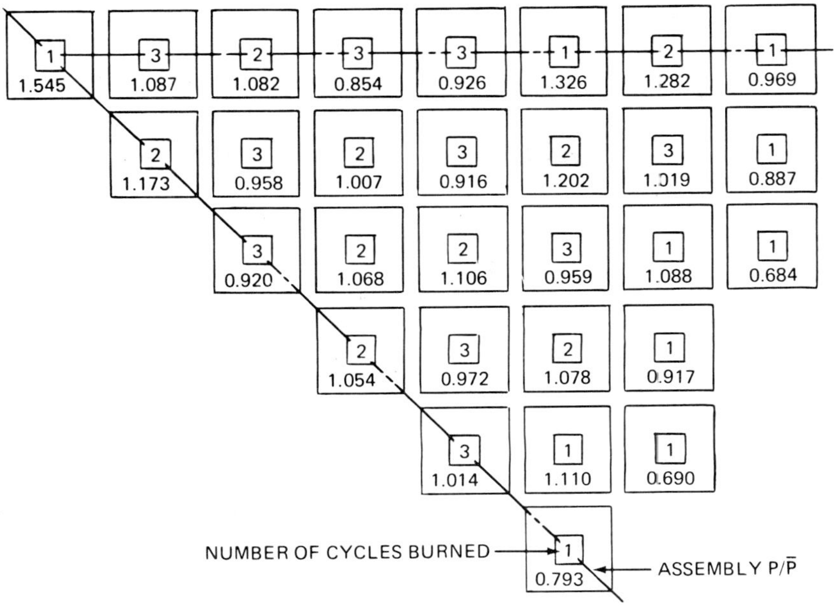

Fig. 7.19 Pressurized-water-reactor core arrangement and power distribution at end-of-life equilibrium-cycle conditions.

pattern, it is not moved again until discharge. Therefore, during each refueling shutdown, three separate fuel batches are handled, each batch amounting to 25% of the core. Figure 7.21 shows this pattern schematically.[21,22]

REFUELING TIME AND PLANT AVAILABILITY

7.77 Since the outage time for refueling affects plant availability significantly, the associated economics is included in the evaluation of various core-management schemes. A reasonably long interval between refueling, approximately 12 months, is desirable for water-cooled reactors because most of the outage time is required for shutting-down and starting-up operations rather than for the actual fuel handling. Typical operational time percentages for a BWR outage[22] of 15 to 20 days are summarized in Table 7.8. Time is provided within the categories listed for normal additional outage tasks and problems. These include in-vessel inspections, fuel-assembly inspections, dry-well seal testing, control-rod inspection, gamma scanning, control-curtain removal, sample coupon work, equipment failures, delays due to lack of visibility, loss of items into the vessel, containment-pressure tests, cold-criticality tests, excessive-contamination problems, personnel-exposure problems, major internal-replacement difficulties, and drive-replacement difficulties.

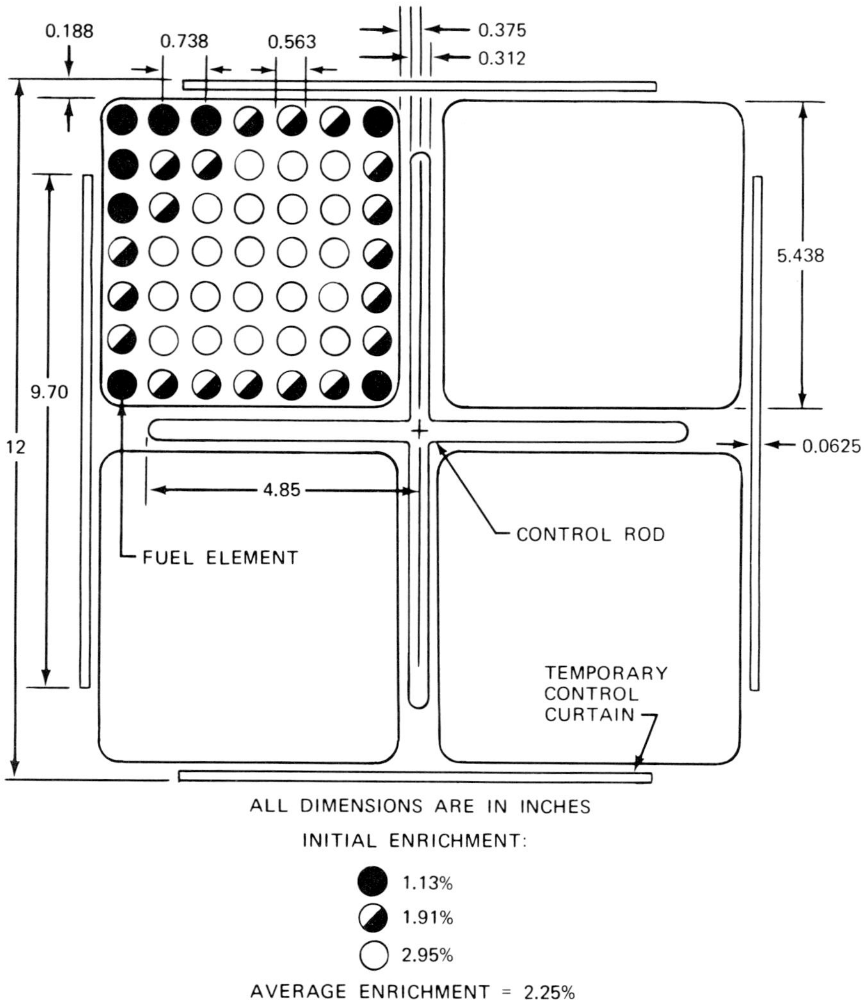

Fig. 7.20 Typical boiling-water-reactor core-lattice unit.

FUEL ECONOMY

INTRODUCTION

7.78 Uranium-requirement projections for light-water-reactor needs in the United States during the next few decades show that the supply of low cost ore is not likely to be adequate unless the energy potential of the fertile component of the fuel can be made available. The parameters affecting fuel economy, which

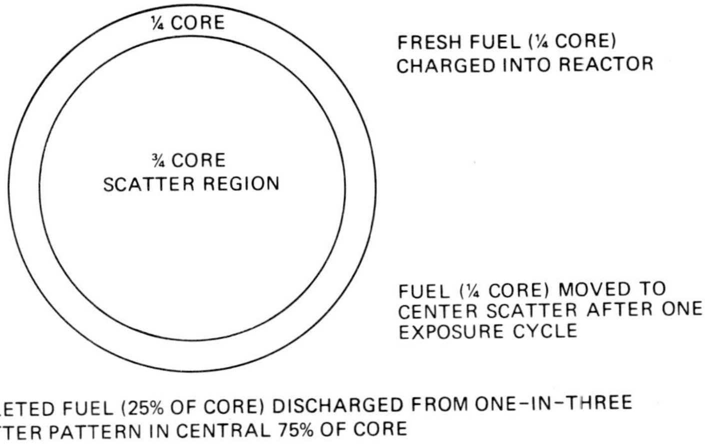

Fig. 7.21 Typical scatter-refueling pattern for a boiling-water reactor.

TABLE 7.8

Boiling Water Reactor Refueling Outage Operations

Operations	Outage time, %
Prepare to open vessel	8
Open vessel and remove dryer and separator	5
Sip fuel (remove leaking assemblies)	9
Unload fuel 25% of core	9
Shuffle fuel 25% of core	6
Load fuel 25% of core	7
Replace in-core flux monitors	10
Replace control rods removed	19
Test new core	4
Prepare to close vessel	6
Close vessel	9
Pre-start-up tests and power stepping	8
Total	100

include possible fuel-use patterns, are therefore important in the long-term analysis of the reactor fuel system. Plutonium recovered from light-water-reactor fuels, for example, could be recycled to reduce the need for new uranium. Various use patterns and associated parameters that might affect the design of a reactor system are discussed, therefore.

7.79 Prediction of the quantities of fuel materials that will be required is helpful. Estimates of future uranium-ore requirements and of plutonium available from reprocessed fuel are based on forecasts of nuclear-power growth through 1980 and beyond. The forecasts, however, have varied by about 30% not only on the basis of when they were made but also among forecasters. In the early 1970s, licensing delays further complicated predictions. The pattern among fuel flows, therefore, is more pertinent than the absolute numbers involved, which are subject to the uncertainty of the forecasts.

TABLE 7.9

Forecast of U. S. Nuclear-Power-Plant Capacity

(Full-Power Operation)

End of calendar year	Mw(e)	End of calendar year	Mw(e)
1974	47,000	1978	108,000
1975	61,000	1979	124,000
1976	74,000	1980	145,000
1977	91,000		

7.80 We shall use an estimate of projected capacities in the middle of the range as a basis. This forecast, for an installed capacity in the United States of 145,000 Mw(e) in 1980, is shown in Table 7.9. The trend is represented by the equation

$$C = 33{,}000 + 12{,}500t + 500t^2$$

where C is the capacity in megawatts and t is the time in years after 1973. Extrapolating according to this relation beyond 1980 gives a capacity of 255,000 Mw(e) for 1985 and 735,000 Mw(e) for the year 2000.

7.81 A picture of ore requirements and recovered fuel[23] was developed from the reactor parameters in Table 7.10. Certain assumptions for processing times and losses were also necessary. For example, for 1973 a period of 30 months was assumed between U_3O_8 procurement and initial full-power operation; this period decreases to 18 months after 1976. Replacement cores require only 12 months since fractional core batches are involved and start-up experiments are shorter.

7.82 It was assumed that procurement of enriched uranium in UF_6 allows an excess of 4% for cold scrap generated during fabrication of fuel elements, that losses during fabrication amount to 1%, and that recovery of the remaining 3% for reuse occurs 12 months after procurement. Twelve months is allowed after irradiation of fuel for cooling, shipment, and chemical processing, with losses of 1.3% for uranium and 1% for plutonium. Recycle of the discharged plutonium

TABLE 7.10

Assumed Design Parameters for Fuel Utilization[23]

Parameter	Boiling-water reactor	Pressurized-water reactor
Thermal efficiency, %	32.5	32.5
Capacity factor, %	85	85
Specific power, Mw(t)/tonne U	22	34
Initial core, average		
Irradiation level, Mwd(t)/tonne U	21,000	26,000
Fresh-fuel assay, wt.% ^{235}U	2.2	2.8
Spent-fuel assay, wt.% ^{235}U	0.8	0.9
Fissile plutonium discharged, kg/tonne U	4.9	5.7
Replacement loadings, typical		
Irradiation level, Mwd(t)/tonne U	28,000	32,000
Fresh-fuel assay, wt.% ^{235}U	2.6	3.4
Spent-fuel assay, wt.% ^{235}U	0.8	0.9
Fissile plutonium discharged, kg/tonne U	5.4	7.4

TABLE 7.11

Annual Uranium Requirements for Forecast U. S. Nuclear-Power Capacity

Year	U_3O_8 fuel, 10^3 tons	
	Without recycle	With Pu recycle
1968–1972	38.8	38.8
1973	14.6	14.3
1974	15.6	14.6
1975	22.6	21.2
1976	26.2	24.3
1977	29.1	26.3
1978	33.5	30.2
1979	37.7	33.7
1980	42.5	37.4
Total (1968–1980)	260.6	240.3

to the reactor as an option was considered and a period of 2 years allowed from the time of discharge to reinsertion in the reactor.

7.83 Annual net requirements for natural-uranium feed to produce enriched fuel for the forecast capacity are given in Table 7.11. These values are based on a diffusion-plant tails assay of 0.2% ^{235}U. Should the assay value be raised, as is likely, increased quantities of natural uranium will be needed. When

TABLE 7.12

Estimated Annual Quantity of Irradiated Fuel from U. S. Water Reactors

Calendar year	Discharged uranium, tonnes	Discharged plutonium, kg	Recovered plutonium (fissile), kg
1973	580	4,200	1,000
1974	870	6,600	2,000
1975	1,100	8,600	4,000
1976	1,600	12,000	6,000
1977	2,100	16,000	9,000
1978	2,500	21,000	11,000
1979	3,000	24,000	12,000
1980	3,600	29,000	16,000
1981	4,200	33,000	19,000
1982	5,000	37,000	22,000

it is not known whether forecast installations will be boiling-water or pressurized-water reactors, the power capacity is assumed to be equally divided between the two types. Also, Table 7.11 shows that requirements are reduced by about 10% if the discharged plutonium is recycled. In 1971, U. S. reserves of U_3O_8 producible at a cost of $8 or less were estimated at 246,000 tons.[24] If the accelerated requirements toward the end of the period in Table 7.11 are considered, these reserves are clearly inadequate if only light-water reactors are used to meet energy needs.

7.84 Table 7.12 shows estimated quantities of discharged fuel on uranium-metal-content and plutonium-content bases, corresponding to the Table 7.9 forecast and the parameters listed in Table 7.10. An estimate of fissile plutonium recovered is also shown. Recovery operations following discharge require from 1 to 2 years. Although such estimates change from time to time, the qualitative picture indicates the trends involved. One important trend is the production of recovered plutonium that can be recycled as thermal reactor fuel or used for fast reactor needs as discussed in the following section.

PLUTONIUM-USE PATTERNS

7.85 Large amounts of plutonium will be contained in the fuel discharged from the reactors forecast for the late 1970s, as shown in Table 7.12. Here again, there is a variation among forecasts. However, whichever forecast is used, several different possibilities exist for the use of the plutonium:

1. The plutonium, together with slightly enriched uranium, could be used in the fabrication of new fuel elements for the thermal reactor in which it was produced.

2. Since its nuclear properties tend to be more favorable in a fast spectrum, the plutonium could be reserved for fast reactor use. Should this approach be chosen, however, it would be necessary to store some of the plutonium for several years until the fast reactors were built.

3. Some of the plutonium could be used to meet all the immediate fast reactor requirements; the remainder would be used in thermal reactors in whatever manner proved optimum.

4. The plutonium could be marketed at a price determined by supply and demand and used in whatever manner the economics dictated.

7.86 Since various modifications and combinations of the preceding options are possible, a description of the possible patterns is more useful here than selection of a pattern that appears to be more favorable than the others. The plutonium price and the factors that determine it are important. A related subject is the behavior of plutonium as a fuel in reactors of different types, particularly in comparison with enriched uranium.* Finally, certain design parameters useful in the analysis of the flow of fuel materials are considered.

7.87 One pattern for the thermal reactor recycle of plutonium is to mix the recovered plutonium with uranium reenriched after reactor discharge. A differential ^{235}U requirement is needed to supplement the plutonium contribution so that the total fissile content is adequate for the desired reactor exposure. This would normally be done batch by batch with the normal core-management scheme for the reactor. Therefore the fissile material is distributed uniformly in a given batch.

7.88 As a rule of thumb, the total fissile requirement can be assumed constant for a given exposure although a slight increase may actually be required to compensate for some reactivity loss from the buildup of ^{242}Pu and other by-products. It should be noted, however, that ^{240}Pu acts as a burnable poison, producing fissile ^{242}Pu, so that the *reactivity swing* required for a given lifetime tends to be smaller than that with uranium fueling. Typical changes in plutonium isotopic composition with successive recycle for a boiling-water reactor[25] are shown in Fig. 7.22. The initial reactivity penalty[26] is indicated in Table 7.13. The detailed interplay of isotopic changes and physics parameters during burnup is quite complex, however, and is not considered here.

7.89 The results of one plutonium-recycle study are given in Table 7.14. Plutonium recycle begins to affect the ^{235}U assay in cycle 5 for the BWR and in cycle 4 for the PWR. In this model some of the fuel elements taken out of the BWR's in early cycles are held and reinserted later; thus plutonium is accumulated for an additional period before it is finally discharged. This method produces an unusually large effect on the ^{235}U assay in cycle 5. Such a

*In this section the term "enriched uranium" refers to an enrichment of 1.5 to 3.5% ^{235}U, such as might be used for light-water reactors.

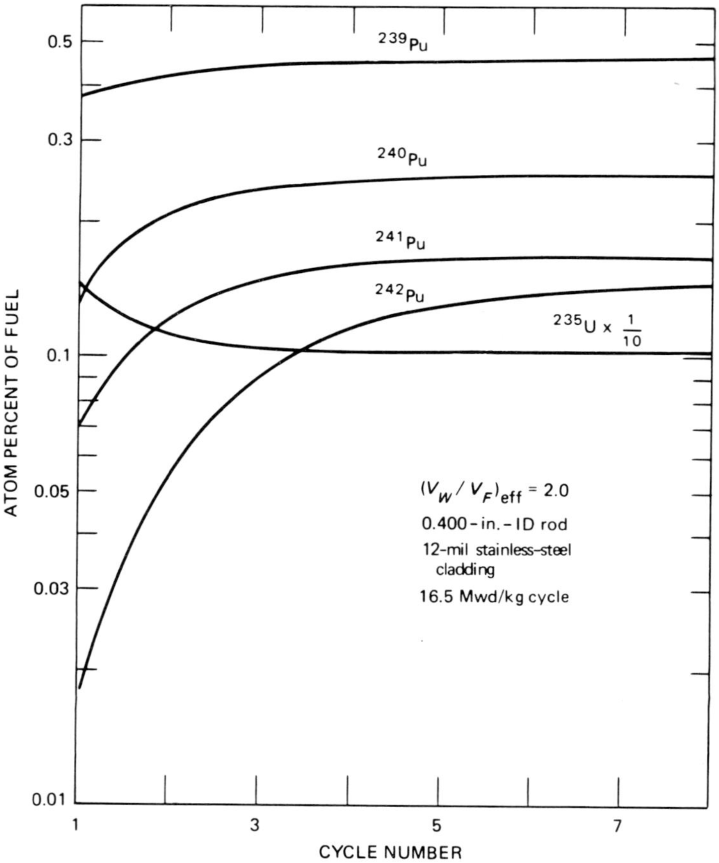

Fig. 7.22 Isotopic compositions of recycled fuel at discharge.

TABLE 7.13

Reactivity Penalty from Fuel By-products

End of recycle number	Worth at 20,000 Mwd/tonne uranium BWR discharge, % Δk			
	^{236}U*	^{237}Np†	^{242}Pu	^{243}Am†
0	0.62	0.13	0.65	0.36
1	0.90	0.59	1.53	0.57
2	1.12	0.73	2.04	0.89

*The ^{236}U concentration is assumed not to decrease in the diffusion plant.

†Neptunium and americium are removed by reprocessing on each recycle.

TABLE 7.14

^{235}U Assay Required for Plutonium Recycle Feed Batch

Cycle	^{235}U, wt.% BWR	^{235}U, wt.% PWR	Cycle	^{235}U, wt.% BWR	^{235}U, wt.% PWR
1	2.2	2.8	6	2.1	2.3
2	2.6	3.4	7	2.2	2.7
3	2.6	3.4	8	2.2	2.7
4	2.6	2.9	9	2.2	2.6
5	1.6	2.9	10	2.2	2.5

procedure is required to phase into the desired core-management scheme. The effect of buildup of plutonium in successive recycles is evident in the continued reductions in the ^{235}U assay of the fresh fuel for the PWR, but buildup occurs beyond the number of cycles shown for the BWR because of the long residence time for fuel. Either reactor type would eventually approach an equilibrium condition. Because of the long out-of-pile time and the fractional core-loading schemes used, however, equilibrium may be reached only after the reactor has operated for about 20 years.

7.90 Another pattern for thermal reactors is to segregate the recovered plutonium in fuel elements that are part of a batch loading and to use enriched uranium for the remainder of the assemblies. Therefore the premium fabrication cost associated with plutonium handling is confined to only a fraction of the core batch, not to the entire batch. However, power-peaking problems might result from such a "spiked" core, and analysis is difficult. Some of the economies of large-scale fabrication are also lost by using two types of fuel elements with a smaller volume of each required. However, a variation is likely to be the preferred scheme during the early years of plutonium recycle. In the so-called self-generated mode, plutonium discharged from earlier batches is used in fuel assemblies containing separate plutonium and uranium rods. Power peaking can be controlled by the pin-arrangement pattern.

7.91 Another scheme, useful when large amounts of plutonium are recovered, is to fuel one reactor with plutonium obtained from a number of other reactors. The opportunity for optimum core design specifically for plutonium fuel is better, and premium plutonium-fabrication charges can be paid only for the pins into which the plutonium has been segregated, as well. On the other hand, fast reactors may be a better use for the plutonium by the time the necessary large amounts become available.

7.92 Since plutonium has a "nominal" value of the order of $10 per gram (fissile), the fixed charges for its use are appreciable. Therefore the economic incentive to minimize the time between recovery of the plutonium from the reactor and its return to productive use as fissile material in a reactor core is considerable. The usefulness of recycled plutonium varies with the type of reactor, being greatest in a fast reactor. Some delay until the plutonium can be accepted in a fast reactor may therefore be economical if the enhanced usefulness is great enough to merit a sufficiently higher price so that the present worth of the expected sale price and storage costs is greater than the "present" thermal reactor price. Factors that influence the plutonium "price" are therefore important in the choice of a use pattern.

Plutonium Price

7.93 The concept of plutonium price is somewhat different from that of plutonium *value*. A market price depends on supply and demand. Demand, in turn, depends on a definite commercial application, which serves as a basis for determining the so-called value. Value is defined here as the price that must be assigned to plutonium to make the fuel-cycle cost using plutonium fuel equal to the fuel-cycle cost obtained if enriched uranium would be used as fuel for a particular reactor. This is the basis of the so-called indifference, or break-even, method of plutonium valuation proposed by Eschbach.[27] In this method, numerical values for each fuel-cycle cost component for both enriched uranium fuel and plutonium recycle fuel must be established, leaving only the price of plutonium as an unknown. Considering the fuel-cycle conditions in each case and analyzing the parameters affecting the result become quite complex, however. For a sharply defined analytical solution for the plutonium value, Eschbach specified that a minimized fuel cost be considered in each case as shown in Fig. 7.23.

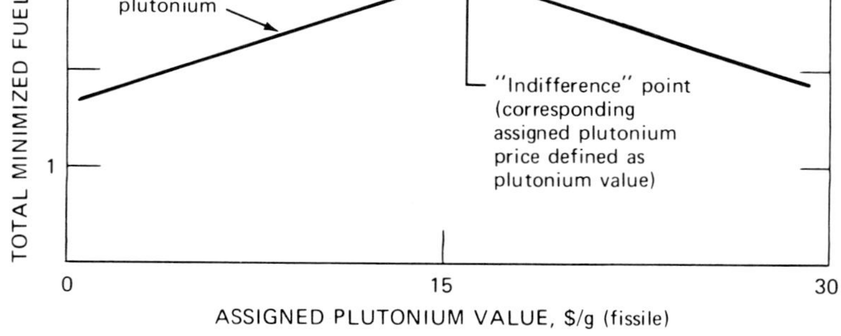

Fig. 7.23 Plutonium value.

7.94 If the market price of plutonium is less than the value for a given reactor operation, the plutonium will be used in that operation. On the other hand, if the actual market price is higher than the value, the plutonium will not be used, or, if produced in the reactor, will be sold rather than recycled. The market price therefore tends to seek a level dependent both on the supply of plutonium and on the quantity and types of reactors using plutonium at a given time.

7.95 In practice, achieving the fuel-cost minimum for the analysis is difficult since optimization involves so many parameters, particularly considering recycle and core-loading-strategy options. For many fuel studies in which a decision among alternate designs is desired, the overall effect of inventory charges is more significant than the plutonium value at any particular step or at any specific composition. A simplifying assumption can then be that the plutonium value is entirely a function of the fissile-plutonium content. The fissile-plutonium value determined at the optimum exposure can then be used as the value at other exposures.

7.96 The ability to separate plutonium chemically from uranium gives plutonium a unique advantage over ^{235}U as an enriching material for ^{238}U because all the unburned residual plutonium is recoverable from spent fuels, and the ^{235}U present, which can also be burned, is available at lower cost than that contained in the equivalent ^{235}U-enriched fuel. Plutonium used to enrich ^{238}U has a value essentially proportional to that of fully enriched uranium, regardless of the burnup conditions.[26] Furthermore, this is also true for any uranium price schedule that is consistent with the mathematics of the present cascade-enrichment process.* This result is surprising since it is often assumed that the value would be proportional to the burnup cost of the ^{235}U, which varies as a function of its enrichment level.

7.97 A plutonium value for a fast reactor system could be determined by the indifference method, with highly enriched uranium as the basis for comparison. Another approach is to solve for the plutonium value when the energy costs for a plutonium-fueled fast reactor system are set equal to those for a thermal reactor using plutonium recycle. Differences in timing could be handled with the present-worth concept. Thus the plutonium value determined would be consistent with the operational decision required for a use pattern.

Plutonium Storage

7.98 If plutonium produced in thermal reactors is stored in anticipation of the higher value in fast reactors, several cost considerations apply. Since storage

*In practice, both the premium plutonium fuel-element fabrication cost and the need for additional fissile-material loading in water reactors compared with ^{235}U tend to reduce the theoretical plutonium value.[28]

costs must be paid from the proceeds of the sale, they tend to reduce the expected premium. Carrying charges on the plutonium investment are the major storage cost contribution. Consideration of such charges is equivalent to "present worthing" (§2.31) the expected sale proceeds or recognizing the time value of money in some other way.

7.99 The actual costs of storage must also be considered. These include costs for a storage facility, which must be built and maintained; containers; security arrangements; and insurance on the stored material. Although little information is available on plutonium storage costs, one estimate[28] cites an annual charge of $0.35 per gram of fissile plutonium.

7.100 During storage of a mixture of plutonium isotopes such as those obtained from the discharge of thermal reactor fuel elements, the decay of ^{241}Pu, which has a 13-year half-life, reduces the fissile content somewhat. The radioactive ^{241}Am formed by this decay also increases fabrication costs. Recovery of the ^{241}Am before fabrication, however, may be economically attractive.

Storage Vs. Recycle

7.101 By balancing the increase in the worth of the plutonium obtained by future use in fast reactors against the costs of storage, one can estimate the maximum economical storage period as a function of various economic parameters. In one study[27] in which a 40% increase in plutonium value was assumed, the economically justified storage period was found to be only 3 to 4 years. A short storage period was found to apply even if the economic-parameter assumptions are varied over the entire realistic range. Furthermore, the effect of increasing demand for fast reactor fuel on plutonium price will probably be quite gradual over a period of years since only a small number of such reactors are expected before the late 1980s. Therefore long-term stockpiling of plutonium does not appear to be economically justified during the 1970s and early 1980s.

PLUTONIUM-RECYCLE PHYSICS DESIGN CONSIDERATIONS

7.102 In discussing plutonium recycle, we should examine the differences in the physics design of the core which may be introduced.[28] Since existing light-water-reactor cores build up appreciable plutonium during their lifetimes, the problems encountered are by no means new. However, the presence in the core of significant quantities of plutonium isotopes with nuclear properties quite different from those of uranium makes the physics analysis difficult.

7.103 The differences between plutonium-fueled and uranium-fueled thermal reactors are due primarily to the thermal- and near-thermal-energy-range resonances that occur in the cross sections of all plutonium isotopes, as shown in Fig. 7.24, and to the role played by the higher isotopes of plutonium.

7.104 For ^{235}U the only significant higher isotope formed is ^{236}U, which builds up slowly and is a weak absorber. Plutonium-239 absorbs neutrons to form ^{240}Pu, which is a strong absorber and which captures a neutron to form ^{241}Pu. Plutonium-241 is a fissile material whose nuclear properties are more favorable than those of ^{239}Pu and which gives rise to the relatively stable parasitic absorber ^{242}Pu on further neutron capture. These various differences affect the nuclear design of reactors fueled with recycled plutonium.

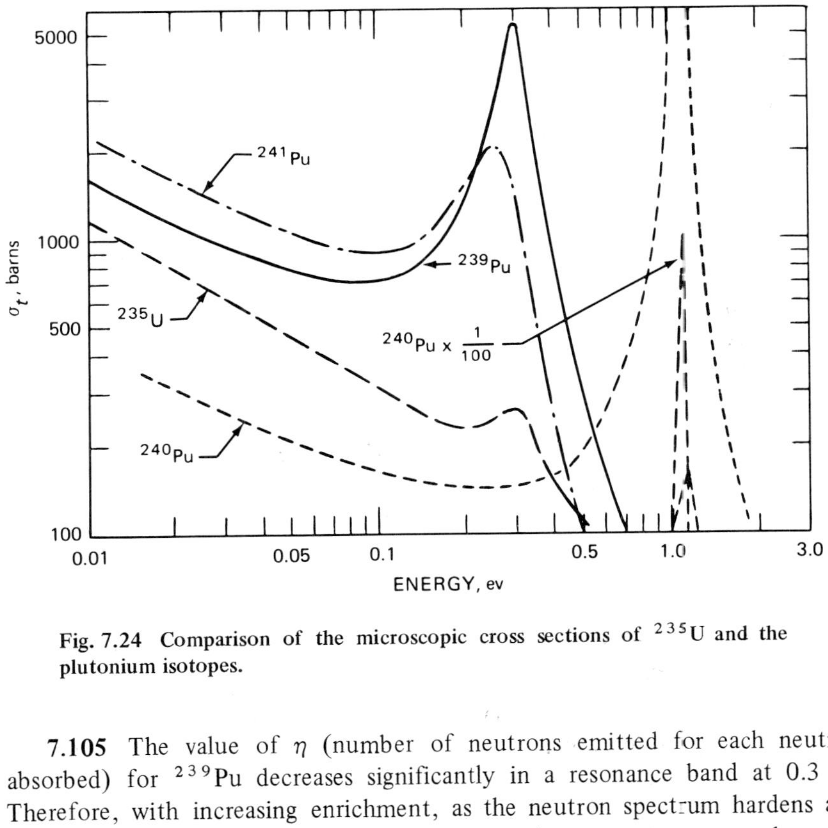

Fig. 7.24 Comparison of the microscopic cross sections of ^{235}U and the plutonium isotopes.

7.105 The value of η (number of neutrons emitted for each neutron absorbed) for ^{239}Pu decreases significantly in a resonance band at 0.3 ev. Therefore, with increasing enrichment, as the neutron spectrum hardens and more neutrons are absorbed in the resonance, the neutron economy becomes worse. Complicating the picture is the fact that the ^{239}Pu ratio $\alpha = \sigma_c/\sigma_f$ varies considerably over the thermal-energy range. For example, spectrum-averaged values for α for both PWR's and BWR's can be as high as 0.55, whereas the 2200 m/sec value is 0.36.

7.106 This variation of multiplication with amount of moderation results in plutonium having less multiplication at a moderating ratio (ratio of moderator atoms to fuel atoms) that is optimum for a ^{235}U lattice. Thus, if a mixed plutonium–uranium fuel is to be substituted for an enriched-uranium fuel in an operating PWR, there may be an economic incentive to increase the moderating ratio. This could be accomplished by reducing the fuel density, reducing the fuel-rod diameter, or changing the basic lattice configuration. An alternative is to use a higher enrichment.

7.107 Since the thermal absorption cross section of plutonium is significantly higher than that of ^{235}U, plutonium presents more competition for thermal neutrons; this reduces control-rod worth somewhat and the worth of a dissolved neutron absorber. Preliminary experiments indicate an approximate worth reduction of 15%. This normally can be handled in the design without problems.

7.108 The higher plutonium cross section also results in relatively rapid burnup. Thus, if a replacement plutonium core is designed for the same initial excess reactivity as its predecessor uranium core, its reactivity lifetime will be shorter. Conversely, if it is designed for the same reactivity lifetime, higher initial reactivity will be needed, perhaps requiring additional reactivity control.

7.109 The delayed-neutron fraction of ^{239}Pu, which is only about one-third that of ^{235}U, must also be considered. However, since the delayed-neutron fraction of ^{241}Pu is only slightly smaller than that of ^{235}U, the effect is reduced as burnup increases. Experience[28] indicates that no design problem is presented.

7.110 In a heterogeneous thermal reactor, a region having excess moderation causes power peaking in adjacent fuel rods. This is particularly true in PWR's, in which there are large fuel elements and no coolant feedback mechanism for suppressing local peaking as in a boiling system. The higher thermal absorption and fission cross sections in plutonium fuel rods aggravate this situation. It has been estimated that, in a typically designed PWR, a water-induced power peak will increase by about 40%. This complicates the in-core fuel-management picture.

7.111 The high cross section of ^{240}Pu contributes several nuclear design advantages.[29] For example, because of its high neutron absorption in under-moderated systems, it can act both as a burnable poison and as a fertile material and therefore reduce reactivity variations during burnup.

7.112 Although some nuclear properties of plutonium isotopes and ^{235}U differ, many of the reactor physics characteristics are not appreciably changed in a slightly enriched core, since the large amount of ^{238}U present has a stabilizing effect in each case. Characteristics such as the fast-fission effect, resonance escape probability, and the Doppler coefficient are therefore affected only in a minor way when plutonium is substituted for some of the ^{235}U in a light-water-reactor core.

7.113 The fuel economy[30,31] of a reactor is affected by many design characteristics. A few generalized parameters, however, serve as guides for the reactor designer. These relate to the amount of fissile material destroyed during the production of energy and to the amount needed for the fuel inventory of the reactor. Design parameters for fuel economy include the conversion ratio, specific utilization, the capture-to-fission cross-section ratio of fissile material, the level of fuel exposures, specific power, and the ratio of total inventory to the in-pile inventory. The doubling time and the breeding ratio are particularly important in considering the fuel economy of fast breeder reactors. Several of these which are useful as criteria are described.

7.114 A fuel-economy parameter in terms of *net amount of fuel required* (in some cases, including out-of-reactor fuel cycle) *per unit of energy produced,* in kilograms of uranium per kilowatt-hour (electrical), is conveniently used to indicate the performance of a given reactor concept. Another parameter is the *net amount of fissile material destroyed per unit of energy produced,* in grams of ^{235}U per megawatt-day. This is the reciprocal of the *reactor specific utilization* (sometimes called static reactor utilization) defined as the total energy produced per unit of net fissile material destroyed.[30] The latter representation shows the effect of alpha, the capture-to-fission fissile-material cross-section ratio, on fuel economy. In terms of natural-uranium requirements, the term "dynamic system utilization," the total energy [megawatt-years (electrical)] produced per tonne of natural uranium, signifies losses in the fuel cycle, which include the diffusion-plant cascade tailings.[30]

7.115 For thermal reactors an important performance index of economy is the conversion ratio, CR:

$$\mathrm{CR} = \frac{\text{atoms of fissile isotope produced}}{\text{atoms of fissile isotope destroyed}} \quad \text{(initial)}$$

$$\mathrm{CR} = \frac{N^{28}\sigma_c^{28} + \epsilon P_1(1-p)N^{25}\sigma_a^{25}\eta^{25}}{N^{25}\sigma_a^{25}}$$

$$= \frac{\sigma_c^{28}}{\dfrac{N^{25}}{N^{28}}\sigma_a^{25}} + \epsilon P_1(1-p)\eta^{25} \qquad (7.24)$$

where P_1 = nonleakage probability from fission energy to resonance energy
p = resonance escape probability
ϵ = fast-fission factor

superscript 25 = ^{235}U
superscript 28 = ^{238}U

The first term in the equation represents the thermal absorption in fertile material, and the second term represents the resonance absorption also in fertile material. This expression for the conversion ratio applies only to the initial period after start-up before plutonium isotopes contribute to the picture. By definition the conversion ratio also depends only on the ratio of fissile to fertile materials and the neutron energy spectrum. However, in practice, the neutrons available for conversion do depend on losses to fission products and other materials effects that depend on the reactor design.

7.116 A higher initial fuel enrichment tends to decrease the conversion ratio. On the other hand, an increase in the resonance capture by ^{238}U, which depends on the size and moderating properties of the reactor, augments the conversion ratio. Therefore, for a given reactor size, the conversion ratio is a function of the neutron energy as well as the properties and proportions of the scattering material. The resonance capture normally increases either as the density of the scattering material increases or as the neutron-absorption cross section increases in the resonance region owing to spectral hardening. Spectral hardening arises, for example, as the system temperature increases after the reactor starts operation or in lattices having a low moderator-to-fuel ratio.

7.117 A high conversion ratio leads to two trends: (1) a higher plutonium concentration at discharge of the fuel and (2) a slower decrease in reactivity with burnup caused by the fissioning of some fraction of the plutonium produced.

7.118 The specific utilization (§7.114) developed from the *effective* conversion ratio R_{eff} is a useful parameter. This quantity is similar to the conversion ratio but applies over the core lifetime of a fuel and therefore is really a more important parameter. Since the energy attainable from the fission of 1 g of fissile material is approximately 0.95 Mwd, the energy attainable from 1 g of fissile material destroyed is

$$0.95 \frac{\sigma_f}{\sigma_f + \sigma_c} = \frac{0.95}{(\sigma_f + \sigma_c)/\sigma_f} = \frac{0.95}{1 + \alpha_{av}} \quad \text{Mwd/g } ^{235}\text{U destroyed} \qquad (7.25)$$

Now, the grams of fissile isotope produced equals the grams of fissile isotope destroyed times R_{eff}. Thus

Net amount of fissile isotope destroyed

$$\begin{aligned} &= \text{grams } ^{235}\text{U destroyed} - \text{grams fissile isotope produced} \\ &= \text{grams } ^{235}\text{U destroyed} - \text{grams } ^{235}\text{U destroyed} \times R_{eff} \\ &= (1 + \alpha_{av}) - (1 + \alpha_{av})R_{eff} \\ &= (1 + \alpha_{av})(1 - R_{eff}) \end{aligned}$$

Therefore the reactor specific utilization, Υ_r, is

$$\Upsilon_r = \frac{0.95}{(1+\alpha_{av})(1-R_{eff})} \quad \text{Mwd/g}\ ^{235}\text{U} \tag{7.26}$$

7.119 If an average value of alpha is taken as 0.5, for example, the energy yield from 1 kg of natural uranium for an effective conversion ratio of 0.7 will be

$$\left(\frac{0.95}{1.5 \times 0.3}\ \frac{\text{Mwd}}{\text{g}\ ^{235}\text{U}}\right)\left(\frac{7.115}{1}\ \frac{\text{g}\ ^{235}\text{U}}{\text{kg U}_{nat}}\right) = 15.02\ \text{Mwd/kg U}_{nat}$$

This compares with an energy yield of 90.12 Mwd/kg U_{nat} for a conversion ratio of 0.95, an approximately sixfold increase.

7.120 Being derived from average values of α, η, and the conversion ratio, the reactor specific utilization, too, is a characteristic of the reactor. The dynamic system utilization, Du, can be expressed as

$$\text{Du} = \frac{(\text{Ex})(\text{Ef})}{(X_{nat})(365)} \quad \text{Mw-years(e)/tonne U}_{nat} \tag{7.27}$$

where X_{nat} = tonnes of natural uranium required per tonne of fuel
Ex = the fuel exposure [Mwd(t)/tonne U]
Ef = the conversion efficiency [Mw(e)/Mw(t)]

7.121 Design parameters for uranium conservation in fast breeder reactors include the *specific power relative to the fuel,* the *breeding ratio,* and the *doubling time,* variables interrelated with one another. The specific power is normally expressed as the thermal power per unit amount of fuel in the core (kw/kg) but can also be expressed in terms of the fissile isotope. Sometimes the out-of-core inventory is also included in the fuel basis.

7.122 Although the breeding ratio is similar in concept to the conversion ratio in that it relates the fissile atoms produced to those destroyed, the definitions used vary. Generally, for the breeding ratio the chemical species formed, such as plutonium, is the same as that destroyed. One definition[32] of breeding ratio represents an integrated average value over the fuel-cycle period:

$$\text{BR}_1 = \frac{\text{fissile atoms produced}}{\text{fissile atoms destroyed}}$$

$$= \frac{\text{final fissile atoms} - \text{initial fissile atoms} + \text{fissile atoms destroyed}}{\text{fissile atoms destroyed}} \tag{7.28}$$

Other definitions of breeding ratio are in terms of individual isotopic nuclear interactions:

$$BR_2 = \frac{\text{neutron captures in } ^{238}\text{U, } ^{240}\text{Pu, and } ^{242}\text{Pu in core and blanket}}{\text{neutron absorptions in } ^{239}\text{Pu and } ^{241}\text{Pu in core and blanket}} \tag{7.29}$$

$$BR_3 = \frac{\text{neutron captures in } ^{238}\text{U and } ^{240}\text{Pu}}{\text{neutron absorptions in } ^{239}\text{Pu and } ^{241}\text{Pu}} \tag{7.30}$$

and even

$$BR_4 = \frac{\text{neutron captures in } ^{238}\text{U and } ^{240}\text{Pu}}{\text{neutron absorptions in } ^{235}\text{U, } ^{239}\text{Pu, and } ^{241}\text{Pu}} \tag{7.31}$$

7.123 Different breeding-ratio definitions have evolved primarily as a result of a desire to use static macroscopic-cross-section data to yield a quantity related to the long-term fuel consumption and generation that will have some relevance in economic studies. A more general definition for a reactor region n has therefore been proposed[33,34] to meet this objective:

$$BR_n = \frac{\sum_i \gamma_i C_{n,i-1}}{\sum_i \gamma_i A_{n,i}} \tag{7.32}$$

where i = a suffix that can take the values 5, 6, 8, 9, 0, 1, and 2, representing ^{235}U, ^{236}U, ^{238}U, ^{239}Pu, ^{240}Pu, ^{241}Pu, and ^{242}Pu, respectively
$C_{n,i}$ = the capture for isotope i of region n
$A_{n,i}$ = the absorption rate for isotope i of region n
γ_i = weighting factors

This breeding-ratio definition can be made equivalent to those previously given by assigning appropriate values to the weighting factors. For ^{240}Pu, ^{241}Pu, and ^{242}Pu, if $\gamma_i = \eta_i/\eta_9$, the value of the breeding ratio becomes insensitive to shifts in ^{240}Pu concentration which otherwise would have an important effect.

7.124 The doubling time is the time required for a breeder reactor to produce excess fissile material equal to the initial total quantity of fissile material in the fuel cycle. Complications arise, however, when attempts are made to write a simple relation for the doubling time in terms of static calculation quantities. A basic problem is the manner in which the fissile atoms formed by breeding can contribute to the formation of a second generation of bred fissile atoms. If this possibility is ignored or the process visualized as occurring only after a time interval longer than that under consideration, the growth is analogous to *simple interest.* Such a process would occur with plutonium formed in the blanket of a fast reactor, which could not be effective as fuel leading to additional breeding until it was recovered from the blanket and loaded into the core, probably after a period of many years. On the other hand, if the bred material can be used as fuel instantaneously, as would be the case for in-core breeding, we would have a growth analogous to *interest compounded continuously*. Behavior in most breeders is likely to be at some point within these extremes. A compound-interest (not continuous) approach can also be applied, of course, to a model in which the bred fuel is made available at the end of a fixed period, then added to the "principal" during the subsequent period, and so on.

7.125 The omission or inclusion of the out-of-pile fissile inventory is another basic point of disagreement among expressions for doubling time, although most do include it. The methods of including it vary, however, with some using a fraction of the in-pile time as a measure of out-of-pile inventory and others using a cycle time (variously defined) together with an inventory in the reactor averaged both in and out of the reactor. Some methods include an estimate of fissile losses sustained during the fabrication and reprocessing operations; others omit these losses. Although the losses must be accounted for in a true doubling time, they are possibly omitted by some because they are not characteristic of the performance of the reactor. Another difference is in the measure of the excess fuel created, which may be expressed in terms of breeding ratios and capture-to-fission ratios, or the difference between initial and final mass balances.

7.126 With the simple interest approach, if the initial total quantity of fissile material is M, the doubling time, T_d, is then given by

$$T_d = \frac{M}{gG} \text{ days} \tag{7.33}$$

where g is the quantity of fissile material consumed in the breeder reactor per day and G is the breeding gain (BR – 1). Of the g grams consumed, $g/(1 + \alpha)$ grams have undergone fission. Since the fission of 1 gram of fissile material per day produces approximately 0.95 Mw of power, the fission of $g/(1 + \alpha)$ grams of fissile material will produce $P = 0.95\, g/(1 + \alpha)$ Mw of power. Then,

$$g = \frac{P(1 + \alpha)}{0.95} \text{ grams} \tag{7.34}$$

Substituting Eq. 7.34 into Eq. 7.33 gives the doubling time, T_d, in terms of $\alpha, P,$ $M,$ and G:

$$T_d = \frac{M}{G[P(1 + \alpha)/0.95]} = \frac{0.95\, M}{GP(1 + \alpha)} \text{ days} \tag{7.35}$$

Such simplified expressions must be used with caution, however, with α properly averaged over the isotopes present and spatial and burnup effects taken into consideration.

7.127 Since the power attainable from the fission of 1 gram of fissile material per day is almost constant, a higher specific power will be obtained by minimizing the out-of-pile inventory. This smaller quantity of out-of-pile inventory, together with a high breeding gain, yields a shorter doubling time.

7.128 If a breeder reactor operates for a year at a plant factor F, the extra fissile material produced, e, as a fraction of the reactor inventory is

$$e = 365F \frac{1 + \alpha_{\mathrm{av}}}{0.95} \frac{SG}{H \times 10^6} \tag{7.36}$$

where S is the specific power and H is the ratio of the total inventory for the reactor to the in-pile inventory.

7.129 A starting point for a compound-interest doubling time is the classical relation for the amount of money, S, obtainable in n years from a principal, P, compounded at an annual interest rate, i:

$$S = P(1 + i)^n \tag{2.1}$$

By analogy the amount of fuel, M, obtainable after n irradiation periods from an initial fuel inventory, M_0, as a result of bred fuel added to the original fuel at the end of each irradiation period will be

$$M = M_0 \, (\mathrm{BR})^n \tag{7.37}$$

A time variable can be introduced by setting the total time required, T, equal to the product of the number of irradiation periods, n, and the time required for each period, t_n. Then,

$$\frac{M}{M_0} = (\mathrm{BR})^{T/t_n} \tag{7.38}$$

For the special case where $M/M_0 = 2$, T will be equal to the doubling time, DT.

$$\frac{M}{M_0} = 2 = (\mathrm{BR})^{\mathrm{DT}/t_n}$$

or

$$\ln 2 = \frac{\mathrm{DT}}{t_n} \ln \mathrm{BR}$$

$$\mathrm{DT} = t_n \frac{\ln 2}{\ln \mathrm{BR}} \tag{7.39}$$

Various other relations have been developed to account for out-of-pile times, losses, etc.[32]

7.130 A short doubling time for a higher specific power and a higher breeding gain implies that more plutonium becomes available in a uranium–plutonium system in a given period of time than in a system having a longer doubling time, with a resulting decrease in the total amount of natural uranium used. The doubling-time parameter is also often used in comparison with a typical growth rate of electrical-energy production needs, which yields a "doubling time" of about 10 years. Such a 10-year doubling time is therefore a design target for breeder reactors, which could then supply sufficient extra fuel for the new reactors necessary to meet the energy demand.

FUEL-DESIGN PARAMETERS

INTRODUCTION

7.131 The designer of a reactor fuel system must consider many parameters that interact in complicated ways. Parameters for core-management schemes, fuel-cycle-operations costs, and fuel economy have been discussed. However, dimensional and lattice specifications in the design of a fuel rod and fuel assembly are also affected by neutron physics, control, safety, materials, thermal–hydraulic, and other parameters. Systems-engineering methods should be applied to the parameter matrix to optimize the core design. Although this is indeed a formidable challenge, some progress has been made, as discussed in Chap. 9. A systematic treatment of all parameters is not appropriate here, however.

LATTICE EFFECTS

7.132 To illustrate the interplay of parameters that the designer must consider, we shall discuss lattice effects on the fuel cycle, along with the fuel-rod diameter and its relation to other design specifications, such as specific power.

7.133 Changes in the lattice of a slightly enriched, water-moderated UO_2 system can affect the reactivity lifetime, the conversion ratio, and ultimately the energy cost. At the beginning of core life, an increase in the fuel-to-moderator ratio can provide an advantage in initial conversion ratio. Other effects during the fuel lifetime, however, must be considered to determine the economic consequences of shifting to a drier lattice.

7.134 The trade-off between initial conversion ratio and reactivity lifetime is shown in Figs. 7.25 and 7.26 for a typical system. As the water-to-uranium ratio decreases and the neutron spectrum therefore hardens, several effects occur. Although eta of ^{235}U (neutrons produced per neutron absorbed) tends to

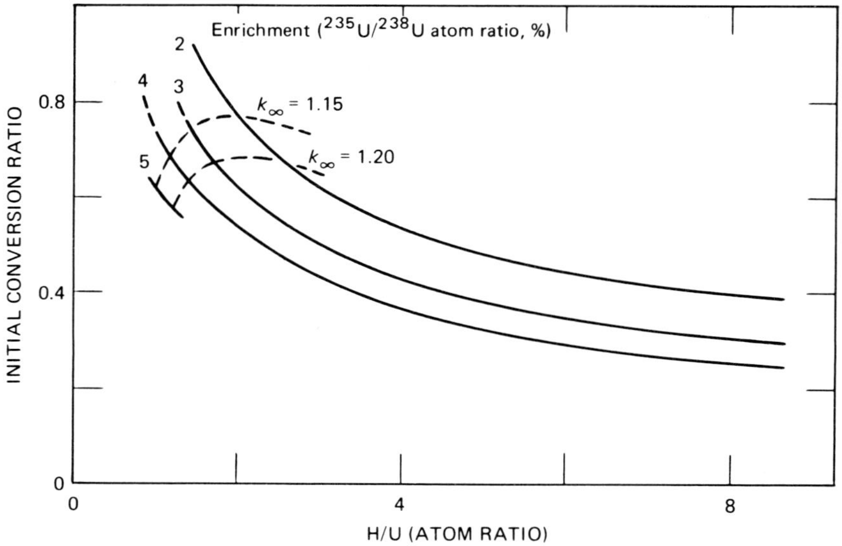

Fig. 7.25 Initial conversion ratio for critical reactors (0.400-in. rod diameter).

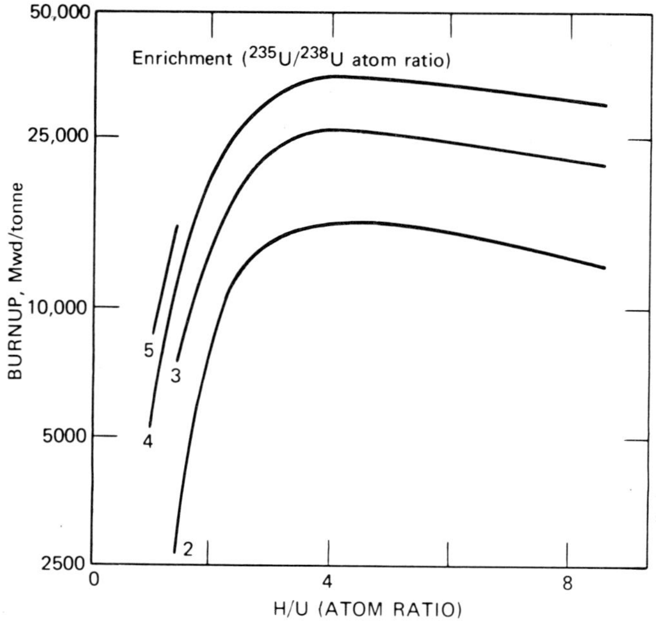

Fig. 7.26 Lifetime for very large reactors (0.400-in. rod diameter).

decrease since alpha for ^{235}U increases, the greater contribution from ^{238}U fast fission increases the total number of neutrons available for absorption in fertile ^{238}U somewhat. Resonance-region absorption is important here and also depends on the pin diameter, which is held constant in Figs. 7.25 and 7.26.

7.135 On the other hand, as the lattice becomes less moderated, neutron leakage tends to increase. As a result a greater enrichment is required to maintain the fuel lifetime. The various isotopic changes and power sharing between the original ^{235}U and the plutonium produced contribute to a complicated picture. In qualitative terms, however, a drier lattice tends to produce more plutonium, which, in turn, is available for power production in situ. This apparent advantage is masked, however, by the need to provide higher enrichment in ^{235}U to obtain the desirable core lifetime. Since the decrease in lifetime with decreased moderation is very sharp, as shown in Fig. 7.26, a compromise moderation limits the initial conversion ratio of such reactors to about 0.6. However, an optimum lattice would depend primarily on fuel costs, thermal hydraulics, and perhaps dynamic behavior.

7.136 The previous discussion considers only a few of the many aspects of changing the degree of core moderation. Diameter effects were included only in a limited way. Although some studies[35] have been made of the effect on fuel cost of the moderator-to-fuel ratio, parameters held constant in such limited studies can affect the results significantly. It is therefore best to note general trends and to carry out detailed analyses of specific cases as needed.

SPECIFIC POWER AND PIN DIAMETER

7.137 An examination of the effect of specific power on fabrication costs identifies a number of important design parameters and their interrelations.[36] The specific power, kilowatts per kilogram of uranium, is a measure of the rate at which energy is obtained from the fuel. Thus changes in the specific power affect the parameters that depend on the *time* required to produce a given amount of energy. If the amount of energy produced (i.e., the burnup) is held constant, the cost of such operations as shipping and processing for a given fuel batch is not affected. On the other hand, when expressed on a unit-energy basis, the fixed-charge component of fuel cost will decrease as the specific power increases since more energy would be obtained in a given time. A high specific power is therefore a desired design objective.

7.138 To increase the specific power, however, for a typical cylindrical fuel element, one must decrease the diameter. Otherwise, the central fuel temperature will be excessive. The fabrication cost *per mass of fuel* then tends to increase, and a slight increase in enrichment may also be necessary to compensate for the probable higher proportion of cladding material in the system. This use of more expensive fissile atoms in the fuel therefore lessens the reduction in fixed charges obtained from the higher specific power.

7.139 Since several parameters affect the system in different directions, an opportunity for optimization exists. The choice between stainless-steel and Zircaloy cladding may also be considered, as well as the effect of changing burnup. The relative effect of these parameters can be systematically considered by expressing them on a functional basis. Since the key to the analysis is interplay between the inventory charges and the fabrication-operations costs, it is useful to express the total fuel cost as the sum of three items: (1) the fabrication cost, (2) an inventory charge, and (3) all other charges.

7.140 To more easily identify the effect of the fuel-rod diameter, one can represent the fabrication cost as the sum of two terms: (1) a constant cost per unit length of fuel rod and (2) a constant cost per kilogram of contained fuel on a uranium basis. The first cost contribution, which is independent of the diameter, is due primarily to operational-labor and fuel-cladding costs, and the second term reflects primarily the cost of manufacturing the ceramic-fuel material to be loaded into the rods (cladding).

7.141 Consider f as the fabrication-cost component independent of the diameter of the rod.

$$f = k_1 \left(\frac{FS}{E_x q_L} \right) \tag{7.40}$$

where F = fabrication expense ($ per foot of fuel rod)

q_L = average heat generation rate [kw/ft of fuel rod (constant over diameters of interest)]

E_x = average burnup (Mwd/kg of uranium in core)

k_1 = conversion constant

The uranium-inventory cost during reactor operation can be given by

$$i = k_2 \left(\frac{r}{\text{PF}} \right) \left(\frac{U}{S} \right) \tag{7.41}$$

where i = uranium inventory cost (¢/million Btu)

r = fractional-uranium-inventory charge rate [(%/year) ÷ 100]

PF = plant factor (ratio of the average plant output to the rated plant capacity)

U = value of uranium inventory ($/kg of uranium)

k_2 = 3.512 {conversion constant [(¢/10^6 Btu)/($/kw-year)]}

7.142 For this simplified model, all other fuel-cycle costs, including the diameter-dependent fabrication component, can be lumped into a contribution, K, so that

$$\text{Total fuel cost} = k_1 \left(\frac{FS}{E_x q_L} \right) + k_2 \left(\frac{r}{\text{PF}} \right) \left(\frac{U}{S} \right) + K \tag{7.42}$$

7.143 This equation can then be differentiated with respect to S. With other parameters fixed and the derivative set equal to zero, an optimum specific power, S^*, which results in a minimum fuel cost, can then be determined.

$$S^* = \sqrt{\frac{k_2}{k_1}\left(\frac{r}{\mathrm{PF}}\right)\left(\frac{q_L E_x U}{F}\right)} \tag{7.43}$$

7.144 The fabrication, f, and inventory, i, contributions to the fuel cost can then be expressed as

$$(f + i)_{\mathrm{minimum}} = 2\sqrt{k_1 k_2 \left(\frac{r}{\mathrm{PF}}\right)\left(\frac{UF}{q_L E_x}\right)} \tag{7.44}$$

at the optimum specific power.

7.145 Owing to the trade-off between inventory and fabrication cost, there is an optimum specific power whose value depends on the actual values used in the relation. With this behavior in mind, the designer should evaluate alternative decisions at the respective optimum condition, not at constant specific power. In considering the relative advantages of stainless-steel and Zircaloy fuel cladding, for example, we see that a different specific power is optimum for each case. Zircaloy offers improved neutron economy at the expense of a somewhat higher fabrication cost per length of fuel rod. An improved neutron economy means that a uranium enrichment somewhat lower than that required for a stainless-steel-clad element can be used to achieve the same degree of fuel burnup. Consequently the ratio of U to F tends to be lower, and the specific power is thus reduced.

7.146 At first glance relaxation of dimensional tolerances appears to reduce the manufacturing expense and hence to reduce the fuel cost, provided the central fuel temperature would not become excessive. For UO_2 fuel, however, the effects of such changes on the internal temperature pattern is difficult to evaluate since the oxide is not dimensionally stable during irradiation (cracks form and there is generally an irregular thermal path to the cladding). If the internal temperature is assumed to be unaffected, a relaxation of pellet OD tolerances could therefore be considered. This question does not arise with vibrationally compacted fuel.

7.147 Relaxing pellet OD tolerance from ±0.001 to ±0.002 in. reduces the fabrication cost only 5%. Similarly, reducing the tubing inside diameter reduces the fabrication cost only a few percent. In fact, for stainless steel a relaxation of the tubing *thickness* tolerances increases the cost slightly since the *nominal* thickness may have to be increased. Consequently greater enrichment may be necessary to compensate for the increased poison. Therefore changes in dimensional tolerances over a practical range do not appear to offer significant reductions in the fabrication cost.[36]

7.148 This analysis is intended as a guide to some of the economic effects of changes in fabrication-design parameters rather than an indication of how fuel-cycle costs can actually be calculated. For example, as Zircaloy costs have decreased during the past few years, the comparison has shifted from a modest advantage over stainless steel for water-cooled-reactor cladding to a definite advantage. Most new designs therefore specify Zircaloy cladding.

FAST REACTORS

INTRODUCTION

7.149 The fuel systems for fast reactors are sufficiently different from those for thermal reactors to justify a separate discussion. Since fast reactors are important primarily because of their potentially high breeding ratio, the breeding gain and parameters mentioned earlier (§7.112), associated with the production of new fissile fuel from fertile material, are of interest to the designer. Some special features of fabrication, core management, and reprocessing also are discussed in the following sections.

7.150 Fast reactor breeding characteristics are affected by the spectrum, which, in turn, depends on the nature of the fuel. A measure of the breeding potential is given by η, the neutrons born per neutron absorbed in fuel, shown in Fig. 7.27 as a function of energy for typical large fast reactor systems.[37] Table 7.15 gives typical core parameters for a 1000-Mw(e) sodium-cooled system.[38] These are mean values from several studies rather than a single optimum design.

7.151 The relative contribution of fuel-cycle cost categories is indicated in Table 7.16. However, they are minor compared with the plant capital cost. Though based on the parameters in Table 7.15 and on reasonable economic assumptions, these fuel-cycle values should be considered as only rough estimates since no experience is yet available. The table does serve as a useful framework for discussion, however. For example, the plutonium credit is proportional to the breeding gain and therefore is not dependent on the specific power. Also, the total energy cost is rather insensitive to the price of plutonium. Although the revenue from the sale of excess plutonium will indeed depend on the price of plutonium, the inventory charge, a contribution in the opposite direction, also depends on the price. Hence a shift in price tends to have a minor effect on the energy cost.

7.152 The combination of breeding ratio, fuel inventory, and other design parameters involves a number of complicated interplays that require study during the design process. A design objective is a high breeding ratio, which is a result of breeding both in the core and in the blanket. Although the proportion of each can vary over a fairly wide range with a satisfactory total breeding ratio being achieved, a high core breeding ratio, perhaps approaching unity, is

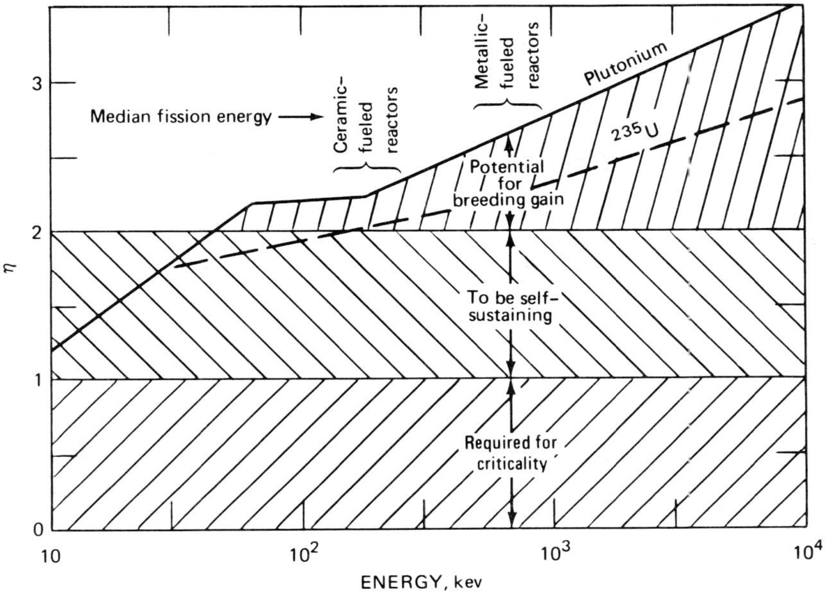

Fig. 7.27 Neutrons born per neutron absorbed in fuel (η) for typical fast reactor systems.

TABLE 7.15

Typical Core Parameters for a 1000 Mw(e) Sodium-Cooled Fast Reactor

Reactor power	2500 Mw(t)
Core power	2090 Mw(t)
Specific power (fissile)	770 kw/kg
Breeding ratio	1.40
Core doubling time*	9.75 years
Core fissile metal	2709 kg
Core fissile fraction	0.169
Average burnup (ceramic fuel)	100,000 Mwd/tonne

*Plant factor of 90%, core inventory only.

normally desirable since the fissile material formed is useful as fuel immediately. Although such a core loses little reactivity with burnup, the lifetime of the core is normally limited from the metallurgical standpoint.

7.153 In general, a large core with low neutron leakage and with a coolant that does not degrade the neutron spectrum by moderation will require a relatively low enrichment. The accompanying high fertile-to-fissile atom ratio results in high internal breeding. However, the atom ratio, as well as the corresponding core breeding ratio, tends to be a dependent variable rather than a

TABLE 7.16

Energy Costs for a Reference 1000 Mw(e) Liquid Metal Fast Breeder Reactor Core

		Mills/kw-hr
Plant capital cost [$400/kw(e)]		8.00
Fuel cycle*		
Fabrication ($300/kg)		0.52
Capitalization (14%)		0.06
Inventory		0.48
Shipping and reprocessing		0.30
Losses		0.02
Plutonium credit		(0.41)
	Total fuel cycle	0.97
	Total energy costs	8.97†

*Plant factor of 90%; 100,000 Mwd/tonne; plutonium; $10/g Pu.

†Excluding operation and maintenance costs.

primary design objective since the designer is also concerned with reactor-size configuration from the heat-removal viewpoint, coolant characteristics, many cost contributions, and perhaps various safety requirements. For example, what might at first appear to be a desirable design direction toward a higher breeding ratio could bring an undesirable increase in inventory requirements should a significant increase in core volume be required. Similarly, other specifications might be shifted in various ways to obtain a practical design, depending on the interplays involved.[39] Some of these considerations are discussed in Chap. 8.

FUEL-DESIGN PARAMETERS

7.154 Parameters affecting the design of fast reactor fuels are similar to those for thermal reactor elements. The very much higher enrichment required for the fast system, however, does introduce some differences. As for thermal reactors, the fuel-inventory cost depends on specific power, whereas the total processing cost (fabrication and reprocessing) is inversely proportional to the plant thermal efficiency and to the burnup level per unit of fuel. For fast reactor fuels, however, the value of the fissile-atom inventory is high, corresponding to completely enriched ^{235}U rather than to the fissile atoms in slightly enriched fuel in thermal reactors. The fuel cost is therefore more sensitive to the specific power. Since specific power should be expressed as the power per unit amount of fuel in the complete fuel cycle, not just in the reactor core, there is a strong incentive to reduce the out-of-core inventory. This relation is shown in Fig. 7.28 for some typical cases.[40] Here the inventory component of fuel-cycle cost is plotted as a function of core specific power on both an in-core and a total fuel

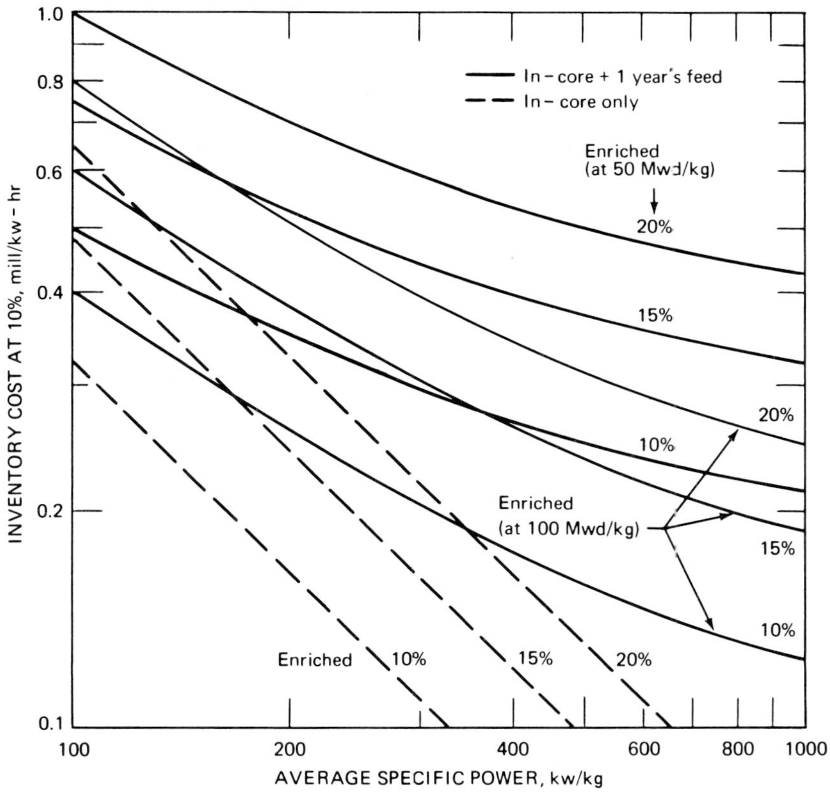

Fig. 7.28 Inventory cost relations to specific power, burnup, and fissionable-fuel concentration. (Fissionable-isotope value is \$10 per gram, and fixed charges are 10% per year; the core loading plus a 1-year throughput are included for the curves with burnup level given).

basis. The quantity in processing and in storage depends on the processing time required and the rate of fuel use. In Fig. 7.28, for burnup levels of 50 and 100 Mwd/kg, a 1-year fuel supply is assumed "out of core." Such graphs are intended primarily to show only trends since the economic assumptions used change significantly with time.

7.155 The interrelations among specific power, burnup, and total processing cost are indicated also in Fig. 7.29. Here total processing cost is the sum of the fabrication and reprocessing charge plus the inventory cost. The numerical values are not significant in the figure since they change as input assumptions are changed. A characteristic of all the curves, however, is to increase rapidly and then to flatten with increasing specific power. Fuel-cycle cost can therefore be significantly reduced by increasing average specific power out at least to the plateau region beyond 300 kw/kg. The occurrence of the plateau region is

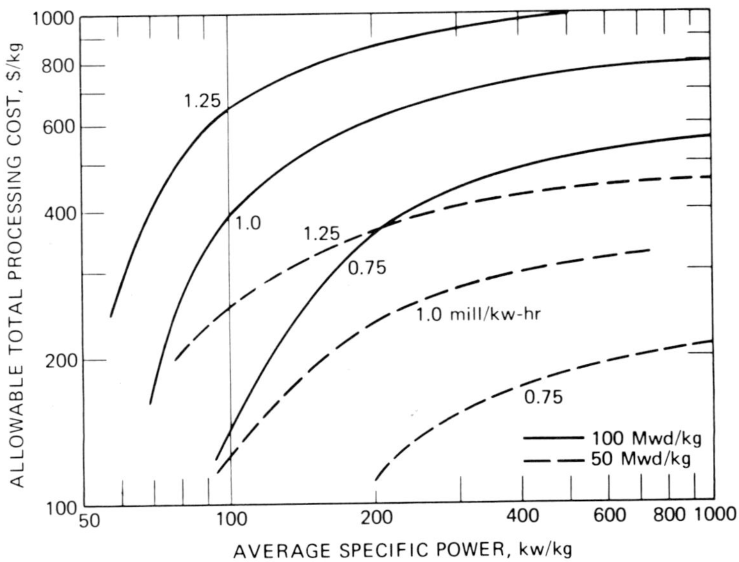

Fig. 7.29 Allowable processing cost for selected fuel-cycle-cost objectives. (All curves are for 15% fissile concentration, $10 per gram value for fissionable isotopes, and 10% per year fixed charges. Curves are for cost of processing plus inventory charges; fuel-cycle costs are less by amount of plutonium credit).

fortunate since, as a result of flux variations, fuel depletion, control effects, etc., part of the fuel in the reactor has a specific power approximately twice the average, and fuel elements must be designed to operate at the maximum.

7.156 This behavior is a result of the reduced relative contribution of inventory charges per kilogram once the specific power becomes reasonably high. An incentive still exists, however, to obtain maximum *exposure* from all the fuel by appropriate fuel management (§7.162).

FUEL REQUIREMENTS

7.157 Most conceptual designs for large fast reactors specify a ceramic fuel, either an oxide or a carbide. Although such fuels may also be used in thermal reactors, some important differences exist. In the oxide, for example, the properties of a binary UO_2–PuO_2 solid solution differ significantly from those of UO_2 itself. Furthermore, the plutonium presence tends to increase both fabrication and processing costs so that a high specific power and large burnup are design requirements as listed in Table 7.15. High power densities produced by the desired high specific power, together with the accompanying large fast-neutron flux and very high heat flux, can introduce material problems.

Mixed-Oxide Fuel

7.158 As for any reactor, the fast reactor fuel designer must know the characteristics of the material and must find ways to solve a host of problems. For example, questions of thermal conductivity, fission-product redistribution, possible diffusion and segregation of plutonium within the fuel as burnup proceeds, interactions with the sodium, and general irradiation effects must all be considered (§4.30).

7.159 An example of the interplay between nuclear, thermal, and materials effects is the possible problem of slumping of molten mixed-oxide fuel in the central region of a fuel pin. Central melting has been observed when the linear heat rate is in the range of 22 to 25 kw/ft. In this case axial transport has been noted even when the melting point has been exceeded for only a very short time. Such transport may have some undesirable reactivity-feedback contributions.[39]

Carbide and Nitride Fuels

7.160 Compared with oxides, carbides have attractive material properties, particularly the thermal conductivity [11.0 Btu/(hr)(sq ft)(°F/ft)], the fission-gas retention, and the fuel density. As a result the rod diameter can be larger to save in fabrication charges, and perhaps also the central fuel temperature can be lower with somewhat fewer problems in fission-gas release and fuel redistribution. A trade-off is involved, of course, in arriving at the proper combination of design specifications. Generally, carbides give a breeding-ratio advantage of about 0.2 compared with oxides. Disadvantages peculiar to carbides include compatibility limitations with claddings and a special requirement for handling in low-oxygen and low-moisture atmospheres. Carbide technology is also not so highly developed as that for oxide fuels.

7.161 Nitrides, which have received very little development, appear to have a favorable combination of physical properties. They also offer certain physics advantages compared with oxide and carbide fuels, primarily a smaller increase in reactivity when sodium is voided because of a greater (n,p) and (n,α) reaction. Many properties, however, are not well known.

Fuel-Burnup Effects[41]

7.162 Internal pressure generated within the pin by fission-gas release from large burnups can be limited by providing a plenum either at the top or at the bottom of the pin or by controlled venting to the coolant. An additional effect, the swelling of the fuel itself and resulting stress on the cladding, is normally controlled by increasing the porosity of the fuel. For example, mixed-oxide solid pellets having about 85% theoretical density may be used. An economic penalty of uncertain magnitude results from such fuel-density reduction, however.

7.163 A major fuel-design challenge is irradiation-induced swelling and creep in stainless-steel cladding and core structural materials. Although theories of swelling are still under development, the growth appears to depend on stress and temperature as well as fluence. The growth also varies among different stainless-steel alloys. For example, reduced swelling, of the order of 5% volume change, has been noted with cold-worked type 316 stainless steel compared with about a 10% change with solution-treated type 304 stainless steel under similar conditions. However, some nickel alloys show promise of much greater stability and thus may provide a means of controlling this effect.

7.164 Since stainless-steel swelling affects not only the fuel element but also the fuel-element support and the core structure, the impact on design is complex. For example, differential elongation could result in assembly bowing, an effect likely to increase reactivity. Various design strategies to accommodate metal swelling must therefore be considered. Spacing between pins and assemblies can be widened, for example, and spacer pads located in low-fluence regions. Reduction in cladding temperature and fuel burnup are other possibilities. Such strategies tend to increase capital cost and possibly also the fuel cost, particularly in a sodium-cooled core. Larger gaps to accommodate growth increase core diameter and the sodium-to-fuel ratio, for example. A higher sodium fraction tends to reduce the breeding ratio. However, a growth in cladding diameter would provide space for fuel swelling, permitting the use of a higher density fuel, thus improving breeding characteristics. The design picture is therefore complicated.

7.165 As an aid to the designer, a number of fuel-performance codes have been developed. Such models also provide a basis for correlating performance data. Two general types of performance codes are used. In one type, such as OLYMPUS,[42] empirical relations are used to describe fuel behavior in terms of observable quantities. In the second category theoretical models of the various microscopic processes that control fuel behavior are used. LIFE[43] is such a code in which models are combined that describe fuel restructuring (§4.42), migration of fuel constituents, fuel swelling from fission products, and cladding swelling. Results obtainable include fuel-temperature distribution, cladding dimensions, fuel-constituent distribution, and regional deformations, all as a function of time and axial position. Each type of code has its own area of application. For example, the simpler empirical codes are very useful for conceptual studies and data correlation, whereas the more complex theoretical codes may be used for detailed design studies.

FAST REACTOR CORE AND BLANKET MANAGEMENT

7.166 The objectives in fast reactor core management are similar to those for a thermal reactor. Uniform power distribution and uniform burnup are desired. At the same time, as many fuel elements as possible should be subject to the maximum exposure. If zones of different fuel enrichment are specified, avoiding power peaks at the zone boundaries is a challenge.

7.167 The possibility of breeding in the core or at least the generation of significant fissile plutonium during irradiation means that reactivity and the flux pattern need not change greatly during burnup. Therefore management objectives are somewhat more easily met than for a thermal reactor. Fast reactor core management has not been studied extensively, however.

7.168 Blanket management is a problem unique to fast reactors. Variables include the residence time in the blanket of the plutonium produced, the power-peaking patterns due to the plutonium growth, and the shutdown schedule for any desired blanket loading, shuffling, and fuel removal. However, since the axial-blanket rods are normally an integral part of the core fuel assemblies, the axial blanket is generally managed in the same manner as the core. The radial blanket is tightly packed; assemblies of depleted uranium oxide fuel rods of larger diameter than that of rods in the core are used. Residence time may also be much longer than that of core assemblies. A steep neutron-flux gradient results in uneven plutonium production and power generation among the rods in a given assembly. There is also a marked difference in average neutron flux between the assemblies closest to the core and those in the next outer radial row. Radial-blanket fuel-management proposals therefore generally provide for *assembly rotation* to improve the uniformity of plutonium production and power distribution per assembly and *shuffling* of assemblies to smooth blanket power distribution. Although various schemes are under study, an *in–out* strategy, in which new assemblies are first loaded into the inner row and then shuffled outward, shows promise.

7.169 In both core- and blanket-management studies the necessary physics calculations are difficult. In a multigroup treatment (§5.65) that includes regions having markedly different isotopic concentrations that change with time, different spectral effects occur. These effects, in turn, require that proper cross sections be generated. Reaction rates in the so-called "transition region" near the core–blanket interface where there is a marked change in neutron spectrum are particularly difficult to describe accurately. Such reaction rates are necessary to develop a model of the growth of bred plutonium in the region where most of the blanket breeding takes place, which is near the core interface.

FAST REACTOR PROCESSING OPERATIONS

7.170 The processing steps for fast reactor fuels are generally the same as those for thermal reactor fuels of the same type (ceramic, metal, etc.). However, a few differences are of interest to the reactor designer.

Fabrication Parameters

7.171 Ceramic fuel pins for fast reactors can be fabricated in much the same way as thermal reactor elements. An important difference, however, arises from the use of plutonium, particularly if the plutonium has been recycled. Higher isotopes of plutonium build up during irradiation. Thus reprocessed plutonium contains significantly active substances that yield neutrons, X rays, and a range of gamma rays. Therefore hand operations must be minimized and shielding provided. In addition, rigid containment is required to prevent ingestion of the highly toxic material. Since fast reactor fuels are much more highly enriched than thermal reactor fuels, the quantities handled must be more carefully controlled to avoid accidental criticality. Plutonium fabrication operations are therefore more expensive than comparable procedures with uranium fuels.

7.172 Another important difference is in the nature of the high-power-density core normally required for a fast reactor. The fuel pins are of small diameter, generally about 0.25 in., and are very close to each other. Fabrication costs per kilogram therefore tend to be high, because of both the small diameter and the close dimensional tolerances required. Fabrication costs of about $300/kg for core and axial blanket are projected for the time when fuel will be fabricated at economical large throughputs, probably in the 1980s. The elements in the radial blanket are, of course, larger in diameter and do not contain plutonium. Fabrication costs of $70/kg have been projected for such elements.[44] A rather wide range of uncertainty exists for such estimates, however, amounting to as much as 50% of the preceding values, depending on the degree of optimism of the estimator.

Reprocessing Parameters

7.173 In reprocessing, fast reactor fuels differ from thermal reactor fuels primarily in that they are subject to much higher burnup and they contain a higher fraction of fissile atoms. Although most parameters for processing do not greatly affect the design of the rest of the system, a discussion of the operations and associated parameters is pertinent here.[45]

7.174 A solvent-extraction process such as Purex, modified to a certain extent to handle fast reactor fuels, can be considered the "traditional" fast reactor purification process. The desirability of reducing out-of-pile process time, and hence out-of-pile inventory, however, has caused alternate processes to be considered. If recovered fuel is recycled to fast reactors, modest decontamination factors are adequate since small fractions of fission products remaining in the fuel do not affect the neutron balance in a fast reactor spectrum. Pyrochemical processes, for example, are of great interest in this category and have received much recent attention. Fluoride volatility, another nonaqueous process applicable to fast reactor systems, shares some of the advantages of the pyrochemical methods but has some problems, such as the recovery of the plutonium.[45] Process details are not considered here.

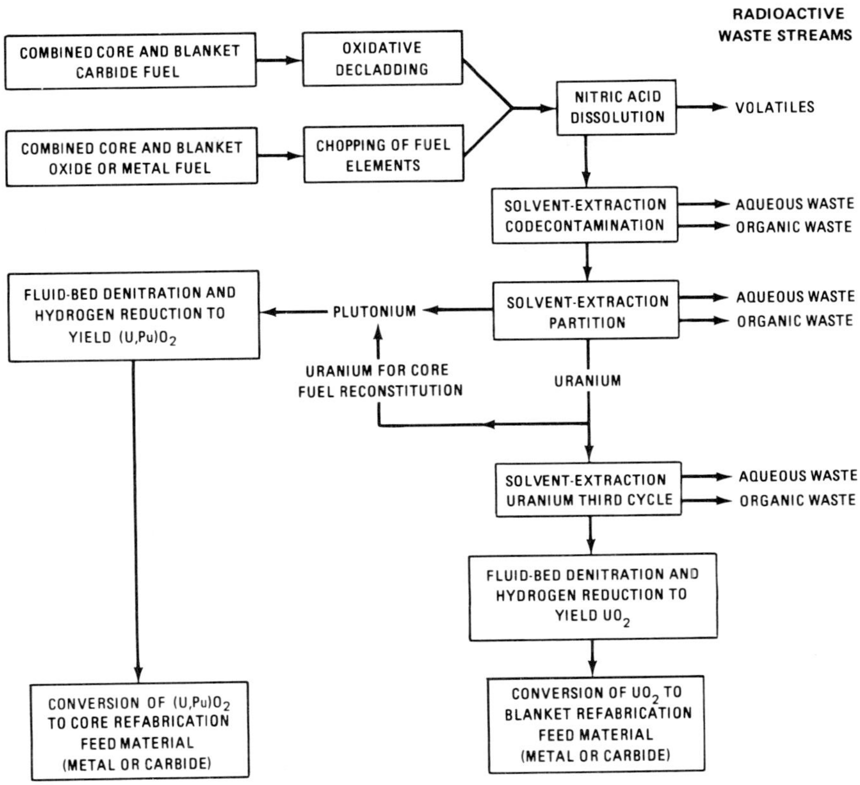

Fig. 7.30 Principal steps in proposed advanced aqueous process for oxide, carbide, and metal fuels.

Aqueous Processes

7.175 The aqueous processes used for fast reactor fuels are essentially the same as those normally used for thermal reactors, as shown in Fig. 7.10. A modified process intended for fast reactor fuels is outlined in Fig. 7.30. Generally, both core and blanket are processed simultaneously to reduce the fissile-element concentration. Carbide fuels must first be converted to oxide as part of the head-end treatment. Since the fissile-atom concentration is still quite high, however, the actinide concentration in the feed solution is normally much lower than that in conventional aqueous process streams. In the step separating uranium and plutonium, uranous ion may be used as a reductant instead of the usual ferrous ion, which would lead to unacceptably high values of iron contamination in the products.

7.176 In the proposal shown in Fig. 7.30, the aqueous products of solvent extraction are converted to oxides by fluid-bed denitration. A dense granular material suitable for fabrication into fuel elements by vibratory compaction results. A disadvantage of the aqueous process is that a substantial amount of low-level liquid waste is produced. Since a degree of remote fabrication is required because of the hazard from actinide isotopes in recycled plutonium, multistage decontamination of the plutonium product, even for the aqueous process, is not really justified. The uranium product, however, can be refabricated by direct-contact methods provided some cooling is carried out.

Pyrochemical Processes

7.177 One example of several pyrochemical processes under development is the salt-transfer process shown schematically in Figs. 7.31 and 7.32. Salt-transport separations depend on the selective transfer of solutes between two liquid-metal solutions equilibrated with the same molten salt. The salt, in effect, serves only as a carrier for the solute transfer between the liquid-metal phases. For rapid transfer of the solute, the salt phase is circulated between the two liquid-metal phases.

7.178 This process has the advantages of requiring a comparatively short cooling period (about 15 days) and of using reagents and solvents not subject to radiation damage. Furthermore, unlike in the aqueous solvent-extraction processes, no large volumes of aqueous-waste materials are produced. Also, there are no neutron-moderating materials, and criticality safety design requirements

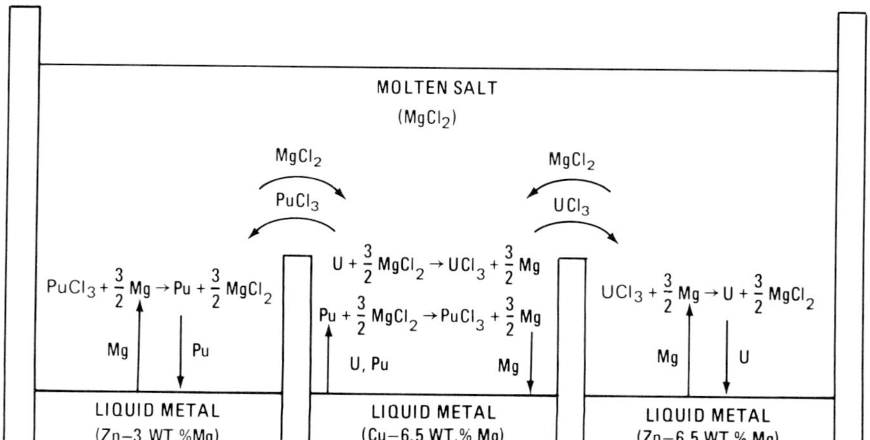

Fig. 7.31 Salt-transport separation of uranium and plutonium.

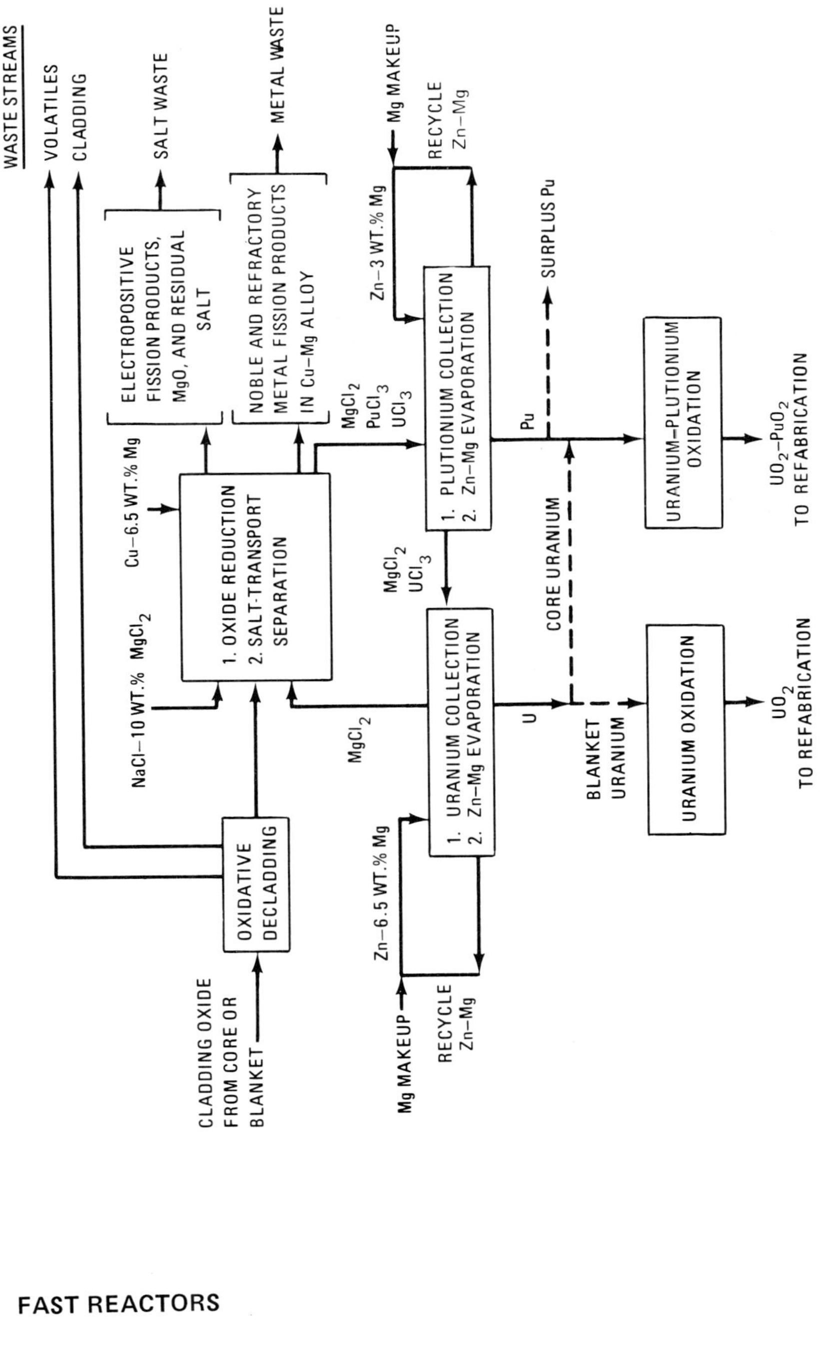

Fig. 7.32 Conceptual salt-transport process.

can thus be relaxed. The equipment is compact, and the radioactive-waste problem is simplified since many of the wastes are collected as solids. On the other hand, only low decontamination is possible, and severe process conditions cause problems in container-material behavior.

7.179 Cost studies have been carried out to identify problem areas and to determine any relative process advantages. In one study[45] three hypothetical plant situations were examined.

1. An on-site processing plant serving a single 1000-Mw(e) reactor plant.
2. A single facility servicing three reactors totaling 3000 Mw(e) installed on the same site in a reactor "park."
3. A central facility servicing reactors having a total capacity of 10,000 Mw(e).

7.180 Estimates are summarized in Table 7.17. There is a saving in shipping expenses for the on-site models. As expected, unit costs tend to be higher in the

TABLE 7.17

Unit-Cost Summary for Processing Fast Reactor Oxide Fuel*

Cost component		Aqueous	Volatility	Pyrochemical
Close-Coupled 1000-Mw(e) Plants				
Processing		0.53	0.35	0.44
Waste disposal		0.15	0.09	0.09
Inventory		0.17	0.12	0.12
Shipping				
1% loss		0.016	0.016	0.016
	Total	0.87	0.58	0.67
Close-Coupled 3000-Mw(e) Plants				
Processing		0.29	0.21	0.25
Waste disposal		0.12	0.08	0.08
Inventory		0.11	0.06	0.06
Shipping				
1% loss		0.016	0.016	0.016
	Total	0.54	0.37	0.41
Central 10,000-Mw(e) Plants				
Processing		0.16	0.14	0.18
Waste disposal		0.10	0.08	0.07
Inventory		0.109	0.079	0.079
Shipping		0.050	0.066	0.071
1% loss		0.016	0.016	0.016
	Total	0.44	0.38	0.42

*All costs in mills/kw-hr(e); plant factor of 80%; 100,000 Mwd/tonne burnup.

smaller plants. A conclusion from the study is that on-site, nonaqueous, reprocessing plants requiring relatively small amounts of risk capital are probably capable of processing spent fast reactor fuel at unit costs very little, if any, higher than costs for larger central plants regardless of the process used in the central plant.

THORIUM FUEL SYSTEMS

INTRODUCTION

7.181 Although considerable emphasis is currently given to the uranium–plutonium fuel system, the Th–^{233}U system provides an alternate path toward a solution to present energy requirements. The most significant advantage of the Th–^{233}U cycle over the ^{238}U–^{239}Pu cycle in thermal reactors is the potential of a higher conversion ratio (CR) (§7.115). In a reactor CR units of bred fuel are produced for each unit of fuel consumed, and the net consumption of nuclear fuel is therefore proportional to 1 – CR. Hence, other things being equal, a typical light-water reactor with a conversion ratio of 0.6 would consume twice as much fuel per unit energy developed as a Th–^{233}U reactor having a conversion ratio of 0.8. Therefore the higher conversion ratio leads directly to a lower depletion charge in the fuel-cycle cost.[46]

7.182 There are similarities and differences in the two isotopic chains, as shown in the simplified comparison in Fig. 7.33. A major difference is the intermediate formation of ^{233}Pa, a strong neutron absorber, which decays slowly. Since the decay of ^{233}Pa yields fissile ^{233}U, reactivity control during shutdown is complicated. In the thermal-neutron spectrum, ^{233}U has a much lower alpha (σ_c/σ_f) and therefore a higher eta [$\nu/(1 + \alpha)$] than ^{239}Pu. This difference in eta, however, is not as large as the difference in alpha might suggest, because ^{239}Pu has a larger ν, or a larger number of neutrons produced per fission. In the thermal and epithermal energy regions, ^{233}U has the potential for a higher conversion ratio than obtainable through the use of plutonium or ^{235}U averaged over a burnup period. Therefore reactivity lifetimes can be longer, and a greater return of fissile material at the end of the fuel cycle may be possible.

7.183 In fast reactors the ^{232}Th–^{233}U cycle can provide a maximum breeding ratio of about 1.20, smaller than that for the plutonium system (~1.60) but still appreciable.[47] However, the practical difference may not be so great since the breeding ratio for large plutonium–uranium oxide systems is about 1.35 whereas thorium systems with breeding ratios close to the maximum may be possible. Thorium metal as a blanket material would be possible since it has greater resistance to radiation damage than uranium.

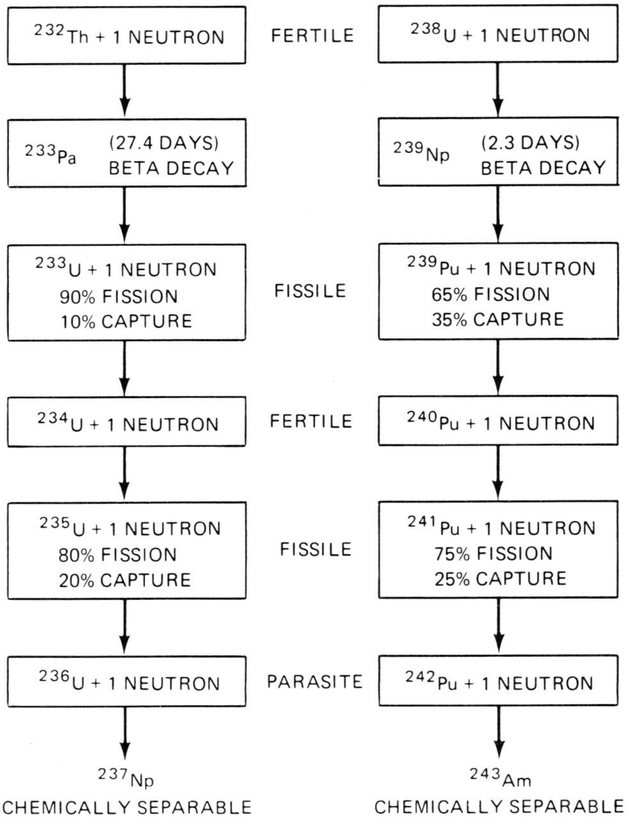

Fig. 7.33 Isotopic buildup in Th–^{233}U and ^{238}U–^{239}Pu systems.

7.184 Thorium has a much smaller fission cross section and a higher threshold energy for fission than ^{238}U. Consequently thorium provides very little fast-fission bonus to the breeding ratio. However, it also contributes a smaller reactivity gain owing to increased fast fission on loss of sodium; hence it should tend to give a less positive sodium-void coefficient than ^{238}U–Pu. Interest in using the thorium–uranium cycle for fast reactors, therefore, is due primarily to favorable safety characteristics. Some work has also been done with mixed ^{233}U and Pu systems in an attempt to combine some of the advantages of each.[47]

FUEL-PROCESSING PROBLEMS

7.185 In thorium systems the radioactivity of daughter products of ^{232}U in recovered thorium and in ^{233}U after the removal of fission products is a challenging problem. In less than a week after high-level decontamination, the gamma activity becomes sufficiently great that fabrication by direct methods can be permitted only on a scheduled-radiation-dosage basis. The magnitude of this problem is directly related to the ^{232}U-concentration buildup that occurs throughout the exposure lifetime of the fuel. This, in turn, is a function of the integrated fuel exposure, including the neutron energy level incident upon the fuel materials since the principal reaction that produces ^{232}U, the ^{232}Th$(n,2n)^{231}$Th reaction, does not occur with neutrons with an energy below 6 Mev. A highly divided fuel in a well-thermalized spectrum will therefore have a lower ^{232}U concentration than a fuel in an epithermal core.

7.186 The recovered fuel can be dealt with in several ways. The thorium product can be stored for a long time to allow decay of ^{228}Th (1.91-year half-life). Uranium could be chemically purified just before the fabrication step. Fabrication could then be done rapidly by using nonremote methods; complicated processes and materials-handling steps would thus be avoided. Such a philosophy was the original basis for the development of the sol-gel process. An alternative is remote fabrication, for which decontamination factors for fission-product removal of 10^2 to 10^3 are acceptable compared with factors of 10^6 to 10^8 for direct methods. This lower level of decontamination results in fission-product activity in the product comparable to that from the uranium daughters in ^{233}U.

7.187 Although this discussion is concerned primarily with fuel cycles, reactor concepts are mentioned as a guide to fuel use. Two interesting examples are the high-temperature gas-cooled reactor (HTGR) and the molten-salt breeder reactor (MSBR). The HTGR is one of several advanced converter-type reactors (DRAGON etc.) being developed. General design features of these concepts are available elsewhere.[48,49]

EXAMPLES OF THORIUM REACTOR CONCEPTS

High-Temperature Gas-Cooled Reactor

7.188 The HTGR is being actively developed as an alternate to light-water reactors in meeting the needs for economical nuclear power. A key factor is the possibility of very low fuel-cycle costs obtained by a high conversion ratio through the use of the Th–^{233}U system, which would also extend available energy resources.

7.189 A typical HTGR core is composed of hexagonal-shape blocks of graphite stacked in a close-packed array of vertical columns. Within the blocks are interspaced both vertical fuel rods, located within fuel channels, and coolant channels, as shown in Fig. 7.34. The fuel consists of two different types of small pyrolytic graphite-coated spheres. One type particle contains fertile material, the other type fissile. Each is of different size to facilitate separation during

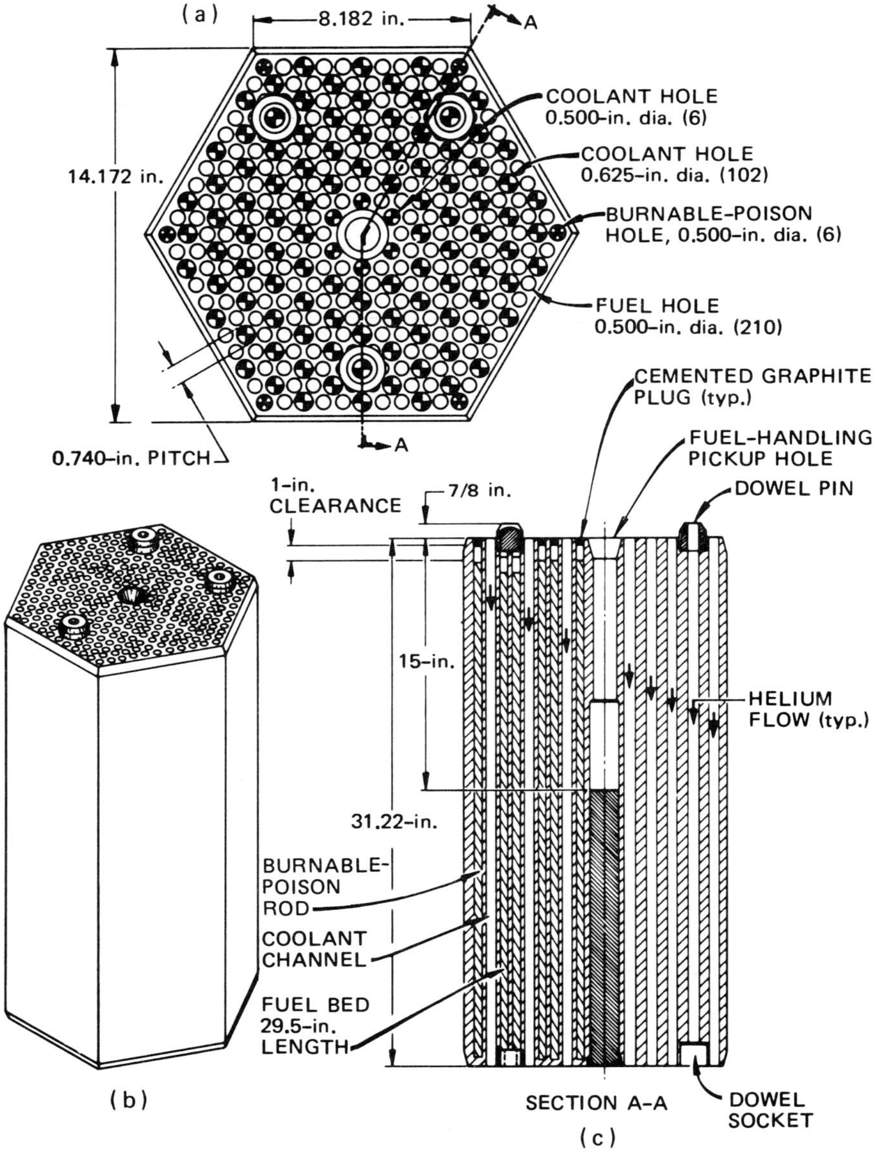

Fig. 7.34 Typical high-temperature gas-cooled reactor fuel element.

processing. Typically a fertile particle has an inner kernel consisting of thorium carbide and the recycled uranium (mostly ^{233}U) carbide. This kernel, about 350 μ in diameter, has a multilayered impervious graphite coating about 130 μ thick. The fissile particle containing the makeup uranium carbide, fully enriched in ^{235}U, has a kernel diameter of about 150 μ and a coating thickness of about 150 μ. Close-packed beds of these particles are bonded together with low-density carbon into short rods typically 0.5 in. in diameter and 2 in. long. The rods are stacked end-to-end in the fuel channels of the hexagonal-block elements to make up the required fuel-column height.

7.190 Because of the size difference, the two types of particles are physically separable. Each type can then be used in a number of different fuel-cycling options. Several modes that have been studied are as follows:

1. Nonrecycle, where the reactor is fed with ^{235}U and the discharged uranium is sold. Makeup ^{235}U is kept separate from thorium and bred uranium.
2. Full recycle, where all the recovered uranium is recycled with ^{235}U makeup. Uranium and thorium are mixed together in one particle.
3. Type I segregation, where only bred uranium from the thorium particle is recycled. The recycle uranium and makeup ^{235}U are combined in a particle separate from the thorium particle. Material recovered from this uranium particle after irradiation is sold or discarded and is not recycled.
4. Type II segregation, where makeup ^{235}U is kept separate from thorium and recycle uranium and its residue is not recycled. Bred uranium is recycled back into the thorium particle, and no bred material is sold.

To accomplish some of these fuel-cycle schemes, several additional types of particles than those mentioned would be required. Such modes require evaluation to determine the best way to establish a breeder economy using the thorium fuel cycle, and present experience is too limited to justify conclusions on the merits of a specific approach.

7.191 A number of methods, generally involving either powder-metallurgical or chemical techniques, have been developed for manufacturing the fuel microspheres. Chemical methods normally are based on the sol-gel process (§7.28), in which oxide microspheres are first produced and then converted to carbide by sintering at high temperature in a graphite bed. Pyrolytic coatings are deposited from either methane or acetylene in a fluidized bed. Much research and development have been devoted to the various processes and to the study of how the many parameters affect the quality of the product. The subject is quite complex and will not be considered here.

7.192 Fuel-reprocessing operations are similar in principle to those used for other fuels. However, the head-end operations must recover the fuel particles intact from the graphite blocks, and a size-classification operation is included to separate the two types of particles used. Actinides are normally decontaminated and separated by solvent extraction using the acid Thorex process, which is similar in principle to the Purex process (§7.37).

Molten-Salt Breeder Reactor

7.193 The Th–^{233}U system does provide a potential for breeding at thermal-neutron energies. One concept* features the dissolution of both fuel and fertile material in a molten-salt fluoride system with the important advantage of continuous fuel loading and reprocessing. An additional feature is the favorable control characteristics inherent in a liquid-fuel reactor system with a high thermal-expansion coefficient that ensures a fast negative reactivity coefficient.

7.194 An important fuel-economy advantage is the low fissile-material inventory both in and out of the core. No fuel fabrication is necessary, and on-site reprocessing requires a minimum quantity of fuel in the out-of-core fuel "pipeline." Thus the doubling time can be about 12 to 15 years, although a breeding ratio of only about 1.07 is expected. The system is quite sensitive to reprocessing efficiency since the loss of only a small amount of fissile material would seriously affect the breeding potential of the concept.

7.195 The conceptual development of the molten-salt breeder reactor has proceeded through a number of stages. An early reference design consisting of a two-fluid core–blanket approach has been superseded by a one-fluid concept that avoids materials problems with graphite, which separated the two fluids. Through core zoning, the one-fluid concept can give almost the same potential performance as the two-fluid design. Important in the new design was theoretical and preliminary experimental evidence that the ^{233}U and possibly rare-earth fission products could be separated from a mixed thorium–uranium fuel salt by reduction extraction with liquid bismuth.

7.196 The fuel for the one-fluid breeder consists of both fissile uranium and fertile thorium as tetrafluorides dissolved in a lithium fluoride–beryllium fluoride carrier salt. This fuel salt is distributed in a core consisting of two zones: a well-moderated inner zone (I) surrounded by an undermoderated outer zone (II). The neutron spectrum in each zone is controlled by adjusting the proportion of salt to graphite, from a salt fraction of about 13% in Zone I to about 37% in Zone II. The overall spectrum is adjusted for the best performance associated with a high breeding ratio and a low fissile inventory. For example, the spectrum in Zone II can be made harder to enhance the rate of thorium-resonance capture relative to the fission rate; thus the flux in that zone is depressed and the neutron leakage reduced.

7.197 The main objectives of fuel processing in the molten-salt breeder reactor are to isolate ^{233}Pa from regions of high-neutron flux during its decay to ^{233}U and to remove fission products from the system. Impurities that may arise from corrosion or maloperation of the reactor system must also be removed

*USAEC research and development on the MSBR was curtailed in 1973 owing to budget limitations. Future commercial application is therefore doubtful.

from the reactor fuel salt. Since the fuel-processing system is an integral part of the reactor system and operates continuously, only a small out-of-pile inventory of fissile material is needed. In most designs the reactor can continue to operate even if the processing facility is shut down but at a gradual decrease in nuclear performance as the poisons accumulate.

7.198 During development of the molten-salt-breeder-reactor concept, several different purification processes appeared to accomplish about the same degree of separation. Development emphasis, therefore, has been given to a system that appears to be simpler than the other candidate processes. This process, shown in Fig. 7.35, consists of two parts: (1) removal of uranium and protactinium from salt leaving the reactor and reintroduction of uranium into salt returning to the reactor and (2) removal of rare-earth fission products from the salt.

7.199 A small stream of fuel salt, taken from the reactor drain tank, flows through a fluorinator, where about 95% of the uranium is removed as gaseous UF_6. The salt then flows to a reductive extraction column, where protactinium and the remaining uranium are chemically reduced and extracted into liquid bismuth containing a reducing agent, lithium, flowing counter-current to the salt. Leaving the column, the bismuth stream contains the extracted uranium and protactinium as well as lithium, thorium, and fission-product zirconium. The extracted materials are then removed from the bismuth stream by contacting it with an $HF-H_2$ mixture in the presence of a waste salt that is circulated through the hydrofluorinator from the protactinium decay tank. The salt stream leaving the hydrofluorinator, which contains UF_4 and PaF_4, passes through a fluorinator, where about 95% of the remaining uranium is removed. Protactinium is allowed to decay in a hold-up tank, and the uranium formed thereby is recovered by additional fluorination.

7.200 The rare earths are removed from the salt stream leaving the top of the protactinium extractor by contacting the salt stream with a stream of bismuth that is practically saturated with thorium metal. This bismuth stream, with the extracted rare earths, is then contacted with an "acceptor salt," lithium chloride. Because the distribution coefficient (metal/salt) is several orders of magnitude higher for thorium than for the rare earths, a large fraction of the rare earths transfer to the lithium chloride in this contactor, whereas the thorium remains with the bismuth. Finally, the rare earths are removed from the recirculating lithium chloride by contacting it with bismuth streams containing high concentrations of lithium.

7.201 The fully processed salt, on its way back to the reactor, has uranium added at the rate required to maintain or adjust the uranium concentration in the reactor (and hence the reactivity) as desired. This is done by contacting the salt with UF_6 and hydrogen to produce UF_4 in the salt and HF gas.

7.202 This process is included as part of a 1000-Mw(e) conceptual design study for a single-fluid molten-salt breeder reactor.[50] Although a number of

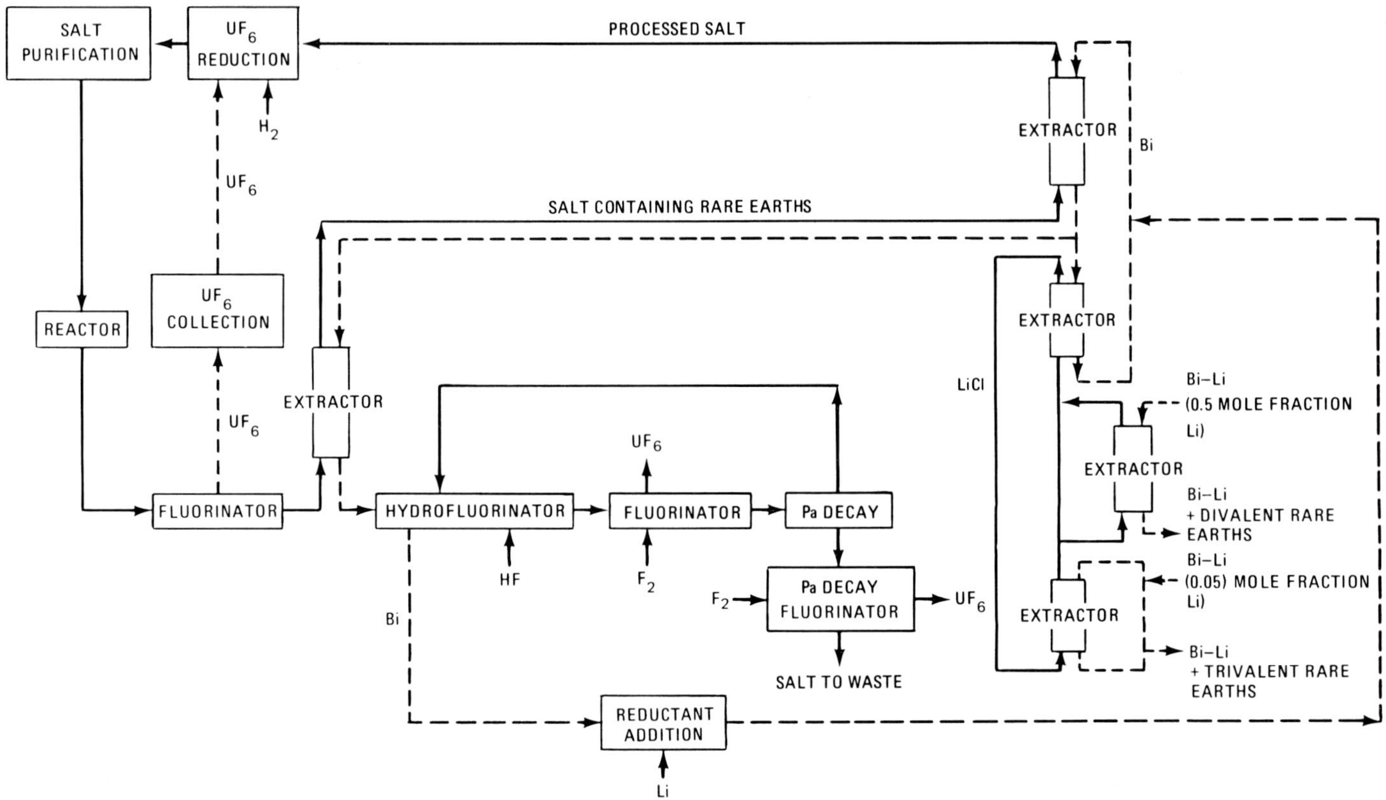

Fig. 7.35 Proposed salt-processing system for molten-salt breeder reactor.

uncertainties exist, the concept was considered technically feasible and economically attractive.

REFERENCES

1. P. J. Fulford and A. Sesonske, In-Core Fuel Management Optimization, *Nucl. News*, **13**(7): 58 (1970).
2. Manson Benedict and T. H. Pigford, *Nuclear Chemical Engineering*, McGraw-Hill Book Company, Inc., New York, 1957.
3. U. S. Atomic Energy Commission, AEC Gaseous Diffusion Plant Operations, USAEC Report ORO-658, Washington, D. C., 1968.
4. U. S. Atomic Energy Commission, Selected Background Information on Uranium Enriching, USAEC Report ORO-668, Washington, D. C., 1969.
5. S. M. Stoller and R. B. Richards (Eds.), *Fuel Processing*, Vol. 2, *Reactor Handbook*, Interscience Publishers, New York, 1961.
6. *Fuel Element Fabrication*, Proceedings of IAEA Symposium, Vienna, 1960, 2 vols., Academic Press, Inc., New York, 1961.
7. A. R. Kaufmann (Ed.), *Nuclear Reactor Fuel Elements*, Interscience Publishers, New York, 1962.
8. T. J. Burke and J. Glatter, Fabrication of Uranium Oxide in *Uranium Dioxide: Properties and Nuclear Applications*, J. Belle (Ed.), Superintendent of Documents, U. S. Government Printing Office, Washington, D. C., 1961.
9. M. W. Rosenthal et al., A Comparative Evaluation of Advanced Converters, USAEC Report ORNL-3686, Oak Ridge National Laboratory, January 1965.
10. C. H. Bloomster, Fabrication Costs for Plutonium Fuel Elements, Part E: The Variation in Fabrication Costs with Some Fuel Design Parameters, USAEC Report BNWL-131E, Battelle–Northwest, February 1967.
11. J. B. Burnham, L. G. Merker, and D. E. Deonigi, Comparative Costs of Oxide Fuel Elements, USAEC Report BNWL-273(Vol. 1 and App.), Battelle–Northwest, July 1966.
12. Argonne National Laboratory, Chemical Engineering Division Semiannual Report, January–June 1967, USAEC Report ANL-7375, October 1967.
13. Royes Salmon, A Procedure and a Computer Code (POWERCO) for Calculating the Cost of Electricity Produced by Nuclear Power Stations, USAEC Report ORNL-3944, Oak Ridge National Laboratory, June 1966.
14. P. J. Fulford, An Analysis of Nuclear Fuel Management Techniques for Utilities, Ph. D. Thesis, Purdue University, 1968.
15. F. H. Warren and N. B. McLeod, Economic Significance of Nuclear Fuel Management, in *Proceedings of the 1965 Conference of the Atomic Industrial Forum*, Atomic Industrial Forum, Inc., New York, 1965.
16. E. A. Mason, An Overall View of the Nuclear Fuel Cycle, *Nucl. News*, **14**(2): 35 (1971).
17. G. J. Bonk, R. L. Crowther, and O. A. Thompson, Interactions Between Nuclear Plant Loading Changes and Nuclear Fuel Cost, *Nucl. News*, **14**(6): 53 (1971).
18. S. Raje and D. H. Petrie, Fuel Management in a Large Chemical Shim Closed-Cycle Water Reactor, USAEC Report WCAP-3269-53, Westinghouse Electric Corporation, May 1966.
19. Consumers Power Company, Preliminary Description and Safety Analysis Report, Palisades Plant, Jackson, Mich., 1966.
20. Sacramento Municipal Utility District, Final Safety Analysis Report, Rancho-Seco Nuclear Generating Station, Unit 1, Sacramento, Calif., 1971.

21. General Electric Company, 1000-Mw(e) Boiling Water Reactor Plant Feasibiltiy Study, Vols. 1–3, USAEC Report GEAP-4476, February 1964.
22. R. W. Lockhart, Feasibility Study of Boiling Water Reactor Nuclear Steam Supply Systems with Capacities up to 10,000 Mw(t), USAEC Report GECR-5155, General Electric Company, February 1967.
23. U. S. Atomic Energy Commission, Division of Operations Analysis and Forecasting, Forecast of Growth of Nuclear Power, USAEC Reports WASH-1084, December 1967, and WASH-1139, January 1971.
24. U. S. Atomic Energy Commission, The Nuclear Industry–1967, Washington, D. C., 1967.
25. E. L. Zebroski, B. Wolfe, P. Greebler, and J. Sueoka, Plutonium Recycle Values in Thermal and Fast Reactors and Expected Patterns of Use of Plutonium, in Plutonium as a Power Reactor Fuel, Proceedings of American Nuclear Society Topical Meeting, Richland, Washington, Sept. 13–14, 1962, F. G. Dawson (Comp.), USAEC Report HW-75007 (Paper 23), General Electric Company, December 1962.
26. R. L. Crowther, E. A. Evans, S. Levy, and W. V. MacNabb, Plutonium Utilization in Boiling-Water Reactor Power Plants, in *Commercial Plutonium Fuels Conference*, Mar. 1–2, 1966, Washington, D. C., pp. 229–244, American Nuclear Society, 1966.
27. E. A. Eschbach, Plutonium Value Analysis, in *Proceedings of the Third International Conference on the Peaceful Uses of Atomic Energy, Geneva, 1964*, Vol. 11, pp. 47–55, United Nations, New York, 1965.
28. Fred Hittman and Marvin Raber, The Importance of Plutonium Recycle, *Nucl. News*, **11**(11): 48 (1968).
29. F. G. Dawson, Physics Design Aspects of Plutonium Recycle, USAEC Report BNWL-232, Battelle–Northwest, May 1966.
30. Combustion Engineering, Inc., and Atomics International, Heavy Water Organic-Cooled Reactor, Quarterly Technical Progress Report, Jan. 8, 1965–June 30, 1965, USAEC Report AI-CE-3, August 1965.
31. J. R. Dietrich, Efficient Utilization of Nuclear Fuels, *Power Reactor Technol.*, **6**(4): 1–38 (1963).
32. Babcock and Wilcox Company, 1000-MWE LMFBR Follow-on Study, USAEC Report BAW-1328(Vol. 1), February 1969.
33. P. Goldschmidt and K. O. Ott, Letters to the Editor, *Nucl. Appl. Technol.*, **9**:450 (1970).
34. K. O. Ott, An Improved Definition of the Breeding Ratio for Fast Reactors, *Trans. Amer. Nucl. Soc.*, **12**: 719 (1969).
35. J. Bengston, J. M. Ravets, and R. J. Rickert, Lifetime Characteristics and Economics of Slightly Enriched Uranium Dioxide in Water Lattices, USAEC Report CEND-145, Combustion Engineering, Inc., September 1961.
36. R. J. Rickert, Economic Consequences of Design Variations of Water Reactor Fuel Elements, in *Fuel Element Fabrication*, Proceedings of IAEA Symposium, Vienna, 1960, Academic Press, Inc., New York, 1961.
37. W. H. Hannum, Status of Fast Reactor Physics Design, USAEC Report LA-DC-7739, Los Alamos Scientific Laboratory, 1965.
38. C. A. Pursel, D. T. Eggen, and J. H. Wright, Status of Fast Breeder Reactor Core Design, in Commercial Plutonium Fuels Conference, Mar. 1–2, 1966, Washington, D. C., USAEC Report CONF-660308, pp. 262–277, American Nuclear Society, 1966.
39. E. L. Zebroski, J. P. Mustlelier, and C. Caldwell, Oxide Fuels for Fast Reactors, in *Fast Reactor Technology*, National Topical Meeting, Detroit, April 26–29, 1965, American Nuclear Society, ANS-100, 1965.
40. C. W. Wheelock and R. J. Herbst, Comparison of Ceramic Fuel Systems for Fast Breeder Reactor Applications, in Commercial Plutonium Fuels Conference, Mar. 1–2,

1966, Washington, D. C., USAEC Report CONF-660308, pp. 278–304, American Nuclear Society, 1966.

41. P. Murray, Fast Breeder Fuel, *Reactor Technol.*, **15**(1): 16-58 (1972).
42. A. Boltax, A Biancheria, B. L. Harbourne, and T. P. Soffa, Mixed Oxide Fuel Pin Performance Analysis Using the OLYMPUS Computer Code, *Trans. Amer. Nucl. Soc.*, **14**(Suppl. 1): 6 (1971).
43. V. Z. Jankus and R. W. Weeks, LIFE-I, a FORTRAN-IV Computer Code for the Prediction of Fast-Reactor Fuel-Element Behavior, USAEC Report ANL-7736, Argonne National Laboratory, November 1970.
44. U. S. Atomic Energy Commission, Chicago Operations Office, An Evaluation of the Atomics International 1000-Mw(e) Fast Breeder Reactor, USAEC Report COO-285, July 1966.
45. M. Levenson, V. G. Trice, Jr., and W. H. Mecham, Comparative Cost Study of the Processing of Oxide, Carbide, and Metal Fast-Breeder-Reactor Fuels by Aqueous, Volatility, and Pyrochemical Methods, USAEC Report ANL-7137, Argonne National Laboratory, May 1966.
46. U. S. Atomic Energy Commission, Division of Reactor Development and Technology, The Use of Thorium in Nuclear Power Reactors, USAEC Report WASH-1097, 1969.
47. David Okrent, Neutron Physics Considerations in Large Fast Reactors, *Power Reactor Technol.*, 7(2): 107–144 (1964).
48. R. G. Wymer (Coordinator), *Thorium Fuel Cycle*, Gatlinburg, Tenn., May 3–6, 1966, AEC Symposium Series, No. 12 (CONF-660524), 1968.
49. Oak Ridge National Laboratory, Proceedings of Gas-Cooled Reactor Information Meeting at Oak Ridge National Laboratory, April 27–30, 1970, USAEC Report CONF-700401, 1970.
50. R. C. Robertson (Comp.), Conceptual Design Study of a Single-Fluid Molten Salt Breeder Reactor, USAEC Report ORNL-4541, Oak Ridge National Laboratory, June 1971.

8 Design Considerations

INTRODUCTION

8.1 Previous chapters describe some considerations affecting nuclear reactor design and indicate the numerous interplays involved. Although systems engineering provides some guidelines for design approaches, design *methods* for the reactor system are not considered here, primarily because approaches vary and design experience plays a major role. However, as discussed in Chap. 1, design methods are systematic, require establishment of specific goals, and provide procedures for meeting these goals. Of major importance in core design and analysis, parameter interplays are discussed in this chapter, together with a systems engineering design approach for a fuel element.

8.2 The beginning engineer can also develop insight into the importance of parameter interplays through *design studies*. In developing commercial nuclear power, it is useful to assess the potential of advanced concepts as well as proposed modifications of commercially proven systems. For this purpose design studies are frequently prepared which examine the features of a proposed concept in the desired detail with the aid of a so-called reference design. The effects of changing various design parameters are frequently studied. The reference design may also sometimes be used to determine the economic potential of the concept and, if desired, the economic effects of changing design parameters. A picture of the "trade-off" pattern between costs and design parameters may then be useful in selecting design values that yield desired low* power costs.

8.3 Also, the student of engineering design receives valuable source material from studies of this nature. Since illustrations of the interplays between parameters and the possibility for trade-offs between economic and other criteria are particularly important, some features of typical liquid-metal fast breeder reactor (LMFBR) studies are examined here. The sodium-cooled fast breeder reactor was chosen primarily because, as an advanced concept of considerable current interest, it offers much more design challenge than present-generation water-cooled thermal reactors. This is particularly true for parametric studies and design trade-off evaluations, which are a major feature of such studies.

8.4 Evaluation, which is an essential step in the engineering design process, enables the designer to review the effectiveness of a design in comparison with established criteria and judge it as acceptable or unacceptable. As part of the examination, information that will help the designer correct any deficiencies is normally developed. This same approach can be applied to the design of a nuclear plant, considering both small subsystems and the larger system as a whole. Thus critical evaluation of the design concept is not really separate from the design task but is carried out as an integral part of it.

SYSTEMS DESIGN APPROACHES

INTRODUCTION

8.5 Systems engineering design approaches, described in Chap. 1, are applicable to design studies and, in fact, to the design of subsystems. Design requirements must be met within the constraints of materials, safety, and economics, and iterative procedures are necessary because of the many interplays between variables. Since a reactor system is very complex, the depth of detail depends on the objectives of the particular study.

8.6 Iteration is also an integral part of engineering design since it provides for evaluation of a proposed solution in terms of established criteria. As indicated in Fig. 8.1 (which is the same as Fig. 1.1), the iteration feedback loop results from the evaluation and analysis decision. In other words, if the evaluation decision is "yes," the design process proceeds to the next step, which is implementation in some form. If the decision is "no," the problem is fed back through the loop for revision or additional development. Since it is likely that

*The word *minimum* is avoided since such a value can be achieved only through a systematic optimization approach as discussed in Chap. 9.

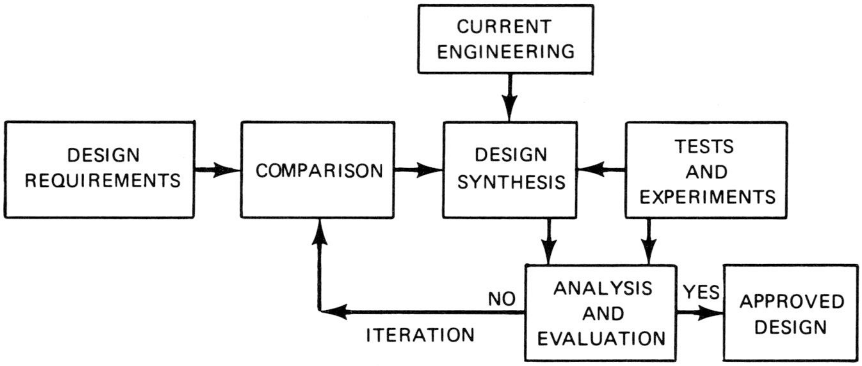

Fig. 8.1 Closed-loop design approach.

new information on deficiencies has been developed as part of the evaluation process, such information provides additional input to the design synthesis step. The new design is again evaluated, with the possibility of additional iteration, until a "yes" decision is achieved. This idealized iteration approach, which lends itself to computer techniques, should be recognized as only a general guide; many modifications are possible to fit it to individual circumstances. For example, information developed during evaluation may indicate that the design is not worthy of revision and should be abandoned.

INTERPLAY OF DESIGN VARIABLES

8.7 Iteration in the design process due to the interplay of variables can be demonstrated by considering the reactor core. Although a number of design requirements apply, we can often simplify the picture by identifying those likely to "control" the design. For example, the ultimate considerations are generally how well the fuel elements perform and whether operating conditions (temperature, stress, etc.) approach the limiting criteria related to failure of the material. In most reactors the *heat-removal* capability from point to point depends on the thermal–hydraulic parameters, and the spatial heat generation depends on the fission rate. As with all nuclear interaction rates, the fission rate, in turn, depends on both the neutron flux and the macroscopic fission cross sections. Therefore the combination of heat removal and heat generation produces a temperature pattern that can be related to limiting criteria.

8.8 Thus the various parameters that contribute to core design generally are not independent but form an interrelated matrix in which the degree of influence of one parameter on another may vary considerably. The core design itself can be considered as the integrated pattern obtained after all individual parametric contributions are brought together systematically.

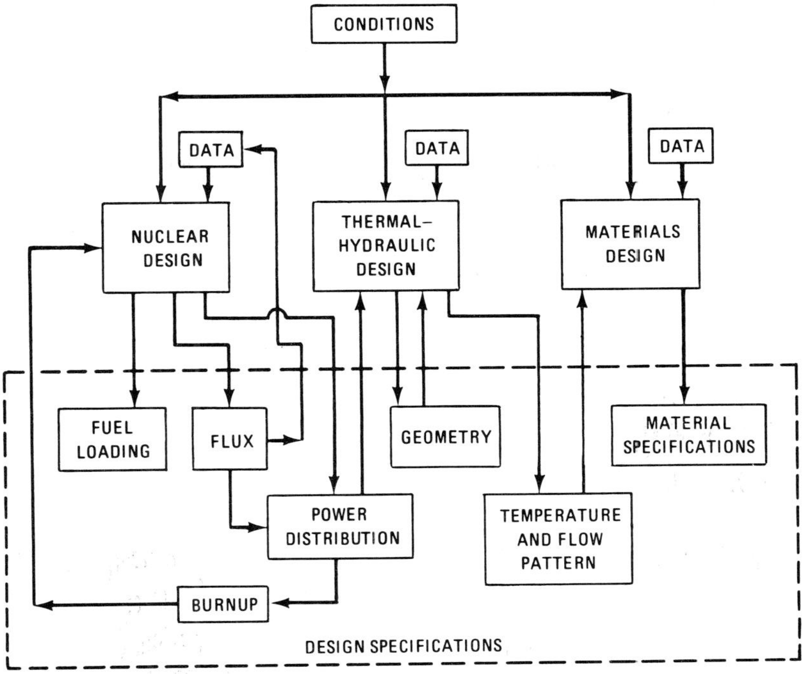

Fig. 8.2 Integration of parametric contributions.

8.9 This integration of core parameters is schematically pictured in Fig. 8.2. Three primary analytical areas are shown: nuclear design, thermal–hydraulic design, and materials design. To simplify the figure, we have omitted the equally important areas of safety design and economics-related design, which also interrelate with the three areas indicated. The desired design specifications (in the area enclosed by the dashed box) are determined by considering all the analytical areas. Fixed conditions or constraints, together with data needed for the analysis (e.g., cross sections, thermal properties, and materials properties), serve as input in each of the analytical areas. The required fuel loading, flux pattern, and power distribution evolve from the nuclear design, and some feedback is provided to both the nuclear data and the thermal–hydraulic design. Some iteration is involved between loading, power distribution, and burnup in developing a fuel-management strategy (§7.62), but this is not shown.

8.10 The coolant temperature, flow pattern, and geometric arrangements evolve from the thermal–hydraulic design. Thermal conditions also contribute some feedback to the materials design, from which the materials specifications evolve. This representation emphasizes some important interdependencies among the design parameters which can be compared with evaluation criteria. A complete picture showing secondary interplays would be much more complex. Some *areas* of design attention are discussed in the following sections.

TYPICAL CORE-DESIGN APPROACH

8.11 The elements of a design approach can best be shown by an example. Figure 8.3 shows a "logic" flow diagram for a typical design approach for a liquid-metal fast breeder reactor fuel element.[1] We emphasize that this is a subsystem of the reactor core and that alternate analysis sequences are possible, perhaps even preferable, since the point of entry into an iteration loop often must be selected arbitrarily. In this example the sequence progresses through nine stages:

1. Thermal analysis.
2. Fuel-element composition and diameter selection.
3. Core sizing.
4. Fuel-cycle economic analysis.
5. Fuel-element structural analysis.
6. Hydraulic analysis.
7. Safety analysis.
8. Fuel-element reliability analysis.
9. Postirradiation handling considerations.

8.12 In this method thermal analysis begins with specification of primary-coolant inlet temperature and mixed outlet temperature. Design specifications that appear as input are normally derived from experience, trade studies (§8.24), or system-optimization studies. These inputs, together with the radial power-peaking factor and the effect of any decision to flatten the coolant outlet temperature, are used to calculate the nominal central-channel outlet temperature.

8.13 Next the axial profile of the hot-channel coolant temperature is computed by using the axial power profile and the engineering factor. The power profile can be estimated with sufficient accuracy since the analysis is relatively insensitive to the uncertainty involved. However, the hot-channel factor is very important and represents a design challenge (§4.87).

8.14 The axial profile of the hot-channel cladding surface temperature is calculated by assuming a surface heat flux and using empirical correlations for the surface heat-transfer coefficient. Cladding material and thickness are tentatively selected (to be verified later in the fuel-element structural analysis), and the temperature profile of the hot-channel *cladding inside diameter* is computed. After the fuel-cladding bonding medium and the bond thickness (or effective thermal conductance) are selected, the hot-channel *fuel-surface-temperature* profile is computed.

8.15 The next series of design decisions concerned with selecting fuel-element composition and diameter are critical to fuel-cycle economics. Design data on the thermal conductivity and limiting temperature of the fuel are used to compute the design-limit linear power rating of the fuel element. The fuel-element diameter is then tentatively selected, considering the trade-off between fabrication costs, which decrease with increasing diameter, and inventory costs, which increase with increasing diameter. Using this initial value of the fuel-element diameter, we calculate the surface heat flux, and the design procedure iterates, considering the flux assumption made in the thermal analysis.

8.16 To size the core, we calculate the number of fuel elements from the core power and length and choose the geometrical array and spacing of fuel elements. These factors roughly establish the core diameter (although additional space must be reserved for control and safety components and, if applicable, for a source, oscillator, internal-core moderator, etc.). The core size and shape are then compared with an acceptable range of sizes and shapes generated from trade-offs involving economics (primarily breeding ratio) and safety (primarily sodium-void coefficient). If the design is unacceptable, the procedure iterates to selection of a different core length. Finally, when core size and shape are firm, the radial and axial power profiles, required early in the thermal analysis, can be recomputed and the subsequent calculations modified appropriately. Coolant velocity, which is needed for the hydraulic analysis, also can be calculated at this point.

8.17 After the core is sized, fuel-cycle economics is predicted to determine whether the calculated cost is less than a maximum allowable value. If the cost is unsatisfactory, iteration (not shown) to a new fuel design is necessary. The fuel-cost calculation uses a suitable computer code that has as input economic data and the design burnup capability. A calculation to determine the burnup with computer codes that describe the deformation of fuel and cladding under irradiation is needed (§4.48), but this area is still being developed.

8.18 The fuel-element structure is analyzed by sizing the gas plenum, considering the combined effects of radiation, pressure, and thermal stress, and evaluating the ability of the fuel element to survive the expected cyclic operation of the core. The fission gas released and the resulting stress picture under planned operating conditions are considered. Swelling and creep of core structural materials receive special attention. Here again, if the results of the analysis fail to satisfy the design specifications, earlier decisions must be reconsidered through logic iteration steps not shown on the diagram.

8.19 The core hydraulic analysis is concerned primarily with determining coolant pressure losses, flow distribution, and related structural-design considerations. Axial-blanket lengths must be chosen before the hydraulic calculations are made to establish the total fuel-element length. The axial blankets are sized primarily from economic studies performed at the "core-system" level. Next we calculate pressure loss across the fuel-element bundle for the reference flow rate, using empirical head-loss coefficients. Over the relevant range of Reynolds numbers, velocity profiles and pressure drops for liquid metal agree with those calculated from generalized correlations developed for water. The calculated pressure loss is compared with an allowable loss determined from the pumping-energy specification. If the loss is excessive, redesign is necessary.

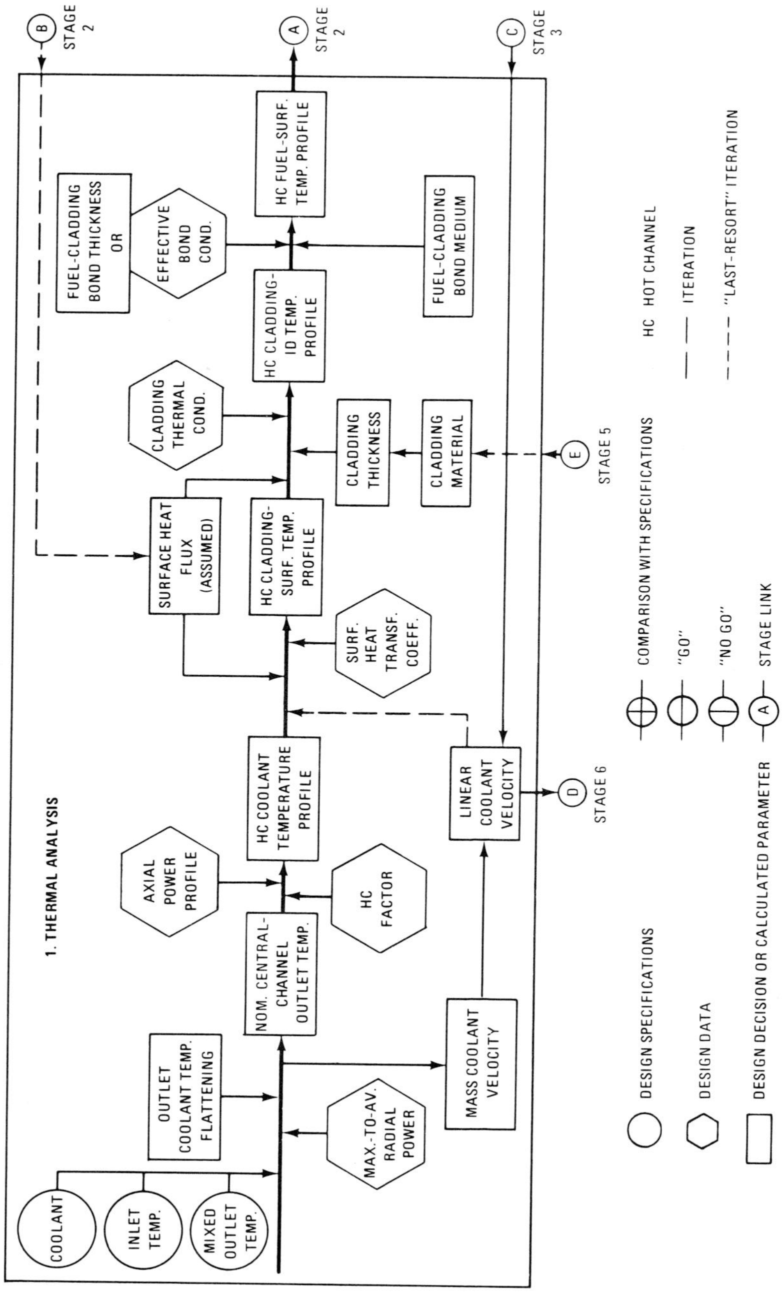

Fig. 8.3 Logic flow diagram for fuel-element design for a liquid-metal fast breeder reactor. Stage 1.

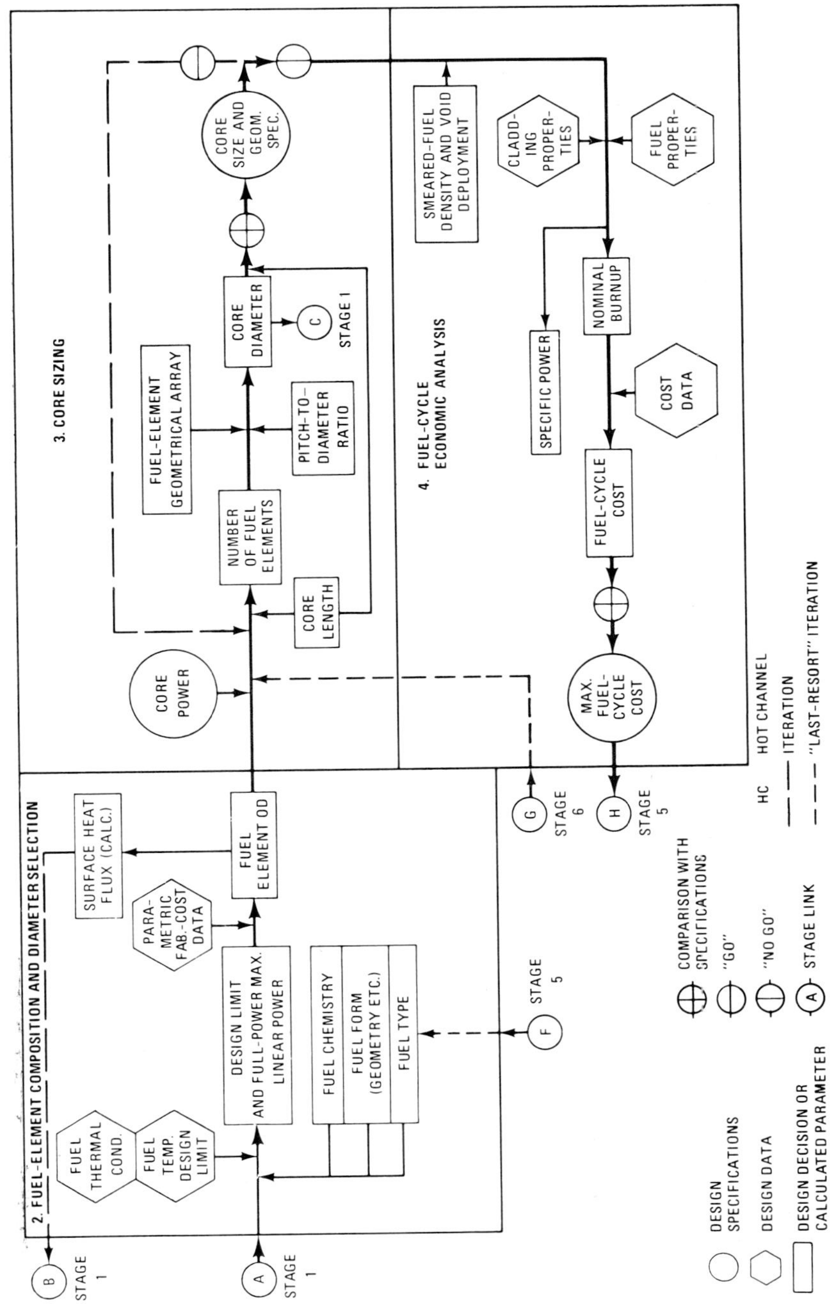

Fig. 8.3 (continued) Logic flow diagram for fuel-element design for a liquid-metal fast breeder reactor. Stages 2 to 4.

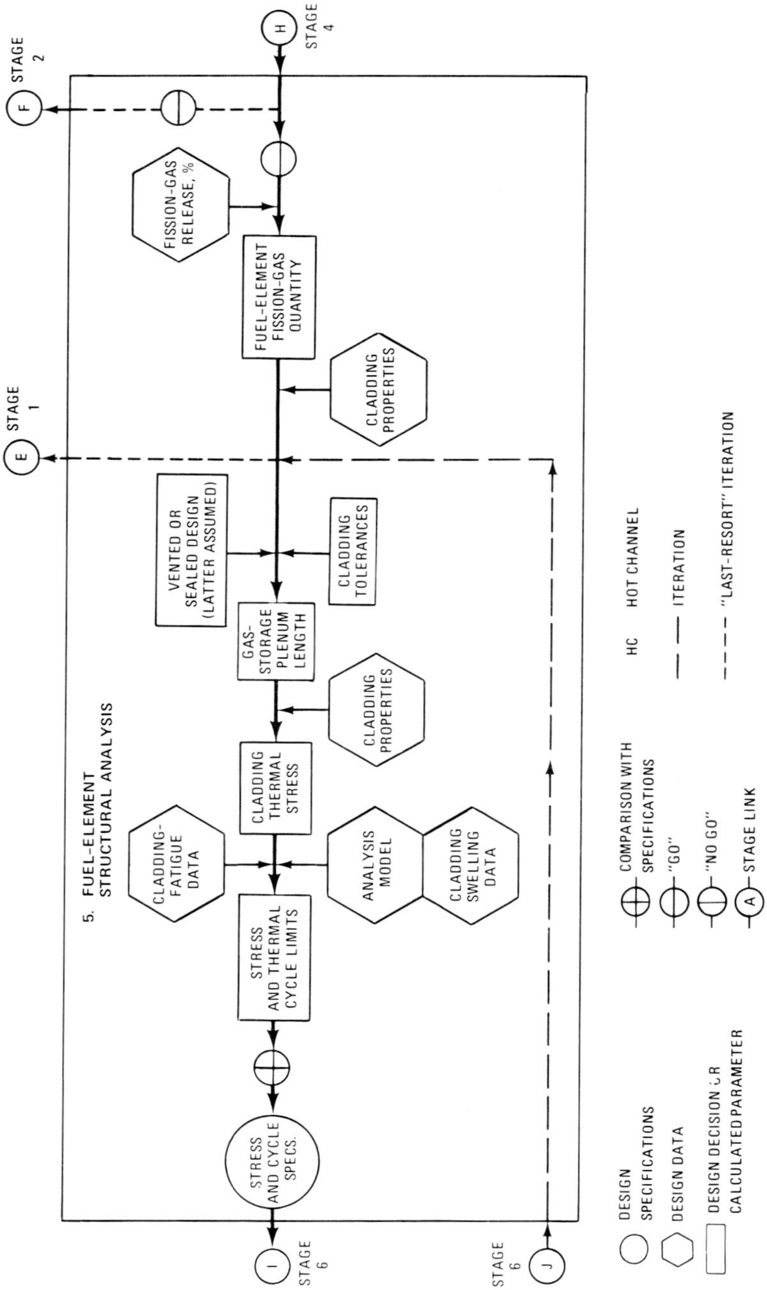

Fig. 8.3 (continued) Logic flow diagram for fuel-element design for a liquid-metal fast breeder reactor. Stage 5.

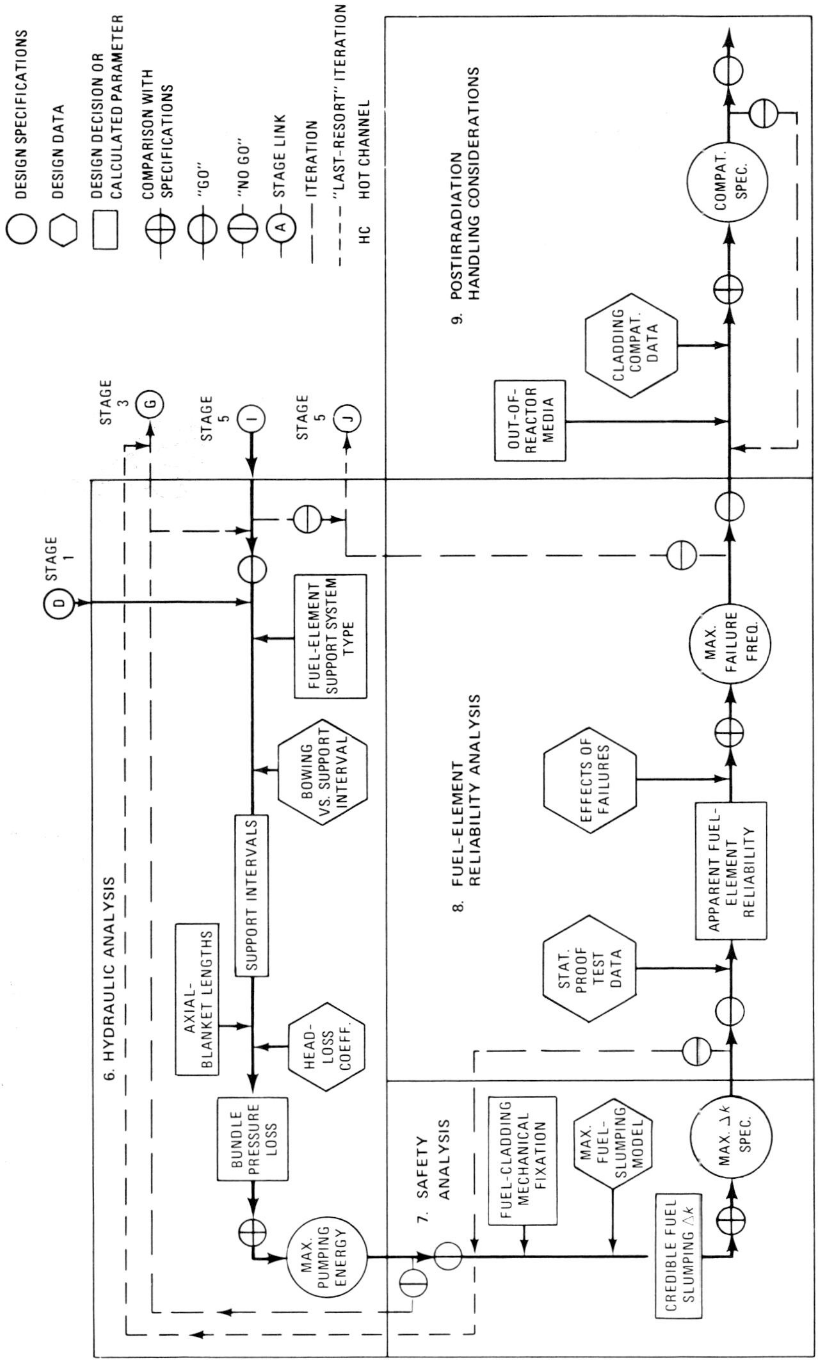

Fig. 8.3 (continued) Logic flow diagram for fuel-element design for a liquid-metal fast breeder reactor. Stages 6 to 9.

8.20 A safety analysis is included since the fuel element contributes heavily to the power coefficient in the operating range. Such reactivity contributions as thermal expansion of the fuel, the possibility of slumping of the fuel into mechanical voids, and other mechanical effects are considered. In fact, the flow diagram gives only an indication of the possible analysis model. Whatever model is used should picture the reactivity behavior; this is then compared with behavior considered acceptable for the design.

8.21 A study of the reliability of the fuel element under operating conditions is another important step in the design sequence. Test data that can be evaluated statistically to provide a failure-probability value are normally needed. The analysis might also include a study of the effects of failure as a basis for an acceptable reliability specification. Since reliability analysis is still being developed, this design step is merely identified here.

8.22 The final stage in the design process concerns postirradiation handling. Before they are reprocessed, fuel assemblies that have been immersed in sodium in the reactor are normally steam cleaned and water washed, a process that could introduce temperature gradients detrimental to the cladding. A review of cladding compatibility under process conditions is, therefore, a part of the design procedure.

8.23 The procedure described is a systematic path toward a design that satisfies all specifications. However, the same goal could be achieved by following other paths. Also, the depth of analysis in each area of attention could vary over a wide range.

ROLE OF TRADE STUDIES

8.24 Often, in developing a design concept, we must explore the effects of changes in parameters and compare different design options to establish a combination of basic specifications that appear to meet the requirements best. Such studies can vary considerably in depth, depending on the detail desired and the level of uncertainty that is acceptable. A procedure used for one phase of the development of a liquid-metal fast breeder reactor illustrates the approach.[2] Figure 8.4 lists the general steps that were followed, together with some factors considered at each step.

8.25 The first step in this approach is to establish requirements for performance, safety, availability, maintenance, operation, design, environmental, and interface features for a subsystem or component. These basic requirements must be met by all alternatives and are not considered design trade variables.

8.26 Alternative designs are then formulated. Many are considered, the number being limited only by the ingenuity of the designers. Then the most promising alternatives that can be studied within the existing budget and schedule are chosen for analysis.

8.27 As analysis of the alternatives proceeds, a conceptual design of each is formulated. The depth of detail required depends on the particular study. The

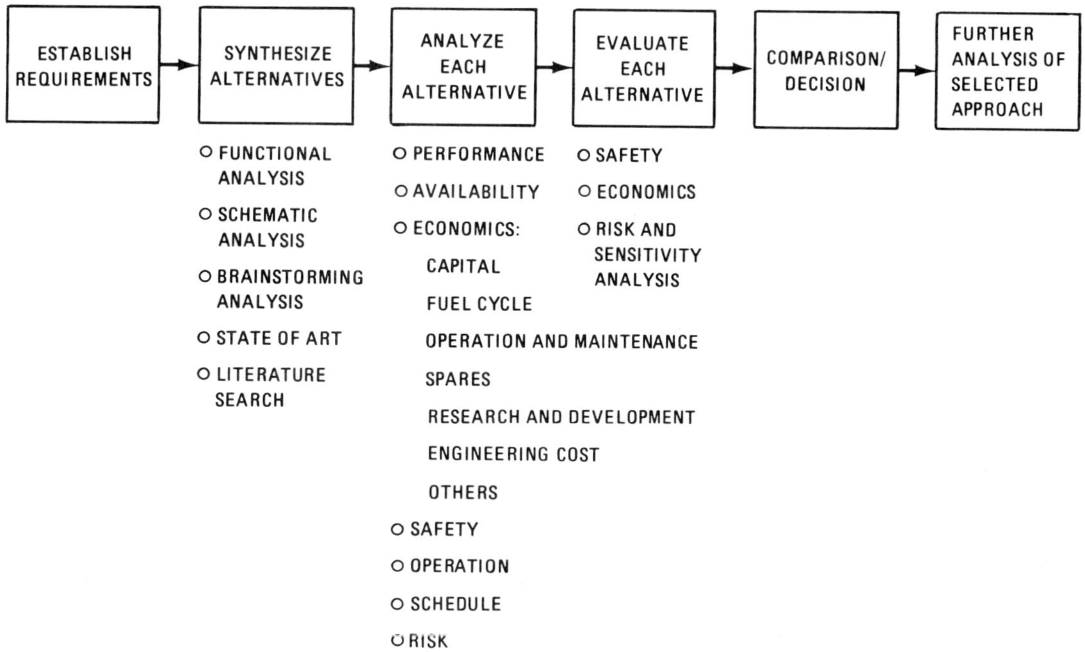

Fig. 8.4 Design-trade-study approach for a liquid-metal fast breeder reactor.

various aspects of the design are analyzed with regard to economics, safety, reliability, availability, operation, maintenance, technical risks, development required, schedule, etc. The design is iterated until it meets all the established requirements. In general, the most extensive iterations are in the safety area since safety requirements are strict and tend to be difficult to meet. This safety iteration goes on in parallel with the trade study and is not completed until all designs are "acceptable" (i.e., meet established safety requirements). Thus safety is not a trade variable but shows up in the economic- and technical-risk penalties required to make each design acceptable.

8.28 When all the design alternatives have been analyzed, all factors are evaluated. The technical advantages and disadvantages of each concept are compared and reduced to three considerations:

1. Safety. All designs must meet the safety criteria. Alternate designs may have safety differences whose weighting is difficult to establish. Such differences are recognized but are not weighted strongly since all designs meet the safety requirements.

2. Economics. Capital cost, fuel-cycle costs, plant availability, operation and maintenance costs, etc., must be normalized to permit a relative comparison of economic factors. For this purpose an economic model is formulated which correlates all these factors.

3. Technical Risks. Technical risks represent the significant problems or uncertainties impossible to evaluate quantitatively in terms of effects on economics, safety, or other requirements. These considerations, should they cause problems, could result in considerable plant downtime, schedule delays, licensing problems, excessive costs, or component development programs. Much engineering judgment is involved in considering technical risks in terms of estimating both the probability of the problem's arising and its consequences. Plant simplicity, available design margins, potential failure modes, past experience, availability of adequate data, backup designs, state of the art, etc., all must be considered carefully in judging the magnitude of these risks.

8.29 The decision on each trade study is reached by comparing the safety, economics, and technical risks, of each alternative design. A firm quantitative evaluation is very difficult to make. In general, the goals of the first-of-a-kind plant are compared with those for a low-risk, proven approach, but the economic penalties must be considered along with the risks. When a large economic penalty is associated with low risks, a higher-risk approach may be allowable if a carefully defined development and test program is set up to minimize the risk.

DESIGN STUDY

INTRODUCTION

8.30 The preceding sections describe design approaches in general terms. Although it is possible to establish a stepwise procedure, as indicated by Fig. 8.1, the importance of parameter interplays can vary from one specific problem to another. A "feeling" for such considerations remains an art acquired by design experience. Some of the results of a specific design study are described in the following sections to provide a picture of the type of insight needed by the designer.

8.31 To show examples of actual system design considerations, we shall discuss some of the features of a design study performed for the U. S. Atomic Energy Commission as part of the Liquid Metal Fast Breeder Reactor development program in the late 1960s. We should emphasize that no attempt is made here to develop a complete picture of the study.

8.32 In this particular study[3] two separate core concepts that differ in assumed safety criteria were compared. In one concept, designated the "advanced design," it is assumed that sophisticated accident-detection and protection devices will prevent complete core voiding and major meltdown of the fuel. In the second concept, called the "conservative design," the remote possibility of a major voiding and meltdown accident is accepted, but the severity is reduced by providing for an enhanced Doppler coefficient in the core and reducing the void reactivity. These two concepts represent different safety-design philosophies. The designs were compared primarily to provide background for a decision on the safety approach for a subsequent reference design for a liquid-metal fast breeder reactor system. The study was carried out in sufficient detail to indicate technical feasibility, costs, and necessary research and development.[4]

8.33 Since the study is a follow on of previous sodium-cooled 1000-Mw(e) fast reactor studies performed by the General Electric Company some features of the reactor concept evolved from a series of prior studies. A "pool" arrangement was chosen in which all primary heat-transfer components are placed within the reactor vessel rather than in a "loop" (the loop concept uses a piped primary system in which the components are separated from the reactor vessel and each other). The reactor is refueled through an open plug into a hot cell rather than through an "under-the plug" mechanism as in the Experimental Breeder Reactor No. II and the Enrico Fermi Atomic Power Plant Unit 1 (Ref. 5).

8.34 Figure 8.5 shows a view of the reactor and refueling cell. The reactor and primary-system equipment are enclosed in a large vessel, and refueling is accomplished through the open plug. For maintenance, the intermediate heat exchangers* (IHX's) and primary pumps, which are mounted on the upper shield deck, can be lifted into a cask to be transported to a maintenance facility.

8.35 The primary coolant is circulated by three vertical, shaft-sealed, centrifugal pumps that draw sodium from the primary-system tank, as shown in Fig. 8.6, at approximately 800°F and pump it directly into and through the

*Six exchangers were arbitrarily assumed in the study described. A three-loop system is used in the reference design, however.[2]

Fig. 8.5 Sodium breeder reactor plant. (Courtesy General Electric Company.)

1 Reactor core.
2 Refueling-cell wall.
3 Shield plug, in raised position.
4 Control-rod drives.
5 Shield-plug lifting screw and guide.
6 Fuel-transfer machine.
7 Fuel-shuffling machine.
8 Fuel-decay tank.
9 New-fuel storage.
10 Intermediate heat exchanger (3 units).
11 Intermediate heat exchanger, shown in raised position.
12 Secondary-sodium inlet.
13 Secondary-sodium outlet.
14 Primary-sodium pump and drive motor (3 units).
15 Primary-sodium vessel.
16 Insulation and outer tank.
17 Steel liner on concrete.
18 Concrete shielding.

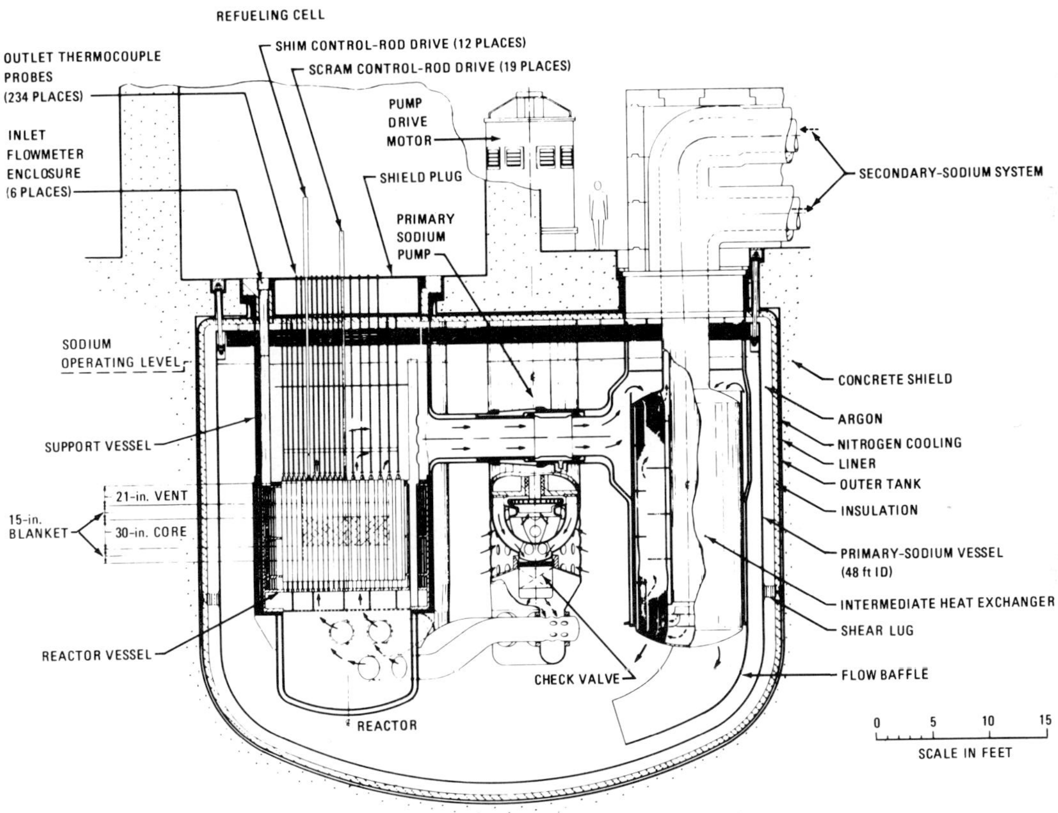

Fig. 8.6 Cross-sectional view of primary vessel and reactor. (Courtesy General Electric Company.)

TABLE 8.1

Variable Design Parameters

Independent parameters	
General	Economic
Core height	Fissile plutonium value
Core volume fraction of BeO (conservative core only)	Working-capital charge rate
Fuel-pin pitch-to-diameter ratio	Fuel-fabrication-cost model
Reactor coolant outlet temperature	Fuel-reprocessing-cost model
Reactor coolant ΔT	Safety
Fuel-pin diameter	Doppler reactivity and sodium-void reactivity limitations (imposed on conservative core only)
Amount of orificing in core	Physics
Type of fuel-channel design	Plutonium alpha value and ^{238}U capture cross sections
Turbine throttle and reheat temperature	Other nuclear data uncertainties
Dependent parameters	
System pressure drop	Fuel-cladding thickness
Net cycle efficiency	Core specific power
Core volume fraction of steel	Blanket thickness
Core peaking factors	Fissile loading and distribution

reactor core, where it is heated to 1150°F. The heated primary sodium is then channeled to one of six IHX's, where it transfers heat to nonradioactive secondary sodium. The primary sodium then leaves the IHX and flows into the primary-system tank. In each of six secondary loops, sodium is pumped from the steam evaporator through the IHX, through the steam superheater and reheater in parallel, and back to the steam evaporator.

8.36 The steam generators are located at a higher elevation than the IHX's and the primary-sodium system to ensure that the static pressure on the nonradioactive sodium in the IHX exceeds that on the radioactive sodium. Consequently no leakage of radioactive sodium can occur. The difference in elevation also aids heat removal if electrical power to the plant is lost since the sodium will circulate by natural convection. Steam at 2400 psia and 950°F is generated in the six once-through steam generators, and it flows to a tandem-compound four-flow 1800-rpm 950°F reheat turbine.

8.37 In the comparison it is desirable to use for each design a combination of parameters that have been optimized for favorable economics. Since complete optimization of all *independent* parameters is a major undertaking, a partial optimization was carried out by varying only independent parameters that seemed to affect the comparison and *dependent* parameters that contribute to significant economic changes. The major independent and dependent parameters are listed in Table 8.1. Several of these are discussed here. Major parameters that were held fixed are listed in Table 8.2 to add to the picture. A listing of plant characteristics would include numerous other parameters, but, since they do not affect the comparison and since we do not intend to discuss the complete design, they are not given here.

8.38 Despite the lack of complete optimization and despite the number of parameters that were fixed, many cases required analysis. Moreover, in most cases analysis consisted of an integrated evaluation of physics, thermal–hydraulics, mechanical design, system design, and economic effects, accomplished with digital-computer techniques.

TABLE 8.2

Fixed Design Parameters

Fuel type	Mixed oxide, gas bonded
Maximum fuel temperature	5207°F
Maximum fuel-cladding surface temperature	1300°F
Fuel-cladding thickness at 0.25 in. outside diameter	15 mils
Maximum reactor pressure drop	200 psi
Maximum coolant outlet temperature	1150°F
Minimum fuel-pin pitch-to-diameter ratio	1.15
Core average burnup	110,000 Mwd/tonne (U + Pu)
Fuel density (hot smear)	9.43 g/cm^3 oxide
Number of core enrichment zones	2
Fuel-plenum-height-to-core-height ratio	1.25
Refueling interval	6 months
Type of plant cycle	Reheat
Steam generator	Once-through parallel-flow reheater–superheater
Number of primary and secondary loops	6
Number of primary pumps	3
Main steam pressure	2400 psia

8.39 Before we discuss the various parametric effects, let us look at the core and blanket arrangement for each concept evolved from the study. As shown in Figs. 8.7 and 8.8, the conservative design has a pancake core with a height-to-diameter ratio of 1 to 6, and the advanced core has a height-to-diameter ratio of about 1 to 2. The relatively-high-leakage, or pancake, concept reduces the positive reactivity effect of a possible loss of coolant. This geometry produces a higher fissile inventory than might otherwise be required, however, with a resulting unfavorable effect on the Doppler coefficient. Thus a moderator, such as BeO, is added to soften the spectrum and enhance the Doppler coefficient. The softer spectrum results in a lower breeding ratio and corresponding higher fuel-cycle costs, however.

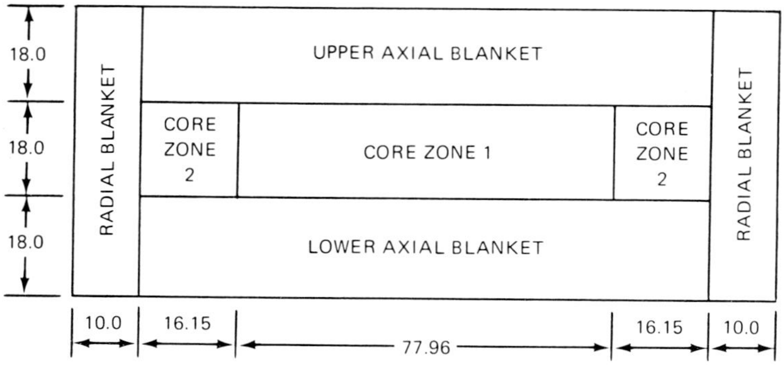

FUEL (UO_2 + PuO_2)	38.6%	BeO	3.0%
SODIUM	38.9%	TANTALUM	0.5%
STEEL	19.0%		

Fig. 8.7 Conservative-design core-and-blanket arrangement. All dimensions are in inches.

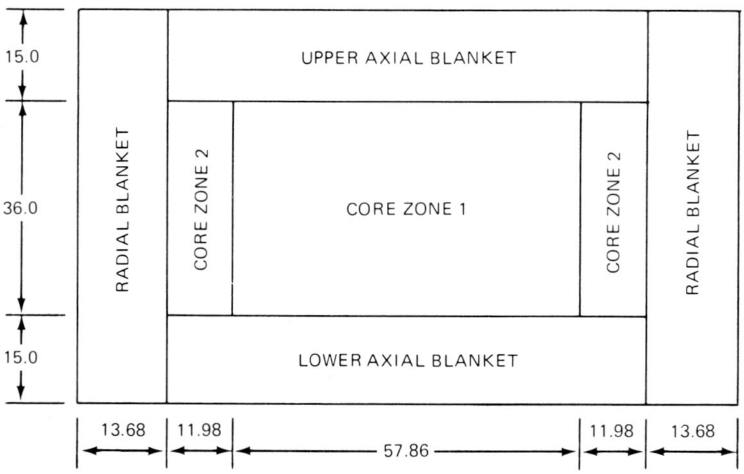

FUEL (UO_2 + PuO_2)	41.7%	STEEL	20.5%
SODIUM	37.7%	TANTALUM	0.1%

Fig. 8.8 Advanced-design core-and-blanket arrangement. All dimensions are in inches.

8.40 Among the various engineering design parameters, those concerned with the core generally play a key role in defining the characteristics of the reactor system. In addition, they serve to illustrate the interrelations between economic, safety, and engineering considerations.

Core Height

8.41 The effect of both core height and volume fraction of BeO on equilibrium-fuel-cycle costs for the conservative design is shown in Fig. 8.9. The 18-in. core height chosen is near the point of minimum fuel-cycle costs within the "allowable" region bounded by an acceptable Doppler coefficient and an acceptable void reactivity. To simplify this particular analysis we have held constant at the values shown in the figure a number of other parameters, e.g., the pitch-to-diameter ratio and corresponding volume fractions. These also affect the core height, as is discussed later. The introduction of BeO has such a strong effect on fuel-cycle costs (the sodium-void limit excludes much of the decision space) that a minimum core height is specified. Furthermore, the shape of the curves indicates that design selection is quite sensitive to *uncertainties* in the sodium-void and Doppler criteria.

8.42 Doppler and sodium-void considerations do not affect the core-height selection for the advanced concept, whose primary objective is a minimum fuel cost. The analysis in Fig. 8.10 shows that taller cores with low pitch-to-diameter ratios tend to give lower fuel costs. In this case, however, the coolant pressure drop through the core becomes the major constraint; a value of 200 psi is considered the practical limit. Thus the value for a 36-in. core height with a fuel-pin pitch-to-diameter ratio of 1.20 is at about the minimum of the 200-psi cost curve.

Fuel-Pin Diameter

8.43 In oxide-fueled sodium-cooled fast reactors, the fuel-pin diameter interrelates with a number of other parameters, e.g., cladding thickness, cladding temperature, fuel swelling, and the approach used for handling of fission-product gases. However, considering fuel-cycle costs and only the trade-off between specific power and fuel-fabrication costs as a function of rod diameter (Fig. 8.11) indicates the economic incentive for examining diameters other than the assumed reference diameter (normally 0.25 in.). A smaller diameter results in a higher specific power (favorable to fuel inventory) but also requires a higher fabrication cost. In Fig. 8.11, for example, curves for both cores suggest an incentive for studying diameters less than 0.25 in.; the curve for the advanced core apparently approaches an optimum. We must remember, however, that such studies are sensitive to the assumptions used for fabrication and other costs.

8.44 The next step is to consider the effect of core pressure drop. Under the constraints of fixed pitch-to-diameter ratio, outlet temperature, and coolant ΔT, the core pressure is a significant function of fuel diameter for the 36-in. advanced core but not for the 18-in. conservative core, as shown in Fig. 8.12. The capital cost of the nuclear steam-supply system is assumed fixed in terms of dollars per thermal kilowatt. An increase in pumping power caused by a larger

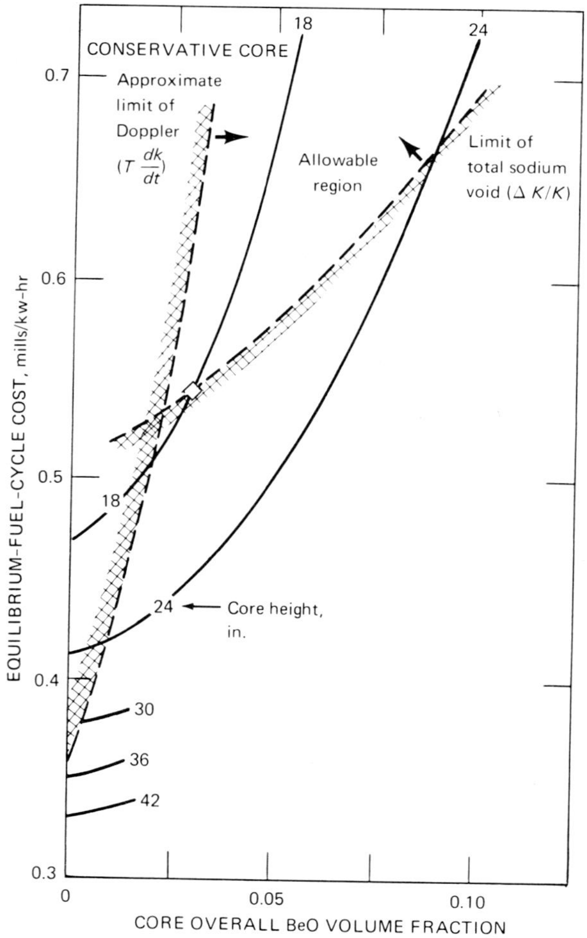

Fig. 8.9 Fuel costs vs. volume fraction of BeO and core height for a 1000-Mw(e) liquid-metal fast breeder reactor, with fissile plutonium at \$10/g and a working-capital charge of 10%/year. Constants: reactor ΔT = 300°F; outlet temperature = 1150°F; fuel outside diameter = 0.25 in.; cladding = 15 mils; core overall volume fractions, coolant = 0.39 and steel = 0.19; fuel-pin P/d_0 = 1.22.

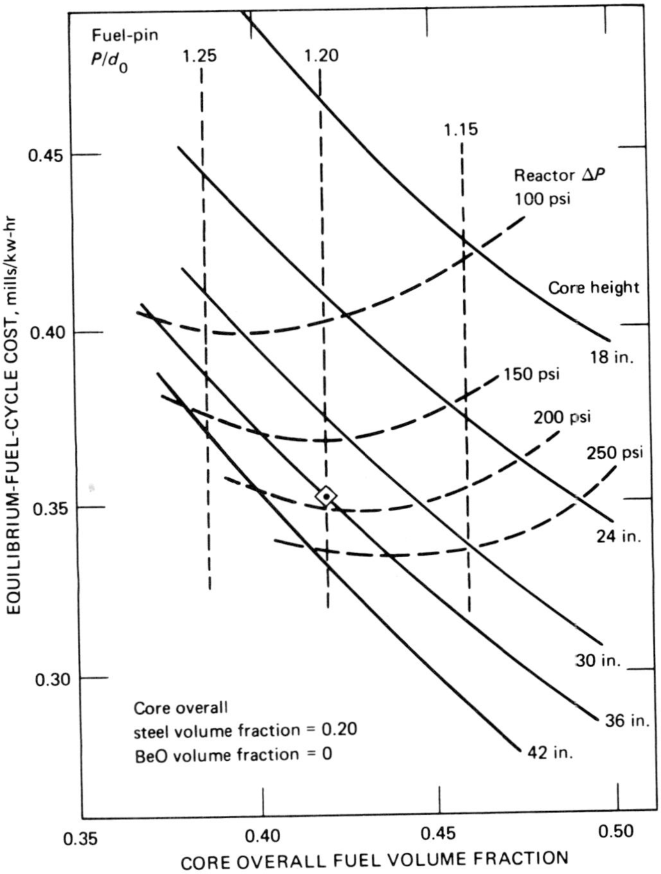

Fig. 8.10 Fuel costs vs. volume fraction fuel and core height for a 1000-Mw(e) liquid-metal fast breeder reactor, with fissile plutonium at \$10/g and a working-capital charge of 10%/year. Constants: reactor ΔT = 300°F; pin outside diameter = 0.25 in.; cladding = 15 mils; peak cladding temperature = 1300°F.

pressure drop will result in less net electrical power produced and, hence, a larger capital cost in terms of dollars per electrical kilowatt. This effect is examined in two ways in Fig. 8.13 for the 36-in. core. The upper solid curve shows the cost behavior, assuming a constant pressure drop maintained by adjusting the pitch-to-diameter ratio, and the upper dashed curve shows the trend given in Fig. 8.11 for a fixed pitch-to-diameter ratio. Differential energy costs are shown by the lower solid curve, which also includes the capital cost effects. This figure indicates that a fuel-pin diameter of 0.23 in. is desirable. Since these trends are not significant for the conservative 18-in. core, a somewhat arbitrary lower limit

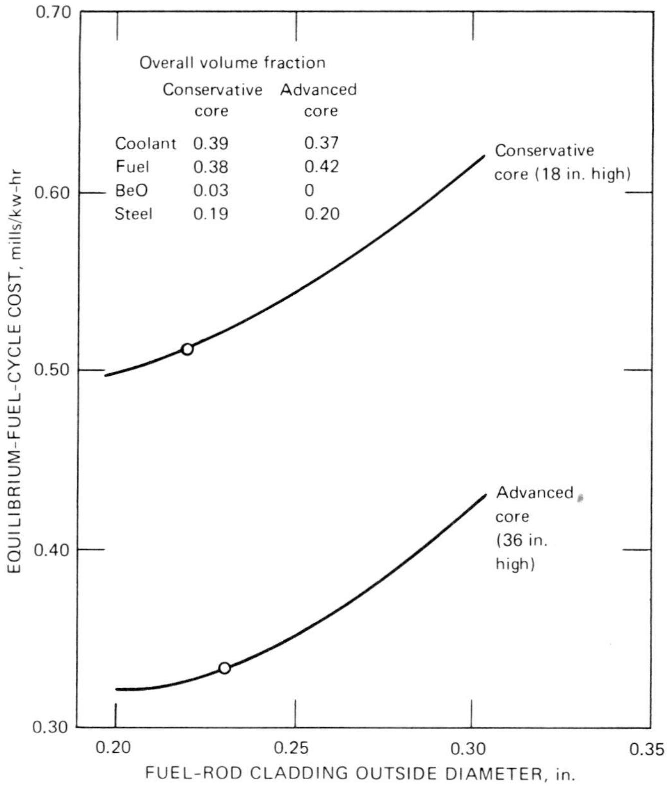

Fig. 8.11 Fuel costs vs. fuel diameter for a 1000-Mw(e) liquid-metal fast breeder reactor, with fissile plutonium at \$10/g, a working-capital charge of 10%/year, recovery costs of \$60/kg (U + Pu), and fabrication costs of ~\$150 to \$250/kg (U + Pu). For both cores, average burnup = 112,300 Mwd/tonne, peak linear power ≈ 18 kw/ft, coolant outlet temperature = 1150°F, and coolant ΔT = 300°F.

of 0.22 in. outer diameter was selected for that case. Since the trends result in rather shallow optimums, however, the influence of other parameters may well cause the designer to select slightly different values of pin diameter.

Core Temperature and Pressure Drop

8.45 The coolant temperature rise through the core is affected by the trade-off between the costs of pumping equipment and heat exchangers. Another factor is the primary-system pumping power, which increases rapidly when both coolant flow rate and pressure drop are increased, as required if the core temperature rise is reduced. Decreasing the core temperature rise, therefore,

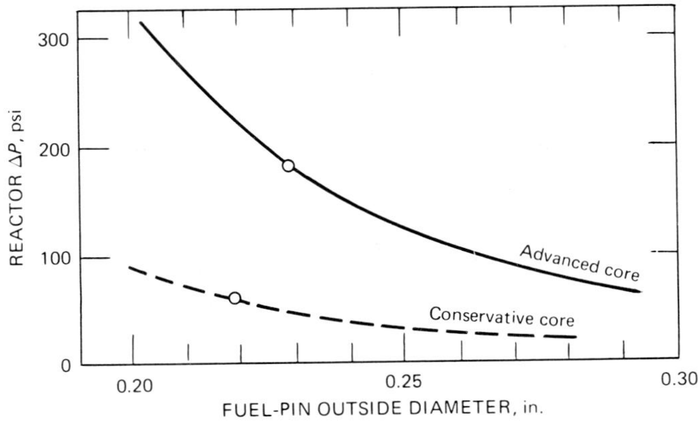

Fig. 8.12 Pressure-drop dependence on fuel diameter for one set of fuel-performance and coolant conditions.

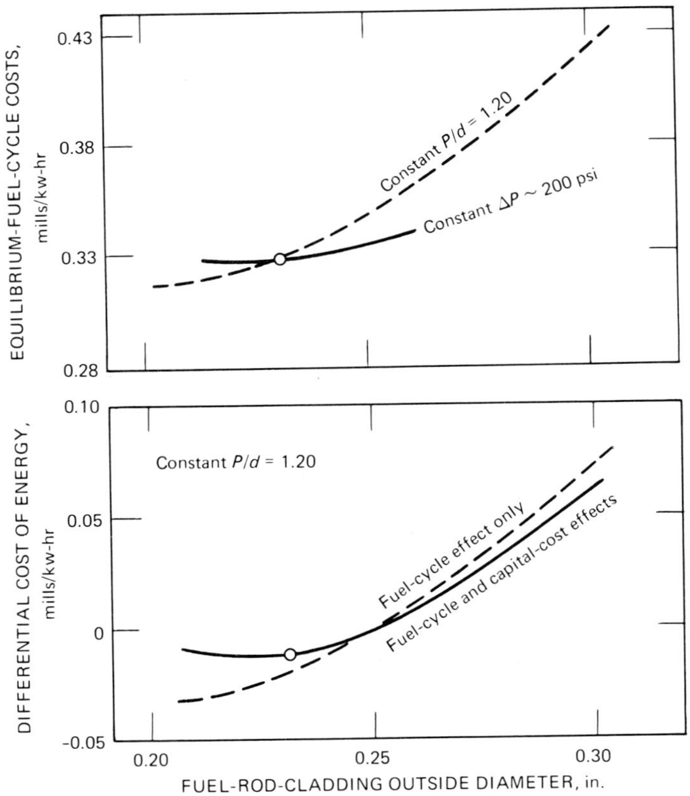

Fig. 8.13 Fuel costs and differential cost of power vs. fuel diameter for 36-in. advanced core design.

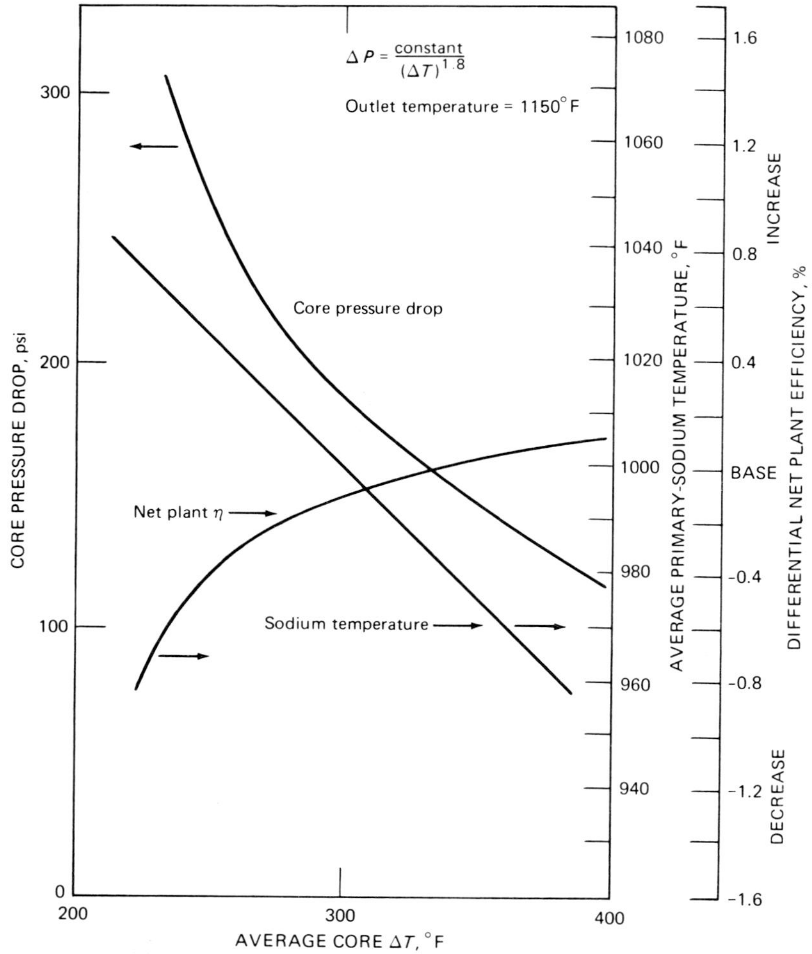

Fig. 8.14 Core ΔP, net plant η, and average sodium temperature vs. core ΔT for 36-in. advanced core design.

increases the pumping power and the cost of pumping equipment but decreases the cost of heat exchangers by increasing the available temperature differential between the sodium heat source and the steam produced. Typical trends for design parameters are shown in Fig. 8.14, and cost contributions are shown on a differential basis in Fig. 8.15.

8.46 Varying the core temperature rise while the outlet temperature remains fixed can be analyzed in a relatively straightforward manner since the change in core inlet temperature affects primarily the boiler section of the steam generator and not the superheater. Therefore optimization of the turbine throttle temperature is not necessary. The interplay of a number of parameters, including secondary effects, can be analyzed by using computer codes. The effects of varying both core outlet temperature and coolant temperature rise are analyzed in Fig. 8.16 for the 36-in. advanced core. Arrows indicate allowable-region boundaries. Since the cladding temperature should be as high as possible

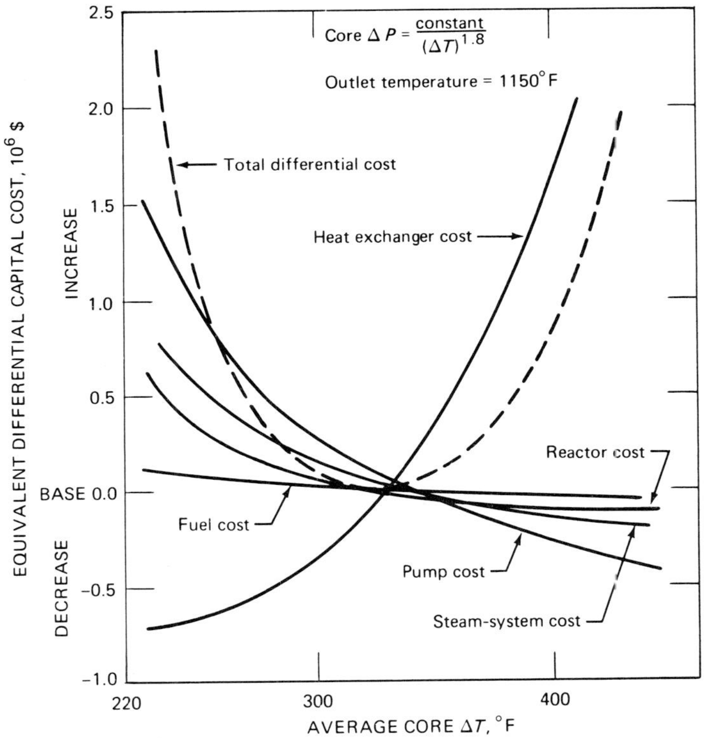

Fig. 8.15. Calculated optimum core ΔT for 36-in. advanced core design.

for minimum power cost, the maximum allowable cladding temperature of 1300°F is used as a boundary. The hydraulic limit of 200 psi pressure drop is also given, but it does not influence the selection of a coolant temperature rise of about 300°F at an outlet temperature of 1150°F.

8.47 A similar analysis for the 18-in. core is shown in Fig. 8.17. The coolant outlet temperature, limited to 1150°F because of materials considerations, becomes the controlling parameter. Thus a coolant temperature rise of about 270°F, corresponding to the minimum on the 1150°F outlet-temperature curve, is selected.

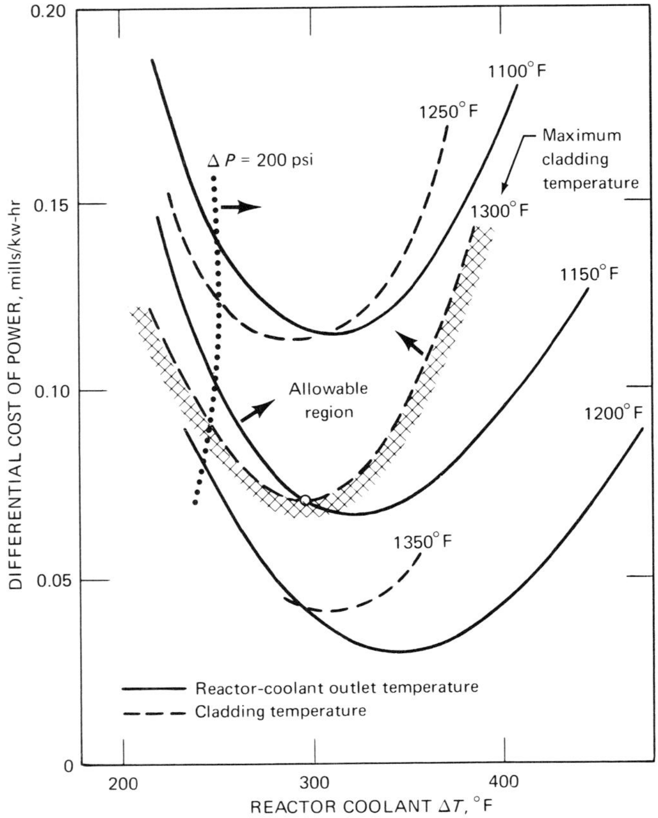

Fig. 8.16 Differential cost of power vs. reactor coolant ΔT and outlet temperature for 36-in. advanced core design.

Fuel-Pin Design

8.48 Though not part of the comparison study, fuel-pin design parameters are quite important for oxide-fueled sodium-cooled fast reactors of this type.[6] Although, as mentioned in §8.46, first considerations indicate that a high coolant outlet temperature is a desirable goal, the picture can change when other variables are considered. For example, as the allowable cladding temperature is increased, both strength and ductility are reduced. Ductility is also reduced by irradiation. If thicker cladding must be used or the fuel burnup must be decreased because of this trend, the cost of the energy produced will be increased. Therefore a trade-off applies between the increased thermal efficiency resulting from the higher coolant exit temperature and the increased heat costs resulting from the higher cladding temperature. Irradiation-induced swelling and

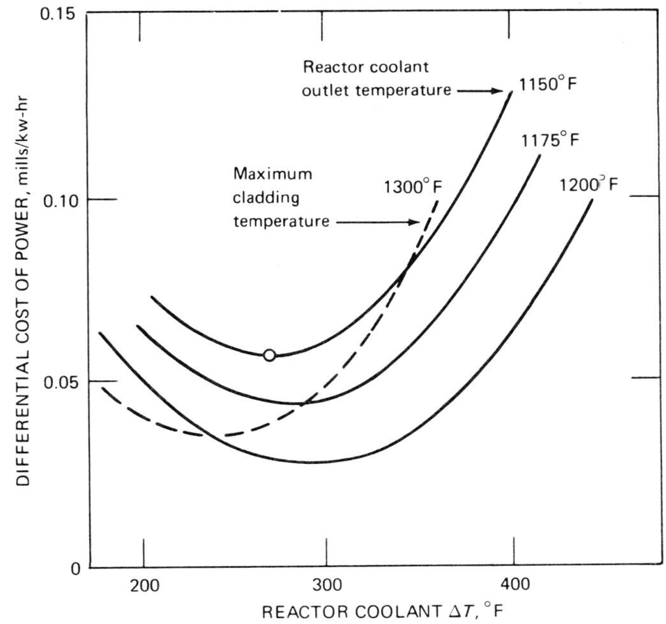

Fig. 8.17 Differential cost of power vs. reactor coolant ΔT and outlet temperature for 18-in. conservative core design.

creep (§7.163) is also of major importance. However, the design studies described in this chapter were completed before its significance was realized and, hence, do not include this effect.

8.49 Even if the effect of internal fission-gas pressure is minimized by using a vented fuel-pin design (§4.44), fuel swelling, which still occurs to some extent with burnup, affects the cladding thickness specification. The swelling can be slightly reduced by lowering the fuel density by perhaps 5 to 10%. Using lower density fuel, however, affects the thermal conductivity and, in turn, the size of the void at the center of the fuel. Adjustments in the nuclear design would also be in order. The combined effect of cladding thickness, fuel density, and burnup on the fuel cost for a typical design is shown in Fig. 8.18. By expressing the results in terms of thermal-energy costs, we need not consider the effect of thermal efficiency.

NUCLEAR-DESIGN PARAMETERS

8.50 Nuclear design and the parameters involved represent a major part of conceptual design study efforts. Such activities are described in Chap. 5 and are not considered at length here. In addition to reactivity requirements, the

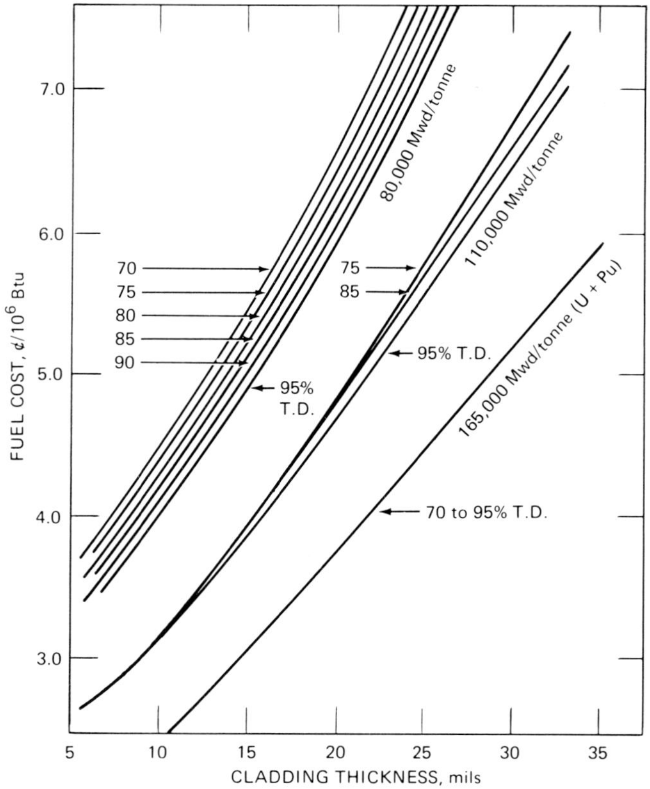

Fig. 8.18 Variation of core power costs with cladding thickness, burnup, and density. T.D. is theoretical density.

designer is concerned with control, core power distribution, fuel burnup patterns, and all the relevant safety parameters. Economic parameters also play a role in selection of the "optimum" nuclear-parameter values. In comparing design studies, we must also bear in mind that differences in such parameters as Doppler coefficient, void effects, and breeding ratio can result from differences in the nuclear data used and the analysis approach followed, as well as in design specifications.

8.51 The nuclear-design parameters having the greatest interplay with the engineering design are generally those concerned with the power distribution in the core as determined by fuel-loading and burnup changes. In the comparison radial power flattening is achieved in both the advanced and the conservative core by varying the fissile plutonium content in two radial core zones of equal volume. Enrichments for the advanced core are 11.3 and 14.5%; the conservative-core enrichments are 15.5 and 18.8%. A power profile for the advanced core

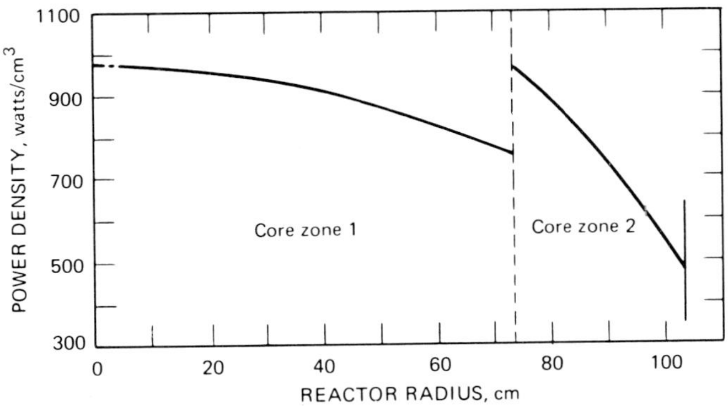

Fig. 8.19 Radial power distribution at mid-burnup of equilibrium cycle.

calculated for mid-burnup is shown in Fig. 8.19. Although the conservative-core radial profile is very similar, its axial peaking factor proves to be about 10% less than that for the advanced core. As a result the overall peak-to-average core power density is 1.38 in the conservative core and 1.50 in the advanced core.

8.52 An important economic advantage results from the lower core fissile inventory and the larger internal-conversion ratio of the advanced concept. Although the difference in breeding ratios for the two cores proves not to be significant because the higher internal conversion in the advanced design is offset by the greater core leakage and blanket conversion of the conservative design, the smaller advanced-core inventory does give a 30% advantage in doubling time.

8.53 One nuclear characteristic that provides some insight into reactor behavior is the neutron-flux distribution. For example, a histogram of the relative neutron flux per unit lethargy averaged over the two core zones is shown in Fig. 8.20. The spectra for the two cores differ only slightly. Therefore comparing safety characteristics involves additional parameters and is relatively complicated, as discussed in §8.60. Thus flux distribution is useful primarily as a starting point for additional analysis.

8.54 In any design study the effect of uncertainties in both nuclear data and analysis methods should be considered. The parameters affecting reactivity are probably of greatest concern to the designer. In addition to questions of safety and heat removal, such parameters can affect the economic potential of a new concept markedly. A large uncertainty in the alpha value for ^{239}Pu, for example, as was the case in 1968, introduced uncertainties in the Doppler coefficient and sodium-voiding reactivity for the two fast reactor concepts discussed here. Additional uncertainty was introduced in the doubling time. When the two designs are compared, the ultimate effect of various nuclear-data uncertainties proves to be somewhat different, partially as a result of spectral considerations, adjustment of the BeO concentration, and the interplay among parameters in each case. Details are not discussed here since they are rather

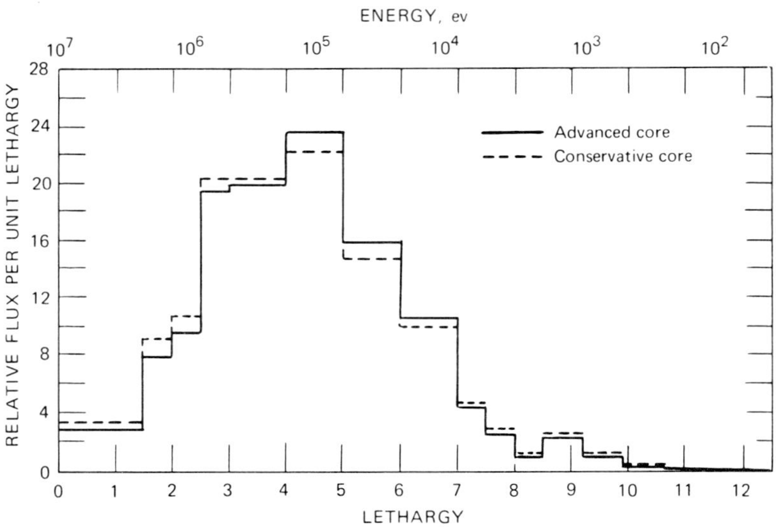

Fig. 8.20 Flux spectra for the advanced and conservative core designs at mid-burnup of an equilibrium cycle.

specific to the concept and the primary objective here is merely to emphasize the necessity of considering such uncertainties.

8.55 Evaluating nuclear-design parameters involves considering not only the uncertainties associated with the nuclear data used but also the suitability of the analytical tools applied. Here the guidance of the specialist is required, particularly for fast reactors. For example, in multigroup calculations variations can be caused by the way group constants are generated from the basic data. Approximations used for flux variations within groups to obtain a flux-averaged pointwise cross section can introduce significant errors.[7] Generally, selecting analytical tools also involves a balance between the need for an accurate description and the computing expense considered appropriate for the objectives of a nuclear-design study. For the accuracy normally needed for preliminary fast reactor design studies for cores larger than 100 cm in radius, diffusion-theory methods are adequate.[8] Results from diffusion calculations of reactivity coefficients (fuel, poison, and sodium) and global flux profiles agree quite well with those from the more expensive transport calculations (§5.68). On the other hand, transport theory must be employed if accurate local flux profiles are desired.

PLANT PARAMETERS

8.56 Among a large number of noncore (or plant) parameters, several interesting parameters related to the thermodynamic or steam cycle are discussed here as examples of questions deserving attention in a design study. Steam conditions, of course, are coupled to the core and the secondary-sodium-coolant temperature pattern since the heat source for steam production is the sodium coolant, as shown in Fig. 8.21, which is a flow diagram and heat balance for the advanced-core-plant steam system. Since the flow pattern is relatively complex and has many possible parameters, optimum conditions are normally determined by digital-computer analysis.

8.57 Shifts in the specifications for steam pressure in the 1800- to 2400-psia range at a constant temperature of 950°F, as limited by heat-exchanger-material considerations, have little effect on the economics of energy production. An improvement of about 2% in cycle efficiency actually results in a smaller cost reduction because of scaling considerations. When pressure is increased, the temperature difference between the sodium coolant and the steam being evaporated is reduced, and a larger heat-transfer area is needed in the steam generator. On the other hand, the higher steam density and improved heat-transfer coefficient work in the opposite direction, with the result that there is little change in heat-exchanger costs.

8.58 If the reactor sodium outlet temperature or the steam temperature is lowered, however, these effects do not continue to compensate. The resulting reduction in temperature-difference driving force across the evaporator tends to dominate the design as the coolant temperature approaches the boiling point of steam, a theoretical limiting condition that would require an infinitely large evaporator.

8.59 An increase in steam superheat temperature from 950 to 1000°F improves cycle efficiency about 1.5%. Again, the temperature spread between the heat source, the sodium coolant, and the steam being heated is reduced, and there is a corresponding need for additional heat-transfer surface. Therefore a trade-off applies between the steam-evaporator cost and other plant costs in response to steam conditions.

SAFETY CONSIDERATIONS

8.60 In the preliminary study effort described here comparing two competing concepts, differences in safety characteristics which might affect the acceptability of either concept are explored. A comprehensive safety analysis would be carried out at a later design stage. The studies of safety characteristics fit into three categories:

1. Analysis of the response of the reactors to reactivity insertions due to severe disturbances not necessarily realizable in an actual plant. This study also provides an estimate of instrumentation requirements, control-system response time, and possible differences in development requirements between the two concepts.

2. Consideration of accidents initiated by local fuel failure and the propagation of damage. This is particularly relevant to the advanced-core

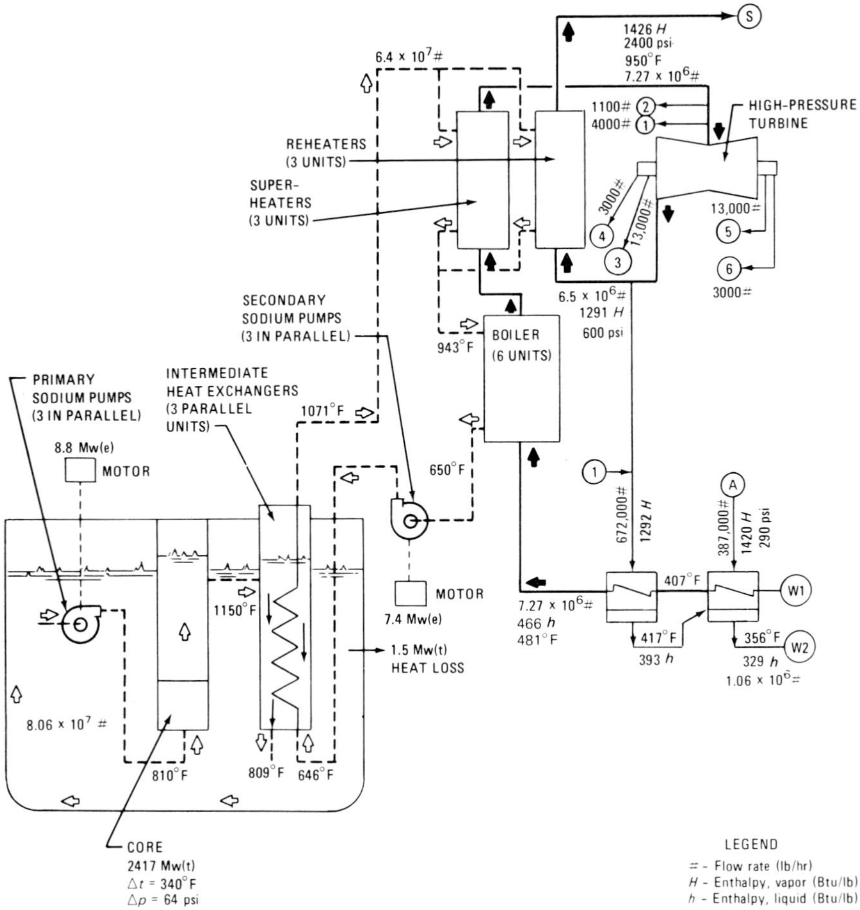

Fig. 8.21 Heat balance for advanced core design. Net plant efficiency is 41.7%.

concept for which large-scale core voiding and compaction are assumed to be not credible because incipient local failures are detected by suitable instrumentation and the reactor is shut down.

3. Design-basis accident analysis and the determination of containment requirements, particularly for the conservative-core concept.

8.61 Under category 1, scoping studies were carried out by two-dimensional perturbation analysis (§5.137) using 1 to 5 dollars/sec reactivity-insertion rates, and the effects of various feedbacks for the two concepts were studied. The amount of reactivity that may be added to the reactor at full power without scram and before cladding failure starts was also determined. The

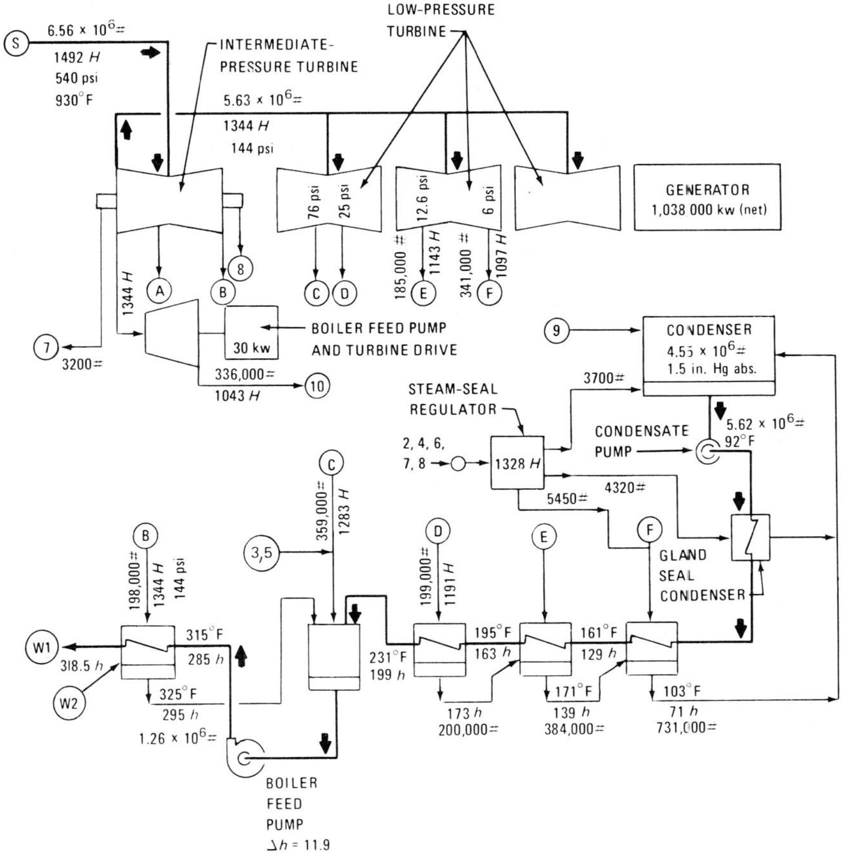

Fig. 8.21 (continued) Heat balance for advanced core design. Net plant efficiency is 41.7%.

feedback picture is somewhat complex but consists primarily of the sodium void, Doppler, and axial fuel expansion. Differences in response between the two core concepts were found to be minor.

8.62 Analysis of coolant flow reduction also fits into the first study category. In the model assumed the reduced flow rate causes a decrease in the frictional pressure drop across the core, with a resulting reduction in pressure acting on the sodium. Thus the tendency for sodium voiding is enhanced as heat is released from the fuel to the coolant, which is flowing at a low rate and hence quickly rises in temperature. Core damage is therefore initiated by sodium voiding. Different options are available in the analytical model for scram and feedback behavior.

8.63 The various flow-reduction cases studied in the core comparison are classified as small (35% or less) or large reductions. For small reductions and the worst possible feedback combination (one-dimensional sodium coefficient and no expansion), the peak power for the advanced core was found to be limited to 114% of the operating power with hot-spot cladding temperature within tolerable limits. For the conservative core the effects are less severe. Therefore it was concluded that small flow reductions appear to be tolerable in both cores, although the resulting transient is more severe for the advanced core.

8.64 For large flow reductions when scram is not assumed, the likelihood of core damage depends on the feedback mechanisms, the fuel axial geometry, and the cover-gas pressure. Feedback behavior for two typical cases is shown in Figs. 8.22 and 8.23 for the advanced and conservative cores, respectively. Voiding occurs in each case at the elapsed time corresponding to the end of the curves shown. A study of the cases analyzed indicates that excessive fuel damage is likely if flow reduction without scram exceeds a threshold level ranging from 35 to 55% for the advanced core and from 35 to 65% for the conservative core. If scram were initiated by the flow reduction, reductions of 75% or higher, occurring in 0.5 sec, could probably be tolerated. The conservative core, therefore, offers no major advantage over the advanced core from the flow-reduction standpoint.

8.65 Analyses of local fuel failure and damage propagation (safety consideration 2) are concerned primarily with describing accident situations rather than comparing the two core designs. Since the fuel pin, fuel assembly, fuel burnup, and operating temperatures do not differ significantly in the two

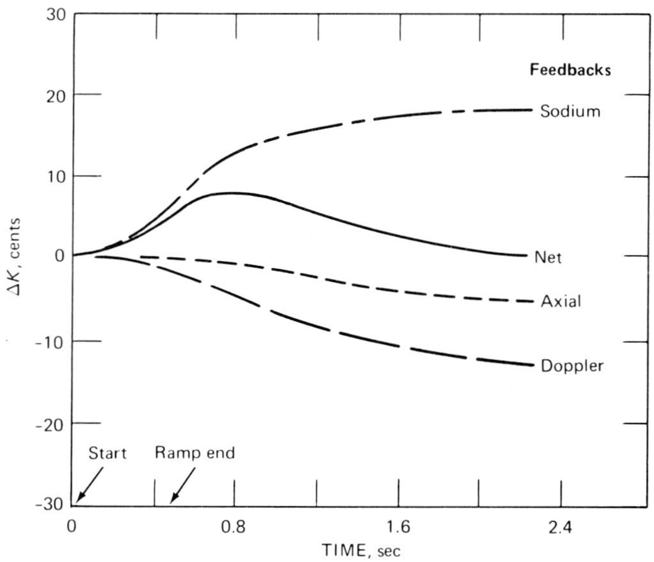

Fig. 8.22 Advanced-core feedbacks for 55% flow reduction and axial-fuel (but no radial-core) expansion.

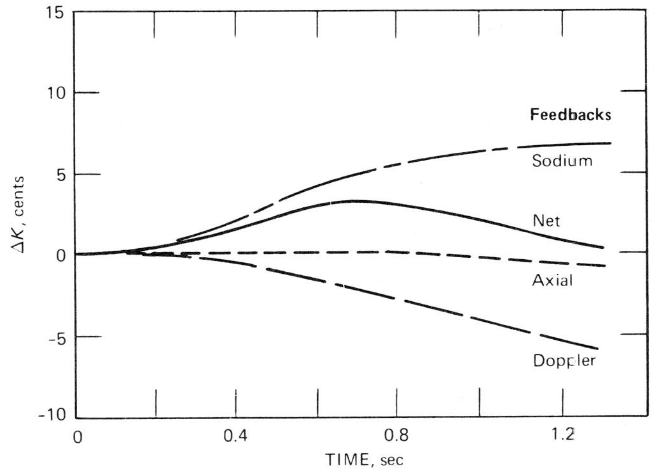

Fig. 8.23 Conservative-core feedbacks for 65% flow reduction and axial-fuel (but no radial-core) expansion.

cores, immediate response to a local blockage is similar. Differences in reactivity feedbacks can become significant if a sodium void forms in a large part of the assembly or if a large fraction of the fuel melts and redistributes itself. Adequate instrumentation for detecting incipient-failure conditions is therefore important. The analyses indicated that limiting damage to one subassembly would be difficult. On the other hand, if reliable flowmeters or temperature detectors are provided on each subassembly, it should be possible to limit propagation to seven subassemblies in the reference design chosen.

8.66 The core-comparison study included an examination of the energy release expected from the prompt critical disassembly of the conservative core to define containment requirements. A modified Bethe–Tait (§6.160) approach was used to determine an effects pattern in response to various initiating events. At the present state of the art, however, a high level of confidence in the results is not available without additional analytical and experimental work. As a result it was concluded that an available work value of 2000 Mw/sec could indeed be chosen for the containment design. However, it is still desirable to rely somewhat on the control and instrumentation systems to achieve scram. Although, for a similar set of assumptions, the energy release for the conservative design is likely to be less than that for the advanced design, it is not certain that in the conservative design every conceivable accident can be contained within the specifications derived.

8.67 The safety-parameter study showed that either concept is acceptable if adequate safeguards are included in the design. A comparison of the two concepts must also include economic parameters, however, since the conservative-core concept is based on a trade-off between presumed favorable safety characteristics and economic performance. Since the study showed that the conservative-core concept still required some reliance on the type of safeguard devices needed for the advanced core, it was concluded that the economic sacrifice inherent in the conservative-core concept was not justified.

ECONOMIC PARAMETERS

8.68 Differences in safety criteria between the two concepts produce technical differences that affect fuel-cycle costs rather than capital costs. Since a trade-off is involved between costs and the design parameters, examination of the economic factors is important to the design study.

8.69 A basis for analyzing the fuel-cycle contributions for the two core concepts is given in Table 8.3, which is a summary of the costs for equilibrium cycles. Fabrication costs for the conservative concept are higher in the core zones than are comparable costs for the advanced concept, the primary reason being that the conservative core contains a larger number of fuel pins of shorter length and a material of higher enrichment. The higher enrichment affects the inventory (fixed-charge) contribution to the fabrication cost.

8.70 The fissile-burnup costs, which include plutonium credit, are much lower in the advanced core because of the higher internal breeding. On the other hand, the high-leakage conservative core shows a high breeding gain in the axial blanket which offsets the advanced-core advantage. The radial blanket for the

TABLE 8.3

Summary of Equilibrium-Fuel-Cycle Costs

Cost item	Core	Axial blanket	Radial blanket	Total
Advanced Core Design (36-in. Height), mills/kw-hr				
Fabrication	0.131	0.023	0.034	0.188
Fissile burnup	0.137	−0.239	−0.267	−0.369
Recovery package	0.056	0.049	0.049	0.154
Working capital	0.263	0.032	0.059	0.354
Total	0.587	−0.135	−0.125	0.327
Conservative Core Design (18-in. Height), mills/kw-hr				
Fabrication	0.168	0.051	0.017	0.236
Fissile burnup	0.378	−0.521	−0.167	−0.310
Recovery package	0.050	0.105	0.039	0.194
Working capital	0.298	0.063	0.030	0.391
BeO	0.018			0.018
Total	0.912	−0.302	−0.081	0.529

taller advanced core is a more effective breeder than that for the conservative core and thus results in some cost advantage. Owing to various compensating contributions, the fissile-burnup charges in the two concepts do not differ greatly. This is consistent with the slight difference in breeding ratio. The closeness of the breeding ratios and the corresponding costs are somewhat surprising, however, when we consider the moderation introduced into the conservative core by the BeO.

8.71 The advanced design also gives some fuel-cycle cost advantages in the "recovery package" and "working capital" categories, primarily because of different refueling-interval and inventory requirements. In fact, there is a cost advantage of about 0.05 mills/kw-hr in each category. Since a number of parameters contribute to the complete picture, the cost-comparison totals depend on the respective parameter values used whether calculated from design specifications or merely assumed.

8.72 The dependence of the fuel-cycle costs on design parameters can, of course, be a basis for selecting optimum design specifications, as described in §8.41 et seq. In addition, the designer can be guided by interrelations that may develop among parameters. For example, the breeding ratio increases with the fuel volume fraction but decreases with the core height as leakage to the various blankets is reduced. On the other hand, enrichment decreases with fuel volume fraction; thus the combined effect of inventory and breeding ratio yields the doubling-time trends shown in Fig. 8.24, with only a weak dependence on core height. Parametric behavior can be determined by calculating a number of systematic variations of the parameters of interest. For a concept comparison

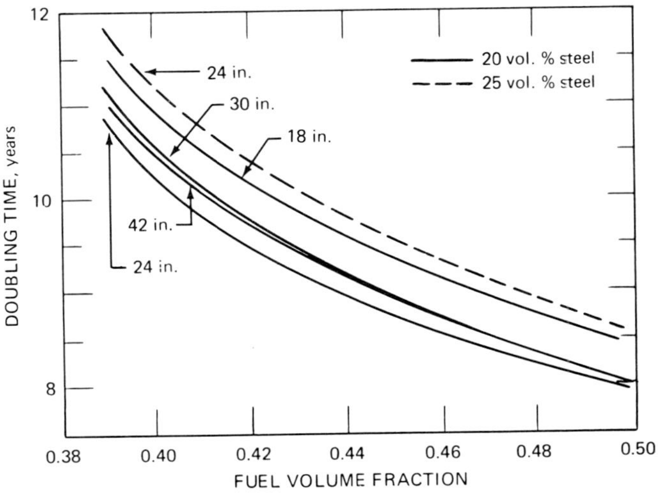

Fig. 8.24 Doubling time as a function of fuel volume fraction and core height.

like this one, it is generally satisfactory merely to consider the results from each reference design. However, in selecting the reference specifications, the designer should recognize the possible influence of parametric design trends on the results.

8.73 The assumptions made for such economic data as plutonium value, unit-fabrication costs, unit-recovery costs, and working-capital charge rate may also affect the results of an economic design study. Although relative costs are not likely to be greatly affected if consistent data are used in concept comparisons, economic data that are either too optimistic or too pessimistic could produce misleading cost estimates when compared with results for other systems. For example, Fig. 8.25 shows the difference in cost between the two core designs as a function of plutonium value for both "optimistic" and "pessimistic" economic data. If the pessimistic data should prove realistic, the conservative concept probably could not compete with other forms of power generation. We also see that the cost range corresponding to the economic data uncertainty is greater than the cost difference between cores determined by using a given set of data for each case.

8.74 Although in this particular comparison fuel-cycle costs are of primary importance, capital-cost effects should also be considered. Consistent with the conservative-core-design philosophy of containing the effects of an excursion, some additional costs are required in the reactor category; these amount to about $1.5 million more than the advanced-concept requirements. Also, the containment structure necessary for the conservative core results in an incremental cost of about another $1.5 million. The extensive additional instrumentation required for the advanced concept, however, results in an increment of almost $3 million. Thus a calculation of the capital-cost differences shows that the additional costs for containing the effects of an excursion in the conservative concept are almost exactly compensated by the additional costs for preventing such an excursion in the advanced concept.

COMPARISON-STUDY CONCLUSIONS

8.75 In addition to comparative behavior, information on the characteristics of each concept was developed in the study. It was concluded that the advanced design should be selected as worthy of additional study in preference to the conservative concept. Major considerations were the high economic potential and the likelihood that devices could be developed which would produce a design with acceptable safety. It was also concluded that probabilistic methodology should be applied to ensure a safe design. Although the conservative-core concept, with its dependence on containment, appeared to require somewhat less development effort, safety instrumentation for reactor scram and the use of probabilistic design to ensure an adequate level of safety

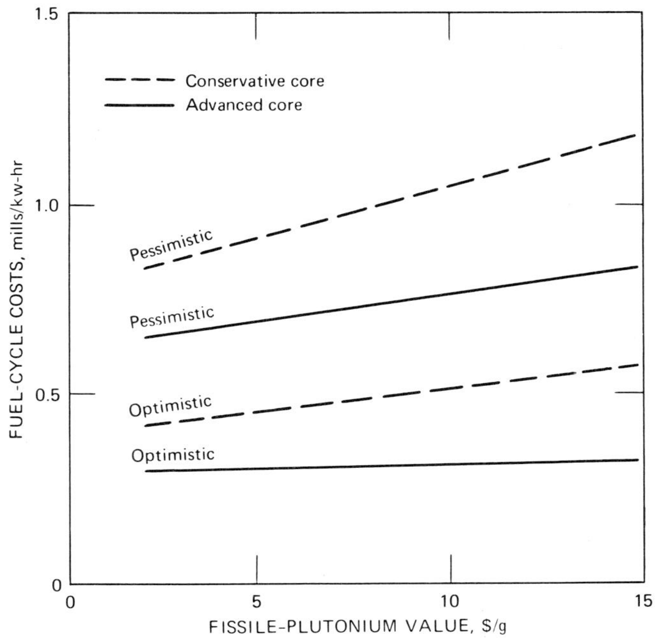

Fig. 8.25 Advanced- and conservative-design fuel-cycle costs determined from optimistic and pessimistic economic data.

Economic criteria	Optimistic approach	Pessimistic approach
Working capital charge rate	10%/year	13%/year
Core-fabrication costs (excluding inventory costs of fissile materials), $/kg (U + Pu)		
Advanced	132	294
Conservative	175	355
Recovery costs (excluding inventory costs of fissile materials), $/kg (U + Pu)		
Advanced	72	140
Conservative	67	140

were still needed. In other words, despite the favorable sodium-void criterion, uncertainties in accident analysis are such that it is still desirable to apply some of the approaches required in the advanced design to ensure safety. As a result, the advanced design, with its greater economic potential, is favored.

EVALUATION AND DESIGN METHODS

INTRODUCTION

8.76 Critical evaluation is an integral step in the design process. In the systems approach, as shown in Fig. 8.1, an initial solution to the design problem is evaluated by comparing it with established criteria, and the design process is iterated if the solution is found to be unsatisfactory.[9] Design efforts often require evaluation of a different sort to determine whether the methods used provide results within the desired confidence limits. We must be sure that all the parameter interplays that might affect the result have indeed been considered. Examples of such considerations were given in the previous discussions. The following sections supplement such examples and provide general perspective on the interrelated role of critical evaluation and design methods.

Range of Evaluation Requirements

8.77 In most practical design situations, decisions are not as easily made as is implied from the systems design approach in Fig. 8.1, i.e., by the straightforward comparison of two numbers, one representing a design parameter and the other the criterion. One problem is the uncertainty of confidence levels in both the design-parameter determination and the criterion value with which it is compared. In fact, the evaluation process may be applied to the analysis confidence level itself. A "no" decision could be based on excessive uncertainty in the design value being considered, and a requirement could be that additional information be developed to improve the confidence level. A second complication is the large amount of interplay among design parameters. The specification under consideration is always the result of compromise. Considerable engineering judgment may be necessary to establish a framework suitable for evaluation, wherein subjective estimates contribute to the "yes–no" decision. In addition, criteria for the evaluation may not be available and may have to be developed, and extra costs may be entailed.

8.78 The need to optimize design parameters may further complicate the procedure. Although proper evaluation of a proposed reactor concept or a new system is made only on an optimized design, funds for the optimization effort are not likely to be available until an evaluation has demonstrated that the additional design expense is justified. As part of the evaluation, it is therefore desirable at least to recognize the likely trends of design parameters as they move toward optimum values and the possible effects on the results.

8.79 Since computer calculation methods are used for most reactor-design efforts, the designer very often must compromise between a desire for results as accurate as technically possible and budgetary limitations on computer expense. This is particularly true for nuclear-design calculations, which can be very elaborate and costly if the most complex methods are used. Thus an important evaluation is to determine whether the calculational models used and the design methods applied to them are sufficiently accurate. Of course, what represents a "sufficient" degree of accuracy varies with the specific calculation and tends to evolve as a result of design practice.

8.80 Applying the "cost-of-the-benefits" criterion can provide some guidelines in establishing criteria for calculational accuracy. Often a monetary value can be assigned to the design uncertainty associated with a particular approach. For example, a small percentage of improvement in core average power due to reducing the uncertainty of such parameters as power peaking or departure from nucleate boiling (DNB) condition can yield substantial economic rewards. The cost of calculational methods needed to reduce the uncertainty may, therefore, be compared directly with the savings anticipated as a result of their use, and an optimum level of calculation effort can be determined.

8.81 In matters of safety analysis and evaluation, it is often difficult to develop an economic basis for the confidence level. Acceptable confidence levels may be prescribed by adopted standards and codes or by licensing agencies. Often, however, the designer can establish confidence criteria for himself by considering the alternatives available to him and determining whether the more sophisticated, and presumably more expensive, analysis method does indeed provide a significantly higher level of confidence in the safety of the system. Also, keep in mind that interplays among parameters are frequently quite complex and thus some variable other than the one under consideration can have a controlling effect on the safety characteristic being considered.

Role of Performance Testing in Evaluation

8.82 Related to design evaluation is performance testing, from which some guidance in selecting design criteria can be obtained. As for many engineering structures, process plants, and power plants, a program of "acceptance tests" is prescribed for the facility after it has been completed to determine whether it meets design specifications and is acceptable to the purchaser. Such tests vary but normally are carefully specified in the purchasing agreement.

8.83 Performance-test codes have been, or are being, developed for various parts of nuclear power plants. These generally evolve from experience and represent the state of the art. Committees of the American Society of Mechanical Engineers (ASME), for example, have developed performance-test codes for nuclear steam-supply systems and for nuclear reactor fuel.* In the code development emphasis was given to pressurized-water and boiling-water

*ASME Performance Test Code Committees 32.1 for Nuclear Steam Supply Systems and 32.2 for Nuclear Reactor fuel.

reactors. For pressurized-water reactors uniform procedures are established for conducting tests to determine the performance of a nuclear steam-supply system as a unit. Such performance codes provide some guidelines for evaluating the engineering design of the system.

Types of Criteria

8.84 Many kinds of criteria can be evaluated. For commercial nuclear power plants, it seems appropriate to consider performance, safety, and economic criteria. These areas are closely related to one another. Performance involves engineering criteria, which tend to interact with economic criteria. Safety criteria must be met if the design is to be acceptable. Since safety evaluation is treated in Chap. 6, engineering and economics evaluations are emphasized here. Because of the interplay between parameters, however, the distinctions between these criteria often are not strong.

Independent Evaluation

8.85 A type of evaluation that is somewhat separate from the process shown in Fig. 8.1 is that carried out by an independent organization to determine the merits of a design submitted by a sponsor seeking financial support for additional work on the concept. In addition to comparison with criteria, this evaluation is likely to verify features of the proposed concept by independent design calculations.

REVIEW AND ANALYSIS OF DESIGN APPROACH

8.86 We have emphasized the importance of reviewing the adequacy of the design method as a feature of the evaluation step in the generalized design process. Recognition by the designer of interplays between parameters is also essential. However, a design review is often concerned primarily with the design considerations themselves. The following sections discuss some of these design considerations from the evaluation viewpoint and summarize previous material.

8.87 Of primary interest in reviewing the reactor system design in terms of engineering criteria is the response of the system as a whole to a disturbance or change of a parameter. This response is the net result of the contributions of the subsystems comprising the system. But the subsystems, which are likely to interact in complicated ways, are normally subject to constraints in their individual response. For example, local power peaking in the core and fuel-temperature limitations are common constraints on parameters such as power density. As a result it is usually necessary to consider each subsystem separately as well as the system as a whole. It is well to point out again, however, that a rigorous systems engineering approach has yet to receive wide acceptance among reactor designers. Thus individual parameters are often given primary attention in the criteria comparison, and secondary consideration is devoted to interplay effects. Interplay effects, however, deserve careful attention because they may operate in opposing directions on a given parameter, and thus the designer may not be sure whether changing a parameter value in a given direction will produce the desired change in the response of the entire system. For this discussion the core and vessel are considered a single subsystem to focus attention on the design variables and their interplay which should be considered in the evaluation.

8.88 As indicated in Fig. 8.2, it is useful to classify core-design efforts into three primary areas: nuclear design, thermal–hydraulic design, and materials design. These areas provide a framework for additional discussion from the viewpoints of design evaluation and economic parameters.

Nuclear-Design or Physics-Analyses Evaluation

8.89 Examining the nuclear-design procedures and establishing a level of confidence are important in the engineering evaluation. Three general objectives are involved:

1. Determination of fuel inventories and lifetime fuel requirements needed for the fuel-cycle cost analysis.
2. Development of the power-distribution pattern in the core for use in the thermal–hydraulic design.
3. Analysis of reactivity coefficients as needed for the safety evaluation.

As discussed in Chap. 5, the approaches used are not independent. The calculations are of two basic types, however. One is concerned with the instantaneous behavior of the system and the other with changes during the operating history of the reactor to determine system fuel requirements.

Analytical Techniques

8.90 A wide variety of computational techniques is available for nuclear design. They vary in requirements from approximate calculations intended for "scouting" surveys to detailed, sophisticated analyses with multidimensional and heterogeneous representation for interpreting critical experiments. Some of these approaches were mentioned briefly in Chap. 5 and, being complex, are not described here; only some general comments are provided for orientation. A nuclear design can be independently evaluated only by a specialist who is able to determine whether the techniques are appropriate for the design objectives. Furthermore, many areas are still in a state of development.

8.91 Reactor manufacturers generally have developed their own calculation procedures, which may be very specific for the types of core lattices in the reactors they market. Evaluation may be somewhat difficult but can be simplified by showing comparisons with either experimental results or calculations performed with generally recognized procedures.

Nuclear Data

8.92 The validity of the nuclear-design calculations depends strongly on the nuclear data used and on the method of reducing the data to the form required by the specific calculation procedure. In multigroup calculations, for example, cross-section data obtained from a code "library" must be reduced to relatively broad group constants by averaging over the neutron spectrum. Therefore a code must be used in an iterative procedure to calculate the spectrum. In the independent evaluation of a nuclear design, therefore, the source of the constants used deserves attention.

Fuel Loading and Depletion

8.93 Initial fuel loading can usually be determined by straightforward eigenvalue calculations using a two-dimensional multigroup diffusion code. In considering depletion, we can use the fluxes developed from this problem to calculate reaction rates for a specified time, at the end of which the eigenvalue problem can be repeated and new fluxes determined, and so on.

8.94 Describing isotopic changes accurately tends to be a challenge since both space and time dependencies are involved and compromises are necessary to keep the calculation from becoming too complex. Uncertainties in depletion-analysis results are of importance primarily in evaluating fuel performance rather than in the initial design. It is true, however, that expected fuel performance can have a major effect on the reactor's economic potential. Some flexibility is normally available in the design and specification of replacement cores. Thus evaluation of the depletion-analysis methods need not be of major concern in the design evaluation of the reactor system if the results presented and the methods used appear to be within the range expected from experience with similar reactors.

Power Distributions

8.95 A detailed temperature "map" of the core depends on the calculation of power-generation distribution. Since the designer is concerned with the relation of power distribution to limiting conditions involving peak fuel temperature or heat flux (§8.100), he needs to determine local neutron-flux distributions under various projected operating conditions at different times during core life to ensure that the limitations are not exceeded.

8.96 The calculation presents considerable challenge. The control rods, part-length rods, burnable-poison rods, and the fuel-burnup distribution are all significant in establishing the power distribution. In addition, thermal–hydraulic and Doppler feedback considerations can have an influence. In pressurized-water reactors, where part-length rods may be used, for example, the behavior of the channel axial power in response to core overpower can affect the acceptability of the specified power density.

8.97 Reactor manufacturers have developed confidence in the analytical procedures and design methods used for water-moderated reactors by comparing them with results from critical experiments and operating reactors. How appropriate such methods may be for new concepts must be evaluated since geometry and other parameters may differ. Unfortunately evaluation is sometimes difficult since some of the procedures are proprietary.

Kinetic Characteristics

8.98 The response of the reactor core to both normal and abnormal changes in operating conditions must be determined as part of the design effort. As described in Chap. 6, the reactivity response is normally expressed in terms of feedback coefficients that are likely to change in value during the life of the core. These include moderator, Doppler, and power coefficients. Another area of nuclear design required for safety evaluation includes control-rod worth and related characteristics, xenon stability, and other reactivity-related inputs needed for accident analyses. This is a large, sophisticated area that for many concepts is still in a state of development.

Thermal–Hydraulic Design

8.99 Thermal–hydraulic parameters include the coolant flow pattern, coolant properties, heat-transfer coefficients, and such considerations as boiling, two-phase behavior, and interchannel mixing. Some of these are indicated in the integration pattern given in Fig. 8.2, and the design interaction for a fuel element for a liquid-metal fast breeder reactor is shown in Fig. 8.3. These illustrations emphasize that the thermal picture of the reactor core is complex and that the considerable interplay among parameters must be considered by the designer.

8.100 The design iteration required can best be handled through the use of digital-computer methods. The calculation flow sheet for one such code, REPP, for water-cooled reactors[10] is shown in Fig. 8.26. Both pressurized-water and boiling-water reactor cores can be designed with reactor power, burnout heat flux, and fuel center-line temperature as the design limits. Pressure drop, coolant temperatures, surface heat fluxes, and fuel center-line temperatures are predicted from input data and a thermal-energy balance at various mesh points in a one-dimensional calculation. Iteration is used to determine geometry and core size for a given power level. Either square or triangular lattice designs can be accommodated.

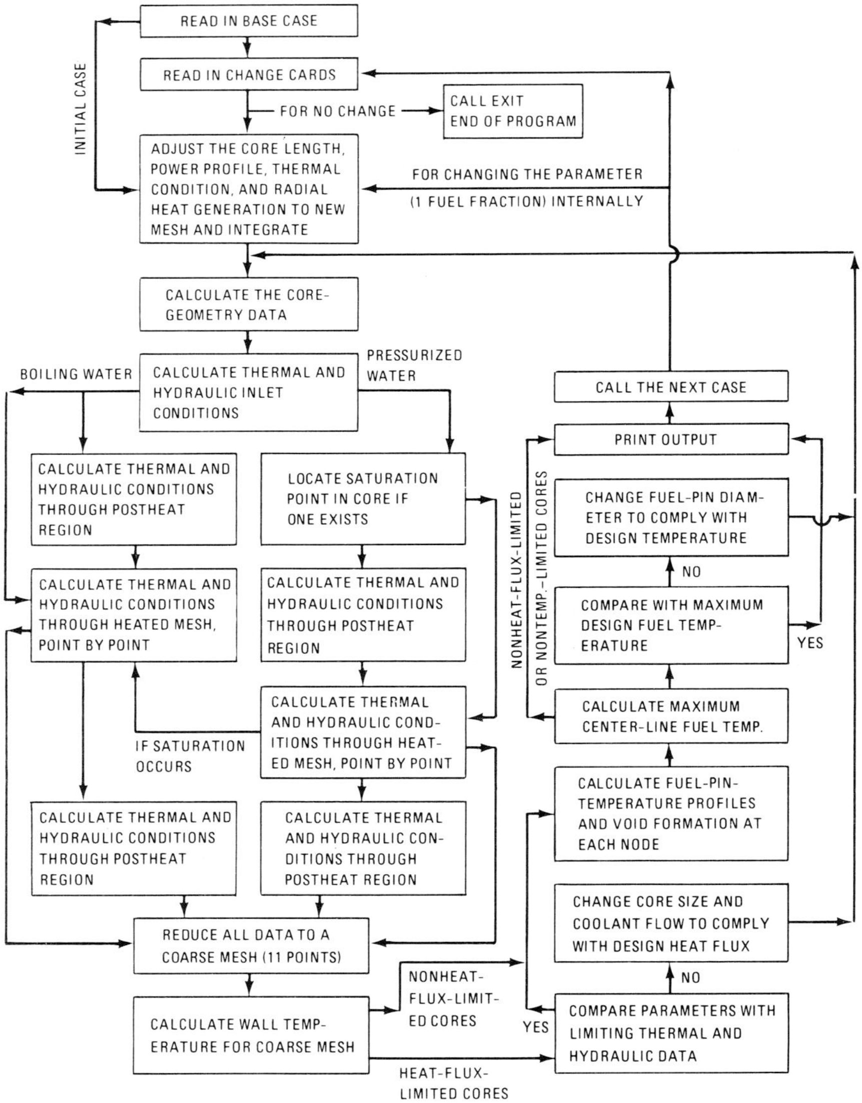

Fig. 8.26 REPP flow chart for water-cooled reactors.

8.101 Some of the features of this program illustrate the type of calculational capability available to the thermal–hydraulics designer. The program includes three major steps:

1. Calculation from input data of an initial core geometry and prediction of thermal and hydraulic conditions throughout the core.

2. Determination by iteration of the core size required so that the thermal–hydraulic parameters compare favorably with design criteria, particularly burnout heat flux, at a given reactor power level and a specified power distribution.

3. Design of a core, including fuel-pin diameter required to perform within specified fuel center-line temperature limits at a specified reactor power level.

After each step is completed, the user can either request the output and call for another case or continue the design process through the next step as indicated in Fig. 8.26, which serves as a "road map" of the procedure.

8.102 In calculating the thermal–hydraulic conditions in the core, we define from 10 to 100 nodes parallel to a coolant channel. At each nodal position both nominal and hot-channel conditions are predicted, based on peaking factors, energy and momentum balance, and nominal conditions. Both one-dimensional conduction and forced convection, including local boiling, are considered; for this the relative power profile (local/average), the relative fuel-pin heat-generation rate (local/average), and a thermal-conductivity profile are used as input parameters.

8.103 The local heat flux at each of the nodes and the energy increase of the coolant are calculated by using the integrated power-profile data and, in turn, the coolant temperature and local-fuel-surface temperature based on a calculated heat-transfer coefficient. A fuel-pin temperature profile perpendicular to the coolant channel can also be determined from integrated thermal-conductivity data and integrated radial-heat-generation data.

8.104 After the thermal–hydraulic conditions are described, the user can transfer control to the output section, concluding the run, or can cause the code to iterate on a core heat-transfer surface area that will be limited by heat flux. In this process a burnout heat flux is calculated for each node by using semiempirical correlations and input data. The position of minimum departure from nucleate boiling ratio* (MDNBR) is located, and the core is sized on the basis of conditions at this location. This is accomplished by transferring control to a sizing routine where the equivalent core radius, the core length, the number of fuel pins, and other parameters are adjusted to satisfy the heat-flux requirements. After the core parameters are modified, control is transferred to the beginning of the program and the process is repeated until the design heat-flux requirements are met at the point of MDNBR. The reactor power level is left unchanged in the process.

8.105 When the core sizing routine is completed, control can be transferred to the output section, concluding the run, or the process can be continued, with a fuel-pin center-line temperature limitation used to determine a fuel-pin diameter that will perform within desired temperature requirements when subjected to hot-channel conditions. This option requires more computer time

---

*The DNB ratio is equal to the local burnout heat flux divided by the local heat flux.

than the previous cases because each time the fuel-pin diameter, and consequently the fuel-pin pitch (as mitigated by the temperature requirement), is modified, control is transferred to the beginning of the program where a new heat-transfer-limited core based on the new rod size is generated.

8.106 In design calculations involving descriptions of physical phenomena, the accuracy obtained is limited by the relations used in the analysis. Thermal–hydraulic calculations are particularly subject to uncertainties since a number of relations, all having some error, are involved. These include relations for heat transfer, the need to locate the point of incipient nucleate boiling, pressure drop under two-phase flow conditions, coolant cross-flow effects, and hot-channel effects. In addition, when comparisons must be made with limiting parameters, uncertainties in the analytical expressions that can be used for the parameters are usually involved. For example, for pressurized-water reactors the W-3 correlation can be used to evaluate DNB conditions (§§4.77 and 4.78). For boiling-water reactors empirical burnout curves (§4.79) can be used. Although these may well represent the best description available, their possible limitations and the potential for improvement should be recognized. In the mathematical models used to represent heat transfer and fluid flow in the core, accuracy is likely to be limited by the uncertainties in the equations used to predict heat and momentum transfer rates. With computers available, the conduction heat-transfer models used are probably fairly complete. Inherent uncertainty, which may be as high as 20%, however, can be found in the convection, flow-distribution, and fluid heat-transfer relations. About a 20% uncertainty is also associated with DNB predictions for water-cooled reactors.

Materials Criteria

8.107 The reactor core must be constructed of materials that will retain their properties for a long period of time in the severe environmental conditions normally existing in the core. Since many different types of materials are used (e.g., for cladding, fuel, and core structure) and since requirements for each material vary, the subject is a complicated one. Checking the suitability of the materials selected is important in the design evaluation. This could be accomplished by comparing properties of the materials with criteria evolved from design specifications. However, a frequent challenge is to estimate the material's performance under reactor operating conditions. In addition to making a straightforward comparison, therefore, the evaluator may consider the evidence available to justify materials-performance predictions, including the confidence limits of available data.

Fuel-Element Evaluation

8.108 In most reactor designs much of the materials evaluation concerns the fuel elements. Cladding and fuel-element temperature distributions are generally determined as part of the thermal–hydraulic analysis. This picture can be compared with criteria for the specific materials selected. Dimensional changes, strength, and corrosion of the cladding are of particular concern. In ceramic fuels the possibility of central melting is usually considered undesirable, particularly for fast reactors, because vertical rods could suffer fuel redistribution or slumping and local hot spots would be likely to develop. Finally, the effect of burnup on temperature-related parameters must be considered. In addition to a shift in power distribution, there are likely to be a lowering of the thermal conductivity of the fuel, with a corresponding higher central temperature, and deterioration of the cladding physical properties, all as a result of irradiation exposure.

8.109 At the high burnup desirable for fast reactor fuels (§8.70), fission gases are released and internal pressure can build up in the fuel pin, with a corresponding effect on the stresses applied to the cladding. In addition, the possibility of some fuel swelling from the accumulation of fission products must be considered.

8.110 The mechanical and chemical behavior of the fuel pin and cladding combination at extremes of temperature, radiation, and operational cycling is another area of review. The fuel cladding might possibly collapse under external pressure. In addition to corrosion by the coolant, the compatibility of the fuel cladding and the fuel coolant should be considered. Vibrational effects are also worthy of attention.

8.111 Often overlooked is the "manufacturability" of the fuel-element system. The system should be designed so that it can be fabricated at reasonable cost with manufacturing tolerances at low levels consistent with the need for low engineering factors.

Other Core Materials

8.112 In addition to the fuel assemblies, the core materials include the internal structure required to support the fuel, the various flow baffles, and the control-rod assemblies. Most reactors have a lower grid structure for supporting fuel and perhaps associated baffles for separating coolant streams. In design evaluation the difficulties involved with such matters as accessibility for removal of foreign objects and the possible removal and replacement of the structure are of concern. The structure must, of course, be properly designed for its intended function and be fabricated of suitable materials.

8.113 The control-rod system is usually considered with the instrumentation and control requirements. From the materials viewpoint, however, the design must provide for dimensional stability over long periods of time under the severe environmental conditions in the core, including neutron irradiation. Design details differ considerably among concepts, with corresponding different materials requirements. In boiling-water reactors, for example, where cruciform B_4C blades are generally used (§6.47), an operating lifetime of about 14 full-power years is usually planned. Lifetime is limited by both loss of rod effectiveness and deterioration of mechanical properties.

Reactor Vessel

8.114 In reactor concepts requiring a thick-walled vessel, the vessel system is a major plant component. Since one or more radial thermal shields are normally needed to reduce the fast-neutron exposure and gamma flux at the vessel walls, the design requirements for the shields are integrated with those for the vessel. In pressurized-water and boiling-water reactors, which have reached a high stage of development, the reactor vessel is designed, fabricated, and tested in general accordance with the American Society of Mechanical Engineers Boiler and Pressure Vessel Code. It is usually fabricated from low-alloy steel clad on the inside with stainless steel to minimize corrosion. Advanced concepts, such as a steam-cooled fast breeder reactor, may not be covered by the ASME code and thus may require special attention. Prestressed-concrete reactor vessels, usually specified for gas-cooled reactors, may also require special analysis.[11]

Noncore Steam-Supply-System Parameters

8.115 In accordance with industrial practice, the responsibility of the nuclear engineer is normally limited to the nuclear steam-supply system, which includes the reactor and the steam-generating equipment required to transfer the energy generated in the reactor to the working fluid of the power cycle. Previously in this section attention has been given to parameters applying to core and vessel design. Since these are the key parameters that define the system, at least at the conceptual stage, they deserve emphasis. Noncore parameters are those associated with the coolant system and steam generators. Coolant circulation rate and temperature pattern, on the other hand, are best considered as part of the core-design specifications. Among the various noncore parameters, those affecting the power-cycle efficiency (e.g., the steam-generator temperature pattern and operating pressure) are probably the most significant at the conceptual stage. Other parameters, such as coolant-pump specifications and material selections, are vital to the finished design but do not greatly influence the potential of the concept and do not interplay strongly with parameters that do. Problems requiring significant research and development effort may possibly arise, however.

8.116 In evaluation it is appropriate to give primary attention to the core-design-related parameters and to review carefully possible design problems in the noncore system. In sodium-cooled fast reactors, for example, the development of a satisfactory steam generator represented a major challenge. Questions of component reliability and quality assurance, which could affect the engineering performance of the design, should not be overlooked.

8.117 Arrangements for refueling, practicality of the component layout, number of coolant loops, and features affecting the ability to provide maintenance at reasonable cost are other important areas in the engineering evaluation. Such plant engineering features as these have an important bearing on the capital cost. Although optimization may not be attempted in a preliminary design study, the features chosen must be acceptable if the corresponding cost analysis is to be valid.

Economic Criteria

8.118 To evaluate the present and/or future economic performance of the concept is an objective of most reactor-design studies. This is usually done by determining power costs under a set of reference economic conditions (§8.73). It is usually helpful to calculate the power costs in terms of capital, fuel-cycle, and operation and maintenance cost components as an analysis guide. Fuel economy can be considered but usually is not a primary consideration in determining the potential of a concept (§7.113). Also important in the cost analysis is to provide a framework for selecting optimum design parameters, as mentioned in §8.37, for fast reactors. Figures 8.27 and 8.28 illustrate for

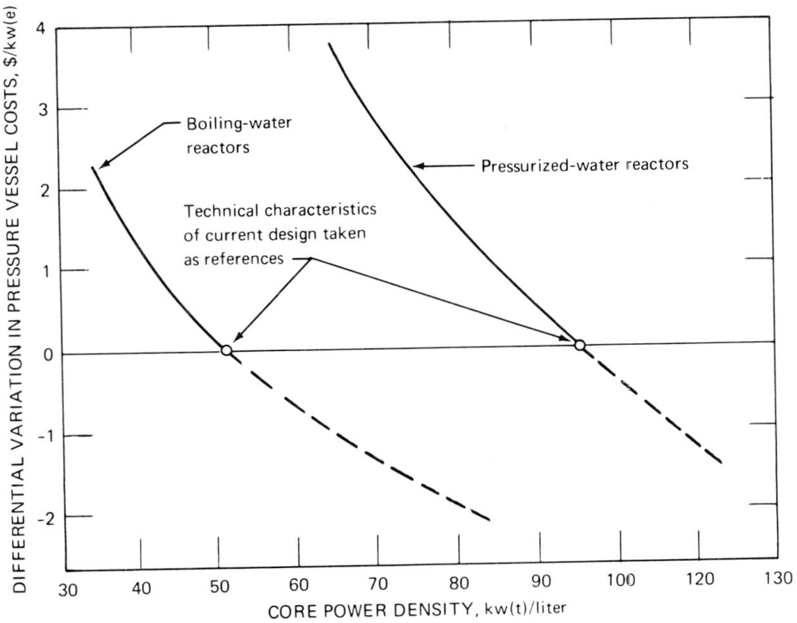

Fig. 8.27 Savings in light-water-reactor pressure vessels as a function of core power density.

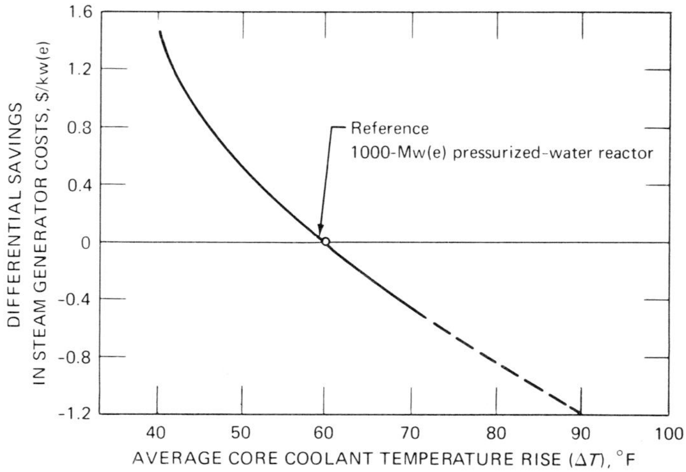

Fig. 8.28 Differential savings in steam-generator costs for pressurized-water-reactor plants.

light-water reactors the influences on steam-generator costs of core power density and coolant temperature rise in the pressure vessel, respectively.

8.119 The reference economic conditions may vary from year to year depending on prices, interest rates, etc. Therefore it is desirable to determine whether conclusions reached are likely to be affected by changes in the reference economic conditions assumed. A parametric study or sensitivity analysis can be carried out by using a computer program to calculate costs and by systematically varying the input reference-economic-condition values, as described in §8.73.

Fuel Costs

8.120 Among the several cost components, the fuel-cycle costs are most difficult to determine, primarily because many parameters that contribute to the fuel cycle are effective throughout the lifetime of the plant. As a result, as described in Chap. 3, a wide range of sophistication is possible in fuel-cost analyses. In a simplified approach a time-averaged cost can be calculated by using assumed values for plant parameters, such as load factor and efficiency; fuel-cycle economic parameters, such as unit-fabrication and separative-work processing costs; and financial parameters, such as the rate of return on invested capital, interest rates, and taxes. If more accuracy is desired, changes in parameters occurring from one year to another during operation can be considered. Some attempt can still be made to develop levelized costs by using present-worth techniques (§2.33) to account for cash flows at different times. At another level of sophistication, we might consider operational parameters of the utility network, e.g., load sharing between plants due to shifts in incremental fuel costs (§7.56). In fact, greater and greater complexity can be developed by considering more and more detail.

8.121 Since attempts to forecast market conditions and various other parameters for long periods of time are subject to inherent uncertainties, there is a limit to the degree of sophistication in fuel-cycle analyses which is meaningful. Since standard approaches have not yet evolved, the evaluator of a design cost estimate must use considerable judgment in determining whether the procedure and its accuracy are consistent with the objectives of the design effort.

Capital Costs

8.122 Capital-cost estimates included with a conceptual design study can be evaluated in several ways. Comparison with cost data or estimates for analogous systems may show some areas of difference which should be investigated. Agreement with other studies does not necessarily indicate confirmation, however, since the same sources may have been used for all estimates. Where possible, independent estimates of major components can be made, perhaps with the aid of manufacturers' bids. Finally, the evaluator's cost experience can provide guidance regarding the reasonableness of the estimate.

8.123 In advanced-concept studies the possibility of uncertainties in the capital costs should be recognized. Some of these can be attributed to uncertainties in design features. When competing designs for the same concept are being considered, differences in component layout or safety features may lead to different results; this in turn, may be interpreted as an uncertainty. Finally, cost projections over a period of years, normally required for advanced concepts, involve considerable uncertainty in the indirect costs that may be appropriate and the economic parameters that should be applied.

REFERENCES

1. Argonne National Laboratory, Liquid Metal Fast Breeder Reactor (LMFBR) Program Plan. Volume 6. Core Design, USAEC Report WASH-1106, pp. 6–241, August 1968.
2. Atomics International, 1000 MWe LMFBR (Liquid Metal Fast Breeder Reactor) Follow-On Study. Task I Report, USAEC Report AI-AEC-12765(Vol. 2), Nov. 12, 1968.
3. General Electric Company, Comparison of Two Sodium-Cooled 1000 MWe Fast Reactor Concepts. Task I 1000 MWe LMFBR Follow-On Work, USAEC Report GEAP-5618, June 1968.
4. A. S. Gibson, C. E. Boardman, and J. H. Germer, Plant Design, System Optimization, and Equipment Selection for an Advanced LMFBR, in Proceedings of the International Conference on Sodium Technology and Large Fast Reactor Design, Argonne, Ill., Nov. 7–9, 1968. Part II. Sessions on Large Fast Reactor Design, USAEC Report ANL-7520 (Pt. 2), pp. 369–387, Argonne National Laboratory, November 1968.
5. A. Sesonske and J. G. Yevick, Descriptions of Fast Reactors, in *Fast Reactor Technology: Plant Design,* pp. 677–746, The M.I.T. Press, Cambridge, Mass., 1966.

6. D. B. Sherer, W. A. Sangster, and F. W. Sciacca, Alternate Core Design Parameters and Economic Trade-Offs for a 1000 MWe Sodium-Cooled Fast Reactor, in Proceedings of the International Conference on Sodium Technology and Large Fast Reactor Design, Argonne, Ill., Nov. 7–9, 1968. Part II. Sessions on Large Fast Reactor Design, USAEC Report ANL-7520(Pt. 2), pp. 358–368, Argonne National Laboratory, November 1968.
7. Ref. 3, p. 317.
8. L. D. O'Dell, R. W. Hardie, and W. W. Little, Jr., A Numerical Comparison of Diffusion and Transport (S_n) Codes for Selected Fast Reactor Configurations, USAEC Report BNWL-992, Battelle–Northwest, March 1969.
9. Morris Asimow, *Introduction to Design,* Prentice-Hall, Inc., Englewood Cliffs, N.J., 1962.
10. R. M. Hiatt and C. Bromley, Jr., REPP: A Thermal Hydraulic Design Code for Water-Cooled Reactors, USAEC Report BNWL-1013, Battelle–Northwest, March 1969.
11. Oak Ridge National Laboratory, An Evaluation of Gas-Cooled Fast Reactors, USAEC Report Wash-1089, April 1969.
12. Jackson and Moreland and Stoller Associates, Current Status and Future Technical and Economic Potential of Light Water Reactors, USAEC Report Wash-1082, March 1968.

Some Optimization Approaches

9

INTRODUCTION

9.1 As part of the engineering design process, optimization is concerned with the search and decision activity needed to obtain a solution most favorable with respect to a given criterion. This can be described in a slightly different way as the process of determining the system parameters and mode of operation which maximizes a prescribed figure of merit.

9.2 This chapter is intended to draw the attention of the designer, who may be uninitiated in optimization procedures, to the many possible approaches that may prove useful to him. Consistent with such a limited objective, only a superficial discussion is given, with mathematical theory at a minimum. Since such theory is relatively sophisticated, the reader must consult other sources to develop the background needed to apply the methods to real problems.

9.3 A wide variety of methods have been developed for determining the optimum solution to a problem. If a mathematical model of the problem can be formulated, the generally familiar methods of the calculus are available for determining maxima and minima.[1] For a single variable the behavior of the first and second derivatives can be used to locate maximum and minimum points on a curve. The methods of partial differentiation can be applied when a function of more than one independent variable is considered. In recent years the more sophisticated of the differential methods have become part of a discipline that can be called *optimization theory*, as discussed later (§9.14).

9.4 For many years before optimization theory evolved, engineers also used graphical methods to find the "best" variable or combination of two variables in a given design, particularly in situations that did not lend themselves to mathematical modeling. In simple cases where an optimum value of one independent variable is desired with respect to a criterion such as cost, various contributing costs can be plotted as a function of the independent variable. The algebraic sum of the ordinates at several values of the independent variable can then be used to plot the total cost as a function of the variable with an optimum obtained at the minimum cost value. Such a procedure is frequently used to analyze fuel costs as a function of burnup, as shown in Fig. 9.1.

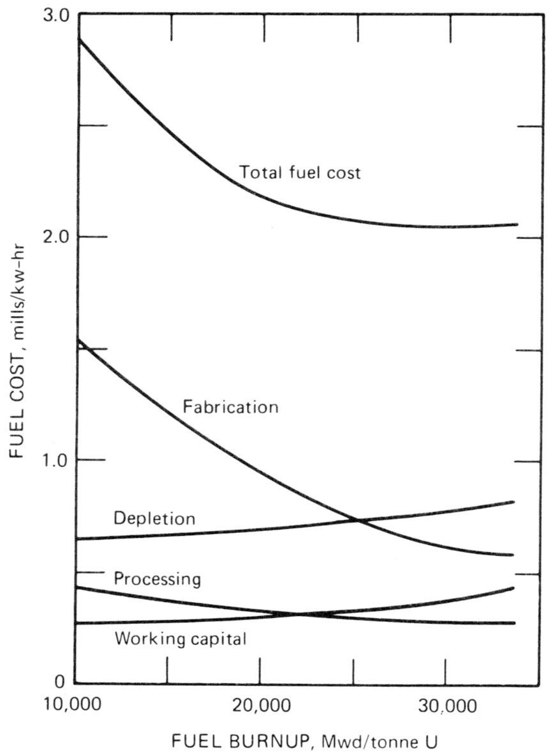

Fig. 9.1 Typical fuel-cycle costs as a function of burnup.

9.5 Another example is shown in Fig. 9.2 from a design study for a flash-distillation water-desalting plant.[2] In such a plant the steam, which serves as a heat source, can be more efficiently used if the temperature differential over the brine heater is reduced and additional heat-transfer surface provided. As a

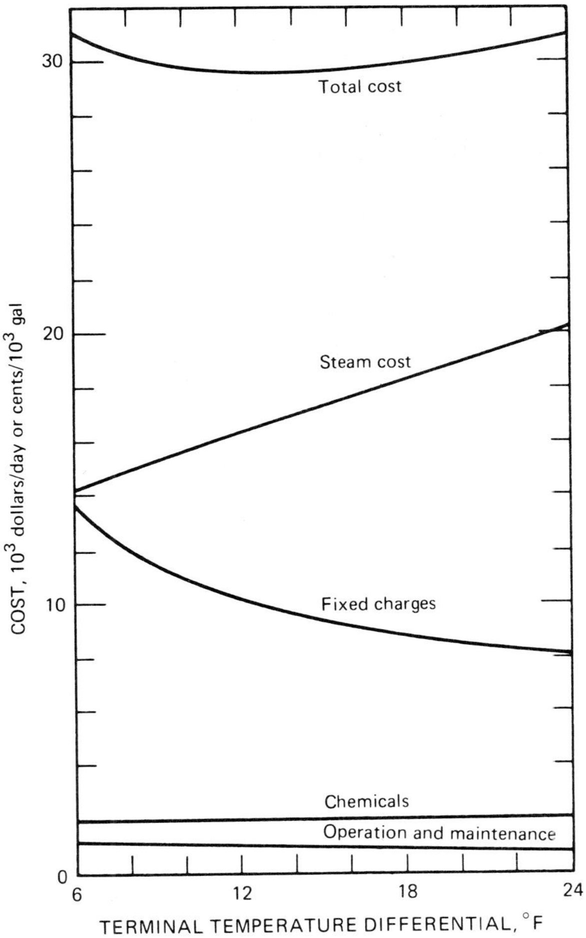

Fig. 9.2 Water-production cost optimization, 10^8 gpd plant.

result a trade-off applies between the cost of steam and the fixed charges associated with the heat-transfer surface.

9.6 If several *independent* design variables are to be determined, *guesses*, based on past experience, would be made of suitable values as an initial step, and then the necessary dependent variables would be determined to establish the design. An optimization process can then be initiated for a criterion, such as cost, by taking the first independent variable and determining system costs corresponding to several assumed values both larger and smaller than the original guess. The total cost can then be plotted vs. the variable, and an optimum value can be obtained from this curve. The dependent variables can then be adjusted. This procedure may be repeated for all independent variables in turn to yield a "first-round optimized design." Since in practical situations the assumed "independent" variables can have some dependency, additional iteration may be necessary to achieve a satisfactory optimum combination of variables.[3] This approach is quite tedious but can be made less so with the digital computer. In fact, variations of this approach using computer programs are still widely used, including some applications to nuclear systems.[4,5]

9.7 For a number of years, therefore, two different types of optimization approaches have been developed. First, classical approaches based on the calculus provided means of optimizing continuous variables with differentiable functions. A geometric background formed the basis of many of the problems considered. Optimization theory from classical calculus was developed primarily before the end of the nineteenth century. A second, much less elegant, parallel effort used graphical and approximate solutions to practical problems that generally could not be solved by the calculus.

9.8 The development of the digital computer and the requirements of World War II led to advances in optimization theory as part of a subject known as *operations research,* in which the characteristics of mathematical models of complex logistic, production, and distribution systems were studied. As digital computers became more sophisticated, the solution of problems containing many variables and the development of iterative optimization schemes became practical. Modern optimization theory therefore evolved as a way of solving complex decision problems. Such search methods as linear programming, nonlinear programming, and dynamic programming were developed.[6]

TERMINOLOGY

9.9 Before we discuss some of these techniques as background for possible nuclear applications, some terms should be defined.[7] The *modeling,* or mathematical representation of the problem, is a key requirement for any optimization approach. In fact, there may be a degree of refinement of the model itself which is optimum in response to conflicting requirements of economy and the cost of sufficiently detailed representation. A minimum or a maximum value is sought for an *objective function* which depends on *state variables* and *parameters** but which may be subject to *constraints.* In other words, we can measure the overall effectiveness of the design model by the objective function, or, as it is sometimes called, the *criterion function.* The functional relations can be quite complex in practical cases, the objective function being determined by a number of dependent variables, each of which in turn may depend on numerous parameters. A major challenge in most

*Variables can be manipulated to achieve the desired objective, but parameters are not controllable, although they do affect the objective.

optimization, therefore, is the setting up of the objective function with the necessary quantification of the variables and parameters.

9.10 The general optimization problem can be formulated as in Eqs. 9.1 and 9.2. It is desired to determine values for *n-decision* variables $x_1,...,x_n$ that satisfy the equalities or inequalities

$$g_i(x_1,...,x_n)\{\leqslant, =, \geqslant\}b_i; i = 1,...,m \tag{9.1}$$

and maximize or minimize the function

$$F = f(x_1,...,x_n) \tag{9.2}$$

The restrictions in Eq. 9.1 are called the constraints, and Eq. 9.2 represents the objective function. Equation 9.1 designates a set of equations and/or inequalities that determine the *region of feasibility* of the problem.

9.11 In most optimization problems limitations on the values that the decision variables can assume are generally given by specifying a constraint set. Any set of decision variables satisfying the constraints is known as a feasible solution to the model. However, the optimum solution is the one which results in the maximum or mimimum of the objective function. When the objective function must also satisfy relations between the independent variables, the optimum lies on a boundary of the feasible region, where standard indirect (§9.15) methods are not valid. It is possible, however, to transform such a problem into a new one having the optimum inside the feasible region (§9.21). Inequality constraints can frequently be handled by the same methods used for equality constraints by reformulating to introduce a *slack variable,* which provides a basis for a class of optimization problems called *mathematical programming.*

9.12 In the formulation of optimization problems, the objective function, or *payoff function,* is also frequently considered a *state vector,* whose value is determined by a set of decision-variable vectors, x_n, and a set of parameter vectors, p_n. The values of x contribute to a decision *policy* that can be adjusted to obtain the best value of the objective function within the limitation of the constraints. In multistage decision processes, to which dynamic programming (§9.49) can be applied, each stage in the progression can be described by a set of state variables.

9.13 In all types of optimization problems, the solution to a model is no better than the model itself. As is true in much of design analysis, therefore, a trade-off occurs between accuracy of representation and the approximations needed for practical analytical solutions. In optimization work a simple model is normally suggested as a first step. It also provides a basis for sensitivity analysis, in which the effects of changing the values of variables, parameters, or constraints can be studied. If parameters and variables apparently having a negligible effect on the objective function can be identified, it may be possible to simplify the solution and still achieve adequate accuracy.

9.14 Optimization procedures can be classified as either *direct methods* or *indirect methods*. Indirect methods are those in which a minimum or a maximum is sought by means of a necessary condition. For example, classical analytical approaches involving differentiation and the behavior of derivatives are indirect approaches. In direct methods a maximum or a minimum is obtained by direct comparison of the values of the function at two or more points. Many systematic numerical procedures have been developed for solving maximum–minimum problems by direct methods. Some of these will be considered here (§9.33). However, optimization theory is a complicated discipline that is described in its literature[7] and cannot be developed here. Only an orientation-level discussion of some methods that have found some application in nuclear reactor design is therefore intended.

INDIRECT METHODS

INTRODUCTION

9.15 The distinction between direct and indirect methods is useful for orientation but is not necessarily absolute, since there is sometimes some overlap, particularly when approximate methods must be used in conjunction with analytical approaches. Classical methods involving the differential viewpoint are generally included in what is known as the Theory of Ordinary Maxima and Mimima[8] and are not considered part of modern optimization theory. Such methods depend on differentiation to yield the characteristics of continuous functions about maxima and minima and are described in most calculus books.

9.16 The possibility of local extrema, which satisfy the conditions but may not be the absolute maximum, can be of concern, however. Consider the geometric representation shown in Fig. 9.3, for example, where contour lines represent the value of a function z, depending on two independent variables y and x.

9.17 The maximum of the function is located at point M at the top of a "peak." If this were a sharp ridge, the derivative of z with respect to x and y would be discontinuous. A second but lower maximum is located at point m, which is "higher" than all points in its immediate vicinity. The highest of all the points in a suitably defined region is called an absolute, or global, maximum; and a point, such as m, that is higher than all the points in a suitably defined small neighborhood is called a local maximum. The derivative of z with respect to x and y at point m is equal to zero. A point at which a function has all its partial derivatives with respect to the independent variables equal to zero is called a

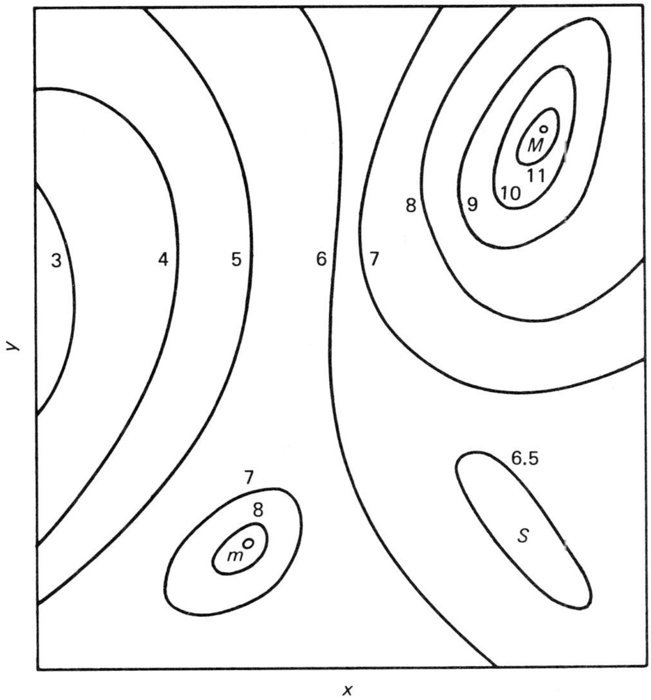

Fig. 9.3 Geometric representation of local extrema. Contour lines represent the value of a function z dependent on independent variables y and x. M, maximum of function. m, second, lower maximum. S, saddle point.

stationary point. A so-called *saddle point* could occur on a plateau such as at S, which is also a stationary point.

9.18 If the sides of the figure are considered as the boundaries of the problem, the minimum of the function z does not occur in the interior of the region, but along the boundary. In fact, depending on the contour structure, local minima (or maxima) could occur in some problems in several different places along different boundaries. The absolute minimum, or maximum, must then be found by comparing such values.

9.19 It is therefore possible to obtain stationary points that are other than optima by differential methods. Several methods are available, however, for distinguishing between various kinds of stationary points. For example, one useful approach involves completing the square on second-order terms of a Taylor expansion.[8]

9.20 Variables not independent but related to one another frequently complicate practical problems. In other words, the problem can be to find an extremum of $f(x_1,x_2)$ subject to a subsidiary condition $g(x_1,x_2) = 0$ that has the effect of a constraint. Although in principle the subsidiary condition may be used as part of a simultaneous equation, a solution of the subsidiary relation in terms of the variables may be difficult. Under such conditions *Lagrange's Method of Undetermined Multipliers* is useful.

9.21 For an extremum problem with a subsidiary condition such as the preceding one, for example, the necessary condition for a stationary solution at a point is

$$\frac{\partial f}{\partial x_1}\frac{\partial g}{\partial x_2} - \frac{\partial f}{\partial x_2}\frac{\partial g}{\partial x_1} = 0 \tag{9.3}$$

which can be rewritten as

$$\frac{\partial f/\partial x_1}{\partial g/\partial x_1} = \frac{\partial f/\partial x_2}{\partial g/\partial x_2} = -\lambda \tag{9.4}$$

where λ is a constant called the Lagrange multiplier. Equation 9.4 may be expressed

$$\frac{\partial f}{\partial x_1} + \lambda\frac{\partial g}{\partial x_1} = 0 \quad \text{and} \quad \frac{\partial f}{\partial x_2} + \lambda\frac{\partial g}{\partial x_2} = 0 \tag{9.5}$$

These relations represent the necessary conditions for a stationary point of the function $f + \lambda g$ without subsidiary conditions. Lagrange multipliers may therefore be used to transform a constrained optimization problem into an equivalent unconstrained one.

VARIATIONAL METHODS

9.22 Some important practical optimization problems involving integrals rather than functions can be solved through the calculus of variations.[6] First, the distinction between variational methods and ordinary maxima and minima should be clarified. As mentioned in §9.15, the problem in ordinary maxima or minima is to find the respective values of each of n-independent variables $x_1, x_2, ..., x_n$ for which a function of these variables has a so-called extremum. Now, consider a *functional* that depends on n-independent functions $x_1(t)$, $x_2(t)$, ..., $x_n(t)$. In the calculus of variations, the problem is to find an extremum of the functional.[9]

9.23 A geometric interpretation is the determination of a curve, specified by the n parametric equations $x_1(t) = x_1$, $x_2(t) = x_2$, ..., $x_n(t) = x_n$, which maximizes a functional of the n functions $x_1(t), x_2(t), ..., x_n(t)$. This curve may be considered to lie either in the n-dimensional space composed of $x_1, x_2, ..., x_n$ or in the $(n + 1)$-dimensional space composed of t, $x_1, x_2, ..., x_n$. Such a curve corresponds to the determination of a point in ordinary maxima and minima in an n-dimensional space which is composed of the n-independent variables $x_1, x_2, ..., x_n$ and which maximizes a function of the n variables.

9.24 An example of a functional is the integral,

$$I = \int_{t_1}^{t_2} F(t,x,dx/dt)\,dt \tag{9.6}$$

To optimize the value of the integral, we need to determine the function $x(t)$. An important simplification is a functional integral that contains no derivatives.

$$I = \int_{t_1}^{t_2} F(x_1,x_2,\ldots,x_n,t)\,dt \tag{9.7}$$

In the variational approach a maximum or a mimimum must satisfy the *Euler* condition. This is analogous to the requirement that the first derivative must vanish in ordinary maxima and minima theory. The Euler condition has the form

$$\frac{d}{dt}(F_{x'}) - F_x = 0 \tag{9.8}$$

or in expanded form

$$F_{x'x'}x'' + F_{xx'}x' + F_{tx'} - F_x = 0 \tag{9.9}$$

where notation F_x indicates a partial derivative of F with respect to x,

$$F_x = \frac{\partial F}{\partial x} \quad \text{and} \quad F_{xx'} = \frac{\partial^2 F}{\partial x \partial x'}$$

and

$$x' = \frac{dx}{dt} \qquad x'' = \frac{d^2 x}{dt^2}$$

9.25 In this case the Euler equations reduce to

$$\frac{\partial F}{\partial x_1} = 0, \quad \frac{\partial F}{\partial x_2} = 0, \quad \ldots, \quad \frac{\partial F}{\partial x_n} = 0 \tag{9.10}$$

These equations represent the solutions of a succession of ordinary minimum problems for every value of t from t_1 to t_2. The solutions define a series of functions $x_1(t)$, $x_2(t)$, ..., $x_n(t)$ that represent the solution to the problem given in Eq. 9.7.

9.26 The preceding problem is a simplification of a more general problem requiring the solution of n simultaneous ordinary differential equations that are usually nonlinear and highly resistant to analytic solution.[3] However, as a general procedure, the Euler equations represent a good starting point. After expressing the functional in Euler equation form, we need a stationary expression that will lead to at least an approximate solution of the Euler equations.

9.27 In addition to the Euler equations, sometimes called Euler–Lagrange equations, several other conditions are useful in determining an optimum by variational methods. Too lengthy for meaningful discussion here, these conditions are mentioned merely for identification. They carry such names as Weierstrass and Legendre–Clebsch and are helpful in distinguishing a relative maximum from a relative minimum. Special techniques are also available for discontinuous solutions.

9.28 Methods based on the calculus of variations have been applied to several types of aerospace optimization problems including the determination of optimum trajectories and the study of aerodynamic shapes.[9] Trajectory problems are consistent with the geometric interpretation of a curve mentioned in §9.23 and can be considered analogous to a nuclear reactor control problem that determines the "best" path from one reactor state to another. Applications of such approaches, however, have been rather limited in nuclear reactor technology. One application in the optimization of the preshutdown power operating program was to minimize the shutdown maneuvering time while keeping the xenon concentration below a prescribed limit.[10]

LINEAR PROGRAMMING

9.29 Mathematical programming is a vast subject with many types of optimization problems and solution methods. Here, programming refers to scheduling or selecting variables leading to the optimum. One very useful method for problems in management and economics is known as linear programming.[11] It is a special case of a more general problem in which the inequality constraints are linear and the objective function is quadratic. In linear programming, the objective function is also linear. As an illustration of linear programming, consider a very simple two-variable problem that lends itself to graphical representation. Most practical problems have many variables but can be handled with digital-computer methods without difficulty.

9.30 This problem concerns resource allocation in which it is desired to maximize an objective function such as

$$y = 2x_1 + 3x_2$$

where y is a profit and x_1 and x_2 represent the units of manufactured products 1 and 2. Here product 1 requires 2, 2, and 4 units of stock items D, E, and F, and product 2 requires 4, 2, and 0 units of these items from an inventory of 20, 12, and 16 of D, E, and F, respectively. This picture can be expressed by the inequality expression representing the regional constraints,

$$2x_1 + 4x_2 \leqslant 20$$
$$2x_1 + 2x_2 \leqslant 12$$
$$4x_1 \leqslant 16$$

9.31 Each of these linear constraints can be represented by a straight line as shown in Fig. 9.4 with the region containing feasible solutions bounded by *all*

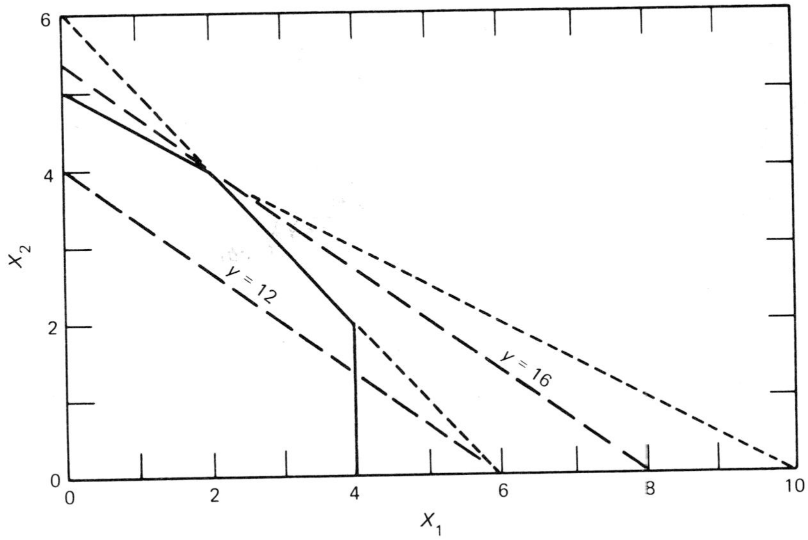

Fig. 9.4 Linear-programming example.

three lines as indicated. The two straight parallel dashed contour lines represent values of y, the profit. The optimum point at $y = 16$ is the "corner" of the feasible region corresponding to a maximum value of y.

9.32 A more general form of solution for more complex linear problems uses an algorithmic or algebraic representation. A systematic process known as the *simplex* method has been developed to test tentative solution points for location within the allowable region and for the presence of an optimum. Since many simultaneous equations can be considered in representing constraints, etc., matrix approaches are normally incorporated into the procedure.

DIRECT METHODS

INTRODUCTION

9.33 Direct optimization methods must be used when an analytic expression cannot be developed for the objective function or when such an expression is too complex for indirect methods. Should this be the case, an *approximation technique* might be used. Here, the objective-function value is determined at many points, and curve-fitting techniques are used to develop an analytical expression that approximates the behavior of the objective function over the region of interest. Indirect methods can then be applied to the analytical expression. However, such approximation approaches are not normally considered *direct methods*, which require the direct use of the performance measure to *search* for the optimum. Direct methods can be classified in two types, *elimination* techniques and *gradient* or *climbing* procedures.

9.34 Before discussing these various possibilities, we should not overlook *exhaustive search,* or exhaustive enumeration. When only a finite number of possibilities need to be considered, it may be practical to evaluate the objective function for every case and choose the optimum directly. The practicality depends on the computing expense involved and the possible difficulty of using more elegant methods. Such a theoretically simple technique is often called the "brute force approach."

ELIMINATION METHODS

9.35 Elimination methods are schemes that narrow down the region containing the optimum to a size wherein trial methods can be used to find a point. They are generally most useful when there is only one independent variable, the so-called one-dimensional case. In addition, the function is usually unimodal and defined in a closed interval. A unimodal function has only one local maximum within the interval although it need not be continuous. A trial evaluation of the objective function in an attempt to locate a maximum is called an *experiment.* Elimination methods therefore involve searching systematically for a maximum by using various schemes to limit the number of experiments required. The objective of such a search plan is to close in rapidly on the optimum and thus obtain "good" values of the criterion early in the search.

9.36 An interesting class of such search methods is based on *Fibonacci* numbers.[12] In a Fibonacci search scheme, each new experiment reduces the interval of uncertainty. The approach is analogous to a population growth sequence, which was the basis of the original development. The series is formed by making each number the sum of the two preceding numbers, i.e., 0,1,1,2,3,5,8,13, ... A selection of spatial intervals for the measurements or experiments based on this number sequence results in an efficient search method for the region containing the optimum value.[12]

GRADIENT METHODS

9.37 Another type of general search approach is known as *gradient, climbing,* or *steepest ascent* methods. The basis is to determine the slope, or gradient, at a selected point, then at another point in the direction of the steepest slope, etc., so that the search progresses efficiently to the maximum point, or optimum. A geographical analogy is helpful in visualizing the picture. For example, the search for the "peak" should begin on some convenient location on the terrain, preferably as close to the top as might be guessed. The slope of the terrain at this point is determined by measuring the elevation at differential distances from the point. We then move in the direction of the steepest slope for a convenient distance and repeat the measurement at that point. This procedure is repeated until the gradient becomes zero, when the optimum point is presumably reached. Although such a point could occur on a "local" peak, saddle point, or ridge, methods are available to test for such possibilities and then proceed to the true optimum.

9.38 A geometric interpretation is useful again in the simplified case of two independent variables x_1 and x_2 in the form of an objective function that determines the dependent variable y. In a three-dimensional coordinate system, the various values of y can be considered to lie on a *response surface* "floating" in space above the base plane described by the x_1 and x_2 coordinate system. A portion of this base plane area including the x_1 and x_2 values of actual interest can be considered the "experimental region," with values of y determined which define the y response surface above the base. The optimization task may then be to determine the peak of the surface. For the slope determination previously described as part of the gradient technique in §9.37, the concept of a plane that is *tangent* to the surface at the point of interest is helpful in developing a general picture. The equation of the plane can be developed by taking small deviations in the x_1 and x_2 values of the point and determining the values of y. Analytically, this amounts to determining the *gradient* of y, or

$$\nabla y \equiv (\partial y/\partial x_1, \partial y/\partial x_2) \tag{9.11}$$

9.39 The tangent plane can be considered an approximation of the response surface in the neighborhood of the point, an approach that may then be generalized to include many independent variables in so-called N-dimensional hyperspace. Multidimensionality does introduce some complications, however. Simple methods effective with only one or two variables are just overwhelmed by the very vastness of multidimensional space. In addition, measures of effectiveness to limit the region of uncertainty cannot be applied in multi-dimensional problems. Special approaches are therefore required.

9.40 A variety of gradient methods that vary in sophistication have been developed and can be applied according to the needs of the problem. Search plans, or strategies, generally combine the need to improve the value of the objective in the move to the "top" with the need to "explore" so that the "jump" to the next point can be made wisely, with the expenditure of a minimum number of steps along the path to the optimum. Examples of such plans are found in standard works.[13,14]

Constraints

9.41 Since most practical problems involve both general and specific constraints that must be satisfied, to obtain useful results, some discussion of constraints and the techniques applicable to gradient methods is in order here. One approach, called the *gradient projection method,* serves as an example. It lends itself to a digital-computer code formulation and has been used to solve various engineering and management problems, including those concerned with gaseous diffusion, desalination, and nuclear reactor operations.[15]

9.42 The general approach in this method is merely to move in a direction that decreases the objective function without immediately striking a constraint rather than seeking a direction that will yield the greatest decrease in the objective function. A particular program given by Cross[15] uses a gradient search procedure, including refinements for recognizing patterns in the response surfaces and boundaries on the variables. The constraint technique, however, should be applicable to most gradient search programs since they operate within the feasible solution space while searching for an optimum. The goal of the constraint technique is to project the gradient so as to obtain a feasible direction and to adjust the points that fail to satisfy the constraints.

9.43 As mentioned earlier, the basic problem is to optimize (maximize or minimize) a function of n variables:

$$F = f(x_1, x_2, x_3, \ldots, x_n) \tag{9.2}$$

which is subject to m constraints:

$$g_i(x_1, x_2, x_3, \ldots, x_n) \geqslant 0 \tag{9.1}$$

where $i = 1, 2, 3, \ldots, m$. Since most techniques for projecting onto constraints are designed for equality constraints, the problem of handling inequality constraints arises. Although slack variables, as found in linear programming, have been suggested, they provide no advantage here.

9.44 Constraints may be handled with Lagrange multipliers to project a vector onto the constrained subspace of interest. In an unconstrained region, by definition n variables exist in an n-dimensional domain. However, when m equality constraints are applied, the dimension of the space is reduced by m. Lagrange multipliers are coefficients λ_i $(i = 1, 2, \ldots, m)$ applied to the partial derivatives of the equality *constraints* with respect to the n variables (§9.21). This application results in a vector that adjusts the gradient to lie in an L-dimensional $(L = n - m)$ subspace. The L, or constrained subspace, is defined as the feasible area in which the m equality constraints are satisfied.

9.45 For example, let $\mathbf{P}$ be the n-dimensional column vector representing the gradient:

$$\mathbf{P} = \nabla f$$

$$P_i = \frac{\partial f}{\partial x_1} \quad (i = 1, 2, ..., n) \tag{9.12}$$

Let **G** be the m by n matrix of partial derivatives of the constraints:

$$\mathbf{G}_{ji} = \frac{\partial g_j}{\partial x_i} \quad (i = 1, 2, ..., n; j = 1, 2, ..., m) \tag{9.13}$$

The Lagrange multipliers are the elements λ_i of an m-dimensional column vector:

$$\lambda = (\mathbf{G}\mathbf{G}^T)^{-1} (\mathbf{G}\mathbf{P}) \tag{9.14}$$

Using this vector, one can compute a new direction, **PP**, by adjusting **P**:

$$\mathbf{PP} = \mathbf{P} - \mathbf{G}^T \lambda \tag{9.15}$$

This new direction, **PP**, will lie along the constraints themsleves if they are linear. However, with nonlinear constraints the new gradient will lie along the hyperplanes, tangent to the constraints at the point being considered.

9.46 After a feasible direction is determined and steps that improve the objective function have been taken, further progress may violate an inequality constraint. When a specified upper or lower linear bound on a variable is violated, the most direct procedure is to reduce the step size. The length of the step is adjusted by a ratio: the distance between the last point and the bound, divided by the distance the variable actually moved. Should more than one variable violate a boundary, the step is scaled small enough to avoid the most imminent and therefore all violations.

9.47 For nonlinear constraints a somewhat different procedure can be used. It depends on a matrix that approximates the partial derivatives of the constraints with respect to the variables. This is only one approach, however. Remember that many different techniques have been developed to handle the different kinds of constraints in gradient problems. The literature should therefore be consulted as needed.[14]

Elimination Methods

9.48 Elimination methods (§9.35) are helpful in reducing the number of necessary trials. In general, some type of *acceleration* of the search is desirable. One type concerns finding and following a "ridge" that reduces the effective dimensionality of the problem. In addition, combinations of tactics are sometimes effective. For example, an acceleration procedure can help achieve the region of the stationary point quickly. Another method, which may consider effects of secondary importance in the search for the region, is then used to explore the region to determine the actual optimum. It is therefore important for the designer to consult the literature to determine promising "climbing" methods for his problem.

DYNAMIC PROGRAMMING

INTRODUCTION

9.49 In the optimization methods previously described, multidimensionality, as required by many practical problems, introduces mathematical and computational difficulties. It is possible, however, to optimize many large systems so that a portion of the problem can be handled at a time. This partial optimization procedure, known as *dynamic programming,* considers a *multistage* process, in which one or more decisions must be made at each stage. In this way many N-dimensional optimization problems can be reduced to N one-dimensional problems or at least to a series of problems of reduced dimensionality. The problem must therefore be planned so that its components can be arranged in series.

9.50 A basis for this approach is the "principle of optimality" as stated by Bellman, "An optimal policy has the property that whatever the initial state and initial decision are, the remaining decisions must constitute an optimal policy with regard to the state resulting from the first decision." The principle tends to be intuitively obvious since, if the remaining decisions were not optimal, the whole policy could not be optimal.

TERMINOLOGY

9.51 Discussions of dynamic programming include some special terms, several of which have been mentioned earlier but are used in a slightly different sense here. For example, in stage-wise problems, a *state variable* transmits information between stages and acts as input to one component of the problem and output from another. *Decision variables* can be manipulated directly at each stage. For example, in a multistage process involving the flow of material, the *state* of the material can be characterized by a set of parameters. For nuclear fuel, for example, the state can be specified by a parameter such as burnup, and a decision variable might be the core location chosen for the fuel on reloading. Decision making is really the essential characteristic of the optimization. The designer, therefore, has several solutions available, but he needs to select the best one.

9.52 Analytically, the system may be conceived as a state vector comprising as many components as necessary to describe the properties of the system. In a general sense the state may therefore be described by an m-component vector. In a multistage process we can consider N stages, each described by an m-component vector. A relation that expresses each component of the stage output state as a function of the input stage and the decisions is called a *stage transformation.*[16]

9.53 A classical illustration of stage-wise decision characteristic of dynamic programming is the transportation network shown in Fig. 9.5. In this simple

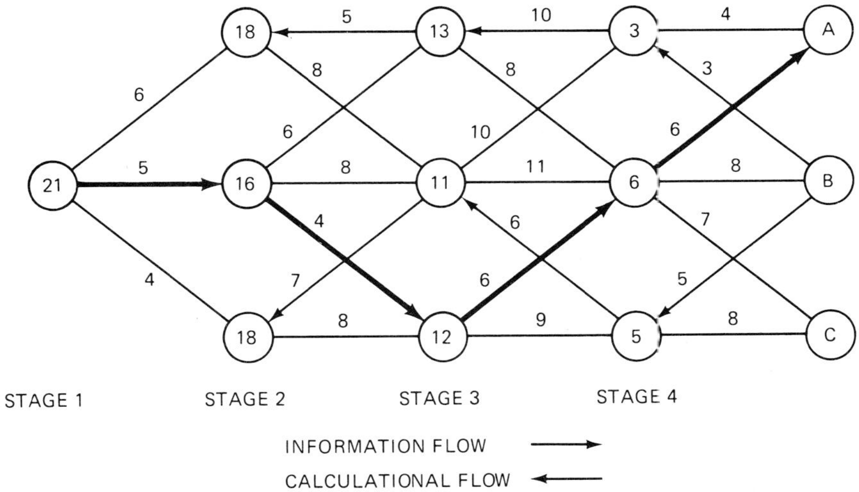

Fig. 9.5 Dynamic-programming problem.

example circles represent locations of cities, and the problem is to travel from one city on the left of the figure to any one of three cities on the right for the least amount of money. Costs between cities are given above the connecting intercity lines. The city locations are the state variables of the system, the state controls or decision variables are the choices of paths to follow, and the four stages are defined by the longitudinal positions of the cities. The information flow is the direction of travel (left to right) in this problem. The decision search begins at stage 4, uppermost city, where the question is asked, "What is the cheapest route from this city to the end of the journey (city A or B)?" The obvious choice is city B, and an arrow and the cost, "3," are stored for later use. The same question is asked at the other two cities at stage 4, and the optimal partial costs and decisions are stored also. The search then proceeds backwards to stage 3, where the same question is asked at each city, and the cost is the sum of the intercity path and the circled cost at stage 4. This process continues until the starting city at stage 1 is reached. The minimum cost of the trip is known immediately then ("21"), and there only remains the retracing of the arrows to find the optimum path (heavy black line).

9.54 By the process of state inversion,[17] the roles of the input and output states can be interchanged, and the problem can be transformed to that of traveling from any of three cities on the east coast to one city on the west coast for minimum cost. The only assumption necessary is that the intercity path cost for traveling from right to left is the same as that for traveling from left to right. The search then starts at the end of stage 1 and proceeds eastward to the end of stage 4. A general algorithm for dynamic-programming problems can be derived from the stage-wise process shown in Fig. 9.6.

Let

$$c(M_n, k_n) = \text{cost of stage } n \text{ of an } N\text{-stage process}$$

where M_n = the stage variable at the beginning of stage n and may be an m-component vector

k_n = the decision variable at stage n and may be a j-component vector

$M_{n-1} = T(M_n, k_n)$, the transformation function

The problem is to minimize the following sum by choosing appropriate values of k_i:

$$C_N = c(M_N, k_N) + c(M_{N-1}, k_{N-1}) + \dots + c(M_1, k_1) \tag{9.16}$$

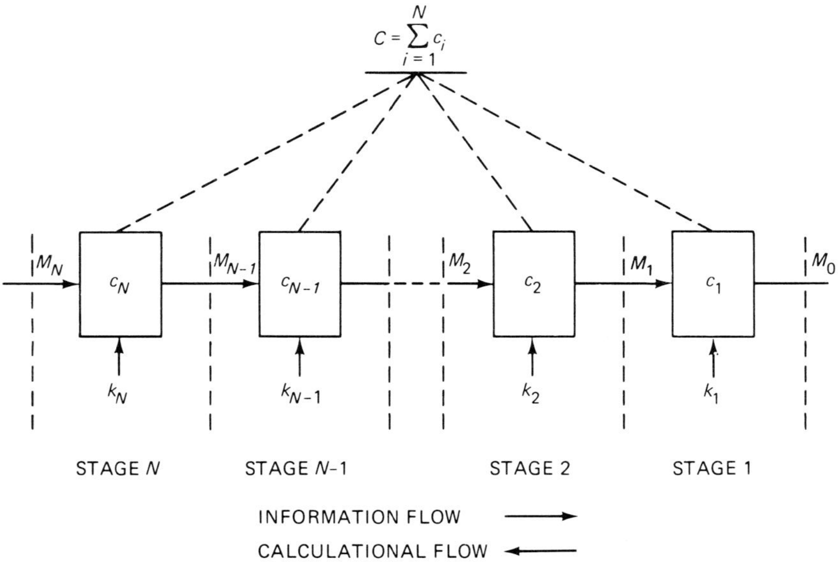

Fig. 9.6 The *N*-stage process for dynamic programming.

9.55 At stage 1 the search process starts and proceeds opposite to the information flow (e.g., time). The total cost, C_N, of the process is a function of the state variable input to the final search transformation, $M_{N-1} = T(M_N, k_N)$, since only when M_N is known can the optimum policy $K(k_N, k_{N-1}, ..., k_1)$ and the total cost C_N be known for the N-stage process (by Eq. 9.16). The minimum total cost is found by searching for those values of k_i which make the total cost a minimum.

9.56 The standard dynamic-programming algorithm has two important disadvantages, however, for nuclear core-management optimization. For a scatter-loaded core with several zones, the state variable vector may have 20 or 30 components. Then the description of the lifetime change in the fuel during a "stage" requires the storing and searching for many values with corresponding high computational costs. Also, since the beginning-of-life state of the reactor is known, it would be desirable to use stage inversion. This process would require, however, the inversion with time of the neutron-diffusion and depletion equations, an operation that has not been developed and may not be possible.

9.57 Both of these drawbacks can be overcome by a modification of dynamic programming, which can be called "Elimination of Similar End States."[18] Figure 9.7 shows the same optimization problem as the one shown in Fig. 9.5, except that the direction of travel and direction of the search are the same. Starting at stage 1, one could explore all possible paths (exhaustive search) and arrive at the optimal path shown by the heavy black line. Or, having found the optimal values for the three cities at stage 2 (the only three possible), one could go to stage 3, uppermost city, and ask the question, "Now that I am here, what city at stage 2 should I have come from so that the total cost to where I am now is a minimum?" The obvious answer is the uppermost city at stage 2, and

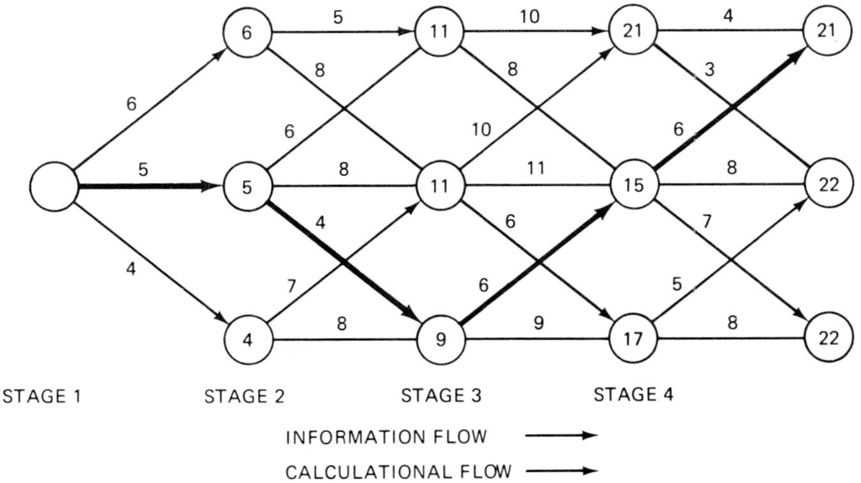

Fig. 9.7 Elimination of similar end states problem.

the total cost is therefore 6 + 5 = 11. The same logic is used for all cities and all stages, and the same optimal cost and path are found as in dynamic programming.

9.58 No comparison of computational efficiency is made for the three methods since, for this example, the number of calculations is approximately the same. For nuclear core management the elimination of similar end states reduces the number of computations enough to make optimization of scatter loading possible, whereas the same problem cannot be solved by dynamic programming.* Information is lost, however, when the exhaustive search and elimination of similar end states are used. Upon considering Fig. 9.5 for dynamic programming, one sees that after the optimization is completed he has the information to start at *any* city in the network and proceed in an optimal manner to city A, B, or C. The information in Fig. 9.7, however, only relates the optimum path from a city on the path. If one strays from the path, he has to work the problem again from the new starting city.

OTHER APPROACHES

POLICY IMPROVEMENT

9.59 For several state variables per stage, formal dynamic-programming procedures become too complex to be practical. Trial decisions followed by adjustments that improve the objective function require less computation. In other words, the entire *policy* is improved in a systematic way.

9.60 The effect of a change in decision value on the state of the stage can be found by a method known as *constrained derivatives.* If the decision variables are systematically perturbed and the resulting rate of change of the objective function is used, the policy improvement and optimization can be accomplished in a stepwise manner. The stage-wise procedure, which involves difference equations, approaches in the limit the differential-step continuous-transition case. The continuous case is useful in automatic-control applications where minimum time or minimum control effort is the objective. An optimization of this general type, using constrained derivatives, possibly is based on the *maximum principle* of the Russian mathematician Pontryagin.

HEURISTIC PROGRAMMING

9.61 *Heuristic methods* or *heuristic programming* refers to approaches seeking a "good" solution rather than an optimal one. A good solution merely defines the regions of the optimum and may be adequate for many practical

*The term "dynamic programming" is sometimes also applied to modified methods, such as the elimination of similar end states, in which the direction of travel and search are the same.

problems wherein the designer can then exercise judgment in selecting a final solution within the region by considering the effect of parameters or constraints that cannot be represented analytically.

9.62 Heuristic methods may also include any optimization approach, normally using a computer algorithm, which is primarily empirical in nature, often with intuition as one of the inputs. With these methods the experienced designer can cut corners for many industrially important problems that would involve prohibitive computation effort if a complete mathematical model were attempted.

9.63 Many problems of this type concern allocation. A classical example[19] is to determine a minimum distance for a round trip that includes each of N cities. Dynamic programming may be used for up to about a 13-city problem of this type. A heuristic approach starts with a random choice of three cities, followed by the selection of a single city from the remaining list. This city is inserted, in turn, between each of the two pairs, and a minimum-length four-city tour is selected. One city from the remaining list is inserted, in turn, between each pair of cities of the previously selected four-city tour, and a five-city tour of minimum length is selected. The procedure is continued until a tour of the desired city length is obtained and is then repeated for a different order of selection. A separate tour therefore results for each set of trials. The desired minimum tour is then selected from the collection accumulated after an arbitrary number of trials, say 100. In many problems of this type, some simplifications can be made as patterns evolve after some trials or may even be obvious when the problem is stated.

APPLICATIONS

INTRODUCTION

9.64 Although some of the methods described have been applied to nuclear engineering problems, commercial usage has been very limited. One reason is that for most applications uncertainties in input data are so large that only simple optimization methods are justified. However, potential use is high as the industry matures, the accuracy of design calculations is improved by experience, and design improvement becomes a major objective. As an indication of the possibilities, some application examples described in the literature are discussed in this section.

9.65 The approaches used can be classified into two areas. One is concerned with the design of systems that do not lend themselves to a stage-wise description. Generally, gradient methods have been used here. Stage-wise problems, particularly those concerned with core fuel management or control lie within the other area and generally use dynamic programming or a related method.

9.66 Optimization methods have been applied to the design of reactor shields of various types, particularly when minimum weight is desired. Variational methods have been used in attempts to synthesize minimum-weight proton and reactor shields.[20] Generally, a procedure was determined for simplified models for the proper ordering of materials in a single laminated shield. This procedure depends on the various attenuation properties and densities of the materials. However, no simple way was developed for determining the optimal thicknesses of these materials in practice.

9.67 Simplified gradient methods[21] have minimized shield weights. However, only a single source and dose point was considered, and the methods for measuring effectiveness were relatively crude. Terney[22] used the optimum gradient technique to determine the optimum dimensions of a shipboard-reactor shield system. He considered both shield weight and cost.

9.68 As a reference Terney used an analytical model that included a symmetrical cylindrical system consisting of a primary shield of water and lead and a secondary shield of heavy concrete, lead, and polyethylene. Design variables include the radiuses and height of the shield sections as well as the thicknesses of the various materials. Radiation dose levels at five specified points constrain the system. An optimal design is achieved when the cost of shield weight change is equal to the worth of the corresponding weight change from the viewpoint of the economics of the ship. The condition for optimality is therefore

$$\frac{dC}{dW} = -K \tag{9.17}$$

where C and W are the shield cost and weight in dollars and tons, respectively, and K, in dollars per ton, is the worth of a 1-ton reduction in shield weight to the ship, a value that depends on the type of vessel. The optimization criterion is therefore to minimize G in the relation

$$G \equiv C + KW \tag{9.18}$$

9.69 The model, which includes five dependent and three independent variables, is considered a constrained-parameter-minimization problem that was afterwards analyzed by the so-called iterative optimum gradient method. In this method, a variation of the one discussed in §9.41, the independent variables are changed at each step in the direction of the gradient and for an amount that gives the maximum change in the objective function. A similar technique has also been used by Nel and Fenech[23] for a gas-cooled reactor-system model that included 15 independent and 48 dependent variables. In each case Nel and Fenech used digital computers to carry out the actual solution; they started with initial estimates of the variables and proceeded until successive iterations yielded an acceptable decrease in the objective function, normally less than 0.01%.

9.70 Another interesting optimization concerned the EURATOM ORGEL Program.[24] A computer code based on *quadratic programming* was used to optimize a nuclear-power-plant design with respect to eight independent variables. In quadratic programming the objective function is of second degree, and the constraints are linear. For this application the cost of electricity was approximated by a function of such variables as core height, lattice, pitch, reflector thickness, coolant velocity, flux ratio, cladding temperature, and coolant outlet temperature. The least-squares method was used to generate the coefficients of the second-degree function from many combinations of numerical values taken systematically in the feasible design region. Optimization was then carried out by partial differentiation of the objective function with respect to each decision variable in turn. In this application the optimized design gave an energy-cost advantage of 0.5 mill/kw-hr over a prior nonoptimized design.

9.71 Lewins[25] used inhomogeneous perturbation theory to derive an optimum distribution of materials in a system such as a reactor lattice design. As mentioned earlier (§9.60), this approach is related to Pontryagin's maximum-principle optimization methods applicable to control problems. As a simple illustration Lewins uses the method to maximize the thermal utilization of a lattice cell in which the relative location of the fissile material and coolant channels may be varied. The process appears to lend itself to automation through the preparation of computer codes for given types of problems.

MULTISTAGE-DECISION-PROCESS EXAMPLES

Introduction

9.72 Optimization problems that can be formulated as multidecision processes can be treated by dynamic programming, as discussed in §9.49. In 1960 Kallay[26] showed how both component-design and system-design problems could be approached in a stepwise decision manner so that dynamic programming would be applicable. In considering spatial distribution of core materials to maximize power density, he divided the core into a number of concentric rings. Then, starting with the outside ring, he was able to consider a poison-allocation problem in a stepwise fashion. Using a recursive stepwise calculation again in another example, he suggested a calculational procedure for selecting a maximum-system-efficiency combination of plant components arranged in series (core, heat exchanger, pumps, etc.). Although Kallay described the structure of the relations and the general approach to be followed, he did not develop actual examples or show results.

9.73 The areas of fuel-management and control-rod programming, which involve operations inherently discontinuous with time, are natural candidates for optimization processes of a multistage-decision-process nature. In core fuel management (§7.62), fuel assemblies are replaced and sometimes rearranged in the core at the end of each depletion period. Since a fuel change normally causes a shift in reactivity, the control-poison pattern also requires a change. As a result, core fuel management and, to a lesser extent, control programming are fertile areas for optimization development. A discussion of several examples therefore provides a picture of the potential of optimization methods of this type.

In-Core Fuel Management

9.74 A review of applicable background for in-core fuel management will be helpful here. From the optimization viewpoint there are a number of possible decisions and constraints. For example, at the point of refueling in a nuclear plant, the designer may be faced with five basic decisions:

1. Fraction of the core to be replaced.
2. Fuel design (H/U ratio, rod pitch, cladding thickness, etc.).
3. Feed enrichment.
4. Physical relocation of all assemblies.
5. Desired length of subsequent operating cycle.

These decisions affect the basic fuel-performance variables of power level and burnup as well as to what degree the fuel is used within the core, subject to the constraints of material and operating limitations plus performance safety. This relation is summarized in Fig. 9.8.

9.75 In general, the power level represents the rate at which energy is extracted from a given fuel investment, and the burnup represents the total energy extracted, regardless of the time involved. The relation between design decisions and these two parameters follows.

9.76 The power level influences the time-value economic components of the fuel cost, such as fissile-atom inventory and working-capital charges. Since plant designs allow for at least a 5 to 10% increase in power level over the

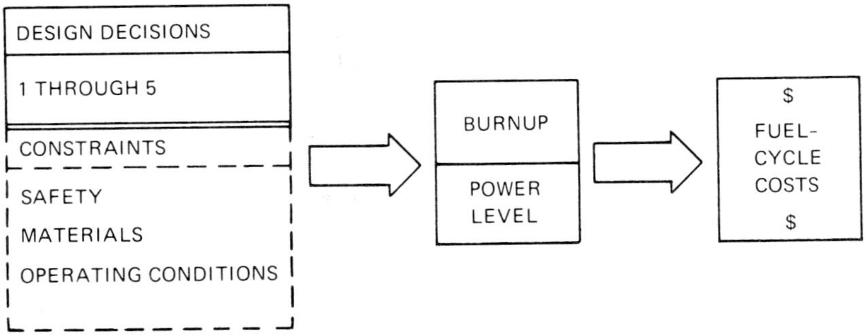

Fig. 9.8 Effect of fuel utilization on fuel-cycle cost.

nominal rating,[27] it is reasonable to seek a higher power level through improved fuel management. If the power level is increased at fixed plant factor, the fixed-cost charges, when expressed on a unit-energy basis, will be reduced. The total charge for a given energy output is likely to be also lowered but depends on decision 5. The nonfuel reactor plant capital charges per unit of energy vary inversely with the average power level. An increase in the average power level could therefore permit a significant saving in capital charges.

9.77 In water-cooled reactors, however, the power level that can be tolerated at any location in the core is limited by thermal–hydraulic parameters, such as fuel-pin central temperature, cladding temperatures, and the possibility of nucleate boiling. These limitations must be considered if the average power level is to be increased by increasing the peak power in the core. A more satisfactory way of raising the average power level, therefore, is by reducing the peak-to-average power density throughout the core by a judicious rearrangement of assemblies to accomplish what is commonly termed "power flattening" (decision 4).

9.78 The fuel burnup influences such significant fuel-cycle costs as the charges expressed in mills per kilowatt-hour for enrichment, fabrication, reprocessing, and uranium and plutonium credits. An increase in the fuel burnup will reduce charges applicable to a fuel batch, such as fabrication and reprocessing, but will increase somewhat the charges for enrichment. These opposing trends therefore create a minimum cost at a specified burnup that tends to be shallow, however, for light-water reactors with an effective range of several thousand megawatt-days per tonne of uranium (Fig. 9.1). This effective minimum range is determined primarily by the unit costs of the fuel-cycle components (enrichment, fabrication, reprocessing, etc.). Although design decisions 1 through 5 have a secondary effect in determining this range, their primary effect is to specify the cycle burnup for all fuel loadings in the core.

Dynamic-Programming Applications

9.79 Dynamic programming is well suited to the discontinuous, stage-wise nature of fuel management. The fuel-depletion equations and decision functions, both of which are discontinuous at the loading points, can be adapted readily to the transformation of the state variables from stage to stage by the search technique in dynamic programming. The optimal policy obtained by dynamic programming is also the global optimum, and constraints are easily handled since the optimum policy obtained automatically satisfies the constraints on the state variable. Computer storage, however, is the most serious difficulty with dynamic programming since storage requirements increase exponentially with the dimension of the state variable. If 10 values, a very modest number, are stored for each component of an n-component state vector, 10^n storage locations are needed. Storage requirements alone in a 32,000-word computer therefore limit the components treated to four.

9.80 In a pioneer effort Wall and Fenech[28] considered the refueling policies for a single-enrichment three-zone 1000-Mw(e) pressurized-water core that would yield a minimum power cost based on fuel burup as the sole criterion with power peaking as the constraint. The model, based on partial-batch refueling, permitted a wide variety of zonal loading strategies to be considered.

9.81 The reactor was described by a three-dimensional state variable that was the specific energy (megawatt-days per tonne of uranium) produced in each of three reactor zones. If 10 values of specific energy or burnup are considered for each of three zones, there would be 10^3 values of the state variable. If 28 decision searches per state variable are assumed, 28,000 calculations would therefore be required for each reloading stage in conventional dynamic programming. Since an average calculation takes approximately 0.01 sec and a typical optimization will have 20 stages, the total computer time would be approximately 93 min per problem. The solution to this problem therefore becomes impractical for n greater than 3.

9.82 Wall and Fenech, therefore, had to use an alternative to dynamic programming and, in general, to limit the values of the state variable to less than 200 per stage. This alternative to dynamic programming is a computational acceleration method of exhaustive search, similar to elimination of similar end states (§9.57).

9.83 The merits of many possible operating decisions and the effects on power costs of changing the constraints were examined in the work. It was recognized, however, that the zonal approach could cause poor power distribution in practical cores of large size. Stover[18,29] therefore extended this approach to scatter-loaded systems.

9.84 In scatter loading, fresher fuel elements are placed next to irradiated elements. Close neutron coupling between newer and older fuel results, and the local reactivity effect of each fuel tends to be balanced. Uniform mixing of fresher and older fuel throughout the whole core therefore produces a more even power distribution than is possible with out–in schemes. Close neutron coupling between fresher and older fuel also increases the average discharge burnup of the fuel over the attainable in out–in schemes in large cores. Furthermore, since the local reactivity of a scatter-loaded zone is less than that of a fresh, batch-loaded zone, the maximum reactivity to be controlled is less than that in most batch-load schemes. These advantages apply, however, only after an equilibrium core condition has been established. Also, local power peaking from heterogeneity of the core could occur when a less depleted fuel element is placed next to a more depleted element. This may present a greater design challenge in a pressurized-water reactor (PWR), which has larger fuel elements and no coolant feedback mechanism for suppressing local power peaking, than in a boiling-water reactor (BWR), where feedback is provided through the formation of voids.

9.85 Stover used scatter loading of a three-zone boiling-water reactor as the reference fuel-loading scheme for his optimization study to maximize discharge burnups. This model was considered sufficiently representative of industrial practice to provide an evaluation of the optimization technique. The optimum decisions to be determined in scatter loading are the volume fractions of each core zone to be replaced by fresh fuel at each reload point in the life of the reactor. A minimum fuel-cycle cost over the life of the reactor was used as the criterion. Once calculated, this series of decisions formed the optimum operating strategy.

9.86 In three-zone scatter loading, however, the dimensionality of the state variable was found to be much greater than that for a three-zone partial-batch scheme. With the assumption that the reactor's performance is well described by the ^{235}U mass at any time in each group of fuel with different irradiation histories, it was necessary to know the zonal ^{235}U mass and corresponding volume fraction for each of these groups. If the minimum fresh-fuel-loading fraction considered was $1/5$, there were five fuel groups in each of three reactor zones for which ^{235}U mass and volume fraction had to be known. The result was a 30-dimension state vector. Such a problem, unsolvable by dynamic programming, was optimized by using elimination of similar end states (ESES).

9.87 By applying the recursion relations of ESES to the optimization of a scatter-loaded reactor, Stover found the state variables at each succeeding stage from the transformation $M_n = T(M_{n-1}, k_n)$ (§9.54), where all combinations of M_{n-1} and k_n are considered. In this manner the number of state variables and thus the number of calculations increased exponentially at each stage. The algorithm therefore rapidly digresses into an exhaustive search which involves all combinations of the state variable and decision variable and which requires excessive computer time and memory. However, the exponential growth can be moderated by eliminating similar end-of-life state variables when they occur. If state variables differ by less than a small number, ϵ, for example, the decision policy that leads to the smallest value of $C(M_n)/E(M_n)$ is chosen, and the corresponding state and decision variables are stored. The state variable resulting from the nonoptimum policy is then discarded.

9.88 The algorithm used by Stover in the computer code MINFUL to optimize scatter loading by ESES was:

1. At each stage consider all possible combinations (policies) of state variable values and decision variable values.
2. Calculate beginning-of-life and end-of-life fuel masses for each policy from previously determined burnup polynomials.
3. Apply constraints and calculate fuel cost and energy production for each policy.
4. Eliminate similar nonoptimum end states whenever possible.
5. Repeat steps 1 to 4 for all state variables at all stages.
6. After the last stage find the minimum fuel-cycle cost and retrieve the corresponding optimum-loading fractions at each stage from the computer memory.

9.89 Stover was able to show that the reduction in dimensionality achieved by the acceleration had little effect on the optimum policy when compared with a standard result obtained by exhaustive search. A nonuniform scatter-loading policy, obtained by allowing the decision values to be different in each zone, was found to be préferable to a uniform scatter-loading scheme.

Control-Rod Programming

9.90 The strategy for planning control poison from one fuel-depletion period to another is another type of design challenge. Terney and Fenech[30] adapted the same general approach used by both Wall and Stover to this problem. In this case power peaking serves as the criterion, and constraints on the rod program include the criticality requirement, allowable rod-motion patterns, and restrictions on the power peaking and burnups. A representative first-generation pressurized-water reactor comparable to Yankee-Rowe* was the basis of the analytical model in the study.

9.91 The approach was used successfully to obtain an optimal control-rod program for minimizing the power peaking, which was a marked improvement over normal operating practices, such as straight-bank or radial-zone withdrawals.

Other Fuel-Management Optimization Approaches

9.92 The development of systematic or "automated" fuel-loading schemes has by no means been limited to those based on dynamic programming, previously described. Several examples of other approaches are mentioned here as possibilities.

9.93 Fagan[31,32] considered fuel-replacement policy options in which the identity of each fuel subassembly is maintained so that shuffling patterns as well as scatter loading could be handled. The methods previously described consider only unit replacement of geometrically defined zones of the core, using average characteristics of the contained fuel assemblies. An important feature of the approach was the representation of all time-dependent effects, such as heat-generation rate, fuel burnup, system reactivity contribution, and fuel composition, by a single empirical time-dependent parameter. This proved to be a reasonable assumption as long as the gross neutron distribution did not change radically with time. The empirical representation of the operating state involving only the fuel location and the time-dependent burnup factor is valid, however, only if fuel assemblies of an existing design are replaced in an established core configuration.

9.94 Nearly identical equilibrium cycles were assumed to occur throughout most of the plant lifetime. Further assumptions were that a single-fuel feed enrichment is used and that the same number of fuel elements are replaced for

---

*Yankee Atomic Electric Co. plant at Rowe, Mass.

each cycle. Under these conditions sequences of equilibrium cycles differing in length by at most 5% were calculated.

9.95 Direct fuel replacement and replacement with shuffling were evaluated by comparing them with this operating-state model. In both cases the refueling scheme was defined by optimizing the operating time for the cycle. For the conditions assumed this proved to be a good approximation of the economic optimization. Although the direct refueling scheme did not give meaningful results, shuffle refueling did show fuel-cycle-cost advantages of approximately 0.05 mills/kw-hr, which amounts to over $300,000 per year for a 1000-Mw(e) plant, corresponding to an increase in operating time for the same amount of new fuel. In this scheme the new fuel always replaced the most highly irradiated elements in the reactor. After the new elements were added to the core, the elements were shuffled until they yielded the maximum operating time consistent with the constraints on power distribution. Since the operating-cycle times for shuffle refueling varied by less than 1.2%, the accuracy of the equilibrium-cycle assumption was ensured.

9.96 All the core-management optimization approaches described here used fuel burnup or the equivalent as the criterion, a parameter which, compared with power peaking, only weakly affects energy costs in commercial light-water reactors once a threshold burnup is achieved. This effect, shown in Fig. 9.1, is due primarily to the present-day values of the compensating contributions to the fuel-cycle cost and is not of a fundamental nature. There is a greater potential for cost savings for commercial reactors if the optimization could be based on the power peaking and burnup used as a constraint rather than the opposite. Once the most favorable power-peaking picture is obtained through fuel management, control-rod management should be included in the analysis to obtain a more inclusive power-peaking optimization, if possible on a three-dimensional basis.

9.97 Considering only the two-dimensional in-core fuel-management problem, the development of a generalized approach to fuel replacement and shuffling based on an accurate representation of the nuclear contributions to the power profile is indeed a formidable task, particularly if the computing effort is to be kept within reasonable bounds. Both spatial and time-dependent parameters are relevant on a noncontinuous basis. Therefore present design techniques rely heavily on engineering judgment. A loading pattern is estimated, and a core nuclear design is calculated by using two-dimensional diffusion codes such as PDQ (§5.73) or equivalent methods. Design goals for power peaking and region-averaged burnup are set on the basis of past experience. If the calculated results do not meet these goals, a new loading pattern is chosen to shift the power profile, and a new nuclear calculation is carried out. Additional pattern selection and calculation cycles are needed as fuel-depletion and -replacement possibilities are considered.

9.98 Fuel replacement introduces shuffling as an important decision option. The objective in shuffling is to remove power peaks and to fill in valleys or to shift the power profile to achieve a more desirable burnup. Shuffling can be planned by using graphical representation with movable indicators for assemblies and searching by hand for a desired pattern, a rather unsophisticated technique, particularly since a favorable pattern must be checked by calculations.

9.99 The assemblies are usually in a very delicate reactivity balance, and this complicates the shuffling. Slight changes in position can cause considerable variations in the power profile. Using past experience, a designer can bring a tightly coupled core, such as San Onofre (initial core design), to within design goals in about ten iterations since all patterns are highly symmetric and the number of patterns satisfying these symmetry constraints are limited. For larger, more loosely coupled cores, the number of choices is likely to be much higher. If burnable poisons, variable assembly enrichments, or PuO_2 is added, the number will probably by even greater. This design process represents an appreciable investment of both man-hours and computer time.[33] In addition, its flexibility is greatly reduced owing to the long lead time required for the design of replacement cores.

9.100 Some "semiautomated" techniques involve a combination of graphical and analytical approaches. A method developed by Mélice[34] for pressurized-water-reactor cores, for example, evolved from an empirical graphical approach based on the power profile. The power profile depends on a reactivity plot characteristic for an optimal design. The so-called "k-profile" of infinite-multiplication-factor distribution was shown to be the basis of several different fuel-reloading optimization options.

9.101 Naft[27,35] developed a direct-search scheme to optimize the fuel-loading pattern of pressurized-water reactors from the initial loading through several reloading cycles that included shuffling. The search, however, does depend on an "initial guess" in the neighborhood of the optimal pattern as input. Such an estimate can normally be made without difficulty by an experienced designer. An optimal pattern is achieved when the peak-to-average radial power is at a minimum and all constraints on assembly and region-average burnups are met.

9.102 The optimization scheme uses a simplified function to represent the two-dimensional power profile. This function is the product of a slowly varying component, which is a one-dimensional analytic solution to the diffusion equation, and a rapidly varying component, which uses the material buckling as a parameter to represent the material properties of the nearest neighbors of an assembly. Through a variational approach several undetermined coefficients were established for each specific case.

9.103 A discrete direct-optimization scheme that treats discrete nonlinear variables through a univariate search operates on the simplified function. Through this search a graded table of possible moves that specifies a direction or pattern for the optimization is created.

9.104 In a test of the optimization procedure using the San Onofre reactor as a model, the initial core power was described by the simplified power function to within about 1%. In addition, the optimization scheme duplicated the initial core-loading pattern as well as a replacement core-loading pattern.

CONCLUSIONS

9.105 Though yet to be widely used in nuclear reactor design, formal optimization procedures appear to have considerable potential, particularly as computer capability improves with time and reactor engineering experience is accumulated. The accurate representation of economic parameters as may be necessary for many applications could be difficult, however. Inflationary trends associated with long lead times needed for design and construction introduce inherent uncertainties.

9.106 Despite present limitations optimization is one of the vital steps in the engineering design process needed to bring concepts to realization. A knowledge of the available procedures is therefore a valuable addition to the professional knowledge of the design engineer. In addition, the needs of optimization theory for an accurate description of the system and the definition of measures of effectiveness encourage rigor and a high level of insight by the designer.

REFERENCES

1. L. A. Pipes, *Applied Mathematics for Engineers and Physicists,* 2nd ed., pp. 313 et seq., McGraw-Hill Book Company, Inc., New York, 1958.
2. A. Sesonske and R. P. Hammond, A Preliminary Evaluation of Fast Oxide Breeder Reactors for Sea Water Conversion, USAEC Report LA-2733, Los Alamos Scientific Laboratory, August 1962.
3. P. H. Margen, *Nuclear Reactor Optimization,* pp. 22–30, Simmons-Boardman Publishing Corporation, New York, 1960.
4. R. B. Winsbro, ORIC: A FORTRAN IV Code for Computing Optimum Plant Designs for the Desalination of Seawater and the Production of Electricity, USAEC Report ORNL-TM-1523, Oak Ridge National Laboratory, August 1966.
5. D. M. Eissenberg and C. C. Burwell, A Survey of Optimum Dual-Purpose Desalting Plants as a Function of Product Ratio Using Alternate Steam Supplies, USAEC Report ORNL-TM-1659, Oak Ridge National Laboratory, July 1967.
6. G. L. Nemhauser, *Introduction to Dynamic Programming,* p. 2, John Wiley & Sons, Inc., New York, 1966.
7. D. J. Wilde and C. S. Beightler, *Foundations of Optimization,* Prentice-Hall, Inc., Englewood Cliffs, N. J., 1967.
8. T. N. Edelbaum, Theory of Maxima and Minima, in *Optimization Techniques,* George Leitmann (Ed.), Academic Press, Inc., New York, 1962.
9. S. E. Dreyfus, *Dynamic Programming and the Calculus of Variations,* pp. 25–26, Academic Press, Inc., New York, 1965.

10. J. J. Roberts and H. P. Smith, Jr., Time Optimal Solution to the Reactivity-Xenon Shutdown Problem, *Nucl. Sci. Eng.*, **22**: 470–479 (1965).
11. Béla Krekó, *Linear Programming,* American Elsevier Publishing Company, Inc., New York, 1968.
12. Reference 7, pp. 235–242.
13. Reference 7, Chap. 7.
14. D. J. Wilde, *Optimum Seeking Methods,* Prentice-Hall, Inc., Englewood Cliffs, N. J., 1964.
15. K. E. Cross, A Gradient Projection Method for Constrained Optimization, USAEC Report K-1746, Union Carbide Corporation, May 30, 1968.
16. Reference 6, p. 22.
17. Reference 7, p. 367.
18. R. L. Stover, Optimized Scatter-Loading of a Nuclear Power Reactor, Ph. D. Thesis, Purdue University, Lafayette, Ind., 1968.
19. P. J. Muder, Heuristic Programming, *Chem. Eng. Progr.*, **65**(12): 20 (1969).
20. A. D. Krumbein, P. S. Mittelman, E. S. Troubetzkoy, F. Nakache, and J. Calnik, Synthesis of Minimum Weight Proton Shields, in Symposium on Protection against Radiation Hazards in Space, Held in Gatlinburg, Tennessee, November 5–7, 1962, USAEC Report TID-7652, pp. 773–793, Oak Ridge National Laboratory, 1962.
21. R. L. Bernick, Application of the Method of Steepest Descent to Laminated Shield Weight Optimization, USAEC Report NAA-SR-Memo 8181, North American Aviation, Inc., Apr. 17, 1963.
22. W. B. Terney, Optimization Technique in Nuclear Engineering, Ph. D. Thesis, Massachusetts Institute of Technology, Cambridge, Mass., 1967.
23. B. M. Nel and H. Fenech, Physics and Engineering Design Optimization of Nuclear Power Plant Using Gradient Methods, *Proceedings of the Third International Conference on the Peaceful Uses of Atomic Energy,* Vol. 5, p. 184, United Nations, New York, 1964.
24. W. Boettcher, F. LaFointaine, and P. Tauch, Recent Optimization Methods, EURATOM Report EUR 2244e, 1965.
25. J. Lewins, Reactor Optimization, *J. Nucl. Energy,* **20:**6 (January 1966).
26. N. Kallay, Dynamic Programming and Nuclear Reactor System Design, *Nucl. Sci. Eng.*, **8**: 315 (1960).
27. B. N. Naft, Development of an Automated Fuel Loading Optimization Scheme for Pressurized Water Reactors, Ph. D. Thesis, Purdue University, Lafayette, Ind., 1970.
28. I. Wall and H. Fenech, The Application of Dynamic Programming to Fuel Management Optimization, *Nucl. Sci. Eng.*, **22**: 285 (1965).
29. R. L. Stover and A. Sesonske, Optimization of BWR Fuel Management Using an Accelerated Exhaustive Search Algorithm, *J. Nucl. Energy,* **23**: 673 (1969).
30. W. B. Terney and H. Fenech, Control Rod Programming Optimization Using Dynamic Programming, *Nucl. Sci. Eng.*, **39**: 109 (1970).
31. J. R. Fagan, An Optimization Analysis for Fuel Replacement in Large Pressurized Water Reactors, Ph. D. Thesis, Purdue University, Lafayette, Ind., 1968.
32. J. R. Fagan and A. Sesonske, Optimal Fuel Replacement in Reactivity Limited Systems, *J. Nucl. Energy,* **23**: 683 (1969).
33. B. Rothleder, Power Reactor Fuel Management Beyond First Cycle Operation, *Trans. Amer. Nucl. Soc.*, **14**: 89 (1971).
34. M. Mélice, Pressurized Water Reactor Optimal Core Management and Reactivity Profiles, *Nucl. Sci. Eng.*, **37**: 451 (1969).
35. B. N. Naft and A. Sesonske, Pressurized Water Reactor Optimal Fuel Management, *Nucl. Technol.*, **14**: 123 (1972).

Index

Numbers refer to sections rather than to pages.

Nuclear Power Reactor Instrumentation Systems Handbook

Volume 1

Editors
Joseph M. Harrer, Argonne National Laboratory
James G. Beckerley, U. S. Atomic Energy Commission

This handbook is for designers and operators of power-reactor instrumentation systems and those concerned with applications and not with invention. The accepted practices used in the design of nuclear reactor instrumentation systems are covered, and the systems used on today's power reactors are described. In some cases the plants discussed are still in the design stage. The requirements (or design bases) to be satisfied by each system and subsystem are given first; then the current practices are outlined and evaluated.

The performance and characteristics of major components of instrumentation systems are presented with a minimum of component design discussion. For example, chapters covering nuclear radiation sensors give performance in detail, but sensor design data are omitted. This follows the basic intent of emphasizing the system aspect of power-reactor instrumentation.

The basic data on neutron sensors, kinetics, computers, process sensors, rod drives, power supplies, installation methods, and quality assurance and reliability are discussed in this volume. Volume 2, which is in preparation, covers applications and includes material on plant protection, standards, and radiation monitoring, and descriptions of four plant designs: BWR, PWR, and sodium- and gas-cooled reactors.

Available as TID-25952-P1 for $7.60 from
National Technical Information Service
U. S. Department of Commerce
Springfield, Virginia 22151

May 1973
315 pages, $8\frac{1}{2}$ by 11 inches
International Standard Book Number 0-87079-005-6
Library of Congress Catalog Card Number 72-600355

Reactor Shielding for Nuclear Engineers

By N. M. Schaeffer
Radiation Research Associates, Inc.

Prepared for the Division of Reactor Development and Technology, USAEC, and published by the Office of Information Services, USAEC

As the number of nuclear power plants on order continues to grow (currently more than thirty per year in the United States alone), the demand for nuclear engineers should also increase, and a new text on reactor shielding is overdue. Shielding technology has matured considerably in the last decade, and shield physics must routinely be translated into shield design. Since the publication in 1959 of *Fundamental Aspects of Reactor Shielding,* by Herbert Goldstein, new generations of computers have become available to exploit techniques heretofore considered too costly, and new measurement techniques have been devised. The energy and angular distributions of neutrons and gamma rays can be followed, both in theory and in practice, throughout their transport histories. Such powerful tools have brought correspondingly large dividends to the shielding community.

These advances and their underlying fundamentals are recorded in this volume, which is intended as a text for a two-semester course in reactor shielding directed at an advanced-undergraduate or a graduate level. (800 pages; 7 by 10 inches; Library of Congress Catalog Card Number 73-600001)

Available as TID-25951 for $13.60 from
National Technical Information Service
U. S. Department of Commerce
Springfield, Virginia 22151